International Fundamentum Science Symposium (iFunS 2018)

IOP Conference Series: Materials Science and Engineering Volume 440

Terengganu, Malaysia
25 – 26 June 2018

Editors:

Aziz Ahmad
Mohd Ikmar Nizam Hj. Mohamad Isa
Wan Mohd Khairul Wan Mohamed Zin
Khairul Anuar Mat Amin
Maisara Abdul Kadir
Mohd Hasmizam Razali

ISBN: 978-1-5108-7574-6
ISSN: 1757-8981

Printed from e-media with permission by:

Curran Associates, Inc.
57 Morehouse Lane
Red Hook, NY 12571

Some format issues inherent in the e-media version may also appear in this print version.

Printed by Curran Associates, Inc. (2019)

For permission requests, please contact the Institute of Physics
at the address below.

Institute of Physics
Dirac House, Temple Back
Bristol BS1 6BE UK

Phone: 44 1 17 929 7481
Fax: 44 1 17 920 0979

techtracking@iop.org

Additional copies of this publication are available from:

Curran Associates, Inc.
57 Morehouse Lane
Red Hook, NY 12571 USA
Phone: 845-758-0400
Fax: 845-758-2633
Email: curran@proceedings.com
Web: www.proceedings.com

TABLE OF CONTENTS

CHEMISTRY

PHYSIC

BIOLOGY

IOP Conf. Series: Materials Science and Engineering **440** (2018) 011001 doi:10.1088/1757-899X/440/1/011001

International Fundamentum Science Symposium (iFunS) 2018

PREFACE

From 25 to 26 June 2018, the International Fundamentum Science Symposium 2018 was held in Kuala Terengganu – the heritage waterfront city, situated at the east coast of peninsular Malaysia. This first ever scientific meeting hosted by the School of Fundamental Science, Universiti Malaysia Terengganu (UMT) and co-organized by Academy Sciences of Malaysia (ASM). With the theme "Exploring Possibilities Through Fundamental Sciences", this symposium was divided to three main sub-themes which were Fundamental Chemistry, Fundamental Physics and Fundamental Biology.

The aim of this symposium was to promote and encourage the fundamental research among the researchers. It is important to note that although fundamental studies do not pursue immediate commercial objectives, nevertheless, findings of fundamental studies may result in innovations, as well as generating solutions to practical problems. In fact, the findings and knowledge of fundamental studies in chemistry, physics and biology will generate new and advanced technologies such as nanotechnology, green technology, biotechnology, tissues engineering and etc. The conference was attended by over 80 specialists from leading universities and research institutes, as well as representatives of the scientific community from United Kingdom, Indonesia and Singapore. More than 60 papers were selected to go through a strict double blinded peer reviewed process. In turn, these selected papers were published in IOP Conference Series: Material Science and Engineering.

We would like to thank Associate Editors and members of reviewers for their kind assistance in reviewing these papers.

The Editors
Prof. Dr. Aziz Ahmad
Prof. (Assoc.) Ts. Dr. Mohd Ikmar Nizam Hj. Mohamad Isa
Prof. (Assoc.) Ts. Dr. Wan Mohd Khairul Wan Mohamed Zin
Prof. (Assoc.) Dr. Khairul Anuar Mat Amin
Dr. Maisara Abdul Kadir
Dr. Mohd Hasmizam Razali

IOP Conf. Series: Materials Science and Engineering **440** (2018) 011001 doi:10.1088/1757-899X/440/1/011001

SCIENTIFIC COMMITTEE

EDITORS
Prof. Dr. Aziz Ahmad
Prof. (Assoc.) Ts. Dr. Mohd Ikmar Nizam Hj. Mohamad Isa
Prof. (Assoc.) Ts. Dr. Wan Mohd Khairul Wan Mohamed Zin
Prof. (Assoc.) Dr. Khairul Anuar Mat Amin
Dr. Maisara Abdul Kadir
Dr. Mohd Hasmizam Razali

ASSOCIATE EDITORS
Prof. (Assoc.) Dr. Mohd Sukeri Mohd Yusof
Dr. Fatimah Hashim
Dr. Lee Oon Jew
Dr. Malinna Jusoh
Dr. Mazidah Mamat
Dr. Md. Uwaisulqarni Osman
Dr. Mohd Sabri Mohd Ghazali
Dr. Nabilah Ismail
Dr. Noor Aniza Harun
Dr. Vigneswari Sevakumaran
Dr. Nor Omaima Harun

ORGANIZER
School of Fundamental Science (PPSA), Universiti Malaysia Terengganu (UMT).

CO-ORGANIZER
Academy of Sciences Malaysia (ASM)

ORGANIZING COMMITTEE

Chairman: Prof. (Assoc.) Dr. Khairul Anuar Mat Amin
Director: Dr. Mohd Hasmizam Razali
Secretary I: Dr. Faizatul Shimal Mehamod
Secretary II: Pn. Sharifah Suryani Al Sayed Ali
Treasurer I: Cik Norazlina Abdul Aziz
Treasurer II: Pn. Suhaila Abu Samah

Scientific Journals & Publications
Prof. (Assoc.) Ts. Dr. Mohd Ikmar Nizam Hj. Mohamad Isa
Prof. (Assoc.) Ts. Dr. Wan Mohd Khairul Wan Mohamed Zin
Dr. Maisara Abdul Kadir

IOP Conf. Series: Materials Science and Engineering **440** (2018) 011001 doi:10.1088/1757-899X/440/1/011001

Invitation & Registration
Dr. Siti Nor Khadijah Addis
Dr. Aima Ramli
Dr. Razifah Mohd Razali
Dr. Nurul Aliaa Idris
Pn. Khadijah Mamat
Cik Nurfarhana Rosli

Inauguration & Protocol
Prof. (Assoc.) Dr. Wan Iryani Wan Ismail
Dr. Hanis Mohd Yusoff
Dr. Vigneswari A/P N Sevakumaran
Pn. Ku Naiza Ku Nordin

Publicity & Relations
Prof. (Assoc.) Dr. Mohd Sukeri Mohd Yusof
Dr. Nabilah Ismail
Dr. Mazidah Mamat

Logistics & Exhibition
Dr. Md Uwaisulqarni Osman
Dr. Noor Aniza Harun
Dr. Ahmad Nazif Aziz
En. Mohd Hanif Ahmad
En. Mohd Yusri Mirang
En. Maswadi Mustafa

Food & Dining
Dr. Suzana Misbah
Pn. Nurul Huda Abdul Wahab
Cik Faridah Jaafar

Certificates & Souvenirs
Dr. Syara Kassim
Dr. Fariza Hanim Suhailin
Pn. Wan Azdayanty Wan Awang

Finance & Sponsorship
Dr. Mohd Sabri Mohd Ghazali
Dr. Khadijah Hilmun Kamarudin
Dr. Lee Oon Jew
Dr. Nora Salina Md Salim

IOP Conf. Series: Materials Science and Engineering **440** (2018) 011001 doi:10.1088/1757-899X/440/1/011001

CONFERENCE PHOTOGRAPH

IOP Conf. Series: Materials Science and Engineering **440** (2018) 011001 doi:10.1088/1757-899X/440/1/011001

IOP Conf. Series: Materials Science and Engineering **440** (2018) 011002 doi:10.1088/1757-899X/440/1/011002

Peer review statement

All papers published in this volume of *IOP Conference Series: Materials Science and Engineering* have been peer reviewed through processes administered by the proceedings Editors. Reviews were conducted by expert referees to the professional and scientific standards expected of a proceedings journal published by IOP Publishing.

IOP Conf. Series: Materials Science and Engineering **440** (2018) 012001 doi:10.1088/1757-899X/440/1/012001

Mechanical and Antibacterial Activities Study of Gellan Gum/Virgin Coconut Oil Film Embedded Norfloxacin

Nur Arifah Ismail, Khairul Anuar Mat Amin, and Mohd Hasmizam Razali*

School of Fundamental Sciences, Universiti Malaysia Terengganu, 21030 Kuala Terengganu, Terengganu Darul Iman, Malaysia.

*Corresponding author email: mdhasmizam@umt.edu.my

Abstract. Biofilm based on gellan gum, was prepared by evaporative casting technique, and its properties were evaluated. The tensile strength (TS) and elongation-at-break (EAB) of the GGVCO-NOR films were found to be 9 MPa and 7.3%, respectively. The swelling properties and water vapor transmission rates of the prepared films were found to be 278% and 212 g m-2 d-1, respectively, after 24 hours. Surface morphology of the film revealed the presence of the pores. From the anti-bacterial study, it was found that the inhibition zone of GG-NOR and GG-VCONOR film were 5.3±0.06 mm and 5.7±0.06 mm against gram-positive (Staphylococcus aureus) and 5.0±0.01 mm and 6.3±0.06 mm againts gram-negative bacteria (Escherichia coli).

1. Introduction

Mechanical and antibacterial properties were considered as an important in the development of qualified wound dressing materials. The mechanical properties were essential to define the strength of the material while the antibacterial properties will prevent the wound from the bacteria infection. Fluoroquinolones group such as norfloxacin is among an effective antibacterial agent against gram-positive and gramnegative bacteria. Norfloxacin (1-ethyl-6-fluoro-1, 4-dihydro-4-oxo-7-(piperazin-1-yl) quinolone-3 carboxilic acid) is a broad spectrum antimicrobial [1] analog of nalidixic acid [2, 3]. Norfloxacin is derived from fluoroquinolones and it is among the rapid growing class of an antimicrobial agent. Norfloxacin has been reported to be more active than other fluoroquinolones against Gram-positive (*i.e.,*

Staphylococcus epidermidis, and *Staphylococcus aureus*) and Gram negative bacteria (*i.e., Pseudomonas aeruginosa* and *Escherichia coli*) [4]. At the same time, norfloxacin had appeared to be the least cytotoxic compared to other fluoroquinolones such as levofloxacin and ofloxacin [5, 6]. Previous study has reported that norfloxacin has potent activity against gram-positive and gramnegative bacteria but limited against anaerobes [2]. Norfloxacin is also used for common and complicated urinary tract infections treatment including cystitis and prostitis [7, 8]. The norfloxacin has been applied in therapy treatment for gonorrhoea infections and was became the first choice drug to treat diseases caused by *Campylobacter, E-coli, Salmonella, Shigella,* and *V. colera* [9]. In the past few years, norfloxacin has become a choice in the drug determination for biological samples and medicine [10-13]. Besides that, norfloxacin is effective in the treatment of gonococcal urethritis and infectious diarrhea [14].

Gellan gum (GG) received great attention, particularly in the field of biomedicine, due to their biocompatibility and biodegradability properties. Gellan gum is produced by *Pseudomonas elodea* and

IOP Conf. Series: Materials Science and Engineering **440** (2018) 012001 doi:10.1088/1757-899X/440/1/012001

consists of a repeating unit of tetrasaccharide: 1,3-linked -D-glucose, 1,4-linked - D-glucuronic acid, 1,4-linked -D-glucose, and 1,4-linked - L-rhamnose. Gellan gum has been explored as stomach-specific delivery [15] and it is also a candidate material for tissue engineering application [16-19]. In addition, gellan gum also has been applied in cosmetics, lotions, creams, make-ups, face masks, hair care products, and various toothpastes. Besides that, previous studies have been reported that gellan gum can be used in the bioremediation of contaminated soils and aquifers [20, 21].

On the other hand, virgin coconut oil (VCO) is one of the recent promising candidates in promoting antibacterial activities. St-Onge and Jones (2002) and Assunção *et al.* (2009) have been reported that the virgin coconut oil has potential to be used as anti-obesity treatment due to its medium chain fatty acid component of lauric acid [22, 23]. The lauric acid will increases the energy expenditure, then directly absorbed and burnt as energy in the liver, resulted in early satiety and lastly leading to weight loss [24]. The lauric acid also responsible to destroyed all the microorganisms, thus promotes collagen production on the skin and directly speeds the healing process to reduce scarring [25]. The virgin coconut oil also reported to have an anti-viral activity due to its medium chain fatty acid was converted to the monolaurin by the body [26].

2. Experiments

2.1. Materials

Low-acyl gellan gum (Gelzan™ CM, Mw ≈ 2-3 x 10^5 Da, product number-G1910, lot number SLBB0374V), and Triton™ X-102 (product number-1001318460, lot number MKBDD4707V) were obtained from Sigma-Aldrich, Malaysia. The virgin coconut oil (VCO) (product number-617488-D) was obtained from BioNeutraceutical (M) Sdn. Bhd., and the commercial antibiotic, Norfloxacin (product number-N9890) was from Fluka, USA. All materials were used as initially received.

2.2. Film Formation

A stock solution of gellan gum (GG) was made by dissolving 1% (w/v) of GG in 100 mL deionized water (18.2 MΩ) with continuous stirring for 2 h at 70 °C. GG-NOR solutions were prepared by adding 0.01% (w/w) norfloxacin (NOR) into the stock solution of GG. Meanwhile, the GG-VCO solutions were prepared by adding 0.3% (w/w) of VCO into the stock solution of GG, followed by 1.0% (w/v) of nonionic surfactant, i.e. Triton X-102. The GG-VCO solutions were then stirred at 500 rpm at 70 °C for 2 h to obtain a stable emulsion. For GG-VCO-NOR solutions, 0.01% (w/w) of norfloxacin were added into GG-VCO solution and stirred for 2 h at 70 °C. The GG, GG-NOR, GG-VCO and GG-VCO-NOR solutions were next deposited onto petri dishes (90 mm x 15 mm) and placed inside Venticell oven at 30 °C for at least 24 h. All films were pre-conditioned in a desiccator (24 °C, 50 % relative humidity (RH)) for at least 2 days prior to testing.

2.3. Characterization of Film

ATR-FTIR spectra were collected using a Perkin Elmer Spectrum 100 FT-IR spectrophotometer with PIKE Miracle ATR accessory (single-bounce beam path, 45 ° incident angle, 16 scans, 4 cm^{-1} resolution) and all spectra were corrected by the Perkin Elmer spectrum 100 software. Stress-strain measurements were obtained using an Instron Universal Testing machine (model 3366) with ± 10 kN grips and the cross-speed set at 20 mm/min. Film thickness (2.0 cm x 6.0 cm) was measured by a hand-held micrometer (Mitutoyo). Young's modulus (E), tensile strength (TS), and toughness (T) were calculated from the slope of the linear part of the stress-strain curve, maximum stress, and through integration of the area under the curve, respectively. Elongation-at-break (EAB) was also recorded. Scanning electron microscopy (SEM) images were obtained by using a JOEL JSM 6360 LA electron microscope. SEM images of cross-sections were obtained by freeze-drying the samples in liquid nitrogen (-160 °C). Then, the samples were fractured at -150 °C. Later coated with Auto Fine Coats (JFC-1600), and imaged by SEM.

2.4. Bacterial study
Gram-positive (*Staphylococcus aureus - S. aureus*) and gram-negative (*Escherichia coli – E. coli*) bacteria were used for the antibacterial assay. The standard growth medium (Muller-Hinton, MH, Difco™) agar was prepared by sterilized with an autoclave (15 min, 120 °C). *S. aureus* and *E. coli* were grown in MH agar and incubated aerobically (37 °C, 24 h). The suspensions were measured by using spectrophotometer (Biomerieux Densicheck Plus, USA) at 600 nm with optical density at 0.5. Inoculants of both *S. aureus* and *E. coli* were spread on sterile petri plates contained the MH agar. Using sterile cotton swab, both bacteria were swabbed over the surface of the agar plates. The film samples and control was gently pressed on the agar. Prior inserting the samples discs with diameter 6 mm, the plates containing bacterial were let to dry. The plates contained film samples and agar with bacteria were incubated at 37 °C for 24 h in triplicates. The plates were examined for the presence of zones of growth inhibition after 24 h of incubation at 37 °C.

3. Results and Discussion
The ATR spectra of GG-VCO-NOR spectrum in Figure 1 shows an absorption band at 3377 cm^{-1} which is refer to –N-H stretching. This absorption band is absent in virgin coconut oil (VCO) and appeared in gellan gum (GG), norfloxacin (NOR), and GG-VCO-NOR spectra. A strong absorption band of alkane groups also have been observed at 2855 and 2924 cm^{-1} in GG-VCO-NOR film due to the interaction of hydrogen bonding formation between the blends [27] as proposed in Figure 2. Besides that, the peak at 1745 cm^{-1} indicated the formation of esterification and conjugation of carbonyl ester group [28] due to the interaction occurred and the particular peak almost absent in pure norfloxacin. Further, the appearance peak at 1615 cm^{-1} in GG-VCO-NOR films proved that ~N-H bending vibration of quinolones group successfully interact in the blends in which the group is absent in pure gellan gum, virgin coconut oil and also in GG-VCO films.

The chemical interaction of ester group stretching vibration at 1246 cm^{-1} also strongly supported the interaction between the blends in GG-VCO-NOR film due to this peak absent in pure norfloxacin and GG-NOR but appeared in GG-VCO-NOR films. Last but not least, the GG-VCO-NOR interaction was supported by the appearance of fluoroquinolones group in the blends at 1111 cm^{-1}. However, this peak was shifted to the left compared to pure norfloxacin films. This might be due to the formation of βketoesters [29]. This is expected due to the chemically interaction between the carboxylic acid group of norfloxacin and the polymer (GG) also might be with the virgin coconut oil. The nitrogen atoms of quinolone groups at ortho position are less of electron rich due to the deficient of flouroquinolone ring. So, it is are not likely involved in the reaction and sterically hinder the reaction also due to the methoxy and piperazinyl groups [28].

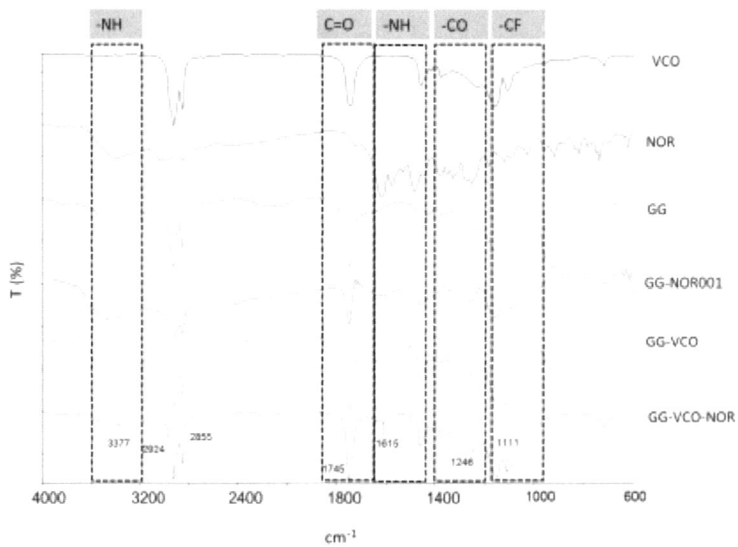

Figure 1. ATR spectra of pure virgin coconut oil (VCO), pure norfloxacin (NOR), gellan gum (GG) film, gellan gum film containing norfloxacin (GG-NOR), gellan gum film with virgin coconut oil (GGVCO) and gellan gum film containing virgin coconut oil and norfloxacin (GG-VCO-NOR).

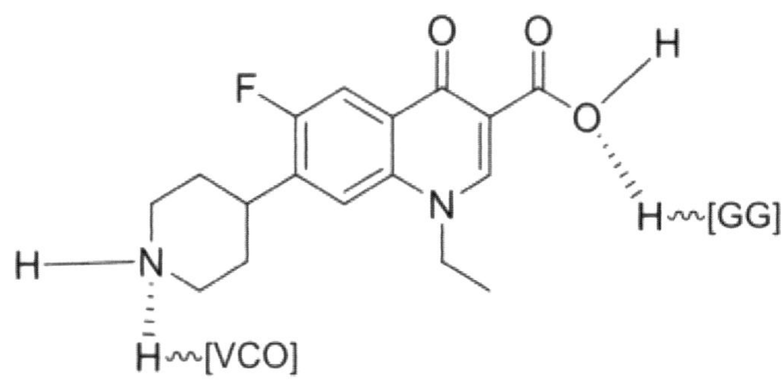

Figure 2. Proposed norfloxacin structured for hydrogen bonding formation between GG and VCO.

The stress-strain diagram of GG, GG-VCO, GG-NOR and GG-VCO-NOR films are shown in Figure 3. The addition of virgin coconut oil (VCO) improved the flexibility or elongation-at-break and inclusion of norfloxacin (NOR) increased the toughness values. The addition of VCO into GG film improved the T and EAB value of composite film to 0.67 J g^{-1} and 13.5 %. Our result show that the reason as to the increase of T and EAB of GG-VCO film was due to the formation of hydrogen bonds by reaction of carbonyl group of virgin coconut oil with gellan gum as proved by ATR spectroscopy. Meanwhile, the addition of norfloxacin in gellan gum films increased the toughness (T) values, however decreased the tensile strength (TS) and Young's modulus (E) of films as summarized in Table 1. The inclusion of norfloxacin in gellan gum films improved the toughness to 0.28 J g^{-1} compared to gellan gum film at 0.26 J g^{-1} which equivalent to 1-fold increase. However, the TS and E value of the GG-NOR films were decreased to 59±11 MPa and 3671±639 MPa; respectively, compared to gellan gum film. This is due to the modification imposed on the composite films lead to weakening the TS of the composite films. For GG-VCO-NOR films, the strain-at-break improved to 7.3 % compared to GG-NOR at 2.1 %, but lowered than GG-VCO at 13.5 % as shown in Figure 3 and summarized in Table 1.

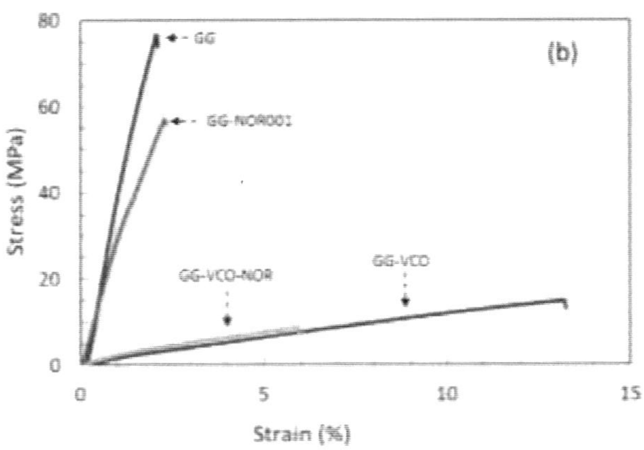

Figure 3. Stress-strain curves of comparison between gellan gum (GG) film, gellan gum film containing norfloxacin (GG-NOR), gellan gum film with virgin coconut oil (GG-VCO) and gellan gum film containing virgin coconut oil and norfloxacin (GG-VCO-NOR).

Table 1. Summary of the tensile strength (TS), Young's modulus (E), toughness (T), and elongation-atbreak (EAB) of gellan gum containing of norfloxacin (GG-NOR) films, gellan gum with virgin coconut oil (GG-VCO) and gellan gum containing virgin coconut oil and norfloxacin (GG-VCO-NOR) film (mean±SD) (n=3)

Sample	**Thick (μm)**	**TS (MPa)**	**E (MPa)**	**T ($J\ g^{-1}$)**	**EAB (%)**
GG	30±0.01	80±12	4600±330	0.26±0.09	2.1±0.2
GG-VCO	80±0.06	14±2	130±19	0.67±0.33	13.5±2
GG-NOR	27±0.01	59±11	3671±639	0.28±0.06	2.1±0.5
GG-VCO-NOR	87±0.01	9±1	195±61	0.25±0.19	7.3±3

In contrast, incorporated both of VCO and NOR decreased the TS and YM values to 9 MPa and 195 MPa; respectively. The addition of virgin coconut oil and norfloxacin into the blends promote the changes of chemical interaction due to the hydrogen bonding formation, thus influenced the mechanical properties. The micelles formation appeared as in the GG-VCO film, then exhibit the flexibility to the materials film. Gellan gum, like many anionic polysaccaharide forms a physical gel by undergoing a random coil-helix to double-helix transition upon cooling [30]. During this transition, the emulsion does not have enough time to settle off, thus micelles formation was produced as shown in Figure 4 (c) and (d) compare to smooth cross section appearance of GG and GG-NOR film (Figure 4 (a) and (b)). The formation of micelles interfered the chemical structure thus enhanced the toughness of the GG-VCONOR films.

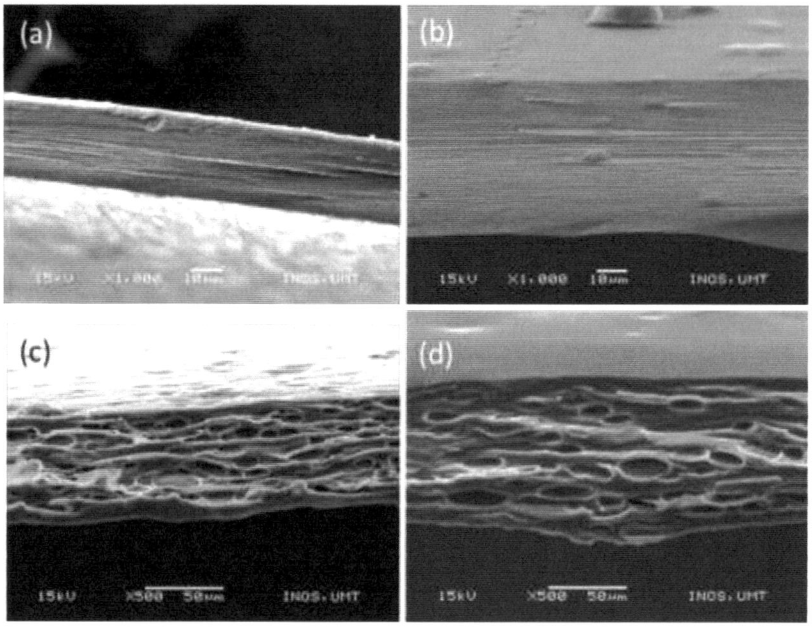

Figure 4. Scanning electron microscopy images of cross-sectional area of (a) gellan gum film, (b) gellan gum film containing norfloxacin (GG-NOR), (c) gellan gum film with virgin coconut oil (GG-VCO) and (d) gellan gum film containing virgin coconut oil and norfloxacin (GG-VCO-NOR).

The antibacterial activity of GG, GG-VCO, GG-NOR and GG-VCO-NOR films were measured by qualitative (disk method) as shown Figure 5. The inhibition zone was recorded after 24 h against *Staphylococcus aureus* and *Escherichia coli* executed by virgin coconut oil and norfloxacin (Table 2). The GG and GG-VCO film shows no antibacterial activities. It has been elaborated that the VCO does not possess antibacterial activity on its own, but rather is induced by its free fatty acids, particularly lauric acid (C12), and small extent of capric acid (C10) and caprylic acid (C8) [31]. Or, in other words, the VCO must be metabolized to release those components and exert its antimicrobial effects [32]. On the other hand, the inhibition zone of GG-NOR about 5.3 mm and 5.0 mm resisted *Staphylococcus aureus* and *Escherichia coli*, respectively (Table 2). The effectiveness antibacterial activity of norfloxacin against Gram-positive and Gram negative bacterial is well known by inhibits deoxyribonucleic acid synthesis intracellularly in bacteria [33-39]. It alters the structure of the enzyme resulting in abnormal change in polypeptide production. Norfloxacin differs from other quinolones, which include nalidixic acid and cinoxacin, by having a fluorine atom at the six position and a piperazine moiety at the seven position. The fluorine atom at the six position provides for increased potency against gram-negative organisms, while the piperazine moiety at the seven position provides for antipseudomonal activity [33]. Inhibition zone of GG-VCO-NOR films incorporated with both virgin coconut oil and norfloxacin improved the antibacterial properties are shown in Figure 5 and Table 2. This is due to the antibacterial properties of virgin coconut oil was enhanced and supported to inhibit the bacterial growth. The inhibition zone of GG-VCO-NOR films (5.7±0.06 and 6.3±0.06 against *S. aureus* and *E. coli*; respectively) were larger compared to GG-NOR001 (5.3±0.06 and 5.0±0.06 against *S. aureus* and *E. coli*; respectively).

.

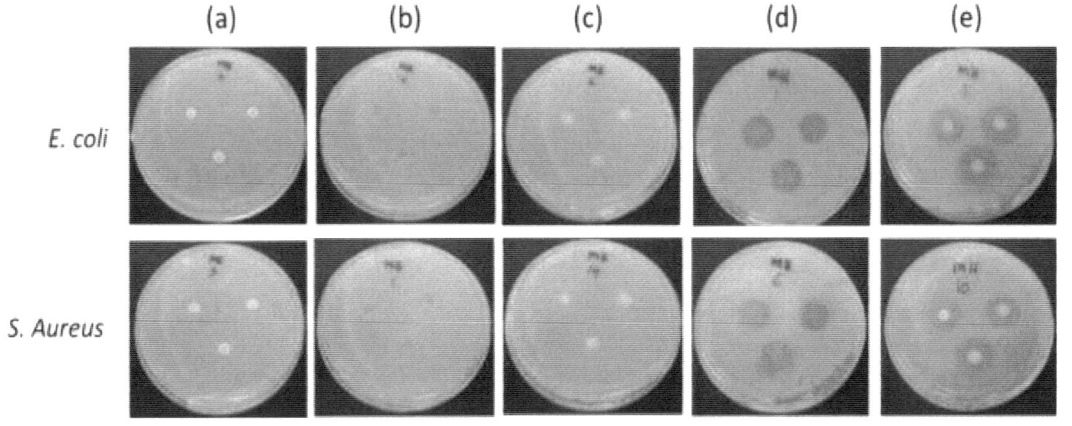

Figure 5. Disk diffusion results of (a) penicillin (positive control), (b) gellan gum film (c) gellan gum film with virgin coconut oil (GG-VCO) (d) gellan gum film containing norfloxacin (GG-NOR), and (e) gellan gum film containing virgin coconut oil and norfloxacin (GG-VCO-NOR) after incubated for 24 h against Escherichia coli and Staphylococcus aureus bacteria.

Material	Inhibition zone (mm)	
	S. aureus	*E.coli*
GG	-	-
Penicillin	-	-
GG-VCO	-	-
GG-NOR	5.3±0.06	5.0±0.01
GG-VCO-NOR	5.7±0.06	6.3±0.06

Table 2. Qualitative results of gellan gum containing different concentrations of norfloxacin (GG-NOR) films, gellan gum with virgin coconut oil (GG-VCO) and gellan gum containing virgin coconut oil and norfloxacin (GG-VCO-NOR) film against the Staphylococcus aureus (S. aureus) and Escherichia coli (E. coli) bacteria as indicated by the zone of inhibition (mm) (mean±SD) (n=3)

4. Conclusion

This work successfully explored the formation of composite films through an evaporative casting technique by using gellan gum (GG), virgin coconut oil (VCO), and norfloxacin (NOR). The mechanical properties of GG films were improved by incorporated of VCO and NOR. The morphological appearance of the GG-VCO films shows the micelles formation due to the presence of surfactant. For the qualitative antibacterial studies, the incorporating of NOR into GG contained VCO was increased the inhibition zone to 5.7±0.06 mm and 6.3±.0.06 mm against *Staphylococcus aureus* and *Escherichia coli*, respectively.

5. References

[1] Gonjaril I D, Jayveera K N, and Kashikar V S 2012 *J. Pharm. Res.* **5** 1620
[2] Wang Y, Zhu L, Dong Z, Xie , Chen X, Lu M, Wang X, Li X, and Zhou W 2012 *Colloids and Surfaces B: Biointerfaces* **98** 105
[3] Shinde A J, More H N, Jadhav S R, Ingole V S, & Kakade H R 2012 *Int. J. Pharm. Biomed. Res.* **32** 90
[4] Fukuda H, Hori S, and Hiramatsu K 1998 *Antimicrobial agents and chemotherapy* **42** 1917
[5] Yamamoto T, Tsurumaki Y, Takei M, Hosaka M, and Oomori Y 2001 *Toxicology in vitro* **15** 721
[6] Alvarez-Lorenzo C, Yanez F, Barreiro-Iglesias R, and Concheiro A 2006 *J. of Control. Release* **113** 236
[7] Nayak K, Khare N K, Sayare A, Ghode P, and Lawrence R M 2012 *Scholars Research Library,*

Der Pharmacia Lettre **4** 192
[8] Chen M and Chu W 2012 *J. of hazardous materials* **219** 183
[9] El-Megharbel S M, Adam A M A, and Refat M S 2012 *Life Science Journal* **9** 1234
[10] Ni Y, Wang Y, and Kokot S 2008 *Spectrochimica Acta Part A: Molecular and Biomolecular Spectroscopy* **70** 1049
[11] Tong C and Xiang G 2006 *J. of fluorescence* **16** 831
[12] Huang K J, Liu X, Xie W Z, and Yuan H X 2008 *Colloids and Surfaces B: Biointerfaces* **64** 269
[13] Cui J, Zhang K, Huang Q, Yu Y, and Peng X 2011 *Analytica chimica acta* **688** 84
[14] Bharath S, Patel N J, Deweswaran R, Bsavaraj B V, and Madhavan V 2012 *Int. J. Pharm. Res. Dev.* **4** 16
[15] Pahwa R, Bhagwan S, Kumar V, and Kohli K 2010 *J. of Pharm Res.* **3** 1312
[16] Tang Y, Sun J, Fan H, and Zhang X 2012 *Carbohydr. Poly.* **88** 46
[17] Ismail N A, Mat Amin K A, and Razali M H 2018 *Materials Letters* **228** 116
[18] Ismail N A, Razali M H, and Mat Amin K A 2017 *AIP Conference Proceedings* **1885** 020045
[19] Ismail N A, Mohamad S F, Ibrahim M A, and Mat Amin K A 2014 *Advances in Biomaterials,* **2014 1**
[20] Johnsen A R and Karlson U 2004 *Applied Microbiology and Biotechnology* **63** 452
[21] Moslemy P, Guiot S R, and Neufeld R J 2002 *Enzyme and Microbial Technology* **30** 10
[22] St-Onge M P and Jones P J 2002 *J. of Nutrition* **132** 329
[23] Assunçao M L, Ferreira H S, dos Santos A F, Cabral C R and Florêncio T M 2009 *Lipids* **44** 593
[24] Mansor T S T, Man Y C, Shuhaimi M, Afiq M A, and Nurul F K 2012 *Int. Food Res. J.* **193** 837
[25] Batool S H 2012 *Basrah J. of Veterinary Res.* **11** 28
[26] Dalmacion G V, Ortega A R, Pena I G, and Ang C F 2012 *Functional Foods in Health and Disease* **2** 290
[27] Sudhamani S R, Prasad M S, and Sankar K U 2003 *Food Hydrocolloids* **17** 245
[28] Sahoo S, Chakraborti C K, and Naik S *2011 Int. J. of Pharm. Sciences and Nanotechnology* **4** 1505
[29] Jiménez-Garrido N, Perello L, Ortiz R, Alzuet G, Gonzalez-Alvarez M, Canton, E., Liu-González M, García-Granda S, and Pérez-Prieded M 2005 *J. of Inorganic Biochemistry* **99** 677
[30] Kirchmajer D M, Steinhoff B, Warren H, and Clark R 2014 *Carbohydr. Res.* **388** 125
[31] Shilling, M, Matt L, Rubin E, Visitacion M P, Haller N A, Grey S F, and Woolverton C J 2013 *J. of medicinal food* **16** 1079
[32] Ogbolu D O, Oni A A, Daini O A, and Oloko A P 2007 *Journal of medicinal food* **10** 384
[33] Qadri S M and Johnson S 1989 *J. of the National Medical Association* **81** 382
[34] Meler J, Grimling B, and Pluta J 2012 *Progress on Chemistry and Application of Chitin and Its Derivatives* **17** 99
[35] Body B A, Fromtling R A, Shadomy S and Shadomy H J 1983 *Euro. J. of Clinical Microbio.* **2** 230
[36] Norrby S R and Jonsson M 1983 Antimicrobial agents and Chemotherapy 23 15
[37] Shungu D L, Weinberg E, and Gadebusch H H 1983 *Antimicrobial agents and chemotherapy* **23** 86
[38] Barry A L, Jones R N, Thornsberry C, Ayers L W, Gerlach E H, and Sommers H M 1984 *Antimicrobial Agents and Chemotherapy* **25** 633
[39] Cohen S P, Hooper D C, Wolfson J S, Souza K S, McMurry L M, and Levy S B 1988 *Antimicrobial agents and chemotherapy* **32** 1187

Acknowledgments
The authors are grateful to Universiti Malaysia Terengganu (UMT) for fnancial assistance under internal research grant (TPM, Grant no. 68006/2012/46) and the Institute of Biotechnology Marine (UMT) for providing the facilities for undertaking this work.

Optimization of NepetaCataria Essential Oil Extraction Yield by Ultrasonic-Soxhlet Extraction Method Using Response Surface Methodology

R M S Marmy[1], N H Norulakmal[1] and G Faridah[1]

1Process Department, Universiti Kuala Lumpur (UniKLMicet), TabohNaning, 78000 Alor Gajah, Melaka, Malaysia.

Abstract. Essential Oil (EO) of *nepetacataria L.* has an attractive benefit due to its potential as alternatives compound of controlling mosquito vectors especially Aedesaegypti. This paper reports on application of response surface methodology (RSM) using Box Behnken design (BBD) to optimize the extraction of EO from *nepetacataria L.* The parameters studied are time of extraction (120-360 min), temperature of sonication (30-50°C) and ratio of sample to solvent (w/v) (30-60 gram/150ml). The extraction was conducted by using soxhlet method assisted by ultrasound using ethanol (polar solvent). Statistical analysis and analysis of variance indicated that the time of extraction and ratio of sample to solvent are significant factors ($p<0.05$) affecting the yield while the temperature of sonication was insignificant ($p>0.05$). For the modelled extraction, the minimum critical values for ethanol solvent was 269.48 minute, 40.18 °C and 33.22 gram/150 ml ethanol. Recovery of oil when 1 gram of raw material was used per 150 ml of solvent was 2.32%. By using High Performance Liquid Chromatography with Ultraviolet detector (HPLC-UV), nepetalactone was detected at time of 12.694 min and 13.941 min using absorption peak 228nm.

1. Introduction

Nowadays, the usage of herbs in daily life become widely spread and common. Herbs been used as medicines to treat illness. Besides, it has been used in a wide range of applications such as dietary supplement, remedies and nutrition. Generally people made herbs beneficial in many manners according to the specified functions. Blumenthal et al. (2006) [1] stated that there are positive growths in total sales of herbal dietary supplements within 10 years from 1996 to 2005. In this study, *nepetacataria L* leaves is choose as the plant material to extract the essential oil. Essential oil (EO) has high potential to be used as medicine to human for many diseases. EO are chemically complex and contain many different substances and compounds. Different types of plants has different chemical characteristic which made essential oil is a reliable source of alternative medical treatment The plant is choose due to its potential as alternatives compound of controlling mosquito vectors especially Aedesaegypti. The aim of this study is to determine the optimization conditions for the extraction of essential oil from nepeta leaves using ultrasonic assisted soxhlet extraction method.

Recently, dengue had becomes one of the major problems to our country, Malaysia and also few other countries in Asia. According to World Health Organization, WHO (2015) about 500,000 people been hospitalized each year due to severe dengue. Unfortunately, a large number of those affected are among children and this problem associated with more serious infections which have more prevalent as 2.5% of those affected die. Dengue is a systemic viral infection transmitted between humans and Aedes mosquitoes. Prevention of this disease almost entirely dependent on various methods of vector control. Control of vectors by insecticides remains the most important method of reducing disease transmission and protection from mosquito bites [2]. Alternative compounds of controlling mosquito vectors are needed which lies in botanical compound that commonly used and known as insect repellent. The usage of current synthetic insecticides for the control of the vector of mosquito has disturbed the natural biological control system and undesirable effects on non-target organisms, fostered environmental and human health concerns [3]. N,N diethyl-z-toluamide (DEET) are the most conventional repellent used which is now still effective. However there is arising concern on the usage of DEET which results in unpleasant odour, can damage plastics and synthetic rubber and has high skin penetration characteristics [4]. Due to the problem for the usage of DEET, new improvement on the repellents and strategies to control mosquito attack is needed. The best alternative is to use bioactive compound which extracted from plant because they gives no harmful effect. Thus, much interest has focused on plant extracts, or plant essential oils, as potential mosquito repellent agents.

A few numbers of botanical extract were tested as alternatives mosquito repellents such as eucalyptus (*Eucalyptus citriodora Hook*), citronella grass (*CymbogonnardusRendle*), thyme (*Thymus vulgaris L.*), clove (*Syzygiumaromaticum L.*) and catnip (*NepetaCataria L.*) [5]. In this study, *nepetacataria L.* commonly known as catnip or catmint is used as herbal plant where historically the oil of *nepetacataria L.* has been used in herbal medicine to treat fever, head and tooth aches, colds, colic and spasms in human [6]. Research of *nepetacataria L.* plants had been widely studied by researcher such as a stable effective fly repellent studied by Zhu, et al. 2012 [7] and the usage of essential oil of *nepetacataria L.* as a natural preservative agent in food products [8]. There are continued interests in studies on behavioural response of NepetaCataria L. towards few species of mosquitoes including AedesAegypti.

Essential oils were obtained from parts of the plant by many different ways mainly by extraction method. Ultrasonic assisted solvent extraction (UAE)is being introduced in the extraction industry. Ultrasound is sound waves that comprise frequency between 18 kHzto 100 kHz. The broad range can be divided into high intensity or called aspower ultrasound (20 – 100 kHz) and diagnostic ultrasound (1-10 MHz)[9]. Ultrasonic waves can easily penetrate opaquematerials, where as many other types of radiation such as visible lightcannot. Since ultrasonic wave sources are inexpensive, sensitive andreliable, this provides a highly desirable way to probe and image the interiorof opaque objects [10]. Ultrasound also achievesgreater penetration of a solvent into a plant tissue and improves the masstransfer. A considerable amount of literature has beenpublished on the success of ultrasound [11]; [12]; [13].Recent developments in the field of extraction have opened up somegreat expectations with a promising result of UAE. UAE utilises acousticcavitation to cause molecular movement of solvent and sample. This methodoffer many advantages such as improved efficiency, reduced extraction time, low solvent consumption, and high level of automation as compared toconventional extraction techniques [14]. Currently, thetypically used ultrasonic systems are either bath and probe type sonicator.Both system are based on electromagnetic transducer. Figure 1 shows theschematic diagram of ultrasonic bath [9].

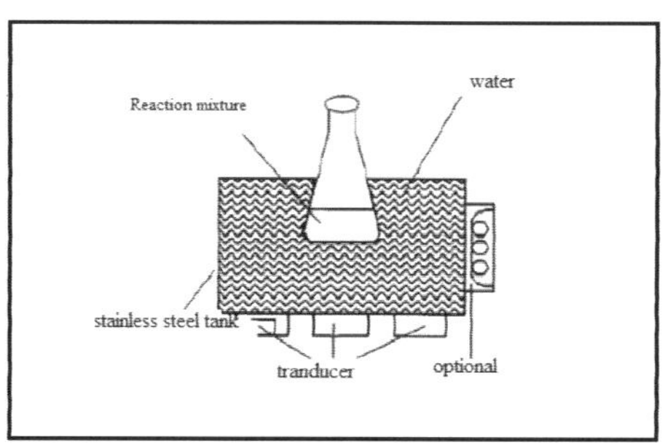

Figure 1. Schematic diagram of ultrasonic bath.

2. Materials and method

The extraction procedure was divided into few subsection where starting with preparation of raw material which is important in order to suit the material structure with the extraction equipment. Preparation of solvent where the two solvent was measured earlier for every run to ease the experiment and avoid any incident to occur if the solvent was being measured right when the solvent is needed. Design of experiment was performed before the start-up of experiment to determine the number of runs and also the matrix of experiment. Prior to extraction of oil, pre-treatment was conducted using Ultrasonic Bath and followed by Soxhlet extractor. Themixture of solvent and oil was separated using rotary evaporator. Selected oil sample was brought to analysis in order to detect the desired constituentwhich was Nepetalactone.

2.1 Selection of design parameters

Thisexperimental design can be generated by Design of experiment method (DOE). It facilitates an arrangement and implementationmethod of the experiment in order to investigate the parameter affecting theexperiment. Once arrangement of method is done, the data obtained arefurther analysed by using statisticasoftware version 10.0 . In this paper, the objectives function forthe optimization of factor affecting the yield of extraction which is to obtainmaximum yield of essential oil. Therefore, three main parameters which aretemperature of the sonication, time of soxhlet extraction and ratio of solventto sample is chosen.

2.2 Box–Behnken experimental design

Experimental design is widely used for controlling the effects of parameters inmany processes. Its usage decreases number of experiments, using timeand material resources. Furthermore, the analysis performed on the results iseasily realized and experimental errors are minimized.In this study, the Box–Behnken experimental design was chosen for findingout the relationship between the yield of extraction and the three variables.Box–Behnken design is rotatable second-order designs based on three-levelincomplete factorial designs. The special arrangement of the Box–Behnkendesign levels allows the number of design points to increase at the same rateas the number of polynomial coefficients.

2.3 Preparation of Raw Material & Drying Process

Sample was prepared by firstly being washed, driedand grinded. The leaves of raw material which is *nepetacataria L.* was obtained on 2nd July 2017 atthe road sideSimpangEmpat, Alorgajah, Melaka. Firstly, after collected theplant material, whole part of plantwere washed to avoid high contamination. The plants were being cut intosmall pieces by using cutter and divided the parts of plant before being keptinside beaker according to the parts. The plant was then left inside oven at 55°C about 6-12 hours to remove moisture content. Then, the dried plants weregrinded by using conventional blender to obtainsample with smaller particles in order to increase surface area and allowhigher yield of extraction. After grinded, the sample was sieved by using sieve to obtain uniform size of plant particles. The sample was the placedinside small beaker to be weighed according to the ratio. Sample wasweighed by using analytical balance stage by stage to obtain better andaccurate reading for weighing. Weigh of the dry sample were recorded forfurther data processing.

2.4 Extraction method

The optimization of the parameter affecting extraction yield will bestudied by using Statistica 10.0 and meanwhile the method that will be used is Box Behnken design (BBD). The extraction method will be done by using Soxhlet ExtractionMethod assisted by Ultrasound Solvent Extraction. The parameter to bestudied is only the extraction time, temperature of extraction and ratio ofsolvent to sample. The solvent will be ethanol. The extraction will be done within 120 minutes,240 minutes and 360 minutes. For the ratio of solvent to sample, 10 gram, 20gram and 30 gram of sample will be used and amount of solvent is fixed to150 ml.

2.5 Analysis

According to Zhang (2009) [15] essential oil yield can be expressed in terms ofthe volume of the oil collected in gram per gram of dry plant material. The essential oil obtained will be analysed by using highperformance liquid chromatography (HPLC) to detect the presence ofNepetalactone.HPLC separation was performed on a YMCPackPro C18 column (150 × 4.6 mm i.d.; 3 μm) with a gradient mobile phasesystem of water containing 0.1% (v/v) formic acid(mobile phase A) and acetonitrile (mobile phase B). The wavelength of UV detection was 228 nm and the flow rate was 1 mL/min.

3. Results and discussion

Box Behnken Method was used as the type of response surfacemethodology to develop polynomial regression equation in order to analyse the correlation between the variables chosen to the yield of extraction. The final empirical formula models for the yield of extraction for Ethanol (Y1) are represented by Eq. 1 whereas Eq. 2 shows the general model ofexperiment before the coefficient of each variable been determined. Thecoefficient of the variables were tabulated and the value was been substituted into Eq.2 in order to obtain the predicted second orderpolynomial model.

$$Y1 = B0 + B1X1 + B2X2 + B3X3 + B12X1X2 + B13X1X3 + B23X2X3 + B11X12 + B22X22 + B33X32 \quad (1)$$

$$Y1 = 156.6942 - 0.7892\ X1 - 4.2160\ X2 - 2.1406\ X3 + 0.0019X12 + 0.0346\ X22 + 0.0783\ X32 + 0.0169X1X2 + 0.0002\ X1X3 + 0.0002\ X2X3 \quad (2)$$

Where, Y is the predicted response, B0 is model constant; X1, X2, and X3 areindependent variables; B1, B2, and B3 are linear coefficients; B12, B13, and B23are cross product coefficients and B11, B22, and B33 are the quadraticcoefficients.Positive signs in all equation indicate synergistic effects. In term of this research variable, increase in time of extraction, temperature of sonicationand ratio of sample to solvent will increase the yield of extraction.Conversely, negative signs in all the equation indicate antagonistic effectswhere yield of extraction will increase only if values of time of extraction,temperature of sonication and ratio of sample to solvent decrease.

The quality of the model developed was evaluatedbased on the correlation coefficient, R^2. The closer the R^2 value to unity themore accurate the response could be predicted by the model. In thisexperiment, the R^2 value for yield of extraction for ethanol was 0.99219. Thisindicated that 99.219%of the total variation in yield of extraction wasattributed to the experimental variables studied. The R^2 value of 0.99219 wasconsidered as relatively high which reflected that predicted values for yield ofextraction would be more accurate to its actual value. The model developedwas successful in capturing the correlation between the variables to theresponses. The plot of experimental value of yield of extraction (%) vs thepredicted values obtained from the model indicated a good fit as presented in Figure 2.

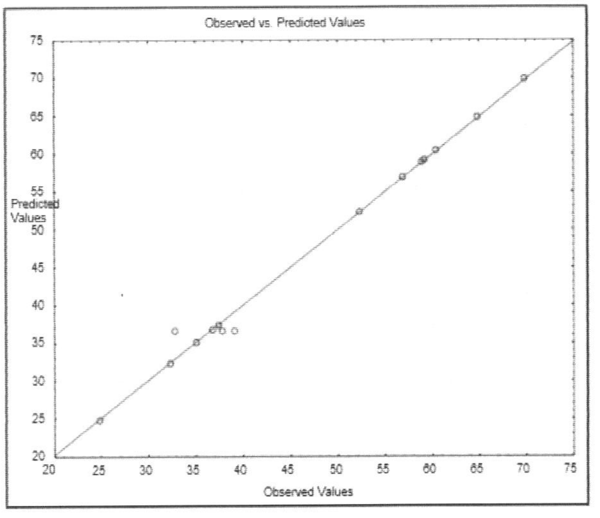

Figure 2 Predicted vs. Actual Extraction Yield (%) for ethanol

3.1 Effect of Parameters and the Interaction Between Parameters on Yield of Extraction for Ethanol

In order to determine the necessity of optimization of the parameter,hypothesis testing was conducted. Statistical analysis heavily relies on test of the hypothesis. It plays animportant role in decision making or analysis. In statistics, hypothesis undertest is referred to as hypothesis null, and associated with alternativehypothesis which is defined to be opposite of the null hypothesis.In this project, the null hypothesis was stated that the yield ofextraction does not affected by the changes of the value of the variablechosen, while the alternative hypothesis was stated that the yield ofextraction was affected by the changes of the value of the variable chosen.For the validation of which hypothesis to be rejected or accepted, anappropriate test statistic which is F distribution test was computed from theexperimental data obtained. Analysis of variance (ANOVA) was performed toobtain the Fishers variance ratio (F-value). The Fishers' variance ratio F-value is the ratio of the mean square owing to regression, to the meansquare owing to error. It is the measure of variation in the data about themean. The mean squares were obtained by dividing the sum of the squares of each of the variation sources, the model and the error variance, by therespective degrees of freedom. The results of the ANOVA are given in Table 1 below.

Table 1 Analysis of Variance (ANOVA) on Yield of Extraction

Sources	Sum of Squares(SS)	Degree of Freedom (df)	Mean Squares (MS)	F-value	R2
Regression (SSR)	2809.656	12	234.138	21.163	0.99219
Error	22.127	2	11.0635	-	-
Total (SST)	2831.783	14	-	-	-

From Table 1 above, the model F value was 21.163 which fall on the criticalregion of the distribution as the F value from F distribution table was 19.4.Hence in this case, dthe null hypothesis was rejected while alternativehypothesis was accepted. On top of this, the optimization of the parameterwas highly suggested to support the alternative hypothesis which stated thatthe yield of extraction was affected by the changes of the value of thevariable chosen.A better view of the interactions can be seen using Pareto chart as shown in Figure 3 which was used to graphically summarize the importanceof each parameter with respect to yield of extraction.

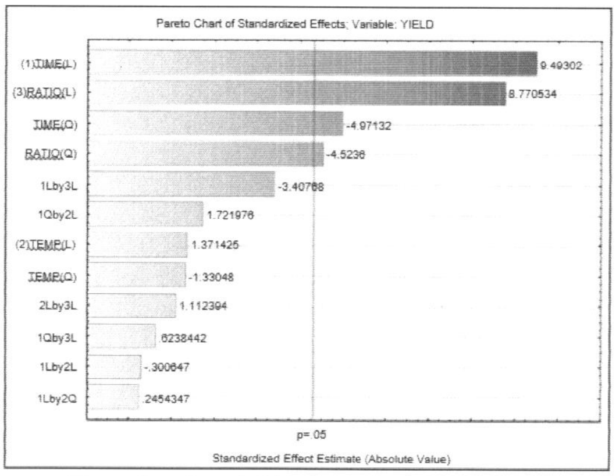

Figure 3 Pareto Chart for the yield of extraction using ethanol as solvent

NOTE :L is linear relation and Q is quadratic relation while the parameter 1, parameter 2 and parameter 3 indicated parameter time of extraction, temperature of sonication, and ratio of sample to solvent

The Pareto chart(Figure 3) above showed all the effect linear, quadratic and also the effect of interactionsbetween parameter. It also displayed a frequency histogram with the lengthof each bar proportional to each of the estimated standardized effect. Thevertical line on the chart functions as indicator whether the effect isstatistically significant within the generated response surface methodology.The bars that extend beyond the lines are statistically significant at 95%confidence level. The sequences of the bar are based on the level ofsignificance of the effect of parameters and the interaction which increasesfrom bottom to top. The least significance bar was at the bottom and thesequence followed by the higher significant parameters.Threedimensional response surface was plotted which can be seen in Figure 4. The use of surface plots was also to explore the effect of changingfactor levels on the response. There are basically three interactions between the parameter to be studied which were the parameter time of extraction vs temperature of sonication, time of extraction vs ratio of sample to solvent, and ratio of sample to solvent vs temperature of sonication. All of those three interactions are not significant based on the p value as stated in the Pareto chart.

Based on the Pareto chart, bar of interactions between parameters did not extend beyond the vertical red line which proven the insignificance of the interactions.

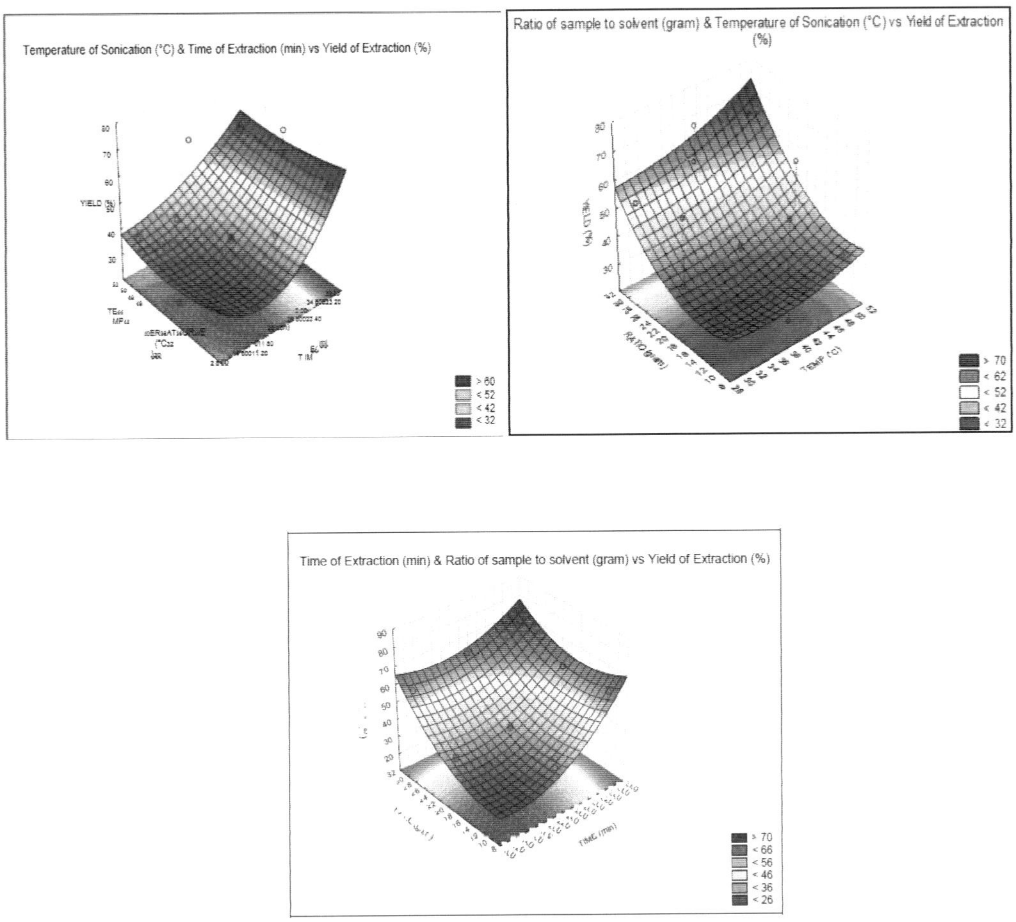

Figure 4 Response Surface Plot of oil yield for ethanol

The effect of single parameters, threedimensional response surface was plotted which can be seen in Figure 4. The use of surface plots was also to explore the effect of changing factor levels on the response. From the 3D fitted surface graph, dark areawith color tone of orange to brown indicates high yield of extraction whilebrighter area with color tone of green to yellow indicates low yield ofextraction. Increment from bright color to dark color shows an increase inyield of extraction due from the effect of all parameters involves.The effect of time of extraction on yield of extraction was illustrated asshown in Figure 4 which reported the interaction between the time ofextraction and temperature of sonication on yield of extraction while the interaction between time of extraction and ratio of sample tosolvent on yield of extraction. As can be seen from both plotted responsesurface, the yield of extraction was high at time of extraction higher than 340minutes. It is apparent from the result which indicate that time of extractiondisplayed a linear effect on the response, where the longer the time ofextraction, the higher the yield of extraction. Therefore, optimum time ofextraction has not been achieved from the response surface when ethanolwas used as solvent.The minimum critical value obtained for the time of extraction was at157.28 minutes which demonstrated that at the beginning of extraction,before reaching the minute of 157.28, the yield was higher. This had proventhe theoretical background for the extraction, on the basis of Fick's secondlaw of diffusion. The osmotic pressure causes the solvent to penetrate theplant matrix from the beginning of extraction, and later at certain level,concentration of solute in plant matrix and the solvent achieved finalequilibrium causes a slow rate of extraction to occur [16].

The change in temperature thus was notsignificant to the yield of extraction. The reason for this insignificance wasbecause the range of temperature or the level of minimum and maximumwas not relevant. Contrary to expectation, the usage of ultrasound expectedto allow changes in extraction condition as mentioned by Romdhane&Gourdon (2002) [17] in their research where decrease in temperature of theextraction process was achieved. Due to this expectation the level of factorswas chosen between 30°C to 50°C only however other factors such as thesize of plant material and polarity of solvent used, possibility of obtainingdifferent results when different type of material had been used is relativelyvery high. The later factor contributed highest influence on the extractionprocess. This evidence is supported by Wang and Weller(2006) [13] who claimedthat co-solvent could give higher yield of essential oil. In addition, the effect oftemperature in the range of 40–66 °C on the yield is negligible, such thatoptimal extraction occurs across the range of temperature from 40 to 66 °C. Therefore, use of ultrasound-assisted extraction is advisable for extraction ofcompounds, which may be altered under Soxhlet operating conditions due tothe high extraction temperature [13].From the graphical representation shown in Figure 4,the plots showed that rising in temperature of sonication does not affect theyield of extraction.The minimum critical temperature obtained from thestatistical analysis was 34.84°C. However the optimum temperature ofsonication need to be determined because too high temperature might givethermal degradation effect throughout the experiment.

Based on graphical representation high yield could be obtain at ratio of range from 30 to 32 gram of sample per150 ml of solvent. This indicated that the yield of extraction increased alongwith the increment of the weight of plant material. A linear effect can beinterpreted hence again, optimum ratio could not be determined at this levelof experiment. In order to obtain higher yield of essential oil, higher amount ofdried plant material is needed however thimble of soxhlet extractor can onlyfit 30 gram of sample. Therefore modification on the parameter ratio can onlybe made if the size of particle was reduced to smaller dimension. At certainextent, the extraction remains unchanged as most of the oil in the plantmaterial had been extracted. According to Sahin and Samli (2012) [18] the yield of extraction decreases with the increase of solventvolume. This is because theconcentration gradient which is driving force is supposed to be higher when alower solid to solvent ratio used, leading to higher diffusion. The optimumratio of solvent to sample need to be determined as in economical point ofview, consuming less solvent or plant material for the extraction is extremelyreasonable and practical.

3.2 Analysis of Nepetalactones

Nepetalactones is the volatile oil produced from this plant. The determinationof this constituent is necessary for this project in order to prove the presenceof nepetalactones. This compound is the desired compound due to its repellence towards for aedes mosquitoes. By using highperformance liquid chromatography (HPLC) it was possible to separate themajor component in the essential oil sample extracted. The method waschosen based on the method used by Wang et al.(2007) [19] whoconducted a comparison of quantification of nepetalactone in NepetaCatariaL. by using HPLC-UV coupled with Ultraviolet and Mass Spectrometric. Based on the analysis, the peakswere observed with maximum ultraviolet (UV) absorptions at 228 nm at flowrate 1mL/min. The detection time at two peak which were12.694 min and 13.941 respectively.

4. Conclusion

Response Surface Methodology (RSM) was successfully been usedin determining the optimum condition for the process of extraction based onthree chosen parameters which were time of extraction(min), temperature ofsonication (°C) and ratio of sample to solvent (gram/150ml). Ethanol andPetroleum Ether were used as solvent and ethanol were proven to befeasible to produce higher yield of essential oil, with several modifications onparameters are needed. Based on regression model, correlation factor (R^2)for ethanol was 0.99219.The optimum condition for ethanol solvent could not be achieved asthe model developed showed a linear effect with minimum critical value. Theminimum critical values for ethanol solvent was 269.48 minute, 40.18 °C and33.22 gram/150 ml. Recovery of oil when 1 gram of raw material was used per 150 ml of solventwas 2.32%. The aim of detection of nepetalactoneto strengthen the evidence of potentiality of this constituent to repel aedesaegypti was proven fruitful with detection of nepetalactone at absorption of228nm at detection time of about 12.694 min and 13.941 min.

5. Acknowledgement

The authors wish to acknowledge full gratitude to the Universiti Kuala Lumpur (UniKL MICET) for funding this work.

6. References

[1] Blumenthal, M., Ferrier, G.K.L., &Cavaliere, C. (2006). Total sales of herbalsupplements in United States show steady growth. HerbalGram.

[2] Roberts, D. R., Chareonviriyaphap, T., Harlan, H. H., &Hshieh, P. (1997).Methods for testing and analyzing excite-repellency response of vectors to insects. J.AmMosq Control Assoc.p13

[3] Hayes, J.B. & Laws, E. R. (1991). Handbook of Pesticide Toxicology. SanDiego' CA: Academic Press

[4] Qiu, H., Jun, H.W., & John, W.M. (1998). Pharmacokinetic, formulation, andsafety of insect repellent N,N-diethyl-3-methylbenzamide (DEET). J Am Mosq ControlAssoc. 12--27.

[5] Barnard, D.R. Bernier, U. R., Furman, K.D., Kline, & D. L. Allan, S.A., (2005).Journal of Medical Entomology 42:306-311.

[6] US EPA. (2010). Refined Oil of Nepetacataria: Hydrogenated Catmint Oil(HCO). Biopesticides and Pollution Prevention Division.

[7] Zhu, J. J. et. al (2012). Nepetalactones from essential oil of Nepetacatariarepresent a stable fly feeding and oviposition repellent. ELSEVIER

[8] Kamiar. et al. (2012). Chemical Composition and Antimicrobial Activities ofEssential Oils from Nepetacataria L. against Common Causes of Food-Bornelnfections. International Scholarly Research Network Pharmaceutics.

[9] Lavilla, I. and Bendicho, C., (2000). Ultrasound Extraction. Academic Press.144-1454.

[10] David, J. and Cheeke, N. (2002). Fundamental and Application of UltrasonicWaves. CRC Press

[11] Pernet, F. and R. Tremblay (2003). Effect of ultrasonication and grinding onthe determination of lipid class content of microalgae harvested on filters. Lipids.1191-1195.

[12] Wang, L. & Weller C. L.(2006). Recent advances in extraction ofnutraceuticals from plants. Trends in Food Science & Technology. 300-312

[13] Cravotto, G., L. et al. (2008). Improved extraction of vegetable oils under high-intensity ultrasound and/or microwaves. UltrasonicsSonochemistry. 898-902

[14] Chen, F., et al. (2007). Optimisation of ultrasound-assisted extraction ofanthocyanins inextractusing highperformanceliquidchromatography-massspectrometry. Ultrasonic Chemistry.

[15] Zhang, Q. A., Zhang, Z. Q. Yue, X. F., Fan, X. H., Li, T. & Chen, S.F. (2009).Response surface optimization of ultrasound-assisted oil extractionfrom autoclaved almond powder. Elsevier
[16] Pawliszyn, J (1997). Splid Phase Microextraction(SPME). The chemical educator.1-7
[17] Romdhane, M and Goron, C. (2002). Investigation in soild-liquid extraction inf;uence f ultrasound. Chemical Engineering Journal.87(1). 11-19
[18] Sahin, S. and Samli, R. (2012). Optimization of olive leaf extract obtained byultrasound-assisted extraction with response surface methodology. Elsevier.
[19] Wang, M., Cheng, K. Wu, Q. & Simon, J. E. (2007) Quantification ofNepetalactone in Catnip(NepetaCataria L.) by HPLC Coupled with Ultraviolet andMass Spectrometric Detection. Wiley Interscience. 157-160.

Synthesis and characterization of carbamate insecticideintercalated zinc layered hydroxide modified with sodium dodecyl sulphate

Z Muda[1], N Hashim[1, 2*], I M Isa[1, 2], N M Ali[1], S A Bakar[2, 3], M Mamat[4], M Z Hussein[5], N A Bakar[1] and W R W Mahamod[1]

[1]Department of Chemistry, Faculty of Science and Mathematics, Universiti Pendidikan Sultan Idris, 35900 Tanjong Malim, Perak, Malaysia.
[2]Nanotechnology Research Centre, Faculty of Science and Mathematics, Universiti Pendidikan Sultan Idris, 35900 Tanjong Malim, Perak, Malaysia.
[3]Department of Physics, Faculty of Science and Mathematics, Universiti Pendidikan Sultan Idris, 35900 Tanjong Malim, Perak, Malaysia.
[4]Pusat Pengajian Sains Asas, Universiti Malaysia Terengganu, 21030 Kuala Terengganu, Terengganu, Malaysia.
[5]Materials Synthesis and Characterization Laboratory, Institute of Advanced Technology, Universiti Putra Malaysia, 43400 Serdang, Selangor, Malaysia.

Corresponding author email: norhayati.hashim@fsmt.upsi.edu.my

Abstract. A novel organic-inorganic nanocomposite based on carbamate insecticide-zinc layered hydroxide modified with sodium dodecyl sulphate nanocomposite (ZLH-SDS-PRO) was synthesized through ion exchange method of propoxur with commercially available zinc nitrate at various concentrations. The intercalation of propoxur into the interlayer of zinc layered hydroxide modified sodium dodecyl sulphate (ZLH-SDS) was confirmed by PXRD analysis with the basal spacing of the pure phase nanocomposite at 33.1 Å. The propoxur anions suggested to be arranged in a bilayer form between the galleries of zinc layered hydroxide. The intercalation also supported by Fourier transforms infrared spectra (FTIR) that showing the resemblance spectra between the host and the resulting nanocomposite. The elemental analysis estimates the loading of propoxur anion in the nanocomposite to be 88.17 %. The surface area analysis revealed mesoporous-type material properties for both the host and resulting nanocomposite. The thermal study shows a significant improved thermal stability of the propoxur anion compared to its pure form. The effect of intercalation of propoxur on the morphology of the host, ZLH-SDS crystals was investigated by field emission scanning electron microscopy (FESEM).

1. Introduction

Propoxur (2-isopropoxyphenyl N-methylcarbamate), a carbamate insecticide widely used in pest control for other domestic animals, mosquitoes, bugs, ants, gypsy moths, and other agricultural pests [1,2]. It was introduced to the market in 1959 and has been widely used against turf, forestry, and household pests and fleas [3]. It has been reported to produce a number of adverse effects including neurotoxicity and adversely affect memory and induce oxidative stress on both acute and chronic exposure [4].

The products of nanochemistry are being been used to create new generations of technologies for curing environmental maladies and protecting public health [5]. Water pollution control, groundwater remediation, potable water treatment, and air quality control are being advanced through nanomaterialbased membrane technologies [6], adsorbents [7], and catalysts [8–11]. Nanomaterial are also inspiring new solutions for providing and using energy that are more environmentally neutral than conventional approaches.

The modern day trend has induced the development of nanocomposite primarily based on natural polymers and inorganic solid mainly clay minerals [12]. Nano clays and layered double hydroxides (LDH) are being developed in this regard [12,13]. Both materials were reported to show excellent biocompatibility, low toxicity, and the potential for controlled release [14–18]. Various chemicals can be intercalated between layers of both materials such as 4-chloro-2-methylphenoxy acetic acid [15], polypropylene [18], 4-aminophenylsulfone [17], poly(methyl methacrylate) [19] and diclofenac [20].

Nowadays, another attention has been focused on zinc layered hydroxide (ZLH). Similarly, with LDH, ZLH structure is based on the mineral brucite ($Mg(OH)_2$) structure and categorised under layered hydroxide salt (LHS) [15]. Among the various two-dimensional layered materials, LHS have been of great interest due to their simple synthetic process and unique anion-exchange properties [21]. The structure of LHS is composed of positively charged hydroxide layers with charge-compensating anions in the interlayer spaces. Positive charges on layers are developed from under-coordination or mixed coordination geometries of interlayer cations. Therefore, counter-anions are required to stabilize the electrostatic charge of layers [22]. These modifications lead to a type of compounds called LHS, with the general formula $M^{2+}(OH)_{2-x}$ $(A^{m-})_{x/m}\cdot nH_2O$ where M^{2+} is the metal cation (e. g. Mg^{2+}, Ni^{2+} Zn^{2+},Ca^{2+},Cd^{2+},Co^{2+}, Cu^{2+}) and A^{m-} is the counter ion.

Occasionally, a few active agents are charged neutral and poorly-soluble pesticides which are difficult to be intercalate into the interlayer of ZLH. The majority of literatures are only focusing to the anionic pesticides. For charge-neutral and poorly water-soluble pesticides, their intercalation is usually dependent on anionic surfactant that forms a hydrophobic region in the gallery [23,24]. The hydrophobic nature and accessibility of the interlayer region of ZLH is helpful for adsorption of target pesticide molecules. In the previous study, Liu et al., [25] have developed a novel method to make chlorpyrifos (CPF) adsolubilize into the interlayer of zinc hydroxide nitrate (ZHN) intercalated with dodecylbenzenesulfonate (DBS). In this study, ZHN is modified with DBS to form a hydrophobic region in the gallery of ZHN–DBS. Liu et al., [25] suggests that DBS forms a tilt monolayer in the gallery when the SO_3 group approaches the layer by three oxygen atoms. The cationic and anionic surfactants interact to form neutral micelles, leaving the ZLH free to capture the desired anion. Since the interlayer spacing in the starting materials for this procedure is already quite large, this method offers particular promise for the incorporation of bulky anions [26].

Therefore, in this study, we discuss our work on the intercalation of propoxur, a carbamate insecticide, into the interlayer of ZLH modified with SDS by an ion exchange method, for the formation of a new organic-inorganic-layered nanocomposite material.

2. Method

2.1 Materials

Sodium dodecylsulphate (SDS) and zinc nitrate ($Zn(NO_3)_2\cdot 6H_2O$) were purchased from Systerm Malaysia. Sodium hydroxide, NaOH was purchased from HmbG Chemicals. Whereas, propoxur pesticide, $C_{11}H_{15}NO_3$ was purchase from Nanjing Essence Fine-Chemical. All solutions were prepared using deionized water. Each chemicals utilised in this synthesis were attained from several chemical providers and used without additional purification.

2.2 Synthesis of ZLH-SDS-PRO

The intercalation compound, ZLH-SDS-PRO was obtained via anion exchange between nitrate ion of precursor ZLH-SDS and anionic propoxur in a mixture solution of water: organic solvent. The host, layered zinc layered hydroxide-sodium dodecyl sulphate (ZLH-SDS) was synthesized by following procedure describe elsewhere [25]. Briefly, ZLH-SDS was synthesized using co-precipitation method by the slow addition of 1.0 M NaOH and 40 mL of 0.5 M of $Zn(NO_3)_2 \cdot 6H_2O$, into a solution containing 40 mL of 0.25 M SDS under magnetic stirring. The pH value was adjusted to 6.5. The slurry was then centrifuged and dried in an oven at 70 °C.

Preparation of propoxur intercalated into the interlayer of ZLH-SDS (indicated as ZLH–SDS-PRO) was achieved by an ion exchange method. Various concentrations of propoxur solution were prepared at 0.0005 M, 0.0010 M, and 0.0025 M. 0.5 g of ZLH-SDS was dissociate in propoxur solution and keep under magnetic stirrer for 2 ½ hours. The slurry was then aged for 24 hours in an oil bath shaker at 70 °C. Then the slurry was centrifuged and the final white solid was dried in an oven for 24 hours.

2.3 Characterization

There are several instruments involved in the characterization of the ZLH-SDS-PRO nanocomposites. The X-ray powder diffraction patterns (PXRD) were obtained using a Power Diffraction Bruker AXS (model D8 Advance, wavelength of 1.5406 Å) with CuKα radiation at 60 kV and a current of 60 mA. The recorded region of 2θ was from 2° to 60° with a scanning rate of 2° min^{-1}. The FTIR spectra were collected in a Thermo Nicolet 6700 Fourier Transform Infrared Spectrometer in the range 400–4000 cm^{-1}. The thermal analyses measurement (TGA/DTG) of the sample was obtained with Perkin Elmer Pyris 1 TGA Thermo Balance with a heating rate of 20°C min^{-1} (N_2 flow rate is 50 ml/min, temperatures of 25-900 °C at a rate of 10 K/min). An inductive coupled plasma optical emission spectrometry (ICP-OES), model Agilent, 720 Axial and (CHNO-S), model Thermofinnigan, Flash EA 1112 was used to study the composition of the samples. The surface morphology of the sample was observed by a field emission scanning electron microscope (FESEM) Hitachi model SU 8020 UHR. Surface characterization of the nanocomposites was carried out by the nitrogen gas adsorption– desorption technique at 77 K using a Quantachrome Autosorb-1and degassed in an evacuated heated chamber at 120 °C overnight.

3. Results and discussion

3.1 PXRD analysis

Figure 1 shows the XRD pattern for the host, ZLH-SDS, propoxur anion and ZLH-SDS-PRO nanocomposite. The XRD pattern of ZLH-SDS showed a basal spacing of 9.8 Å corresponding to NO_3^-, and the peak at 33.0 Å has the same pattern with LDH-SDS that previously reported by [27]. The resulted ZLH-SDS is well crystallized without impurities like ZnO which indicated that a complete reaction during the co-precipitation between zinc nitrate and SDS. Upon dissolving zinc nitrate hexahydrate in water, the zinc species undergo hydrolysis reactions (Equation 1 and 2) as previously reported by Moezzi et al., [28].

$$Zn(NO_3)^2 \rightarrow Zn^{2+}(aq) + 2(NO_3)^-(aq) \quad (1)$$
$$Zn^{2+}(aq) + H_2O \rightleftharpoons ZnOH^+(aq) + H^+(aq) \quad (2)$$

Zinc hydroxide species can actually further transform into ZnO nanoparticles. However, in the presence of dodecyl sulphate ions, these zinc hydroxide species could cause charge–assemblies to form zinc hydroxide-dodecyl sulphate layered nanosheets. The formulas below (Equation 3 and 4) represent the main reactions that were previously reported by Liang et al. [29].

$$Zn(OH)_{2-x}^{x+} + x/4O_2 \rightarrow ZnO + (1 - 1/2x)H_2O \quad (3)$$
$$Zn(OH)_{2-x}^{x+} + xDS^- \rightarrow Zn(OH)_2 - xDS_x \quad (4)$$

The x-ray diffraction pattern of ZLH-SDS-PRO shows a series of basal peaks indicated by (003), (006), (009) in Figure 1. The intercalation of 0.0005M, 0.0010 M, and 0.0025 M, of propoxur pesticide into layered ZLH-SDS exhibited a diffraction pattern characterization with a sharp diffraction peak centred at 33.1 Å at low angle 2θ degree. By comparing the XRD pattern of the resulting with the pure propoxur, it can be observe that the resulting ZLH-SDS-PRO is in pure form without any impurities of propoxur anion. This is due to due to successfully intercalation of propoxur between interlayer of ZLH-SDS well-ordered stacking layer. The sharp and intense intercalation peak reveal that the nanocomposite is high crystalinity [30]. The intensity of the intercalation peak of ZLHSDS-PRO nanocomposite decrease as the concentration of propoxur increases from 0.0010 to 0.0025 M indicates that the poorer crystalinity of nanocomposite at higher concentration of anion. This is due to the presence of the large amount of propoxur in interlayer and this is somehow interferes with the crystal growth during regeneration [31]. For further characterization, sample of 0.0010 M propoxur was then chosen as the phase pure well-ordered nanocomposite material and used for further characterizations.

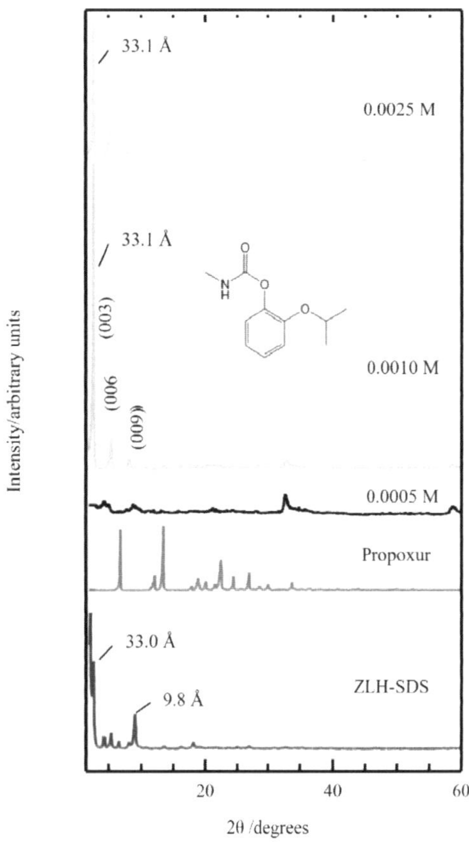

Figure 1. PXRD pattern for ZLH-SDS, propoxur anion, and ZLH-SDS-PRO nanocomposite with concentration of 0.0005 M, 0.0010 M and 0.0025 M of propoxur.

3.2 FTIR analysis

As shown in Figure 2, the propoxur spectra present adsorption peaks at 3309 cm^{-1} indicate the N-H for secondary amine group. The peaks appear at 1708 and 1253 cm^{-1} is due to stretching vibration of C=O and C-N group respectively. The C-O stretching vibration appears at 1197 and 1114 cm^{-1}. The C-H bending vibration peak appear at 1375 and 742 cm^{-1} while, C-H stretching vibration appear at 2963 cm^{-1}. Whereas, peak at 1496 cm^{-1} corresponds to the C-H stretching vibration for aromatic group. The list of FTIR peaks are presented in Table 1.

The FTIR spectra of ZLH-SDS shows a strong and sharp absorption band around 3500–3700 cm^{-1} corresponds to the stretching vibrations of the OH group of the free water molecule [32]. Whereas, a strong and broad absorption band centred at 3461 cm^{-1} attribute to the OH vibration. This band has a broad base due to hydrogen bonds established with the hydration water molecule [33]. Another bending vibration at 1637 cm^{-1} is the H-O-H bending of water molecule in the interlayer of ZLH-SDS. Two main doublet absorption bands appear in the region of 2850-2950 cm^{-1} which is due to stretching vibration of aliphatic group and another band at 1350-1480 cm^{-1} is due to bending vibration of aliphatic group.

While, the stretching vibration at 1210-1240 cm^{-1} is due to the presence of sulphate group in SDS (Hussein, Zainal, & Ming, 2000). A strong absorption band at 1364 cm^{-1} is due to the presence of nitrate as the salt used for the source of metal ions is zinc nitrate.

The FTIR spectra of ZLH-SDS-PRO nanocomposite resemble the ZLH-SDS spectra. A medium and broad band at 3507 cm^{-1} is attributing to the stretching vibration of the hydroxyl group of the water molecule in the interlayer. While, a peak at 1635 cm^{-1} reveal that there is a free water molecule in the interlayer of nanocomposite. A sharp doublet absorption of ZLH-SDS-PRO nanocomposite band resemble the peak in ZLH-SDS layer at 2913 and 2846 cm^{-1} which is due to stretching vibration of the C-H group that present in the nanocomposite. Whereas, the peak at 1375 and 742 cm^{-1} which is also resembles the peak in propoxur, is due to bending vibration of the aliphatic group, C-H. The bands appeared at 1202 and 1082 cm^{-1} is assigned to the asymmetric and symmetric stretching vibration of S=O [27]. The stretching vibration of S=O for ZLH-SDS-PRO particles are shifted to lower frequencies, indicating the configuration variation of the OSO_3^- functional group [27]. Those down-shift conforms to the losing of S=O bond strength, referring the existence of a hydrogen bond within ZLH interlayer and sulphate group (S=O H–O–Zn) as well as the electrostatic attraction [27]. The absence of N-H absorption peak in resulting ZLH-SDS-PRO spectra that should appear around 3350-3310 cm^{-1} indicating that propoxur molecule has been successfully loaded into the ZLH interlayer in an ion form. After the intercalation, propoxur maybe undergo elimination mechanism in hydrolysis reaction [35]. The intercalation also confirm by the disappearance of nitrate absorption peak which indicate the successful of ion exchange between nitrate ion and propoxur anion.

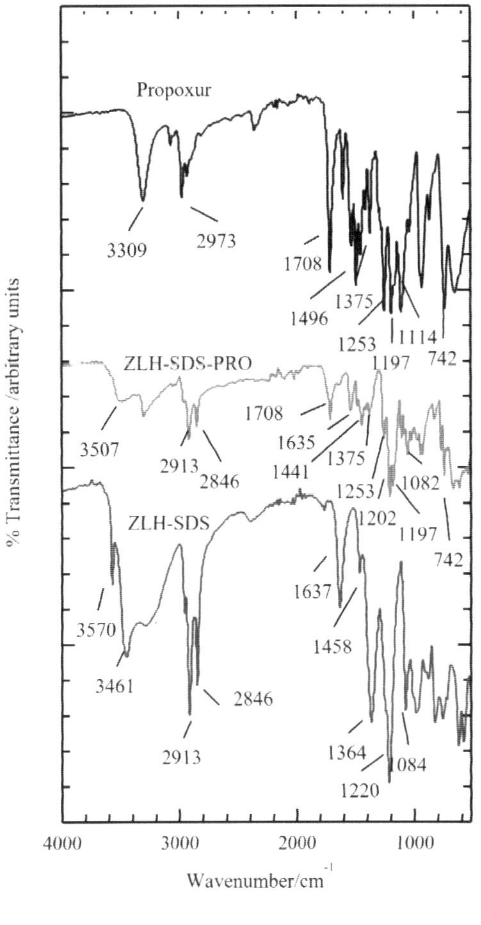

Figure 2. The FTIR spectra for ZLH-SDS, propoxur anion, and ZLH-SDS-PRO nanocomposite

Table 1. FTIR bands for ZLH-SDS, propoxur anion, and ZLH-SDS-PRO nanocomposite

Characteristic group	SDS	ZLH-SDS-PRO	Propoxur
ν (O-H), H-bonded	3461	-	-
ν (O-H) in the interlayer; H_2O	3570	3507	-
ν (H-O-H) in the interlayer; H_2O	1637	1635	-
ν (N-H)	-	-	3309
ν (C = O)	-	1708	1708
ν (C-O)	-	1197	1197 1114
ν (C-N)	-	1253	1253
ν (C-H),stretching	2846 2913	2846 2913	2973
ν (C-H),bending	1458	1375 742	1375 742
ν (C = C),stretching, aromatic	-	1441	1496
ν (N-O)	1364 1637	-	-
νas (S=O)	1220	1202	-
ν_s (S=O)	1084	1082	-

3.3 Spatial orientation

A proposed orientation of ZLH-SDS-PRO nanocomposite is shown in Figure 3. SDS and propoxur are aligned 90° perpendicular to the ZLH planes. Considering that propoxur anions and dodecyl sulphate ion electrostatically attracted to ZLH plane, the arrangement of the intercalated propoxur is proposed in bilayer arrangement as shown in Figure 3 (obtained by ChemDraw software). The interlayer height of ZLH-SDS-PRO nanocomposite is estimated by the basal spacing obtained from PXRD analysis which is 33.1 Å. Regarding the size of the propoxur molecule and interlayer space height of ZLHSDS-PRO nanocomposite, it is reasonable to conclude that the propoxur molecules are intercalated into the hydrophobic region formed by dodecyl sulphate molecules in the interlayer space of ZLH, due to the increase of the ZLH-SDS-PRO nanocomposite basal spacing, as reported elsewhere [27]. Since the layer thickness is 4.8 Å and the Zn^{2+} moiety of the lattice from the basal spacing is 2.6 Å, the interlayer height of the nanocomposite is estimated to be 23.1 Å.

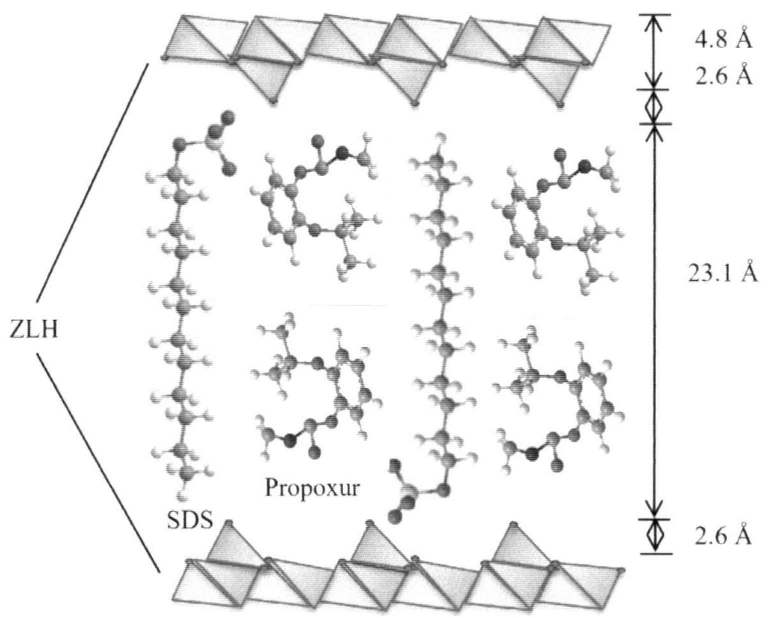

Figure 3. The proposed orientation of propoxur and SDS anions intercalated within the ZLH interlayer gallery resulting in ZLH-SDS-PRO nanocomposite estimated by Chemoffice software.

3.4 Elemental analysis

Elemental and compositional analysis of ZLH-SDS and ZLH-SDS-PRO nanocomposite are listed in Table 1. CHNO-S analysis of ZLH-SDS shows that 2.46 % of N, 24.8 % of C, 5.48 % of H, and 3.38 % of S. ICP-OES analysis shows high percentage of Zn (38.87 %) for pristine ZLH-SDS used in this work. After intercalation process, about 56.49 % of carbon is detected. The increase in percentage of C indicates successful intercalation of propoxur in the interlayer of ZLH [15]. The percentage of N in the resulting nanocomposites is 5.90 % and the percentage of S is 0.43 %. The S element represents the dodecyl sulphate ion in the interlayer of ZLH. The previously PXRD analysis confirmed that there was no nitrate ion after intercalation process. Therefore, the N element detected in the resulting nanocomposite came from the amine group in the propoxur anion, which also confirmed the successful intercalation of propoxur in the interlayer of ZLH. From CHNO-S data, the calculated percentage propoxur anion intercalates between the interlayer of ZLH region is 88.17 %. Whereas, the ICP-OES analysis detects 38.87 % of Zn in the resulting nanocomposite.

Table 2. Compositional data for synthesized ZLH-SDS and ZLH-SDS-PRO nanocomposite

Sample	N (%)	C (%)	H (%)	S (%)	Anion (% w/w)	% Zn
ZLH-SDS	2.46	24.80	5.48	3.38	-	24.14
ZLH-SDS-PRO	5.90	56.49	6.45	0.43	88.17	38.87

3.5 Thermal analysis

The thermogravimetric analysis, TGA-DTG obtained for propoxur, ZLH-SDS-PRO nanocomposite and ZLH-SDS are reported in Figure 4. In Figure 4 (a), the thermal study shows that the maximum temperature of pure propoxur was observed at 174 °C with 99.8 % of weight loss.

For ZLH-SDS-PRO nanocomposite, two stages of decomposition were observed in Figure 4 (b). The first stage is due decomposition of propoxur at 182 °C with weight loss of 63.5 %. Meanwhile the second stage is due to decomposition of SDS which can be observed at 218 °C with 4.2 % weight loss.

The ZLH–SDS display two stages of weight loss (Figure 4 (c)). The first stage corresponds to the removal of intercalated water at 107 °C with 3.2 % of weight loss [36]. Whereas, the decomposition of SDS takes place at 198 °C with a 37.0 % weight loss. From Figures 4(a) and (b), the maximum temperature of the decomposition of propoxur in the resulting nanocomposite increased from 174 °C to 182 °C. Therefore, the thermal study of ZLH-SDS-PRO nanocomposite affirm that the propoxur anion that intercalate in the ZLH-SDS interlayer is thermally stable compared to its prior form which is corresponding to the electrostatic interaction between anions and ZLH layer [37].

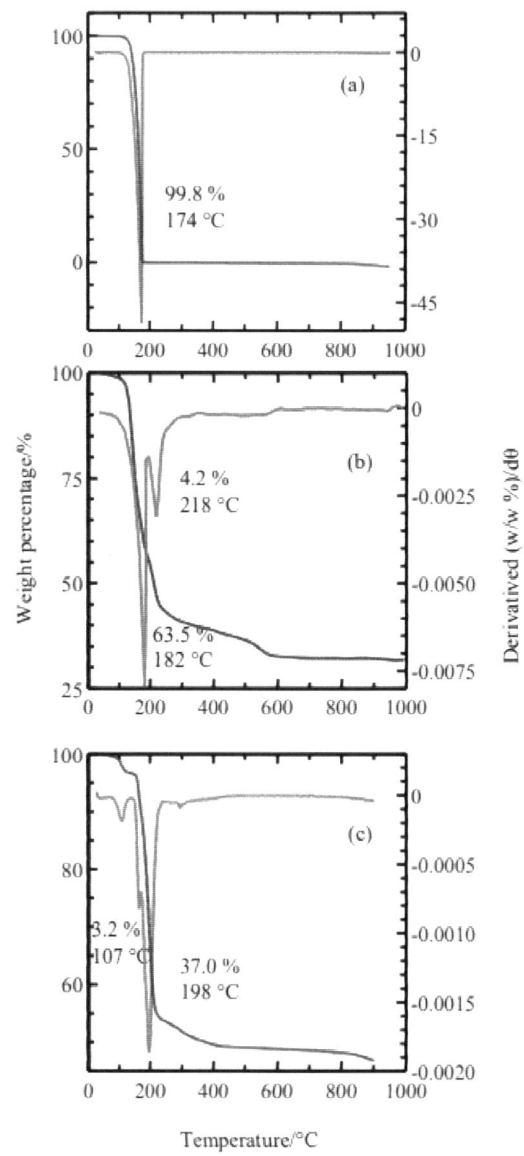

Figure 4. Termogravimetric curve for propoxur anion (a), ZLH-SDS-PRO nanocomposite (b), and ZLH-SDS (c).

3.6 Surface morphology analysis
Figures 5 (a) and 5 (b) illustrate the morphology of ZLH-SDS and ZLH-SDS-PRO at 10k magnification. The FESEM image of ZLH-SDS resemble thin flake-like nanosheets that are expected to be layered materials [38]. The edge of ZLH-SDS sheets is quite fractured with cracked surface. After the intercalation, 3D self-assemble nanoflower-like particle was observed with a size of a few microns. Miao et al., [21] has reported that, the nanoflower-like particle resulting from stirring in the intercalation process. The flower structure is composed of nanosheets which estimated to be 1.5-3.0 μm thickness of the nanosheets. The sheets are in stacked plate, sharp edges, and smooth surface.

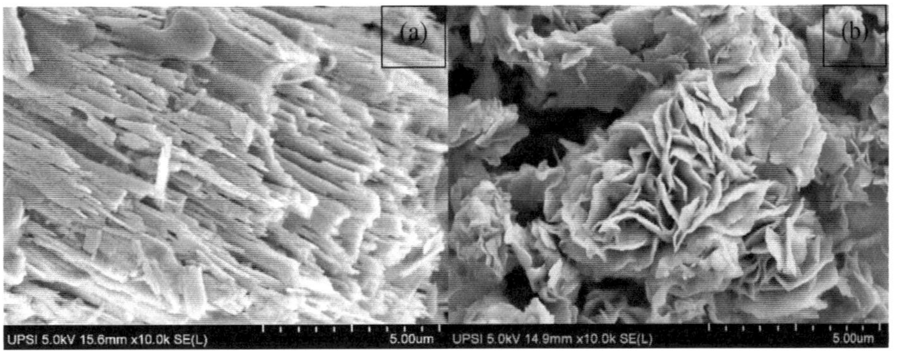

Figure 5. FESEM images of ZLH-SDS (a), and ZLH-SDS-PRO (b) 10k magnification nanocomposite

3.7 Surface area analysis
The comparison of surface area and porosity of ZLH-SDS and ZLH-SDS-PRO nanocomposite are given in Table 3. The intercalation of propoxur anion has decreased the BET surface area from 7.027 m^2g^{-1} for ZLH-SDS to 2.407 m^2g^{-1} for ZLH-SDS-PRO nanocomposite. This is attributed to the increase in the particles size and decrease of the pore volume [39].

As shown in Figure 6, both the host and resulted nanocomposites possessed a relatively narrow distribution of pores with an average diameter of 27.83 nm to 16.88 nm respectively, strongly suggesting the formation of a mesoporous structure. This is indicated by the closing of the pores from insertion of propoxur anion into the pores of the ZLH-SDS layer [20,40]. The nitrogen-adsorption isotherm for ZLH-SDS and ZLH-SDS-PRO nanocomposites are of Type IV isotherm with H3 hysteresis loops based on International Union of Pure and Applied Chemistry (IUPAC) classification [41] indicating mesopores-type materials [42]. The adsorption slowly increases in low relative pressure in the range of 0.0-0.6 (for ZLH-SDS) and 0.0-0.5 (for ZLH-SDS-PRO) which represents of the surface monolayer adsorption process. Further increases of the relative pressure resulting a rapid adsorption of the adsorbent due to the first multilayer adsorption. The presence of H3 hysteresis loop in layered ZLH-SDS and ZLH-SDS-PRO nanocomposite reveals that the pores exhibit slit-shaped pores with non-uniform shape and sizes. ZLH-SDS-PRO nanocomposite showed very much lower maximum adsorption compared to the host, ZLH-SDS which is due to the high capacity of propoxur anion intercalate in the interlayer of ZLH.

The Barret-Joyner-Halenda (BJH) pore size distribution for ZLH-SDS and ZLH-SDS-PRO nanocomposite is shown in Figure 7. The pore size distribution of ZLH-SDS and ZLH-SDS-PRO is almost the same features which centred at around 14 and 21 nm. In general, the pore size distributions do not show the formation of regular mesopores in the layered ZLH-nanocomposites, but could reveal the contribution of irregular mesoporous interparticulate void space [37,43].

Table 3. Surface properties of ZLH-SDS and ZLH-SDS-PRO nanocomposite

Samples	Specific BET surface area (m^2/g)	Average pore diameter (nm)	Classification
ZLH-SDS	7.027	27.83	Mesoporous
ZLH-SDS-PRO	2.407	16.88	Mesoporous

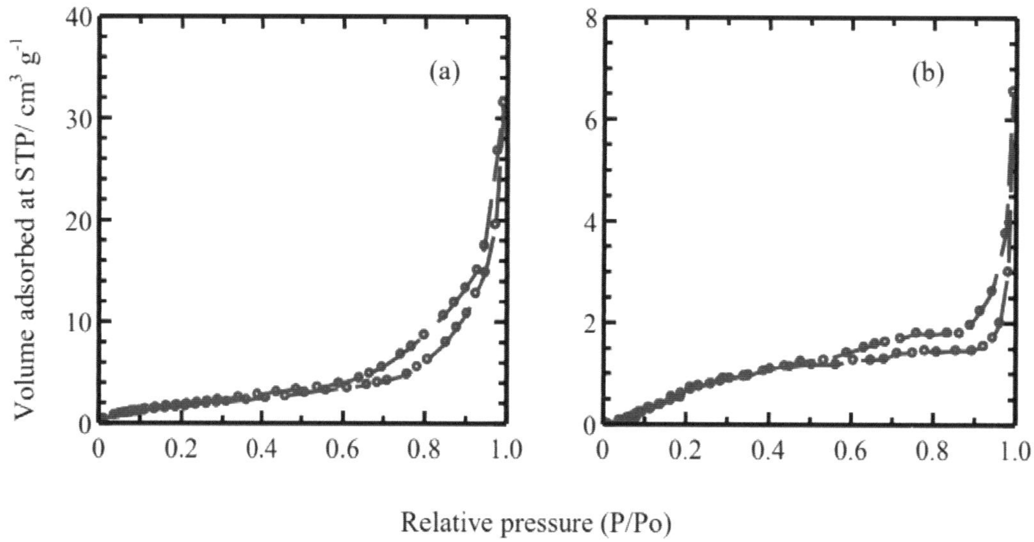

Figure 6. Adsorption–desorption isotherms of nitrogen gas for ZLH-SDS (a) and ZLH-SDS-PRO (b)

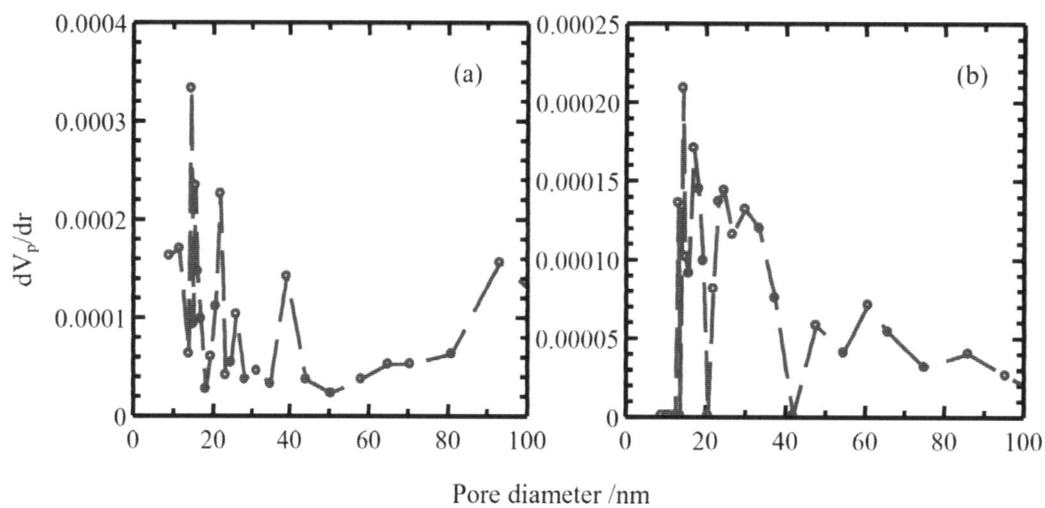

Figure 7. BJH desorption pore size distributions for ZLH-SDS (a), and ZLH-SDS-PRO (b)

4. Conclusion

In the present work, a carbamate insecticide, propoxur was successfully intercalated into the ZLH interlayer by ion exchange method. The modification of ZLH with SDS surfactant gives a great impact in order to intercalate the poor water soluble insecticide like propoxur into the interlayer of ZLH. A well-crystallized nanocomposite was obtained with basal spacing 33.1 Å. The FTIR spectra of ZLHSDS-PRO show a resemblance with ZLH-SDS spectra and propoxur spectra which support the intercalation of propoxur into interlayer of ZLH. The intercalation also has enhanced thermal stability of propoxur compared to its pure form. Both ZLH-SDS and the resulting nanocomposite show Type IV for the nitrogen adsorption–desorption isotherms with BET surface area 7.027 m^2g^{-1} and 2.407 m^2g^1, respectively. This nanocomposite also confirmed to exhibit mesoporous type material with remarkably flower-like surface morphology. The successful intercalation of propoxur proves that ZLH can be an excellent host for propoxur with SDS modification. Further work will be carried out with the controlled release formulation of this nanocomposite in order to study the release behaviour of this nanocomposite in selected medium.

5. References

[1] Pandey M R and Guo H 2014 Evaluation of cytotoxicity, genotoxicity and embryotoxicity of insecticide propoxur using flounder gill (FG) cells and zebrafish embryos *Toxicol. Vitr.* **28** 340–53

[2] Zafiropoulos A, Tsarouhas K, Tsitsimpikou C, Fragkiadaki P, Germanakis I, Tsardi M, Maravgakis G, Goutzourelas N, Vasilaki F, Kouretas D, Hayes A and Tsatsakis A 2014 Cardiotoxicity in rabbits after a low-level exposure to diazinon, propoxur, and chlorpyrifos. *Hum. Exp. Toxicol.* **33** 1241–52

[3] Tomlin C 1994 The pesticide manual; A world compendium, 10th edn. British Crop Protection Council

[4] Mehta K D, Garg G R, Mehta A K, Arora T, Sharma A K, Khanna N, Tripathi A K and Sharma K K 2010 Reversal of propoxur-induced impairment of memory and oxidative stress by 4′-chlorodiazepam in rats *Naunyn. Schmiedebergs. Arch. Pharmacol.* **381** 1–10

[5] Andrievski R A and Khatchoyan A V 2016 *Nanomaterials in Extreme Environments Fundamentals and Applications* (Switzerland: Springer Int. Publ)

[6] Pendergast M M and Hoek E M V 2011 A review of water treatment membrane nanotechnologies *Energy Environ. Sci.* **4** 1946–71

[7] M. Rozi Mat Dris, Chan Kok Sheng, M. Ikmar Nizam Isa M H R 2012 a Study of Cadmium Sulfide Nanoparticles With Starch As a *Int. J. Technol.* **1** 1–7

[8] Razali M H, Mohd Noor A F, Mohamed A R and Sreekantan S 2012 Morphological and structural studies of titanate and titania nanostructured materials obtained after heat treatments of hydrothermally produced layered titanate *J. Nanomater.* **2012**

[9] Razali M H, Ruslimie C A and Khairul W W 2013 Modification and performances of TiO2 photocatalyst towards degradation of paraquat dichloride *J. Sustain. Sci. Manag.* **8** 244–53

[10] Razali M H, Noor A F M and Yusoff M 2017 Hydrothermal synthesis and characterization of Cu^{2+}/F–Co-doped titanium dioxide (TiO_2) nanotubes as photocatalyst for methyl orange degradation *Sci. Adv. Mater.* **9** 1032–41

[11] Ahmad A, Razali M H, Mamat M, Mehamod F S B and Anuar Mat Amin K 2017 Adsorption of methyl orange by synthesized and functionalized-CNTs with 3-aminopropyltriethoxysilane loaded TiO_2 nanocomposites *Chemosphere* **168** 474–82

[12] Devi N and Maji T K 2011 Neem seed oil : Encapsulation and controlled release - search for a greener alternative for pest control *Pesticides in the Modern World-Pesticides Use and Management* pp 191–232

[13] Choy J H, Choi S J, Oh J M and Park T 2007 Clay minerals and layered double hydroxides for novel biological applications *Appl. Clay Sci.* **36** 122–32

[14] Dasgupta S 2017 Controlled release of ibuprofen using Mg Al LDH nano carrier *IOP Conf. Ser. Mater. Sci. Eng.* **225**
[15] Salleh N M, Mohsin S N, Sarijo S H and Ghazali S A I S M 2017 Synthesis and physicochemical properties of zinc layered hydroxide-4-chloro-2- methylphenoxy acetic acid (ZMCPA) nanocomposite *IOP Conf. Ser. Mater. Sci. Eng.* **204** 1–5
[16] Rehab A and Salahuddin N 2005 Nanocomposite materials based on polyurethane intercalated into montmorillonite clay *Mater. Sci. Eng. A* **399** 368–76
[17] Zulfiqar S, Ahmad Z, Ishaq M and Sarwar M I 2009 Aromatic-aliphatic polyamide/montmorillonite clay nanocomposite materials: Synthesis, nanostructure and properties *Mater. Sci. Eng. A* **525** 30–6
[18] Paiva L B de, Morales A R and Guimarães T R 2007 Structural and optical properties of polypropylene-montmorillonite nanocomposites *Mater. Sci. Eng. A* **447** 261–5
[19] Unnikrishnan L, Mohanty S, Nayak S K and Ali A 2011 Preparation and characterization of poly(methyl methacrylate)-clay nanocomposites via melt intercalation: Effect of organoclay on thermal, mechanical and flammability properties *Mater. Sci. Eng. A* **528** 3943–51
[20] Heraldy E, Suprihatin R W W and Pranoto 2016 Intercalation of diclofenac in modified Zn/Al hydrotalcite-like preparation *IOP Conf. Ser. Mater. Sci. Eng.* **107** 1–7
[21] Miao J, Xue M and Feng Q 2006 Hydrothermal synthesis of layered hydroxide zinc benzoate compounds and their exfoliation reactions 474–80
[22] Bae H S and Jung H 2012 Electron beam mediated simple synthetic route to preparing layered zinc hydroxide *Bull. Korean Chem. Soc.* **33** 1949–54
[23] Pavan P C, Crepaldi E L, De A. Gomes G and Valim J B 1999 Adsorption of sodium dodecylsulfate on a hydrotalcite-like compound. Effect of temperature, pH and ionic strength *Colloids Surfaces A Physicochem. Eng. Asp.* **154** 399–410
[24] Dekany I, Berger F, Imrik K and Lagaly G 1997 Hydrophobic layered double hydroxides (LDHs): Selective adsorbents for liquid mixtures *Colloid Polym. Sci.* **275** 681–8
[25] Liu J, Zhang X and Zhang Y 2015 Preparation and release behavior of chlorpyrifos adsolubilized into layered zinc hydroxide nitrate intercalated with dodecylbenzenesulfonate *ACS Appl. Mater. Interfaces* **7** 11180–8
[26] Auerbach S M, Carrado K a. and Dutta P K 2004 *Handbook of Layered Materials*
[27] Zhao J, Fu X, Zhang S and Hou W 2011 Water dispersible avermectin-layered double hydroxide nanocomposites modified with sodium dodecyl sulfate *Appl. Clay Sci.* **51** 460–6
[28] Moezzi A, Cortie M and McDonagh A M 2013 Formation of zinc hydroxide nitrate by H+catalyzed dissolution-precipitation *Eur. J. Inorg. Chem.* **2013** 1326–35
[29] Liang C, Tian Z, Tsuruoka T, Cai W and Koshizaki N 2011 Blue and green luminescence from layered zinc hydroxide/dodecyl sulfate hybrid nanosheets *J. Photochem. Photobiol. A Chem.* **224** 110–5
[30] Sarijo S H, Ghazali S A I S M, Hussein M Z and Sidek N J 2013 Synthesis of nanocomposite 2-methyl-4-chlorophenoxyacetic acid with layered double hydroxide: Physicochemical characterization and controlled release properties *J. Nanoparticle Res.* **15** 1536
[31] Costa F R, Leuteritz A, Wagenknecht U, Auf der Landwehr M, Jehnichen D, Haeussler L and Heinrich G 2009 Alkyl sulfonate modified LDH: Effect of alkyl chain length on intercalation behavior, particle morphology and thermal stability *Appl. Clay Sci.* **44** 7–14
[32] Yang W, Ma L, Song L and Hu Y 2013 Fabrication of thermoplastic polyester elastomer/layered zinc hydroxide nitrate nanocomposites with enhanced thermal, mechanical and combustion properties *Mater. Chem. Phys.* **141** 582–8
[33] Arizaga G G C, Gardolinski J E F D C, Schreiner W H and Wypych F 2009 Intercalation of an oxalatooxoniobate complex into layered double hydroxide and layered zinc hydroxide nitrate *J. Colloid Interface Sci.* **330** 352–8
[34] Hussein M Z, Zainal Z and Ming C Y 2000 Microwave-assisted synthesis of Zn-Al-layered double hydroxide-sodium dodecyl sulfate nanocomposite *J. Mater. Sci. Lett.* **19** 879–83

[35] Bakhti H and Hamida N Ben 2014 Kinetic study by uv spectrophotometry of isoprocarb degradation in aqueous medium *J. la Soc. Chim. Tunisie* **16** 35–43
[36] Nabipour H and Sadr M H 2015 Controlled release of diclofenac, an anti-inflammatory drug by nanocompositing with layered zinc hydroxide *J. Porous Mater.* **22** 447–54
[37] Hashim N, Muda Z, Isa I M, Ali N M, Bakar S A and Hussein M Z 2018 The effect of ion exchange and co-precipitation methods on the intercalation of 3-(4-methoxyphenyl)propionic acid into layered zinc hydroxide nitrate *J. Porous Mater.* **25** 249–58
[38] Liu J and Zhang Y 2016 Inorganic-organic nanohybrid materials of layered zinc hydroxide nitrate with intercalated salicylate: Preparation, characterization and Uv-blocking properties *Int. J. Nanosci.* **15** 1–9
[39] Hussein M Z, Al Ali S H, Zainal Z and Hakim M N 2011 Development of antiproliferative nanohybrid compound with controlled release property using ellagic acid as the active agent *Int. J. Nanomedicine* **6** 1373–83
[40] Hashim N, Hussein M Z, Yahaya A H and Zainal Z 2007 Formation of zinc aluminium layered double hydroxides-4-(2,4-dichlorophenoxy)butyrate nanocomposites by direct and indirect methods *Malaysian J. Anal. Sci.* **11** 1–7
[41] Sing K S W, Everett D H, Haul R a. W, Moscou L, Pierotti R a., Rouquérol J and Siemieniewska T 1985 Reporting physisorption data for gas/ solid system with special reference to the determination of surface area and porosity *Pure Appl. Chem.* **57** 603–19
[42] Wang J, Mei X, Huang L, Zheng Q, Qiao Y, Zang K, Mao S, Yang R, Zhang Z, Gao Y, Guo Z, Huang Z and Wang Q 2015 Synthesis of layered double hydroxides/graphene oxide nanocomposite as a novel high-temperature CO_2 adsorbent *J. Energy Chem.* **24** 127–37
[43] Hussein M Z, Ghotbi M Y, Yahaya A H and Abd Rahman M Z 2009 The effect of polymers onto the size of zinc layered hydroxide salt and its calcined product *Solid State Sci.* **11** 368–75

Acknowledgement
The author is grateful to the support for the research provided by UPSI under GPU Grant no. 20170188-101-01. ZM thanks UPSI for all funding and support for this research.

IOP Conf. Series: Materials Science and Engineering **440** (2018) 012004 doi:10.1088/1757-899X/440/1/012004

Parameters study of lipase-transesterification reaction of ethyl caffeate and glycerol in deep eutectic solvent (DES)

P N M A Bakar, F N Gonawan and A H Kamaruddin[1]

[1]School of Chemical Engineering, Universiti Sains Malaysia, 14300, Nibong Tebal, Penang, Malaysia

Corresponding author email: chazlina@usm.my

Abstract. The toxicity of organic solvent to human and environment has triggered researchers to find an alternative green solvent for the past 20 years. Deep eutectic solvent (DES) is the newly emerging green solvent that is being utilized in biocatalysis. The boiling point of DES is higher than conventional organic solvent thus enables for lipase to perform at its optimum temperature. However, the transesterification reaction in DES is limited due to its high viscosity. Therefore, water was added to DES media to reduce its viscosity. This study reported on the use of choline-based DES for lipase-catalyzed transesterification of ethyl caffeate and glycerol. The transesterification reactions in DES comprised of choline chloride and urea at molar ratio 1:2 resulted in high conversion of ethyl caffeate (over 90%) under the following parameters: enzyme loading (1250 Unit Activity); amount of water (20 vol%) and agitation speed (200 rpm). This work could provide an insight to the potential of DES as a promising alternative solvent for the production of glyceryl caffeate esters due to high conversion, faster reaction time and reduction of dangerous organic solvent consumption.

1. Introduction

The toxicity and harmful effect that the organic solvents possess has boosted researchers to find alternative solvents for synthetic chemical reaction. In addition, one of the green chemistry principles outlined that benign solvent should be used [1]. The most outstanding substitute solvents so far are ionic liquids (ILs) and deep eutectic solvents (DESs). Having non-volatility and thermal stability properties, the latest generation of ionic liquid usually comprised of organic cations such as pyridinium, imidazolium, pyrrolidinium, and phosphonium; and anions like methylsulphate, tetrachloroalumine, methansulfonium and hexafluorophosphate [2]. Despite having great physical properties suitable for synthetic chemical reaction, ILs requires expensive starting material components [3], as well as its common starting material components used, are toxic to the environment according to investigation made by toxicologist, refuting the claim that it is a green solvent [4]. Furthermore, the impurities found in ILs required for purification process of ILs is not economical for further industrial application [5]. The emergence of DESs overcomes the drawback of ILs as it owns physical properties quite similar to ILs. DESs consisting of hydrogen bond acceptor and hydrogen bond donor components to create a liquid with a low melting point than the single components resulting from intermolecular hydrogen bonding between those components [2, 6]. In contrast to ILs, the starting materials of DESs are more environmental-friendly and can be easily prepared from readily available components. The most employed hydrogen bond acceptor, the choline-based quaternary ammonium salt is usually mixed with

IOP Conf. Series: Materials Science and Engineering **440** (2018) 012004 doi:10.1088/1757-899X/440/1/012004

amide, urea, carboxylic acids, alcohols, sugars or organic acids to form the DES at a specific molar ratio [6, 7]. DES is utilized in organic synthesis, biocatalysis, nanotechnology, biotransformation, electrochemistry, extraction, analysis and gas adsorption [6, 8]. Currently, lipase-catalyzed reaction for the production of biodiesel is the most widely employed enzymatic reaction that applies DES as the solvent. For instance, in the production of biodiesel from soybean oil [9], rapeseed oil [10] and yellow horn seed oil [11]. DES also has been utilized as a solvent for the production of phenolic compound involving the reaction between methyl *p*-coumarate and methyl ferulate with 1-octanol [12].

A hydroxycinnamic acid such as caffeic acid (CA, 3,4-dihydroxycinnamic acid) is a phenolic compound that is produced by plant as a secondary metabolite. A plentiful amount of CA is widely distributed in nature for example in fruits, vegetables, oils, and tea. CA exhibits several biological activities like antioxidant [13, 14], anticancer [15] and anti-HIV [16]. Unfortunately, the solubility of CA in a polar and non-polar media demonstrated unsatisfying result thus modification into caffeic acid esters is required to enhance its solubility. The most widely synthesized caffeic acid esters using lipase are alkyl caffeates [17, 18] and caffeic acid phenethyl esters (CAPE) [19]. The first glycerol-based caffeate ester was previously produced in solvent-free system using lipase enzyme [20] and in ionic liquid which functioned both as catalyst and solvent [21]. Glycerol-based caffeate ester has high solubility in water as compared to CA (1.76 mg/ml in water at 20 °C) [20]. At the present time, the production of glycerol-based caffeate esters in DES using lipase has not being reported. The addition of water will improve the performance and activity of lipase in DES media since water is providing hydration to the enzyme. In this work, the lipase-catalyzed transesterification reaction of ethyl caffeate and glycerol to produce glycerol-based caffeate esters using a commercial lipase was carried out. This study aims to investigate the effect of reaction parameters that affect the conversion of ethyl caffeate in DES consists of choline chloride-urea. The outcomes of the investigation are important for better understanding of the potential of DES as a solvent for further application in food, cosmeceutical and pharmaceutical industries.

2. Materials and method

2.1. *Enzyme and materials*

Commercial lipase containing Novozym 435 was purchased from Novozymes (Bagsvaerd, Denmark). Choline chloride (ChCl), glycerol (purity 99.5%) and 4Å molecular sieve were from Sigma-Aldrich, Co. (United States). Urea (U) and caffeic acid (CA) were purchased from Acros Organics (New Jersey, United States). Ethyl alcohol (ethanol), anhydrous sodium sulfate and glacial acetic acid (HPLC grade) were purchased from Fisher Scientific (M) Sdn. Bhd. (Malaysia). Sodium carbonate and methanol (HPLC grade) were from Merck (Germany). Diethyl ether was from Fine Chemicals Corporation (PTY) Ltd (Cape Town, South Africa).

2.2. *DES preparation*

Hydrogen bond acceptor, choline chloride was dried under vacuum oven at 60 ºC as it is hygroscopic. ChCl and U were weighed immediately at molar ratio 1:2 so that only minimum air exposure occurred. The mixture was heated on hot plate with magnetic stirrer (11-100-49SH, Fisher Scientific) at 100 ºC with 300 rpm rotation speed until a colourless liquid is obtained (typically 2 hours).

2.3. *Synthesis of ethyl caffeate*

Ethyl caffeate (EC) synthesis was altered slightly from the method previously described by Pang and Fiuza [17, 18]. Caffeic acid (2.5 g) was refluxed with 200 ml of ethanol and 2.5 ml H_2SO_4 for approximately 8 h with the addition of molecular sieve to absorb any water produced from the reaction. The rotation speed was set at 400 rpm (MS-DMS, PLT Scientific). The mixture was then neutralized with 10 vol% Na_2CO_3 solution. The extraction of ethyl caffeate was done utilizing diethyl ether (3x150 ml) and the organic phases collected were then combined and washed with distilled water (3x200 ml) until a clear yellowish solution was acquired. Later, the solution was dried with anhydrous $MgSO_4$ followed by solvent removal under reduced pressure. The residue was filtered and evaporated until yellowish solid formed.

2.4. Lipase-catalyzed transesterification of ethyl caffeate and glycerol
Enzymatic transesterification of ethyl caffeate and glycerol was carried out according to the procedure previously described by Sun and co-workers [20], with slight modifications. Ethyl caffeate and glycerol (molar ratio 1:50) were dissolved in DES in tight closed jar and the reaction was initiated by adding Novozym 435 at 40°C in orbital shaker (Infors HT Ecotron). At a certain period of times, the aliquots taken were diluted with methanol, filtered using 0.20 μm syringe filter and injected to HPLC. The aliquots were assessed with HPLC (Shimadzu Corporation, Japan) using a chiral column (Regis Technologies, Inc., Illinois, United States) connected to a UV detector set at a wavelength of 325 nm. The isocratic elution was performed using a mobile phase of methanol: water and 0.5% acetic acid (80:20, v/v). All experiments were conducted in triplicates.

3. Results and Discussions
3.1. Effect of lipase loading amount
Novozym 435 with unit activities ranged from 250 U to 1500 U was chosen to monitor the effect of enzyme loading on the lipase-catalyzed transesterification in DES. Final EC conversion increased from 250 U to 1250 U due to the increase in active site for substrates to bind. However, the conversion rate started to decline at 1500 U (see Figure 1). The movement of substrates and enzymes were restricted because of slow mass transfer in the system when more enzymes were added in addition with the high viscosity of DES. Furthermore, enzyme aggregation occurred due to its tendency to attract with each other when in excess resulting in reduction of binding of substrates to the enzyme [17, 22]. Enzyme with unit activity of 1250 U was employed for further investigation.

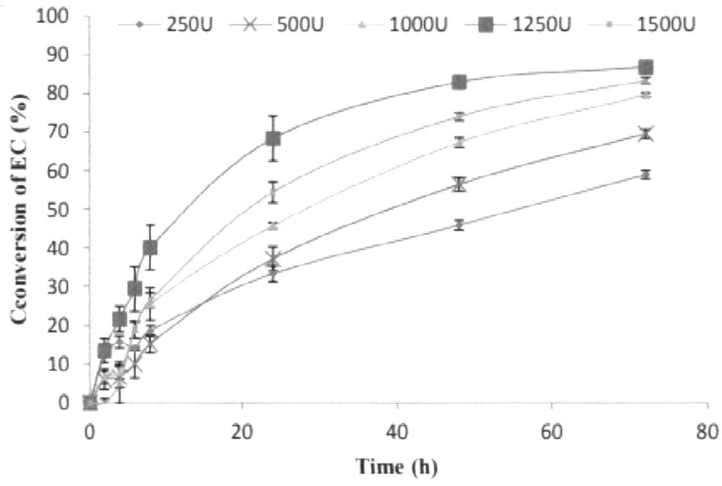

Figure 1. Effect of lipase loading amount on EC conversion. Reaction conditions: molar ratio of ethyl caffeate to glycerol is 1:50; agitation speed, 150 rpm; temperature, 40 °C.

3.2. Effect of water content
The high viscosity of DES led to a practical problem as such the transferring process of the solvent was extremely time-consuming [23]. Diluting it with water could reduce the viscosity. The water content in the lipase-catalyzed transesterification in DES was investigated by varying the amount of water from 1 vol% to 40 vol% of the total volume of the mixture. The conversion achieved a plateau at 20 vol% with conversion over 98% within 4 hours (see Figure 2). Water is essential for enzyme to stay hydrated and

maintain its natural conformation in order to fully function. The addition of water was found to increase the conversion and yield for the lipase-catalyzed transesterification of methyl ester with several alkanols in DES as reported by Durand and his co-workers [24]. Since water can cause hydrolysis reaction of ethyl caffeate to caffeic acid thus 20 vol% was chosen for the next parameter investigation to minimize hydrolysis reaction. Moreover, a higher concentration of water in DES could rupture the choline chloride-urea bond and making the components appeared as free forms in the DES liquid, resulting in the disturbance of DES properties [23, 25].

3.3. Effect of agitation speed

Mass transfer limitations could also be overcome by investigating the effect of agitation speed on the lipase-catalyzed transesterification reaction in DES. Figure 3 depicted the effect of agitation speed on the initial rate of lipase-catalyzed transesterification in DES investigated at four different speeds, namely 100, 150, 200 and 250 rpm. The initial rate showed an increment when the agitation speed increased due to the high collision between substrates and enzymes. However, at higher agitation speed than 200 rpm, the initial rate remained steady. The previous study reported that the initial rates were unchanged as the agitation speed was higher than 200 rpm. They deduced that no external diffusion factor when the agitation speed was higher than 200 rpm [26]. EC conversion as illustrated in Figure 4 showed that no significant change can be seen making 200 rpm the chosen agitation speed.

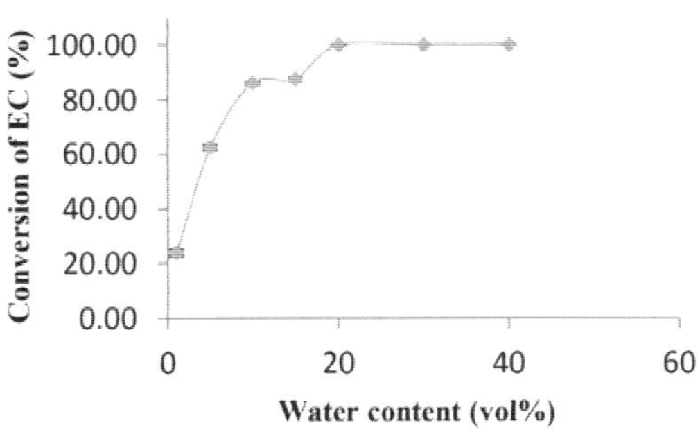

Figure 2. Effect of water content on EC conversion. Reaction conditions: molar ratio of ethyl caffeate to glycerol is 1:50; agitation speed, 150 rpm; temperature, 40 °C; lipase loading, 1250 U.

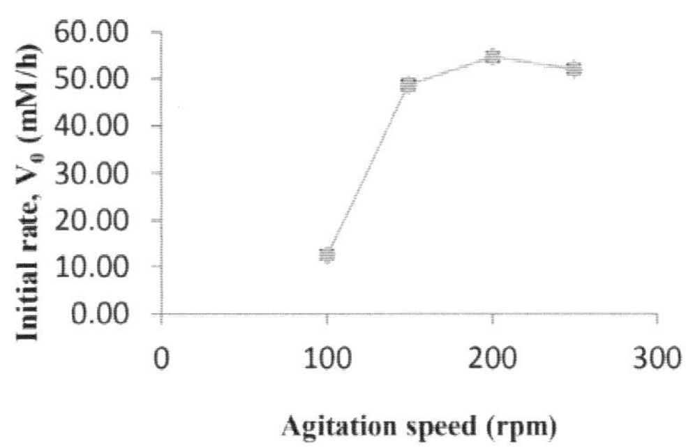

Figure 3. Effect of agitation speed on initial rate. Reaction conditions: molar ratio of ethyl caffeate to glycerol is 1:50; temperature, 40 °C; lipase loading, 1250 U; water content, 20 vol%.

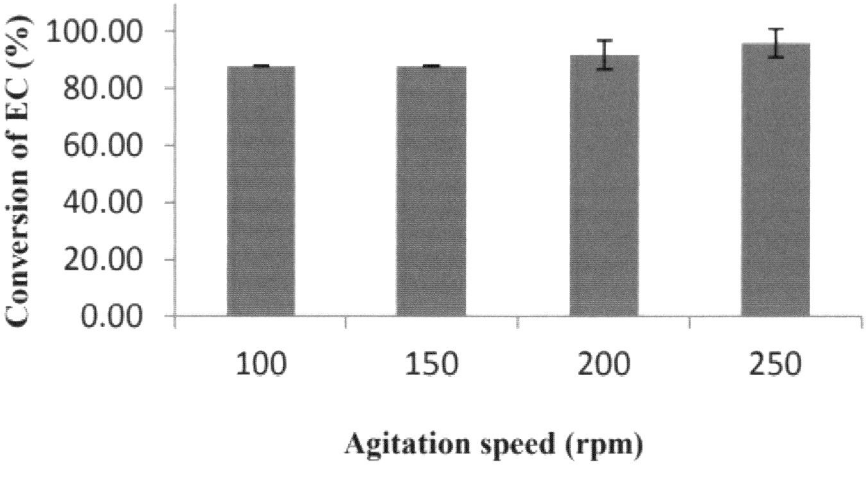

Figure 4. Effect of agitation speed on EC conversion. Reaction conditions: molar ratio of ethyl caffeate to glycerol is 1:50; temperature, 40 °C; lipase loading, 1250 U; water content, 20 vol%.

4. Conclusions

In the present study, it was found that lipase-catalyzed transesterification of ethyl caffeate and glycerol can be carried out in DES. The overall conversion of more than 90% was achieved using 1250 Unit Activity of Novozym 435 with 20 vol% water and agitation speed of 200 rpm in 1:2 molar ratio of choline chloride: urea. It is suggested that the addition of water has reduced the viscosity of DES and reducing the mass transfer limitation during the reaction. Nevertheless, water should be controlled to minimize the hydrolysis reaction and to prevent total rupture of choline chloride-urea bond in DES. A study on the optimization of the reaction parameters, kinetic and thermodynamic analysis should be carried out before it could be applied to industry. This study provides valuable information on the prospect of DES as a solvent which highly biodegradable and possesses no threat to the environment.

5. References

[1] Anastas P T and Williamson T C 1998 *Green Chemistry: Theory and Practice* (New York: Oxford University Press)
[2] Kudłak B, Owczarek K and Namieśnik J 2015 *Environ. Sci. Pollut. Res.* **22** 11975-92
[3] Cvjetko Bubalo M, Jurinjak Tušek A, Vinković M, Radošević K, Gaurina Srček V and Radojčić Redovniković I 2015 *J. Mol. Catal.* B*: Enzym.* **122** 188-98
[4] Zhao D, Liao Y and Zhang Z 2007 *Clean. (Weinh)* **35** 42-8
[5] Durand E, Lecomte J, Baréa B, Piombo G, Dubreucq E and Villeneuve P 2012 *Process. Biochem.***47** 2081-9
[6] Smith E L, Abbott A P and Ryder K S 2014 *Chem. Rev.* **114** 11060-82
[7] Dai Y, van Spronsen J, Witkamp G-J, Verpoorte R and Choi Y H 2013 *Anal. Chim. Acta.* **766** 61- 8
[8] Paiva A, Craveiro R, Aroso I, Martins M, Reis R L and Duarte A R C 2014 *ACS Sustain. Chem. Eng.* **2** 1063-71
[9] Zhao H, Zhang C and Crittle T D 2013 *Journal of Molecular Catalysis B: Enzymatic* **85-86** 243-7
[10] Huang W, Tang S, Zhao H and Tian S 2013 *Ind. Eng. Chem. Res.* **52** 11943-7
[11] Zhang Y, Xia X, Duan M, Han Y, Liu J, Luo M, Zhao C, Zu Y and Fu Y 2016 *J. Mol. Catal.* B*: Enzym.* **123** 35-40

[12] Durand E, Lecomte J, Baréa B, Dubreucq E, Lortie R and Villeneuve P 2013 *Green Chem.* **15** 2275-82
[13] Gülçin İ 2006 *Toxicology* **217** 213-20
[14] Khan F A, Maalik A and Murtaza G 2016 *J. Food. Drug. Anal* **24** 695-702
[15] Tyszka-Czochara M, Konieczny P and Majka M 2017 *Int. J. Mol. Sci.* **18** 462
[16] Fabrice B and Philippe C 2005 *Curr. Med. Chem.* **12** 1811-8
[17] Pang N, Gu S-S, Wang J, Cui H-S, Wang F-Q, Liu X, Zhao X-Y and Wu F-A 2013 *Bioresour.Technol.* **139** 337-42
[18] Fiuza S M, Gomes C, Teixeira L J, Girão da Cruz M T, Cordeiro M N D S, Milhazes N, Borges F and Marques M P M 2004 *Bioorganic. Med. Chem.* **12** 3581-9
[19] Widjaja A, Yeh T-H and Ju Y-H 2008 *J. Chin. Inst. Chem. Eng.* **39** 413-8
[20] Sun S and Hu B 2017 *Food Chemistry* **214** 192-8
[21] Sun S, Hou X and Hu B 2017 *J. Mol. Liq.* **248** 643-50
[22] Sun S and Hu B 2017 *Biochem. Eng. J.* **124** 78-87
[23] Dai Y, Witkamp G-J, Verpoorte R and Choi Y H 2015 *Food Chem.* **187** 14-9
[24] Durand E, Lecomte J, Baréa B and Villeneuve P 2014 T *Eur. J. Lipid. Sci. Technol.* **116** 16-23
[25] Gutiérrez M C, Ferrer M L, Mateo C R and Monte F d 2009 *Langmuir* **25** 5509-15
[26] Zhang J, Wu J and Yang L 2004 *J. Mol. Catal.* B*: Enzym.* **31** 67-72

Acknowledgments
This present study is generously funded by Fundamental Research Grant Scheme (FRGS), Ministry of Higher Education, Malaysia with a grant no. (203.PJKIMIA.6071342). The authors would like to greatly thank Universiti Sains Malaysia (USM) for the facilities provided.

IOP Conf. Series: Materials Science and Engineering **440** (2018) 012005 doi:10.1088/1757-899X/440/1/012005

Characteristics of adsorption isotherm and kinetic study for newly prepared Co^{2+}-imprinted polymer linkage with dipicolinic acid

N F Yusof[1], F S Mehamod[2], M A Kadir[2] and F B M Suah[3]

[1]School of Fundamental Science, Universiti Malaysia Terengganu, 21030 Kuala Nerus, Malaysia.
[2]Advanced Nano Materials Research Group, School of Fundamental Science, Universiti Malaysia Terengganu, 21030 Kuala Nerus, Malaysia.
[3]School of Chemical Sciences, Universiti Sains Malaysia (USM), 11800 Minden, Pulau Pinang, Malaysia.

Corresponding author emails: fshimal@umt.edu.my

Abstract. Co^{2+}-complex imprinted polymer (Co-IIP) was prepared by ionic imprinting technology for selective removal of Co^{2+} ions from aqueous solution. Co^{2+} which act as a template ion established a coordination linkage with dipicolinic acid before being copolymerized with functional monomer (methacrylic acid) in the presence of cross-linker (ethylene glycol dimethacrylate) and initiator (2,2′-azobisisobutyronitrile). The surface features and functional groups present in imprinted polymer were characterized by scanning electron micrograph and infrared spectra. The non-imprinted polymer (NIP) was produced simultaneously to serve as controlled polymer. From the isotherm study, adsorption of Co^{2+} ions by ion imprinted polymer (IIP) followed Langmuir isotherm model, which indicates the sorption can be described by monolayer adsorption by give the maximum adsorption capacity was 19.57 mg g^{-1}. This value was comparable to the experimental value which was 18.48 mg g^{-1} and better affinity compared to NIP. The adsorption kinetics fit a pseudo-second-order and adsorption equilibrium time was around 30 minutes. The present work has successfully synthesized Co-IIP particles with good potential in recognition of Co^{2+} ions.

1. Introduction

Molecular imprinting is a study of molecular recognition which is widely exploited in a numbers of applications. It is a technique for making selective binding sites towards the analyte (template) in polymeric matrices. The concept of molecular imprinting is analogous to lock and key hypothesis proposed by Emil Fischer, a Nobel laureate who explained enzyme-substrate interaction. It was early attributed by Polykov in 1930s to investigate the silica matrix for various additives extraction and it was continued develop by Linus Pauling in 1940s. This discovery had resulted in intensive research on imprinting technology. The imprinting process can be illustrated into three main stages [1]; 1) mixing the functional monomer with the target molecule or template in the suitable porogen according to predetermined ratio. Pre-polymerization occurs between template and monomer via covalent or noncovalent bonding; 2) the growing polymer chain will be created as the free radical polymerization takes part in the solution. Cross-linkers are added and polymerize to achieve highly cross-linked polymers. The structure of adducts will froze and fixed around the template molecule; 3) lastly, the removal of template from the polymer thus leaves the specific binding site corresponding to the template molecule. Due to the complementarity of recognition sites to template in term of size, shape and chemical functionality orientations as well as contribution from environment, rebinding of subsequent template is thus allowed.

Rebinding recognition in imprinting involves the sorption process between template molecule and polymer. When this molecular species is in contact with a solid surface of polymer, there will be an equilibrium establishment. In the liquid phase, monitoring the coverage of the surface in relation to the concentration at a fixed temperature is described as adsorption isotherm. The equilibrium adsorption isotherm is an important assessment for the theoretical determination design of adsorption system. Ion imprinted polymers (IIP) is a branch of molecular imprinting study by changing the template molecule to inorganic ions. Commonly, metal ion is dominates used as template molecule for imprinting purpose. The first ion imprinting was introduced by Nishide and co-workers by cross-linked poly(vinylpyridine) resin towards several metal ions [2].

In the present study, IIP were applied as an adsorbent material for selective recognition of Co^{2+} ions from aqueous medium. Cobalt ion was chosen as a target metal ion as it was environmentally toxic and enters the environment mostly from human activities such as industry of mining and pigment [3] plastic and catalytic process [4] alloys industry [5] and many more. It can form both bivalent and trivalent compounds. Human can be exposed to cobalt in variety of pathways such as drinking water, breathing air and food supply. For this reasons, separation of cobalt ion is needed to overcome the excessive exposure towards it. Because of their importance from an environmental viewpoint, numerous methods have been produced in sequestration of cobalt ion from aqueous solution such as salting-out phase separation [6] extraction by almond green hull [7] biosorptive removal [8] solvent extraction [9] and many more.

Therefore, one of the greatly developed techniques for the preparation of metal extraction is by applying the molecular imprinting technology. It is a technique for the preparation of polymeric materials that are capable of molecular recognition for selective separation and chemical analysis, amongst other applications [10]. The application of IIP using cobalt ions as target analyte was investigated as one of the alternative way to improve any drawbacks of the existing methods and produce less costly procedure in the extraction of cobalt ions in aqueous media. The imprinting effect of Co-IIP was study to get the information on adsorption behavior in terms of equilibrium isotherms and kinetic motion of Co-IIP towards Co^{2+} ions, which will be extended to further development of technological appliances especially for environmental concern.

2. Methodology

2.1. Chemicals

Methacrylic acid (MAA), ethylene glycol dimethacrylate (EGDMA), 2,2'-azobisisobutyronitrile (AIBN), methanol, ethylenediamminetetraacetatic acid (EDTA), hydrochloric acid (HCl) and dipicolinic acid (DPA) were purchased from Sigma-Aldrich. Cobalt (II) chloride (Hexahydrate) was supplied from Scharlau. All chemicals were purified before use and AIBN was re-crystallized from methanol.

2.2. Synthesis of Imprinted Polymers

The IIP was synthesized by bulk polymerization method. The template was first prepared by react $CoCl_2$ with dipicolinic acid to form a complex before being polymerized with monomer. The Co^{2+}-dipicolinate complex, MAA, EGDMA and AIBN was dissolve in 5.00 ml of methanol to allow the polymerization process. The ratio used for template: monomer: cross-linker was 1: 4: 20. The solution was degassed for 10 minutes and then placed in an oil-bath at 60°C for 24 hours. NIP was prepared in the same manner as controlled polymer but in the absence of Co^{2+}-complex. The resultant polymers were then ground and sieved using acetone to obtain the particles with sizes less than 25 μm. It was then collected after

IOP Conf. Series: Materials Science and Engineering **440** (2018) 012005 doi:10.1088/1757-899X/440/1/012005

sedimentation from acetone. The template, unreacted monomers or diluents were leached out *via* continuous stirring with EDTA and HCl in 9:1 ratio for 24 hours. Figure 1 shows the proposed schematic reaction of Co-IIP.

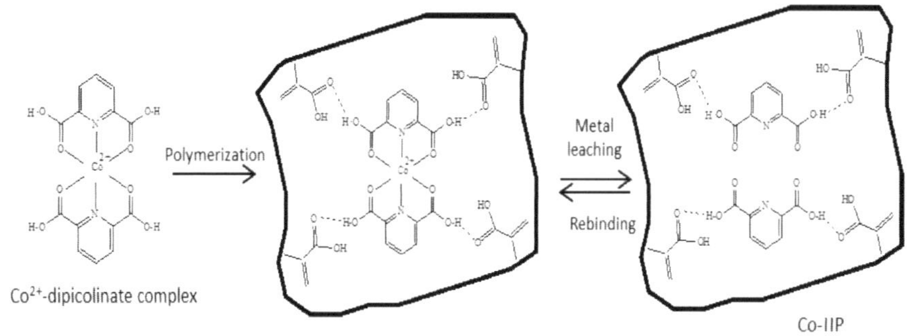

Figure 1. Schematic representation of Co-IIP imprinting process.

2.3. Characterization Techniques
The surface and morphological properties of IIP and NIP monolith size was characterized using Scanning Electron Microscopy (SEM) model JEOL JSM-6360LA. Samples were prepared and placed on metal stubs and coated with thin layer of Au. The textural characteristic was viewed under 5000x magnification. The chemical structure of the imprinted polymer was determined using Fourier Transformed Infrared (FTIR) spectrometer from Perkin Elmer with spectrum 100 as the software. Attenuated Total Reflectance (ATR) method was used in order to get the FTIR spectra. The spectra were recorded in the range of 4000 to 400 cm^{-1} wavenumbers.

2.4. Binding Studies
Equilibrium batch rebinding experiment were studied using Atomic Absorption Spectroscopy (AAS). 25 ml of 10 ppm of Co^{2+} solution was added to the different mass of polymers (1, 5, 10, 11, 12, 13 and 15 mg). Each mixture was shaken for different time (5, 15, 30, 60, 90 and 120 minutes) to facilitate the adsorption of Co^{2+} ions onto polymers. After the binding process was complete for predetermined time, the solution was filtered and analyzed by using AAS to identify the concentration of free Co^{2+} in solutions. The data obtained was used to study the adsorption properties of Co-IIP and NIP by adsorption isotherm modelling. The adsorption capacity was calculated using the Equation (1).

$$q_e = \frac{(C_i - C_f)V}{W} \quad (1)$$

where q_e is the amount of Co^{2+} adsorbed (mg g^{-1}), C_i is the initial concentration of Co^{2+} ions solution (ppm), C_f is the final concentration of Co^{2+} ions solution (ppm), V is the volume of Co^{2+} ions solution used (L) and w is the weight of polymer applied to the solution (g).

3. Results
3.1. Characterization of Polymers
Co-IIP and NIP were successfully synthesized using free radical polymerization. The Co-IIP and NIP were characterized by FTIR analysis. The FTIR spectra for Co-IIP and NIP were recorded as shown in Figure 2, it is obvious that Co-IIP and NIP possessed similar backbone structure. The interaction between template and monomer present in Co-IIP gave changeable peaks from NIP spectrum to Co-IIP spectrum.

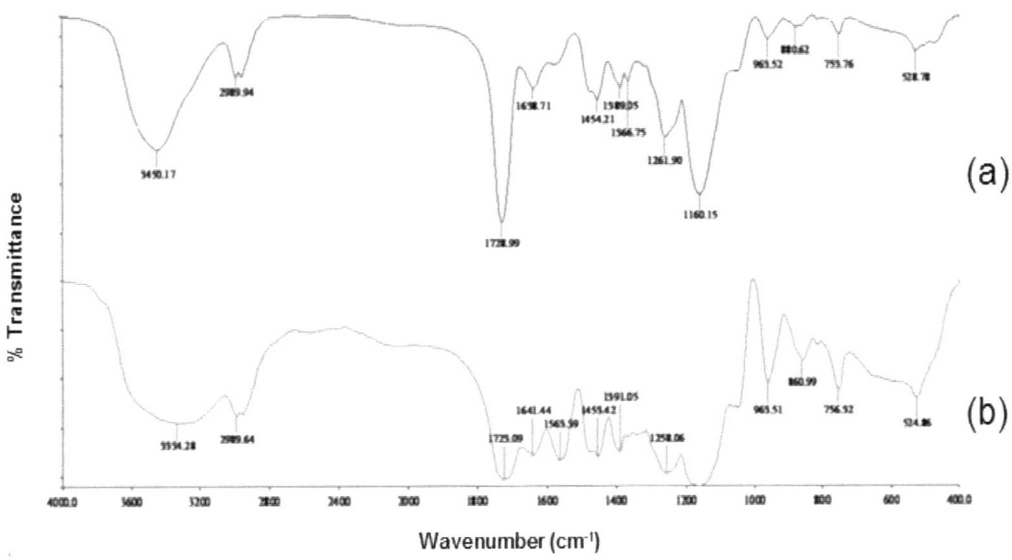

Figure 2. FTIR spectra for (a) NIP and (b) Co-IIP.

From bulk polymerization, polymers were obtained in monolith form; irregular shape and size. Particles with size range of 1-25 μm were collected and used in this study. Figure 3 shows the SEM micrographs for Co-IIP and NIP. The surface morphology for Co-IIP was rougher compared to NIP, which was smoother and cleaner surface.

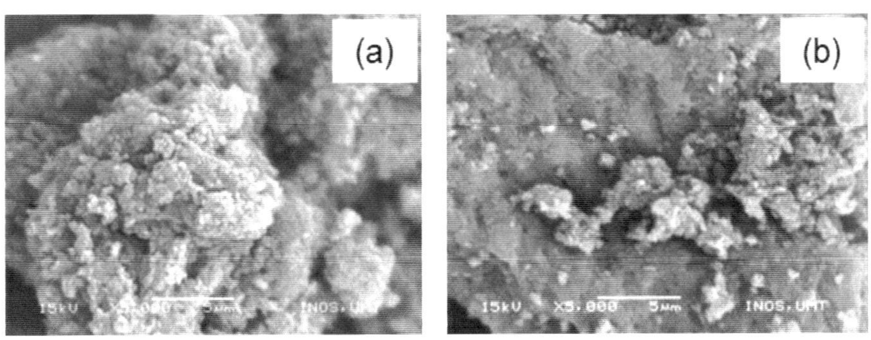

Figure 3. SEM micrograph for (a) Co-IIP and (b) NIP.

3.2. Adsorption Isotherm

The adsorption isotherm for Co-IIP and NIP were investigated to illustrate the interactive behavior between Co^{2+} ions and polymeric cavities. Series of calculation were carried out to investigate the adsorption capacity, as well as describing polymers' surface properties and affinity to Co^{2+} ions. For the effect of polymer dosage (Figure 4), it was observed that when the dosage of polymers were increased from 1 mg to 12 mg, the removal efficiency of Co^{2+} ions increase as expected. The removal efficiency of Co-IIP increased from 10.13% to 94.94%, as compared to slight increased from 3.58% to 36.66% for NIP.

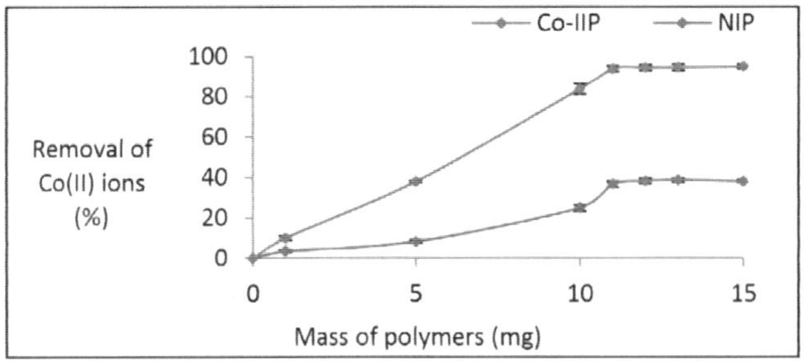

Figure 4. Effect of polymers dosage on the adsorption of Co^{2+} onto Co-IIP and NIP.

Linearized graphs of Langmuir [11] and Freundlich [12] were produced by fitting all the equilibrium data for both polymers in order to get the applicability of isotherm models. Langmuir model of adsorption is described as the Equation (2).

$$\frac{C_e}{Q_e} = \frac{1}{Q_o} C_e + \frac{1}{Q_o b} \tag{2}$$

where C_e (mg L^{-1}) is the equilibrium concentration, Q_e is the adsorption amount at equilibrium (mg g^{-1}), Q_o is the maximum adsorption capacity (mg g^{-1}) and b is Langmuir adsorption constants (L mg^{-1}). Langmuir isotherm can be plotted as C_e/Q_e against C_e, $1/Q_o$ as slope and $1/Q_ob$ as intercept.
The second isotherm model used in this study is Freundlich model and a linear equation can be described as in Equation (3).

$$logQ_e = logK_F + \frac{1}{n} logC_e \tag{3}$$

where K_F is the Freundlich constant while n is the Freundlich exponent. Freundlich isotherm can be plotted as log Q_e against log C_e; log K_F becomes the intercept of the line and $1/n$ becomes the slope of the line. The constant K_F indicates adsorption capacity, while $1/n$ shows the strength of adsorption during adsorption process.

Linearized forms of Langmuir and Freundlich isotherm were plotted as shown in Figure 5 and Figure 6, respectively. It was found that the adsorption of Co-IIP and NIP fitted better with Langmuir isotherm than Freundlich isotherm by given higher correlation coefficient, R^2 values. Therefore, it can be concluded that the sorption of Co^{2+} ions by Co-IIP and NIP were followed the Langmuir isotherm model. The information on isotherm parameter was calculated from the straight line equation and tabulated in Table 1.

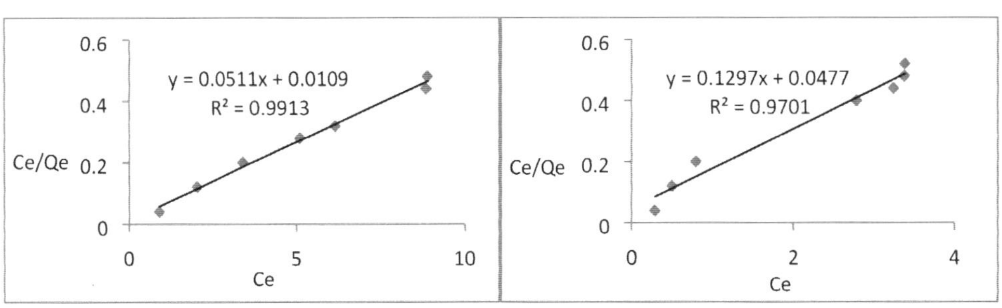

(a) (b)

Figure 5. Langmuir plot for adsorption of Co^{2+} ions by (a) Co-IIP and (b) NIP.

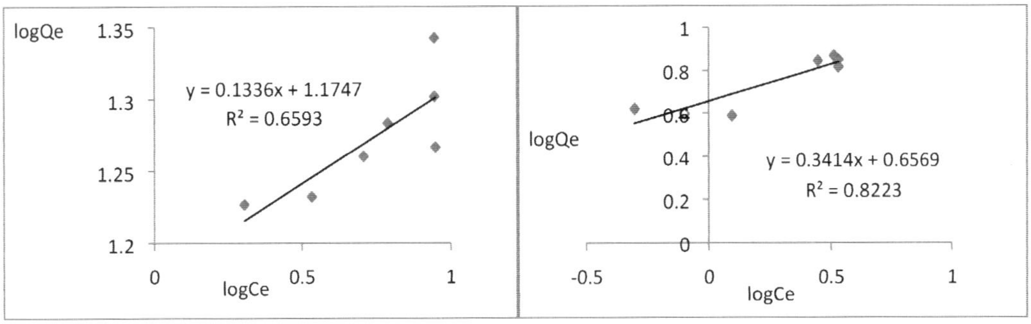

(a) (b)

Figure 6. Freundlich plot for adsorption of Co^{2+} by (a) Co-IIP and (b) NIP.

The separation factor, R_L is a dimensionless constant was calculated to understand the essential features of the Langmuir equation by using Equation (4).

$$R_L = \frac{1}{1 + bC_o} \quad (4)$$

where C_o is initial concentration of Co^{2+} solution and b is the Langmuir constant. R_L indicates the nature adsorption of Co-IIP and NIP. The adsorption process is unfavorable if $R_L > 1$, linear if $R_L = 1$, favorable if $0 < R_L < 1$ and irreversible if $R_L = 0$ [13, 14]. Result obtained for Co-IIP and NIP in this study showed that R_L values were in the range of favorable.

Table 1. Isotherm constant for Co-IIP and NIP.

Polymers	Langmuir				Freundlich			
	Q_o (mg g^{-1})	b (Lmg^{-1})	R_L	R^2	$1/n$ (mg g^{-1})	n (L mg^{-1})	K_F (mg g^{-1})	R^2
Co-IIP	19.56	4.68	0.0208	0.99	0.13	7.48	0.07	0.65
NIP	7.41	4.17	0.0233	0.97	0.34	2.92	4.53	0.82

3.3. Kinetic Studies

The adsorption kinetic was studied to describe the underlying mechanism of the adsorption process and the rate at which Co^{2+} ions are adsorbed onto the surface of polymers. The adsorption capacity curves of Co-IIP and NIP was illustrated as shown in Figure 7. As it can be seen, the adsorption process involved two stages. Initially, the adsorption rate was rapid increase for the first 30 minutes with an adsorption capacity of 8.86 mg g^{-1} and 3.38 mg g^{-1} for Co-IIP and NIP respectively. Then, after 30 minutes, adsorption rate remained constant and achieved equilibrium state because of the saturation of polymeric matrices occupied by Co^{2+} ions. Co-IIP showed much higher adsorption capacity as compared with NIP. Therefore, the Co-IIP was said to perform more efficient than NIP in adsorbing Co^{2+} ions. The equilibrium data were analyzed using pseudo-first-order and pseudo-second-order rate model. Results were obtained in triplicate analysis with errors based on ±1 standard deviation.

IOP Conf. Series: Materials Science and Engineering **440** (2018) 012005 doi:10.1088/1757-899X/440/1/012005

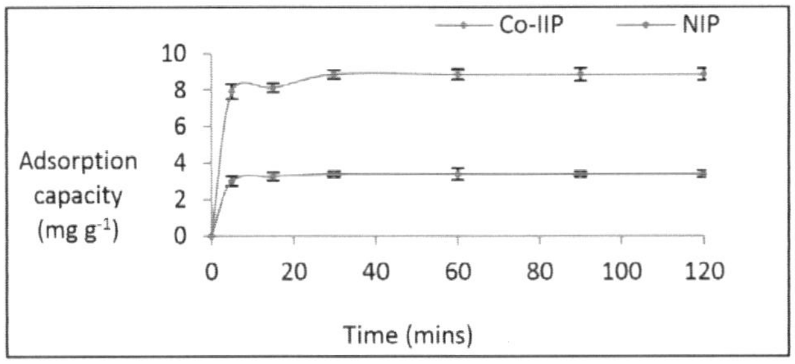

Figure 7. Adsorption capacity of Co^{2+} by Co-IIP and NIP.

The kinetic equation of pseudo-first-order represent as in Equation (5) [15].

$$\ln(q_e - q_t) = \ln q_e - k_1 t \tag{5}$$

where k_1 is the adsorption rate constant of pseudo-first-order kinetic (min^{-1}), t is the contact time between adsorbate and adsorbent, while q_e and q_t are the amount of Co^{2+} ions adsorbed at equilibrium and at any time (mg g^{-1}) respectively. Meanwhile, pseudo-second order kinetic model was described by using the kinetic Equation (6) [16].

$$\frac{t}{q_t} = \frac{k_2}{h} + \frac{t}{q_e} \tag{6}$$

where $h = k^2 q^2_e$ is the initial adsorption rate, while k_2 is the pseudo-second-order rate constant. The parameter q_e here is referred as $q_{e,cal}$ to show the calculated amount of Co^{2+} ions adsorbed at equilibrium from the plotted graphs.

In order to ensure the suitability of kinetic models of polymers, both normalized standard deviation value, Δq (%) and relative errors (%) were calculated as described as the equations below.

$$\Delta q(\%) = \sqrt{\frac{[(q_{e,exp} - q_{e,cal})/q_{e,exp}]^2}{N-1}} \times 100\% \tag{7}$$

$$Relative\ error\ (\%) = \frac{|q_{e,cal} - q_{e,exp}|}{q_{e,exp}} \times 100\% \tag{8}$$

where N is the amount of data points fitted to the plot, $q_{e,exp}$ and $q_{e,cal}$ (mg g^{-1}) are the experimental and calculated adsorption capacity respectively. The lower Δq (%) gives best fitness of polymer to the kinetic model.

The graphs for pseudo-first-order and pseudo-second-order were plotted (Figure 8 and Figure 9, respectively) and the parameters were calculated as shown in Table 2. The kinetic of adsorption exhibited by Co^{2+} ion as divalent metallic ion was described by pseudo-second-order kinetic by give higher R^2 values for both polymers.

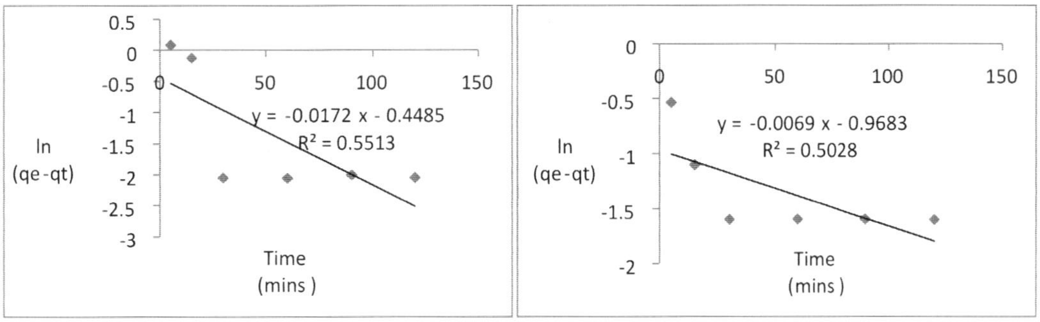

(a) (b)

Figure 8. Pseudo-first-order plots for adsorption of Co^{2+} ions by (a) Co-IIP and (b) NIP.

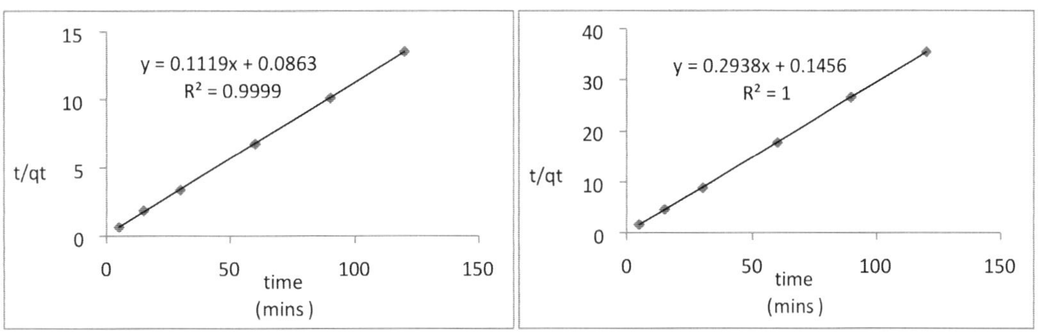

(a) (b)

Figure 9. Pseudo-second-order plots for adsorption of Co^{2+} ions by (a) Co-IIP and (b) NIP.

Table 2. Kinetic constant for Co-IIP and NIP.

Polymer	Pseudo-first-order					Pseudo-second-order					
	$q_{e,cal}$ (mg g^{-1})	k_1	R^2	Δq (%)	RE (%)	$q_{e,cal}$ (mg g^{-1})	k_2	h (g.min mg^{-1})	R^2	Δq (%)	RE (%)
Co-IIP	0.63	0.0172	0.55	41.50	92.7 9	8.93	0.14	11.58	0.99	0.34	0.76
NIP	0.37	0.0069	0.50	39.78	88.7 8	3.40	0.59	6.86	1.00	5.31×10^{-4}	0.54

4. Discussion

For FTIR analysis, as expected for the similar backbone FTIR structure was due to the same monomers were used in the production of the polymers [17] and high level of cross-linking agent used. Based on Figure 2, a few peaks was shifted from NIP to Co-IIP; for example, the O-H stretching band shifted from 3450 cm^{-1} to 3334 cm^{-1}, C=O stretching band shifted from 1728 cm^{-1} to 1723 cm^{-1} and C-O stretching band shifted from 1160 cm^{-1} to 1161 cm^{-1}. These shifted peaks can be reasoned of the formation of dative bond between Co^{2+} ion and dipicolinic acid. Co-IIP exhibited a relatively rougher surface due to the imprinting effect of Co^{2+} ions onto the surface, which resulted in cavities compared to NIP. IIP with rougher surfaces also facilitates the mass transfer rate of metal ions toward the polymer surface and consequently improves its adsorption capacity [18].

For the adsorption study, the higher performance of Co-IIP could be reasoned by the possession of specific cavities to allow the adsorption of Co^{2+} ions onto the surface, as in agreement with SEM images showed slightly rougher surface in Co-IIP compared to NIP. The further increase in dosage of Co-IIP and NIP after 12 mg produced almost plateau values which indicated that equilibrium was achieved. This could be attributed to saturation of cavities for both polymers. Results based on average values for triplicate analysis with error based on ±1 standard deviation.

The equilibrium data obtained was found to be appropriate towards the Langmuir isotherm. This isotherm assuming that the recognition sites of outer surface of adsorbent are homogeneous and the adsorbate is adsorbed at well-defined sites, leading to monolayer coverage of the adsorbate surface. Thus, this can be assumed that Co^{2+} ions were adsorbed at a fixed number of well-definite recognition cavities which each cavity can only uptake one Co^{2+} ions. These cavities are considered energetically equivalent and there is a distance between cavities to ensure that no interaction between adsorbed Co^{2+} ions to adjacent sites [19].

The maximum adsorption capacity, Q_o obtained (Table 1) from Langmuir isotherm for Co-IIP was 2.6 times higher than NIP, with the value of 19.56 mg g^{-1} and 7.41 mg g^{-1} respectively. This proved the presence of high amount of microspores with higher specificity and affinity on the Co-IIP surface than NIP, allowing more amounts of Co^{2+} ions adsorbed into the cavities. The maximum adsorption capacity (experimental) obtained for Co-IIP and NIP was 18.47 mg g^{-1} and 3.58 mg g^{-1} respectively. Thus, the values calculated, Q_o for Langmuir isotherm model is comparable to the experimental values.

In the pseudo-second-order theory, the rate-limiting step is chemisorption process between Co^{2+} ions and cavities. This process involves the valence force that allows the exchange or sharing of electrons to occur. With response to this research, Co^{2+} ions can interact with two types of adsorption cavities, either Co^{2+} ions was imprinted by active sites in the polymers or they can result in the formation of inclusion complex with functional monomers, (MAA) during the molecular recognition stage (Wang et al., 2011). The adsorption rate of Co^{2+} ions onto the cavities was evaluated by looking at the initial adsorption rate, h. Co-IIP was found to be faster in adsorbing Co^{2+} ions relative to NIP with the initial adsorption rate of 11.58 g.min mg^{-1}, which was 1.7 times faster than NIP (Table 2).

The accuracy and validity of kinetic models were further compared with normalized standard deviation value (Δq) and percentage relative errors (RE) to strengthen the data that both polymers are follows pseudo-second order kinetic model. The value of $q_{e,exp}$ obtained for Co-IIP and NIP were 8.86 and 3.38 mg g^{-1} respectively. Both polymers fitted better to pseudo-second-order as the difference between $q_{e.cal}$ and $q_{e,exp}$ was small. This model also showed the relatively low for Δq and RE value as compared to pseudo-first-order kinetic model.

5. Conclusion

The potential of Co-IIP for the effective removal of Co^{2+} ions was successfully demonstrated. It has been shown that Co-IIP showed better performance than NIP by successfully adsorbing 94% Co^{2+} ions in the aqueous solution. Batch rebinding experiment were revealed that the adsorption isotherm was followed Langmuir isotherm which produce the homogenous polymer surface. For kinetic study, pseudo-second-order kinetic models was successfully fit the data thus revealed the chemisorption process occurred in the system.

6. References

[1] Komiyama M, Takeuchi T, Mukawa T and Asanuma H 2003 *Molecular Imprinting: from fundamentals to applications* (Wiley-VCH Verlag GmbH & Co. KGaA) p i-xii

[2] Nishide H, Deguchi J and Tsuchida E 1977 *Journal of Polymer Science: Polymer Chemistry Edition* **15** 3023-3029

[3] Faroon O, Abadin H, Keith S, Osier M, Chappell L, Dimond G and Sage G 2004 Toxicological profile for cobalt. US Department of Health and Human Services AfTSaDR, Atlanta: Agency for Toxic Substances and Disease Restistry
[4] Khodakov A Y, Chu W and Fongarland P 2007 *Chemical Reviews* **107** 1692-1744
[5] Guo W, Chen R, Liu Y, Meng M, Meng X, Hu Z and Song Z 2013 *Colloids and Surfaces A: Physicochemical and Engineering Aspects* **436** 693-703
[6] Habib A, Wahiduzzaman M, Rashid H -O, Islam A, Ferdoushi F K and Alam A S 2008 *Pakistan Journal of Analytical Environment Chemistry* **9** 6-10
[7] Ahmadpour A, Tahmasbi M, Bastami T R and Besharati J A 2009 *Journal of Hazardous Materials* **166** 925-930
[8] Egila J, Dauda B and Jimoh T 2010 *African Journal of Biotechnology* **9** 8192-8198
[9] Lee M, Kim S, Choi Y and Chae J 2011 *Materials Transaction* **52** 1211-1215
[10] Mehamod F S, KuBulat K, Yusof N F and Othman N A 2015 *International Journal of Technology* **4** 546-554
[11] Sime R J 2000 *Infohost. Nmt. Edu* 7
[12] Mittal A, Kurup L and Mittal J 2007 *Journal of Hazardous Materials* **146** 243-248
[13] Malkoc E and Nuhoglu Y 2007 *Separation and Purification Technology* **54** 291-298
[14] Sathish T and Vinithkumar N V 2015 *Applied Water Science* **5** 153-160
[15] Yuh-Shan H 2003 *Scientometrics* **59** 171-177
[16] Ho Y S and McKay G 1999 *Process Biochemistry* **34** 451-465
[17] Othman N A, Daik R, Suah F B M and Mehamod F S 2016 *International Journal of Applied Chemistry* **12** 661-674
[18] Yilmaz V, Arslan Z, Hazer O and Yilmaz H 2014 *Microchemical Journal* **114** 65-72
[19] Yavuz H, Say R and Denizli A 2005 *Materials Science and Engineering C* **25** 521-528

Acknowledgment
We would like to thank for Ministry of Higher Education Malaysia for project funding supported under Fundamental Research Grant Scheme (FRGS-Vote No. 59263).

IOP Conf. Series: Materials Science and Engineering **440** (2018) 012006 doi:10.1088/1757-899X/440/1/012006

Preparation of Gellan Gum (GG) Film: The effect of GG, Calcium Chloride ($CaCl_2$), Glycerol Concentration and Heat Treatment

Nur Arifah Ismail, Khairul Anuar Mat Amin, and Mohd Hasmizam Razali*

School of Fundamental Sciences, Universiti Malaysia Terengganu, 21030 Kuala Terengganu, Terengganu Darul Iman, Malaysia.

*Corresponding author email: mdhasmizam@umt.edu.my

Abstract. The aim of this work was to investigate the effect of gellan gum (GG), calcium chloride ($CaCl_2$) and glycerol concentration on physical appearance of the film. It was found that, after heated at 50 °C for 24, hours the films were successfully produced using low amount of GG (0.50, 0.75, and 1.00 g) with effective concentration of $CaCl_2$ (5 mM) and glycerol (50 % (w/w)). The film was not well dried at 40 °C while at 70 and 100 °C, the film formed was easily to rupture due to the overheated. At 1.5 g GG the film cannot be produced due to the formation of highly viscos and sticky formulated solution. The effective glycerol concentration was within the range of 30-50 % (w/w, based on GG weight). At 60 and 70 % (w/w) of glycerol, the films started to distort at the edge due to high concentration of the plasticizer.

1. Introduction

Gellan gum is a type of microbial polysaccharide that is produced from fermentation of the *Pseudomonas elodea*. It is composed of a tetrasaccharide repeating unit of two β-D-glucose, one β-Dglucuronic acid and one α-L-rhamnose [1, 2]. Native gellan contains a substitution of an ι-glycerate group at C_2 and a partial substitution of acetate group at C_6 of 1,3-β-D-glucose. However, the commercial product based on gellan gum has much lower sol-gel transition temperature and higher in gel strength compared to native product. The commercial gellan gum product has been completely deacetylated and de-esterified by alkali treatment [3-6]. The remarkable properties which has been displayed by the gellan gum gel compared to the other polysaccharide gels are that they are heatresistant and less pH-sensitive [4]. Therefore, gellan gum has been widely used in food industry as a thickening agent [7, 8]. Moreover, gellan gum is one of the most promising in situ gelling polymers in the human body and applicable for biomedicine technology, such as drug delivery vehicles, protein immobilization media and wound dressing materials [9-14]. However, the applications of GG gel have their own limitation due to the poor stability in physiological conditions and low in mechanical strength. Thus, researchers focused on synthesizing GG film with good mechanical properties to extend their applications on food packaging, edible film and coating. It can be obtained by incorporating the crosslinker and plasticizer into the GG during the formulation. It is important to note that GG carries a net negative charge from the carboxylic groups present in the glucaronic acid residues which is the determining factor to the ability of GG hydrogel to transform into film after sufficient heat treatment.

Thus the presence of cation as cross-linker to the negative charge of GG is important for gelation process and film forming [15, 16]. The GG chain will be rapidly cross-linked to form the three-dimensional network with addition of cations. K^+, Na^+, and Ca^{2+} are among the commonly used cations to form strong and durable GG film. Somehow, divalent Ca^{2+} is widely used as can form stronger gels than monovalent cations [17]. Besides the cross-linker, plasticizer is another important additive in the film formulation especially for biodegradable films [18-20]. The addition of plasticizers can improve the mechanical properties of edible films and various plasticizers, such as polyols, have been employed to increase the flexibility and workability of these films. Among the plasticizers, glycerol is one of the most broadly used in film-making techniques, and it has been successfully employed in the production of polysaccharide-based edible films [21-24]. It is a water soluble, polar, and low molecular weight non-volatile substance, which makes glycerol a suitable plasticizer to be used with a compatible water-soluble polymer [22]. However, to our best of knowledge there are no specific investigations about the effect of plasticizer concentration on the properties of GG film so far. Therefore, the purposes of this work were to: (1) develop a novel edible film based on GG; and (2) determine the effect of heat treatment, GG content, and concentration of cross-linker ($CaCl_2$) and plasticizer (glycerol) on the physical appearance of the resulting films.

2. Experiments

2.1 Materials

Low-acyl gellan gum (Kelcogel®, CP Kelco, lot number 5C1574A), Glycerol (lot number STBC1888V) were purchased from Sigma-Aldrich, Malaysia; as a plasticizer to reduce the brittleness of film. Calcium Chloride ($CaCl_2$, lot number K47117278604, Mw 110.98 g/mol) was obtained from Merck, Malaysia. All materials were used as received.

2.2 Preparation of GG Film

Gellan gum (GG) solution was prepared by dissolving GG (0.00, 0.50, 0.75, 1.00 and 1.50 g) into 100 mL distilled water under continuous stirring for 2 h at 70°C. (30, 40, 50, 60, and 70 % (w/w), relative to GG) of glycerol and (2, 5 and 10 mM) calcium chloride ($CaCl_2$) was added into the gellan gum solution. Then, the mixture was stirred for 15 minutes and the resulting suspension was casted onto an acrylic plate. After removing bubbles, the samples were dried for 24 h (at room temperature, 40, 50, 70 and 100°C) in an oven. All films were preconditioned in a desiccator (27 °C, 50% relative humidity (RH)) for at least 2 days prior to testing.

Table 1. Testing parameter of the experimental design

Type	Testing Parameter
Heat temperature, °C	Room temperature, 40, 50, 70, and 100
Concentration of gellan gum, g	0.00, 0.50, 0.75, 1.00, and 1.50
Concentration of $CaCl_2$, mM	2, 5, 10
Concentration of glycerol, % (w/w)	30, 40, 50, 60, and 70

*percentage weight relative to gellan gum

3. Results and Discussion

Figure 1 shows the produced samples when gellan gum suspension was dried at different temperature for 24 h. After drying at room temperature (RT), the samples were presented in hydrogel form (figure 1(a)). Hydrogel can be described as a solution that shows a behaviour similar to gels at low shear stresses however act as fluids at higher shear stresses [27]. Hydrogel resulted from gellan gum is not a real gel but a fluid or paste-like material which shows soft texture with a yield stress. It provides an appropriate environment for a proper texture formation and inhibition of sedimentation or suspension of particles in liquid gel structures [28]. This hydrogel is classified as physical gels category because in these gels joining points are formed through physical reactions such as hydrogen bonds, hydrophobic bonds and cross-sectional joints by the aid of cations [29]. The random coils would form double helices structure with neighbouring chains through hydrogen bonds and van der Waals forces [29]. Subsequently, the aggregation would occur to generate a continuous there dimensional (3D) network giving rise to a gel.

Heat treatment was carried out at higher temperature because the existence of -OH groups is important for changes occurred in solution during heating, as it is involved in reactions with other sugars [31]. At 40 °C, the GG film was formed however it is not well dried thus the film was stacked on the casting plate and cannot be peel off (figure 1(b). At this temperature, the gellan gum which was initially in the membrane suspension liquid, is fully stretched and displaced, thus the integration among them is promoted to produce the film. The formation of film is an initial stage of gellan gum gelatinization accompanied by formation of double helical structures and during heating process strands are formed from the structures which aggregate in the presence of cation to form a three-dimensional hard and brittle gelatin network [32]. After being heated at 50 °C for 24 hours, free standing GG film was formed (figure 1(c)). Meanwhile, at 70 and 100 °C, the produced films were torn apart and ruptured attributed to overheat as the degeneration of film-forming materials may occur, resulting in larger molecules break within the polymer chain (figure 1(d), 1(e)).

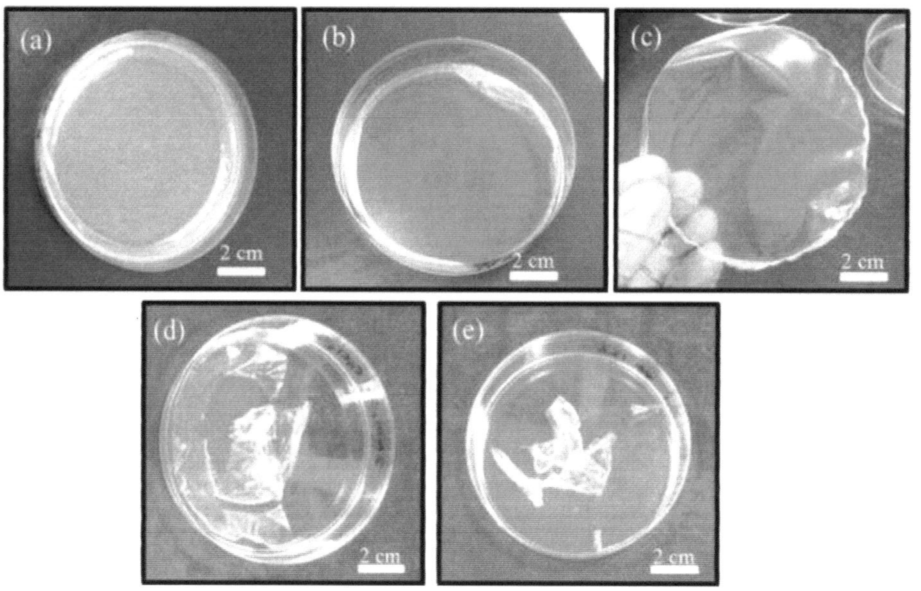

Figure 1. Samples after drying at (a) room temperature, (b) 40 °C, (c) 50 °C, (d) 70 °C, and (e) 100 °C for 24 hours.

The effect of GG mass which were 0.00, 0.50, 0.75, 1.00 and 1.50 g on the physical appearance of films was examined. Films are not able to be formed when the formulation without the gellan gum, presumably due to the lack of intermolecular interactions upon film dehydration. When 0.50, 0.75 and 1.00 g of gellan gum was used, a films was successfully produced as shown in figure 2(a), 2(b) and 2(c), respectively. Among the produced films, 1.00 g GG is the best film as nice spherical shape was observed without shrinkage and distortion as compared to 0.5 and 0.75 g GG films. At the highest amount of GG studied (1.50 g), the film cannot be produced due to the formation of highly viscos and sticky suspension as observed visually (figure 3(a)) and determined by mildly rubbing between two fingers (figure 3(b)). In general, the stickiness of gel was increased with the increasing of polymer concentration as it leads to more intense aggregation of the polymer chains and stronger physical and chemical crosslinking [33]. Thus, at higher polymer concentrations more inter-chain cross linking occurs leading to a significant increase in viscosity of gel.

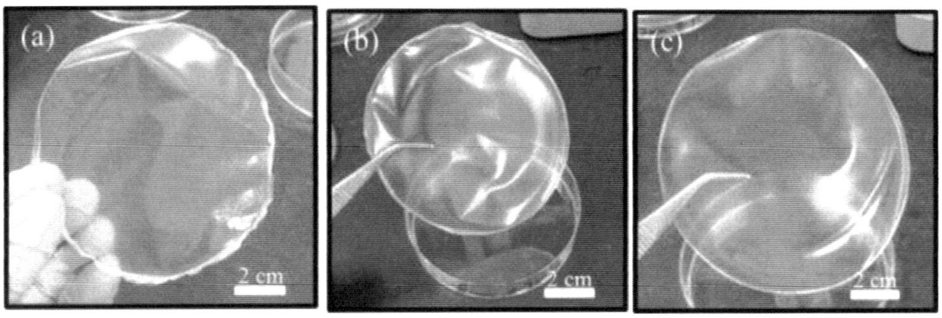

Figure 2. The produced film using (a) 0.50 g, (b) 0.75 g, and (c) 1.00 g of GG.

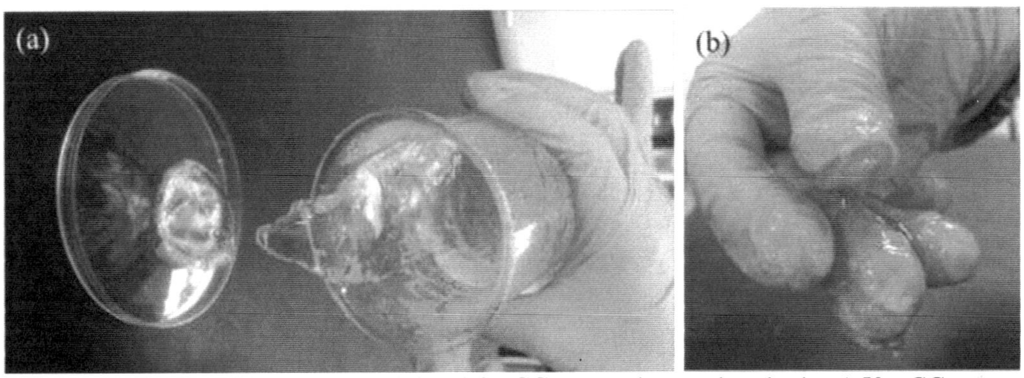

Figure 3. (a) Highly viscos, and (b) sticky GG suspension produced using 1.50 g GG.

The gelation mechanism of GG suspension depends on the nature of cations present as cross-linker. The divalent cation cross-linker promotes the gelation much more effectively than the monovalent cation as they can form ionic bonds directly with the carboxylic acid groups at different GG chains, resulting in the aggregation of the double helices by the inter-chain bridging [34, 35]. One of the capabilities of GG is forming a physical gelation by addition of physical cross-linker (Ca^{2+}). The physical gelation mechanism actually occurred at two stages which is when the temperature of hot GG solution is decreased, the molecules of GG undergo a conformational thermo reversible change from random coils structure to double helices structure [36, 37]. This occurs in GG either with or without additions of ions. However, with addition of cations, the gelation was follow up with the aggregation of the double helical elements with water through hydrogen formation to form three-dimensional (3D) network [38]. In this research the cross-linker (Ca^{2+}) function was observed at 2, 5 and 10 mM. As shown in figure 4(a)-4(c), the films were successfully produced at all studied concentration. It can be observed that increasing the concentration of cross-linker resulted in a significant decrease of the solubility of the biofilms. This indicates that the Ca^{2+} actually promoted to the crosslinking formation between the polymer chains, thus increasing the concentration of Ca^{2+} makes the intermolecular bonds more cohesive and arranged, thereby reduce the solubility of the films.

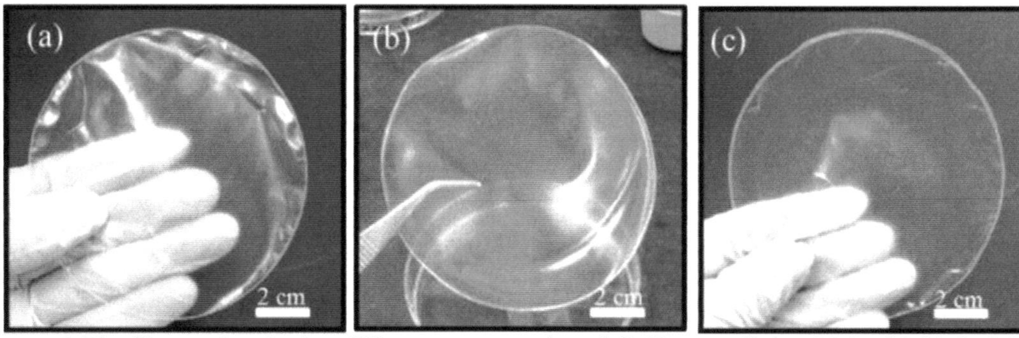

Figure 4. The film produced using different concentration of $CaCl_2$ cross-linker (a) 2 mM (b) 5 mM, and (c) 10 mM.

In film formulation, the plasticizer was used to enhance the film flexibility and decreased brittleness. There are verities of plasticizers available commercially in the market such as polyethylene glycol 400 (PEG), sorbitol, propylene glycol (PG), ethylene glycol (EG), and glycerol. However in this study glycerol was selected due as it is the most effective plasticizers for GG films fabrication as recorded in other previous study [39]. Figure 5 shows the produced film using different concentration of glycerol. Good films were produced at low concentration of glycerol used in film formulation especially at 50 w/w %. They are in good shape as well. However, the shape of the films can be altered depending on the casting plate used in the experiment. At higher concentration of glycerol (60 and 70 w/w %), a defect can be observed on the produced film as they were distorted attributed to the limited dispersion capability of the GG solution at higher glycerol concentration. When the plasticizer concentration exceeds polymer compatibility limits, it will cause the phase separation as well as physical exclusion of the plasticizer [40]. The other problems that often encountered in the formation of the films at higher glycerol concentration are the films tend to be sticky and they will be difficult to handle either for characterization or application.

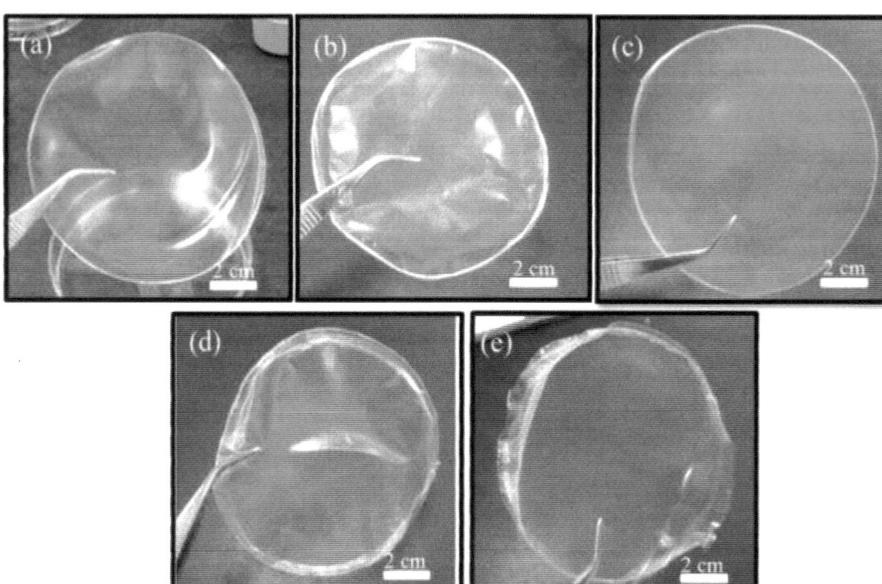

Figure 5. The produced films at different concentration of glycerol (a) 30 % (w/w), (b) 40 % (w/w), (c) 50 % (w/w), (d) 60 % (w/w), and (e) 70 % (w/w).

4. Conclusion

This study demonstrated that GG could be used as a film-forming material. Good film was produced at 1.00 g GG with 5 mM $CaCl_2$ cross-linker and 50 w/w % glycerol as plasticizer after dried at 50 °C for 24 hours. Based on the visual and preliminary characterisations, it can be concluded that glycerol is an appropriate plasticizer for GG films. Overall, the results of the present study indicates that GG with the effective $CaCl_2$ and glycerol could be regarded as an appropriate formulation for films synthesis.

5. References

[1] Jansson P, Lindberg B, and Sandford P 1983 *J. Carbohydr. Res.* **124** 135
[2] O'Neill M, Selvendran R, and Morris V 1983 *J. Carbohydr. Res.* **124** 123
[3] Chandrasekaran R. and Thailambal V 1990 *J. Carbohydr. Poly.* **12** 431
[4] Mazen F, Milas M, and Rinaudo M 1999 *Int. J. Biol. Macromolec.* **26** 109
[5] Rinaudo M 2001 *Food Hydrocolloids* **15** 433
[6] Morris V In A. M. Stephen (Ed.) ed. 1995, New York: Marcel Dekker. 341
[7] Morrison N, Clark R, Chen Y, Talashek T, and Sworn G 1999 *Springer* 127
[8] Omoto T, Uno Y, and Asai I 1999 *Springer* 123
[9] Coviello T, Dentini M, Rambone G, Desideri P, Carafa M, Murtas E, Riccieri F, and Alhaique F 1998 *J. of Control. Release* **55** 57
[10] Ferrance J 2007 *J. of Chromatogr. A* **1165** 86
[11] Nanjawade, B K, F Manvi, and A Manjappa 2007 *J. of Control. Release* **122** 119
[12] Ismail N A, Razali M H, and Mat Amin K A 2017 *In AIP Conference Proceedings* **1885** 020045
[13] Ismail N A, Mat Amin K A, and Razali M H 2018 *Materials Letters* **228** 116
[14] Ismail N A, Mohamad S F, Ibrahim M A, and Mat Amin K A 2014 *Advances in Biomaterials,* **2014** 1
[15] Smrdel P, Bogataj M, and Mrhar A 2008 *Sci. Pharm.* **76** 77
[16] Smrdel P, Bogataj M, Zega A, Planinšek O, and Mrhar A 2008 *J. of Microencapsulation* **25** 90
[17] Moxon S and Smith A 2016 *Int. J. Biol. Macromol.* **84** 79
[18] Bertuzzi M, Vidaurre E, Armada M, and Gottifredi J 2007 *J. Food Eng.* **80** 972
[19] Rivero S, García M, and Pinotti A *2010 Innov. Food Sci. & Emerging Tech.* **11** 369
[20] Vanin F, Sobral P, Menegalli F, Carvalho R, and Habitante A 2005 *Food Hydrocolloids* **19** 899 [21] Ahmadi R, Kalbasi-Ashtari A, Oromiehie A, Yarmand M, and Jahandideh F 2012 *J. Food Eng.* **109** 745
[22] Ghasemlou M, Khodaiyan F, Oromiehie A, and Yarmand M 2011 *Food Chem.* **127** 1496
[23] Khazaei N, Esmaiili M, Djomeh Z, Ghasemlou M, and Jouki, M 2014 *Carbohydr. Polym.* **102** 199
[24] Piermaria J, Bosch A, Pinotti A, Yantorno O, Garcia M, and Abraham A 2011 *Food Hydrocolloids* **25** 1261
[25] ASTM. Designation D882-12. 2012, Philadelphia, PA: Americ. Soc. for Testing and Materials.
[26] ASTM. Designation E96-95. 1995, Philadelphia, PA: American Soc. for Testing and Materials.
[27] Guvendiren M, Lu H and Burdick J 2012 *Soft Matter* **8** 260
[28] Young N 2002 *Special Publication-Royal Society of Chemistry c* **278** 226 [29] Williams P and Glyn P, Cambridge: Woodhead. 2000, ISBN 1-85573-501-6.
[30] García M, Alfaro C, Calero N, and Muñoz J 2011 *Biochem. Eng. J.* **55** 73
[31] Banik R and Santhiagu A 2006 *Biotechnology Letters* **28** 1347
[32] Ogawa E, Matsuzawa H, and Iwahashi M 2002 *Food Hydrocolloids* **16** 1
[33] Dai L, Liu X, Liu Y, and Tong Z 2008 *Eur. Polym. J.* **44** 4012
[34] Kanesaka S, Watanabe T, and Matsukawa S 2004 *Biomacromolecules* **5** 863
[35] Miyoshi E and Nishinari K 1999 *Phys. Chemistry and Industrial Application of Gellan Gum* 68

[36] Miyoshi E, Takaya T, and Nishinari K 1996 *Carbohydr. Polym.* **30** 109
[37] Chandrasekaran R, Millane R, Arnott S, and Atkins E 1988 *Carbohydr. Res.* **175** 1
[38] Morris E, Gothard M, Hember M, Manning C, and Robinson G 1996 *Carbohydr. Polym.* **30** 165
[39] Yang L and Paulson A 2000 *Food Res. International* **33** 563
[40] Donhowe I and Fennema O 1993 *J. of Food Processing and Preservation* **17** 247

Acknowledgments
The authors are grateful to Universiti Malaysia Terengganu (UMT) for providing the facilities to carry out this project and Malaysia Ministry of Higher Education for the financial support vote FRGS 59358 and MyBrain15 (MyPhD).

Effects of PEG-polysaccharide coating on electrospun PLA nanofiber

Asrizuan Jauri[1], Saiful Izwan Abd Razak[1*]

[1]Medical Devices and Technology Group, Faculty of Biosciences and Medical Engineering, Universiti Teknologi Malaysia, 81310 Skudai, Johor, Malaysia

[*]saifulizwan@utm.my

Abstract. This paper reports the preparation of electrospun polylactic (PLA) nanofiber coated with carrageenan (CRG) and polyethylene glycol (PEG). CRG was utilized to increase the hydrophilicty while PEG was used to create strong bonding between the two polymers. The nanocomposites films (PLA/CRG-PEG) were successfully prepared. The presence of CRG improves the hydrophilicity of the nanofiber mat (22.45^{o}) while the addition of PEG increased the mechanical strength. This new approach would enable the applications of electrospun PLA for more hydrophilic environment, especially for medical devices or materials

1. Introduction

Nowadays, electrospinning has become one of the most versatile, cost-effective, and ease of handling technique to engineer advance materials to be used in biomedical applications. Apart of being, flexible in the spinning of a wide variety of polymeric fibres, it is also able to produce continuous fibres. The products gain from this method was consistent in term of its fibre diameter [1]. In tissue engineering and wound dressing, electrospun nanomembranes act as tissue scaffolds which improve cells growth and proliferation. In case of damaged tissues, the nanomembranes scaffolds with seeded cells can be fixed to patient's body to repair the damage [2].

Poly(lactic acid) (PLA) is known as one of the most widely used synthetic polymers due to its non-toxicity which is naturally present in human body. PLA is thermoplastic, high strength and high modulus polymer [3]. Due to these characteristics, PLA has become one of the most famous material in electrospinning. However, there are some major limitation of PLA such as hydrophobic nature and poor ductility which hinder its practical used as substitute materials in tissue regeneration [4].

In order to overcome this problem, CRG has been attached to the surface of the PLA to increase its hydrophilicity. CRGs are polysaccharides derived from red seaweeds that are well known in the food industry. They are highly soluble in water and very popular gelling agents [5]. Research on CRG hydrogels in biomedical and pharmaceuticals are still on-going. Even though CRG have been extensively explored particularly in food packaging applications [6-8], recent studies have shown that CRG exhibit potentials to be formulated as dressings for drug delivery to wounds [9] and possible enhanced bioactivity for bone tissue engineering [10].

Another material that plays a crucial role in this study is polyethylene glycol (PEG). PEG has been extensively studied for biomedical application such as drug delivery and tissue engineering [11]. PEG has the ability to attach a variety of reactive functional group to the terminal sides of PEG polymers

[12]. Hence, we hypothesized that PEG can promote good binding between PLA and CRG. PLA was electrospun and coated with CRG and CRG/PEG in order to increase the hydrophilicty.

2.1 Experimental

2.1 Preparation of electrospun PLA/CRG-PEG

The electrospun PLA was fabricated by using electrospinning method at a feed rate of 1ml/hour and applied voltage of 12.5 V. 2g of PLA was used for each batch. The CRG (0.1, 0.3, 0.5g) and PEG (wt% of 1, 2, 4, 6) were dissolved together and the solution was poured on the surface of the electrospun PLA. They were heated to the glass transition temperature (T_g) of PLA for 3 days.

2.2 Mechanical testing

The tensile test was performed on the electrospun nanofibre samples using a computer controlled Instron machine with crosshead speed of 1mm/min at a standard laboratory atmosphere of 23°C. Thicknesses of the specimens were measured and the experiment was conducted in triplicate.

2.3 Water contact angle

Wetting degree of the electrospun nanofibre was determined by water contact angle using sessile drop method. The contact angles were calculated from images captured by the camera of contact angle measuring system, VCA Optima, AST Products, Inc.

3. Results and discussion

3.1 Tensile strength

Tensile strength of the neat electrospun PLA nanofibre and its modification were shown in Table 1. Presence 0.1g of CRG on the PLA surface revealed a lower tensile strength compared to that of neat electrospun PLA. This mainly caused by the poor coverage of the CRG on the PLA cavities that lead to poor load transfer. Meanwhile 0.3g of CRG seemed to increase the tensile strength while higher amount of CRG (0.5g) decrease the nanocomposites' tensile strength. The reduction of mechanical strength may due to the detachment of the CRG from the nanofiber surfaces. Thus 0.3g of CRG was utilized for the next stage.

Table 1. Tensile strength of neat electrospun PLA nanofibre and PLA/ (0.1,0.3,0.5)g CRG

Samples	Ultimate tensile strength (MPa)
PLA	3.35 ±0.05
PLA/0.1g CRG	3.24 ±0.07
PLA/0.3g CRG	3.50 ±0.02
PLA/0.5g CRG	3.18 ±0.05

Tensile strength of PLA/0.3g CRG with the addition of different composition of PEG are shown in Table 2. It can be observed that the tensile strength increased with increasing percentage of PEG. This shows that addition of PEG promote the binding between PLA and CRG. Though higher amount of

PEG was not prepared, we predict that the tensile strength to be reduced at a certain PEG concentration.

Table 2. Mechanical properties of PLA/0.3g CRG with the addition of PEG

Samples	Ultimate tensile strength (MPa)
PLA/0.3g CRG/1% PEG	4.85 ±0.05
PLA/0.3g CRG/2% PEG	5.76 ±0.10
PLA/0.3g CRG/4% PEG	5.88 ±0.20
PLA/0.3g CRG/6% PEG	6.71 ±0.32

3.2 Water contact angle

The water contact angle of the specimens can be seen in Figure 1. Figure 1a shows that the contact angle for neat electrospun PLA nanofiber was at 120.88 $\pm$ 5.22° and the contact angle for PLA/0.3CRG/6PEG was at 22.45 $\pm$ 3.25 °. Contact angle is determined by the physiochemical nature of the material tested. When polar substances are being used, it is plausible to have a partial rearrangement of the functional groups towards the surface during casting. In this case, diols and carbonyls of PLA should create a three dimensional network with the cellulose -OH, and the non-polar part of the structure may be reoriented to the presence of CRG and PEG, the polarity of the PLA increased due to the presence of more -OH group, thus resulting in lower value of contact angle. Nanofiber patch with lower contact angle would provide preferable environment when in contact with body fluid.

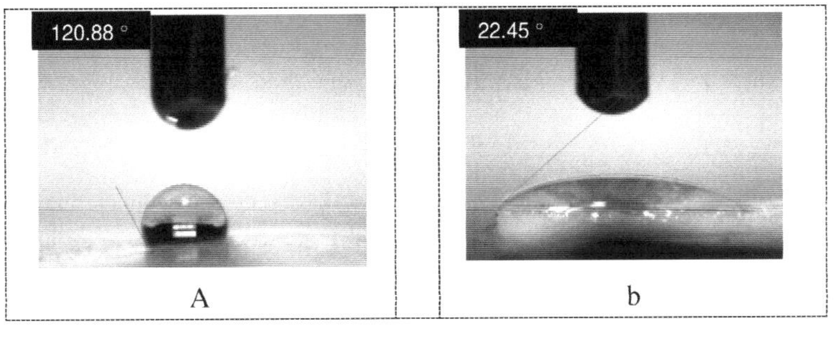

Figure 1. Water contact angle of a) neat electrospun PLA nanofiber and b) PLA/0.3CRG/6PEG

4. Conclusion

In conclusion, the incorporation of CRG has successfully improved the tensile strength and the hydrophilicity of the electrospun PLA nanofiber. The presence of PEG has increase the bonding between the PLA and CRG. The modified electrospun PLA nanofiber has the advantages being mechanically reliable for biomedical applications. In addition, cost-effective approach utilizing natural resources based on renewable synthetic polymer and polysaccharide might open up new frontiers in biomedical materials.

Acknowledgement
The authors would like to thank Universiti Teknologi Malaysia and the Ministry of Higher Education in Malaysia for providing financial assistance for this research.

References

[1] Lasprilla-Botero J, Álvarez-Láinez, M, & Lagaron, J. M. 2018. The influence of electrospinning parameters and solvent selection on the morphology and diameter of polyimide nanofibers. Materials Today Communications 14, p 1-9.

[2] Liang W, Hou J, Fang X, Bai F, Zhu T, & Lang, M. 2018. Synthesis of cellulose diacetate based copolymer electrospun nanofibers for tissues scaffold Applied Surface Science 443, p 374-381.

[3] Lesage F., Roman S, Pranpanus S, Ospitalieri S, Zia S, Jimenez J, & Deprest J. 2018. Modulation of the early host response to electrospun polylactic acid matrices by mesenchymal stem cells from the amniotic fluid European Journal of Pediatric Surgery 28, p 285-292.

[4] Abdal-Hay A, Hussein K H, Casettari L, Khalil K A, & Hamdy A S. 2016. Fabrication of novel high performance ductile poly (lactic acid) nanofiber scaffold coated with poly (vinyl alcohol) for tissue engineering applications Materials Science and Engineering:C 60, p 143-150.

[5] He H, Ye J, Zhang X, Huang Y, Li X, & Xiao M. 2017. κ-Carrageenan/locust bean gum as hard capsule gelling agents Carbohydrate Polymers 175, p 417-424.

[6] Sedayu B B, Cran M J, & Bigger S W. 2018. Characterization of semi-refined carrageenan-based film for primary food packaging purposes Journal of Polymers and the Environment p 1-8.

[7] Farhan A, & Hani N M. 2017. Characterization of edible packaging films based on semi-refined kappa-carrageenan plasticized with glycerol and sorbitol Food Hydrocolloids 64, p 48-58.

[8] Thakur R, Pristijono P, Golding J B, Stathopoulos C E, Scarlett C J, Bowyer M, & Vuong Q. V. 2017. Amylose-lipid complex as a measure of variations in physical, mechanical and barrier attributes of rice starch-ι-carrageenan biodegradable edible film Food Packaging and Shelf Life 14, p 108-115.

[9] Zepon K M, Marques M S, da Silva Paula M M, Morisso F D P, & Kanis, L A. 2018. Facile, green and scalable method to produce carrageenan-based hydrogel containing in situ synthesized AgNPs for application as wound dressing International Journal of Biological Macromolecules 113, p 51-58.

[10] Kavoli S, Mirzaie M, Feizi F, Rakhshan V, Arash V, & Bijani A. 2017. Local injection of carrageenan accelerates orthodontic tooth movement: A preliminary experimental animal study. International Orthodontics 15, 588-599.

[11] Grossen P, Witzigmann D, Sieber S, & Huwyler J. 2017. PEG-PCL-based nanomedicines: A biodegradable drug delivery system and its application Journal of Controlled Release 260, p 46-60.

[12] Khalid A, Abdel-Karim A, Atieh M A, Javed S, & McKay G. 2018. PEG-CNTs nanocomposite PSU membranes for wastewater treatment by membrane bioreactor Separation and Purification Technology, 190, pp 165-176.

Synthesis and characterization of hydrophilic polymer nanoparticles using n-isopropylacrylamide (NIPAM) via emulsion polymerization technique

Nurul Syafika Azmi[1], Nur Nasyita Kamaruddin[2], Syara Kassim[1,2] and Noor Aniza Harun[1,2]

[1]Advance NanoMaterials Research Group, School of School of Fundamental Science, Universiti Malaysia Terengganu, 21030 Kuala Nerus, Terengganu, Malaysia. [2]School of Fundamental Science, Universiti Malaysia Terengganu, 21030 Kuala Nerus, Terengganu, Malaysia.

Abstract. Poly(N-isopropylacrylamide) (PNIPAM) nanoparticles were successfully synthesized via emulsion polymerization techniques using different critical micelle concentrations (CMCs) of sodium dodecyl sulphate (SDS) surfactant, temperature and time of polymerization. The effects of surfactant concentrations on emulsion polymerization of PNIPAM nanoparticles have been discussed. Potassium persulfate (KPS) and N,N'-methylenebisacrylamide (MBAA) were used as initiator and cross-linker, respectively throughout the emulsion polymerization. The formation of PNIPAM nanoparticles was confirmed by Fourier transform infrared spectroscopy (FTIR) with absorption peaks observed at 2997 and 2930 cm^{-1} for C-H stretching of CH_3 and CH_2 groups, 1459 cm^{-1} for 2° amide C=O stretch , 3310 cm^{-1} and 3275 cm^{-1} for N-H stretching band. The particle size and morphology of PNIPAM nanoparticles were determined using scanning electron microscopy (SEM) where large aggregation of PNIPAM nanoparticles were observed with average diameters in the range of 20 – 50 μm and the appearance of pore structure on the hydrogel surface. Thermal stability of PNIPAM nanoparticles were obtained by thermogravimetric analysis (TGA) where it showed the percentage of mass loss at certain temperature.

1. Introduction

The studies of nanoscience and nanotechnology have advanced greatly in recent years and large amounts of research have focused in determining the properties and potential functions of nanomaterials. Nanotechnology is the study of combination of science, medical, engineering and technology at the nano-scale level ranging from 1 nm to 100 nm in size [1]. Polymeric nanoparticles (PNPs) are nanosized particulate carriers which either can entrap or encapsulate the drug substances, antigens, proteins and DNA vaccines and deliver it to the desired site of action [2]. Biodegradable polymers have received lots of attentions due to their biocompatibility, possibility for surface modifications and sustained release of the drug. Hydrophilic biodegradable polymer is a synthetic hydrolytically degradable polymer which has been remarkable developed in the application of biomedical and tissue engineering applications [3].

Amongst numerous synthetic biodegradable polymer, poly(N-isopropylacrylamide) (PNIPAM) which obtained from N-isopropylacrylamide (NIPAM) monomers offer huge interest among researcher especially in medical applications on account of their intelligent and unique stimuli-responsive properties. PNIPAM has distinctive characteristic in which its chains comprise both of hydrophobic and hydrophilic domains that can undergo rapid changes in their microstructure from a hydrophilic to a hydrophobic state cause by the alterations of the external stimuli. This external stimuli including heat, pH, ionic strength, magnetic and electric field, light, ultrasound and chemical species causes the disturbance to the environment structure of PNIPAM. PNIPAM has a thermo-sensitive effect which is related to lower critical solution temperature (LCST). Upon heated above the LCST around 32 °C, liquid PNIPAM has turn into a gel formation structure (Dong *et al.*, 2017). The hydrogen-bonding interactions between the hydrophilic groups of PNIPAM and water are dominant below LCST, enforcing a coil-like conformation in PNIPAM resulting in water soluble PNIPAM [4, 5].

Since polymer nanoparticles have attracted the interest of many research groups and have been utilized in an increasing number of fields during the last decades, the properties of PNPs have to be optimized depending on the particular application. In order to achieve the properties of interest, the mode of preparation plays a vital role. Thus, it is highly advantageous to have preparation techniques at hand to obtain PNPs with the desired properties for a particular application [6]. In this study, PNIPAM nanoparticles were synthesized via emulsion polymerization technique, a typical emulsion polymerization formulation comprises of monomer, water, surfactant and a water soluble initiator. Previous work reported by Jagadeeshbabu *et al.* (2011) showed a thermo responsive macro porous PNIPAM hydrogel was synthesized using free radical polymerization [7]. The reaction was optimized by varying the reaction temperature, monomer, cross-linker and initiator based on the strength and swelling characteristics of the hydrogel. Synthesis of NIPAM using soap-free emulsion polymerization has been prepared by Chen *et al.* (2013) [8]. The polymerization was performed in the absence of surfactant, with the help of the initiator (KPS) and cross linker (MBA). The significance of this study was to determine the structure and physical properties of PNIPAM at lower LCST and cloud point temperature (CPT).

Surfactant has been a major role in emulsion polymerization as it become primary role to lower the interfacial tension so that the emulsification of reactive vinyl monomers and the formation of stable colloidal dispersions of nano-size polymer particles are allowed. Sodium dodecyl sulphate (SDS) is an amphiphilic molecule which is commonly anionic type surfactant that broadly utilize in the production of free radical polymers [9]. Taking into this direction, we are intrigued to explore the effects of different critical micelles concentration (CMC) of SDS in emulsion polymerization of PNIPAM nanoparticles. Hence, in this work, PNIPAM nanoparticles were synthesized via emulsion polymerization technique. The influence of different concentrations of SDS surfactant ranging from below, equivalent and above critical micelles concentrations (CMC) towards the emulsion polymerization of PNIPAM were thoroughly investigated.

2. Methodology

2.1 Materials

N-isopropylacrylamide (NIPAM) monomer, sodium dodecyl sulphate (SDS), N,N'-methylenebisacrylamide (MBAA) and ammonium peroxodisulfate (APS) were used as received. All the reagents and solvents used in this study are commercially available from Sigma Aldrich. Distilled water was used throughout the experiment.

2.2 Emulsion polymerization of PNIPAM nanoparticles

Emulsion polymerization was carried out in a 250 mL two-neck round bottom flask composed of magnetic stirrer, a reflux-condenser, nitrogen gas inlet and thermometer. A mixture of N-isopropylacrylamide (NIPAM) monomer, sodium dodecyl sulphate as surfactant, and N,N'-methylenebisacrylamide (MBAA) as cross-linker were mixed together in distilled water (3 mL). Distilled water act as dispersion medium and nitrogen gas was bubbled through the reaction mixture for 15 min and stirred at 450 rpm/min. The polymerization was occurred in the oil bath at temperature of 70 °C and initiated by addition of aqueous solution potassium persulfate (KPS) as initiator (0.0018 g) in 1 mL distilled water. The polymerization was completed after 2 h and the reaction was terminated by cooling to room temperature.

2.3 Characterizations of PNIPAM nanoparticles

2.3.1 Fourier transmission infrared (FTIR)

The solid sample of PNIPAM nanoparticles was prepared in the form of potassium bromide (KBr) pellet. The PNIPAM nanoparticles were mixed and crushed together with KBr powder in the ratio of 1:7. The pellet then was placed on a plate and clamped in the right position. The FTIR spectra recorded using Perkin Elmer 100 series with sampling range from 450 to 4000 cm^{-1} and 16 times of scan number.

2.3.2 Scanning electron microscopy (SEM)
The hydrogel PNIPAM nanoparticles were mounted on SEM holder and coated with gold by Auto Fine Coater (JEOL) to prevent accumulation of electrostatic charge on the surface of samples. The morphology of polymer nanoparticles were observed using Oxford Instrument attached to JSM-6360 LA Analytical Scanning Microscope at acceleration voltage of 15 – 20 kV.

2.3.3 Thermogravimetric analysis (TGA)
The dynamic weight loss tests were conducted on TA Instruments 2050 thermogravimetric analyser (TGA). All tests were conducted in an N_2 purged (25 mL/min) using sample weights of 5 – 10 mg over a temperature range of 20 °C – 600 °C with scan rate of 10 °C/min.

3. Results and discussion

3.1 Emulsion polymerization of Poly(N-isopropylacrylamide) PNIPAM nanoparticles
The synthetic hydrophilic polymer nanoparticles using N-isopropylacrylamide monomer *via* emulsion polymerization were successfully performed using sodium dodecyl sulphate as anionic surfactants at different critical micelles concentration (CMC) of below, equivalent and above CMCs. The effect of surfactant concentration towards particle size and formation of polymer nanoparticles (PNPs) were studied and the parameters involved in this study were summarized in Table 1. PNIPAM nanoparticles prepared at different SDS concentration of below, equivalent and above CMC (4 mM, 8.2 mM and 10 mM, respectively). PNIPAM nanoparticles were obtained in the form of white fluffy solid right after polymerization and transformed to the solid hydrogel after dried at room temperature for 1 day. The formation of PNIPAM nanoparticles were then confirmed by FTIR spectroscopy.

Table 1. PNIPAM nanoparticles prepared *via* emulsion polymerization at different concentrations of SDS.

Polymer nanoparticles (PNPs)	Concentration of SDS (mM)
PNIPAM_1	4 (below CMC)
PNIPAM_2	8.2 (equivalent CMC)
PNIPAM_3	10 (above CMC)

3.2 Fourier transform infrared spectroscopy (FTIR)
Figure 1 illustrates the IR spectra of PNIPAM nanoparticles which prepared at different SDS concentration of 4 mM (PNIPAM_1), 8.2 mM (PNIPAM_2) and 10 mM (PNIPAM_3). The presence of broad peaks at 3567 cm^{-1} belongs to N-H stretching. The presence of CH stretching for CH_3 and CH_2 groups absorption peaks were observed in the range of 2930 cm^{-1} and 2997 cm^{-1}. The band at region of 1420 – 1460 cm^{-1} belongs to the characteristic of secondary amide C=O stretching. The absorption peaks for NIPAM monomer are identical to the PNIPAM nanoparticles with narrower peaks were observed [10]. It is noteworthy to mention that all PNIPAM nanoparticles that prepared at different SDS concentrations show similar pattern of FTIR spectra confirming the formation of PNIPAM nanoparticles, and this observations also proved that different concentration of SDS do not affect the structure of PNIPAM nanoparticles.

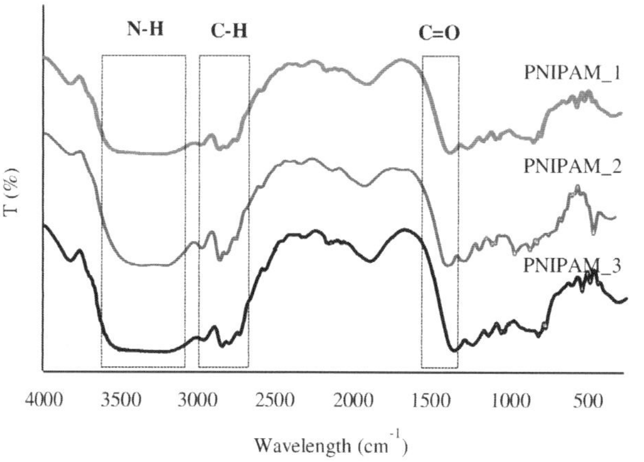

Figure 1. The FTIR spectra of PNIPAM 1, PNIPAM 2 and PNIPAM 3 prepared at below, equivalent and above CMC of SDS surfactant.

3.3 Scanning electron microscope (SEM) analysis

Scanning electron microscopy (SEM) was then performed to investigate the morphology of the PNIPAM nanoparticles. It was assumed that the morphology of pure PNIPAM hydrogels have a morphology consisting of relatively flat foils over which large craters (size around 100 nm) are visible. Even at higher magnifications, PNIPAM hydrogels exhibit a non-porous structure [11]. The SEM photographs (Figure 2) of PNIPAM nanoparticles showed most similar morphology as reported previously.

PNIPAM_1 (Figure 2a) prepared below CMC shows compact with non-porous surface structure, while, PNIPAM_2 prepared at condition of equivalent CMC showed the existence of micro and nano-pore were formed on the surface of the hydrogel (Figure 2b). In contrast, the morphology of PNIPAM 3 (Figure 2c) which prepared above CMC showed flaky nonporous structures. Therefore, it can be suggested that different concentrations of SDS can affect the morphology of PNIPAM nanoparticles.

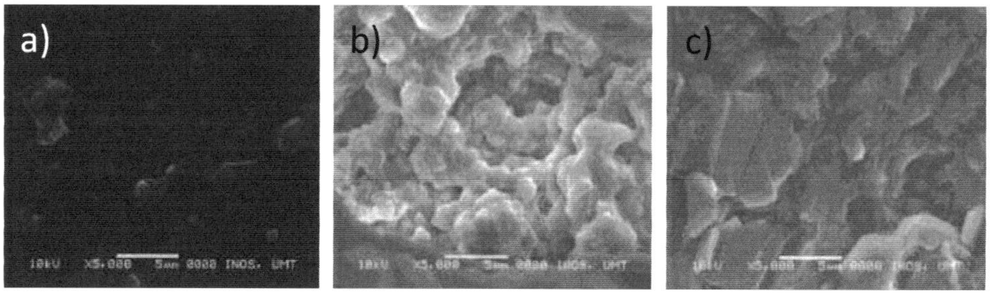

Figure 2. SEM images of (a) PNIPAM_1 (b) PNIPAM_2 and (c) PNIPAM_3

The average particle size of PNIPAM nanoparticles were determined by randomly measured the polymer particles obtained from SEM images and it was found that average particle size of PNIPAM nanoparticles in between 20 – 50 µm with large aggregations were observed. The aggregation of polymer samples were expected to be observed under SEM on account of nature of sample preparation for SEM and during drying process of the samples.

3.4 Thermogravimetric analysis (TGA)

Thermal stability of polymer depends on the polymer microstructure and TGA can be utilized to determine the polymer degradation characteristics [12]. PNIPAM nanoparticles were characterized by using thermogravimetric analysis (TGA) to determine the thermal stability of nanoparticles by measuring weight loss of polymer nanoparticles as function of temperature. To confirm that the mass loss due to water lost from the hydrogel, and was not associated with decomposition of polymer itself, the effect of composition on thermal degradation of the PNIPAM nanoparticles was examined. The samples were heated from 30 °C to 600 °C under a nitrogen atmosphere at heating rate of 10 °C/min. Figure 3 showed the TG curves of PNIPAM_1, PNIPAM_2 and PNIPAM_3 versus NIPAM monomer. From the graph, the first step in a range of 30 - 120 °C is due to the water loss. At 120 - 375 °C, it can be said that it is related to the thermal degradation of PNIPAM nanoparticles in which the degradation of the NIPAM monomer start at 120 °C while the degradation for PNIPAM_1, PNIPAM_2 and PNIPAM_3 are found to be at 369 °C, 372 °C, and 369 °C with percentage weight loss are 75 %, 63 % and 79 %, respectively. The comparison of thermal stability for PNIPAM nanoparticles samples shows that PNIPAM_2 with surfactant concentration equivalent CMC is the most stable during the thermal degradation region on account of higher thermal degradation temperature with the lowest percentage of mass loss.

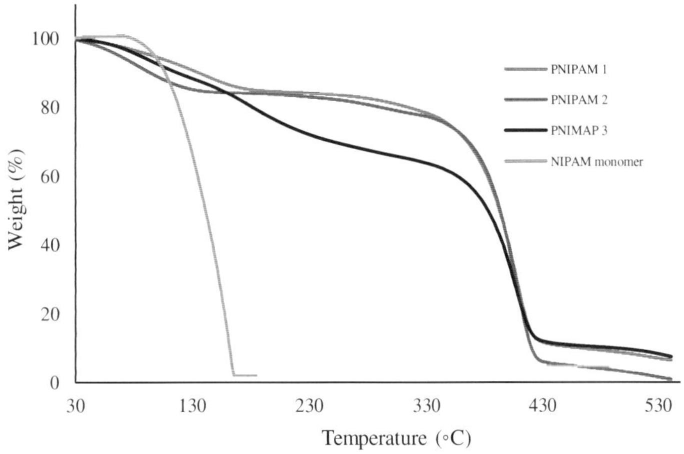

Figure 3 TG curves of PNIPAM 1, PNIPAM 2, and PNIPAM 3

4. Conclusions

Synthetic hydrophilic poly(N-isopropylacrylamide) (PNIPAM) nanoparticles were successfully produced *via* emulsion polymerization technique at different CMC of SDS surfactant. The formation of PNIPAM nanoparticles were confirmed by FTIR where the absorption spectra showed the similar peaks in all samples of nanoparticles. Scanning electron microscope (SEM) images showed different morphology of PNIPAM nanoparticles prepared at different surfactant concentrations with the existence of aggregation of PNIPAM nanoparticles. The comparison from all the TGA curves of PNIPAM nanoparticles indicates that PNIPAM_2 was more stable in terms of thermal degradation based on the highest onset temperature. It can be concluded that the different concentrations of surfactant does give significant effects towards the properties of PNIPAM nanoparticles.

Acknowledgement
We thank the Ministry of Higher Education, Malaysia and Universiti Malaysia Terengganu for the funding (FRGS 59344) and generous support.

5. References

[1] Bera A and Belhaj H 2016 *J. Nat. Gas Sci. Eng.* **34** 1284 – 1309.
[2] Sharma R, Agrawal U, Mody N and Vyas S P 2015 *Biotechnol. Adv.* **33(1)** 64 – 79.
[3] Yin R, Zhang N, Wu W and Wang K 2016 *Mater. Sci. Eng. C Mater. Biol. Appl.* **62** 137 – 143.
[4] Singh R, Deshmukh S A, Kamath G, Sankaranarayanan S K R S, Balasubramanian G 2017, *Comput. Mater. Sci.* **126** 191 – 203.
[5] Pelton R 2010 *J. Colloid Interface Sci.* **348(2)** 673 – 674
[6] Rao J P and Geckeler K E 2011 *Prog. Polym. Sci.* **36** 887 – 913
[7] Jagadeeshbabu P E, Kumar R R S and Maheswari B 2011 Colloids Surfc *A Physicochem. Eng. Asp.* **384** 466 – 472.
[8] Still T, Chen K, Al Sayed A M, Aptowicz K B and Yodh A G 2013 *J. Colloid Interface Sci.* **405** 96 – 102
[9] Salager J L, Moreno N, Anton R E and Marfisi S 2002 *Langmuir* **3** 607-611
[10] Rwei S P, Anh T H N, Chiang W Y, Way T F and Hsu Y J 2016 Materials **9(12)** 1003
[11] Uva M and Atrei A 2016 *Gels* **2(4)** 30
[12] Gałka P, Kowalonek J, and Kaczmarek H 2014 *J Therm Anal. Calorim* 115 1387–1394

IOP Conf. Series: Materials Science and Engineering **440** (2018) 012009 doi:10.1088/1757-899X/440/1/012009

Computational Study towards Schiff Base Derivatives Length as a Potential Recognition Layer in Electrochemical DNA Sensor

Hanis Mohd Yusoff[1,2], Norhafiefa Hassan[1,2] and Ku Halim Ku Bulat[2]

[1]Advanced Nanomaterials Research Group, School of Fundamental Science, Universiti Malaysia Terengganu, 21030 Kuala Terengganu, Terengganu, Malaysia
[2]School of Fundamental Science, Universiti Malaysia Terengganu, 21030 Kuala Terengganu, Terengganu, Malaysia

Corresponding author email: hanismy@umt.edu.my

Abstract. Electrochemical sensing of DNA has been known to be the easiest and sensitive way reported so far. This study focuses on the effect of recognition layer's length in E-DNA sensor, a theoretical approach has been carried out to four different length of Schiff bases derivatives which involved four different length of R group attached (3, 6, 9 and 12 alkane chain). This study has been carried out using density functional theory (DFT) by using Gaussian09 software package 6-31G (d,p) basis set for all the calculations involved. Structure drawing was performed with GaussView 5.0. After optimization, energy, dipole moment, HOMO, LUMO, hardness (η), softness (σ) and energy gap. Dipole moment value is an important value to determine the optimum length of recognition layer. Preliminary result showed that the length does not play an important role in E-DNA sensor.

1. Introduction

Schiff base compound is known for its low-cost preparation and normally formed by the condensation reaction of ketones or aldehydes with primary amines [1,2,3]. Schiff base compound has been extensively applied in pharmaceutical and biological studies [4]. This has sparked our interest to apply Schiff base compound as linker in electrochemical DNA (E-DNA) sensor with four different alkyl chain as the R group. To date there is no study involving Schiff base molecules in E-DNA sensor.

There are many methods to detect DNA. Electrochemical DNA sensor is the easiest, sensitive and selective method reported so far [5]. However, a suitable linker also known as recognition layer is needed to be attached to a suitable substrate such as gold, platinum, silver or indium tin oxide (ITO). Furthermore, DNA probe must be sufficiently immobilised [6]. All molecular based biosensor relies on highly specific recognition events which typically involve probe sequence immobilised within the recognition layer. The length of linker may also affect DNA immobilisation.

Reaction scheme for Schiff bases molecules derivatives can be seen in Figure 1. COOH group will bind with gold surface substrate while the other R group will bind directly with phosphate group from DNA part. Phosphate group involved, normally taking consideration all four nucleotides which are adenosine, cytidine, guanosine and thymidine. Since this is a preliminary study, this article only reported on the adenosine part. There are many computational chemistry methods and also theoretical studies about the determination of biological and chemical activity [7]. GaussView 5.0 has been used to do all the structure drawing. Calculations were performed with Gaussian 09 software package program [8]. In this study, quantum-chemical descriptors are suitable to determine the ranking of physical descriptors. Figure 2 shows the complete Schiff base derivate attached to adenosine nucleotide via the oxygen atom using GaussView 5.0. One side of R group attached to a substrate while the other R group is attached to adenosine nucleotide. This side of R group has been varied the length which involves alkane chain which contain $(CH_2)n$ where n= 3,6,9 and 12 carbon chains to see the effect of recognition length towards the effectiveness of DNA detection.

In this study, quantum-chemical descriptors are suitable to determine the ranking of physical descriptors. Figure 2 shows the complete Schiff base derivate attached to adenosine nucleotide via the oxygen atom using GaussView 5.0. One side of R group attached to a substrate while the other R group is attached to adenosine nucleotide. This side of R group has been varied the length which involves alkane chain which contain $(CH_2)n$ where n= 3,6,9 and 12 carbon chains to see the effect of recognition length towards the effectiveness of DNA detection.

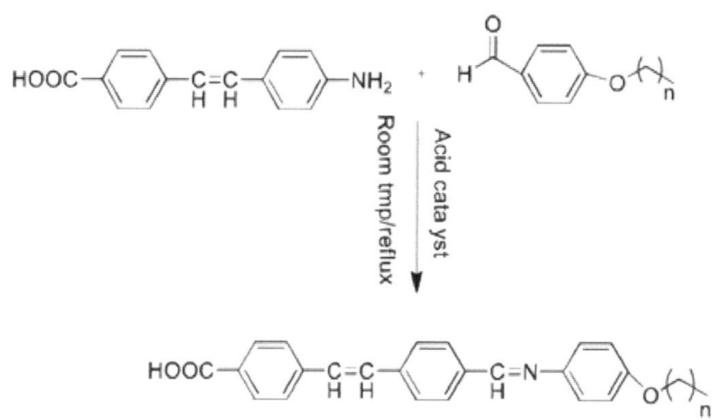

Figure 1. Reaction scheme to synthesise Schiff base derivatives with various R group which involve $(CH_2)n$ where n = 3, 6,9 and 12 carbon chain

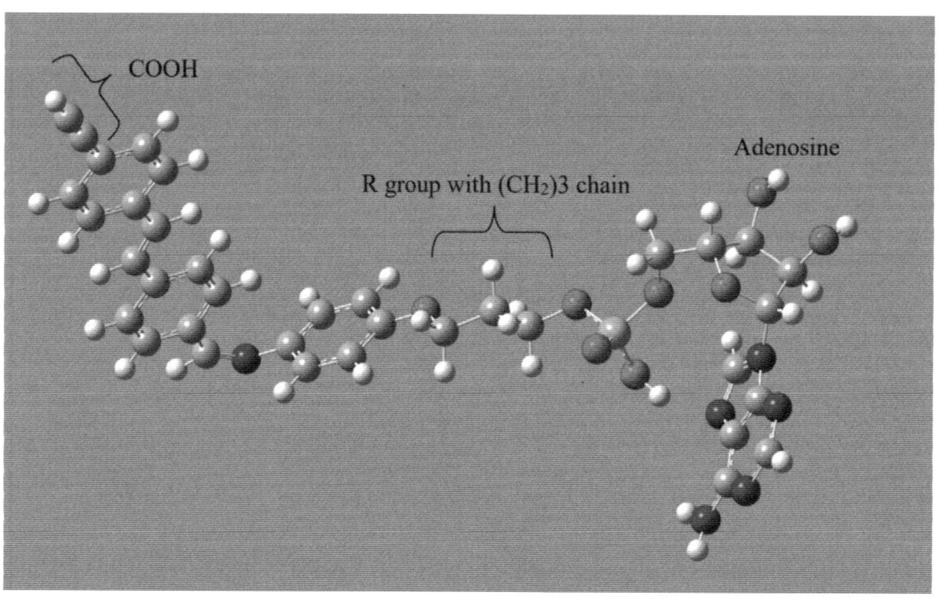

Figure 2. The overall molecule of Schiff base derivatives bind with adenosine with two R groups, COOH (fixed) and alkane chain length and this image represent R group with $(CH_2)3$ chain

2 Computational Method

The structure drawings were done using GaussView 5.0 and all calculations were performed with Gaussian 09 software package program [7]. There are many computational chemistry methods as well as theoretical studies about the determination of biological activity and chemical activity [8,9]. Quantum-chemical descriptors are very suitable to determine the ranking biological activity. Quantum-chemical descriptors are very suitable to determine the ranking biological activity. These parameters are the highest occupied orbitals (HOMOs) and the lowest unoccupied molecular (LUMOs), the energy gap between LUMO and HOMO (ΔE_{GAP}), dipole moment (μ), hardness (η) and softness (σ). Hardness and softness descriptors can be related with biological activity ranking and can determine the stability of chemical system [10]. To analyse these biological activity ranking one of the density functional theory (DFT) methods, named Becke, 3-parameter, Lee-Yang-Parr (B3LYP) was selected and the rest of the atoms were selected as basis sets. The calculation to determine the biological activity were calculated by using Eqs 1-4.

$$\Delta E_{GAP} = E_{LUMO} - E_{HOMO} \quad (1)$$

$$\mu = (E_{HOMO} - E_{LUMO}) / 2 \quad (2)$$

$$\eta = (E_{LUMO} - E_{HOMO}) / 2 \quad (3)$$

$$\sigma = 1 / \eta \quad (4)$$

3. Results and Discussion

All calculated parameters are represented in Table 1. In computational chemistry, normally Hartree-Fock (HF) is used as a method of approximation for the determination of the wave function and the energy of a quantum in a stationary state. HF is also known as self-consistent field method (SCF) assuming each particle is subjected to the mean field created by all other particles. E_{SCF} in table 1 shows the central starting point for most methods that describe the many-electron system more accurately.

Table 1. calculated quantum chemical descriptors at B3LYP method

	$(CH_2)3$	$(CH_2)6$	$(CH_2)9$	$(CH_2)12$
*E*SCF (au)	-2664.6088	-2782.5581	-2900.5080	-3018.4574
HOMO (eV)	-0.19381	-0.19161	-0.19174	-0.19144
LUMO (eV)	-0.07719	-0.07655	-0.07667	-0.07673
*E*GAP (eV)	3.1734	3.1309	3.1312	3.1214
μ (Debye) Dipole moment	1.8296	4.1622	1.7605	3.959
η (eV) Hardness	0.05831	0.05753	0.05754	0.05735
σ (eV^{-1}) Softness	17.1497	17.3822	17.3807	17.4353

IOP Conf. Series: Materials Science and Engineering **440** (2018) 012009 doi:10.1088/1757-899X/440/1/012009

HOMO is an important quantum chemical descriptor and mainly associated with electron donating ability of molecule. The higher the value of HOMO, the higher the tendency of electron transfer to acceptor molecule [10]. The ranking of biological activity of the derivatives are as follows:

$$(CH_2)3 > (CH_2)9 > (CH_2)12 > (CH_2)6$$

While the LUMO value indicates that the electron-accepting ability of the molecules is higher. The binding ability of the compounds are:

$$(CH_2)3 > (CH_2)12 > (CH_2)9 > (CH_2)6$$

It can be seen that from the energy gaps (E_{GAP}) between HOMO and LUMO for the compounds are as follows:

$$(CH_2)3 > (CH_2)9 > (CH_2)6 > (CH_2)12$$

The lower value of E_{GAP} explains the eventual charge transfer interaction taking place within the molecules. The lower value of this descriptor shows that the molecule is more active [11].

On the other hand, chemical hardness (η) is associated with the stability and reactivity of a chemical system. This descriptor measures the resistance to change in the electron distribution or charge transfer. Hard molecules have a big E_{GAP} value and of molecules have small E_{GAP} [9-11]. Biological molecules are known as soft molecules. This means that soft molecules can easily interact with biological molecules or human body. Biological activity is proportional to the increase of the softness value and vice versa. Based on the softness values obtained, the sequence of biological activity are:

$$(CH_2)3 > (CH_2)9 > (CH_2)6 > (CH_2)12$$

Another important descriptor is electronic chemical potential (μ) also known as dipole moment which is defined as the negative of electronegativity of a molecule. Based on the biological activity obtained the ranking for dipole moment of the molecules are:

$$(CH_2)6 > (CH_2)12 > (CH_2)3 > (CH_2)9$$

From dipole moment descriptor obtained, the optimum recognition layer can be seen at Schiff base molecules with 6 carbon chains attached to adenosine, while other descriptors are almost similar for each chain length. This preliminary result indicates that there is a certain length which is suitable towards the electrochemical DNA system. This indicates dipole moment position and interaction may play a role for recognition layer's length in E-DNA sensor. However, the future experimental study is much needed and will be carried out to be compared with theoretical results obtained.

4. Conclusion

DFT is a promising tool to study theoretical aspect of recognition layer's length in electrochemical DNA sensor. From the value of biological activities obtained the length of recognition layers does not play a significant role. The best length obtained was from Schiff base molecules with 6 carbon chain group attached to adenosine. However further study considering other nucleotides which are cytidine, guanosine and thymidine will be further investigated. Furthermore, experimental work will be carried out by synthesizing all the molecules involved to be compared with theoretical results.

5. References

[1] Karamunge K G and Vibhute Y B 2013 *J. Phys: Conf Ser.* **423** 1-6
[2] Sashikala S and Syed Shafi S 2014 *Pharm. Lett.* **6** (2) 90-97
[3] Nahid S, Zeinab G and Saba H 2012 *Bioinorg. Chem. Appl.* 2012 1-9

[4] Hassan K, Majid R, Misagh M-M, Koray S, Nefise D and Huseyin U 2015 *Inorg. Chim. Acta.* **432** 243-249
[5] Gregory D T, Michael G H and Jacqueline K B 2003 *Nat. Biotechnol.* **10** 1192-1199
[6] Jiao W, Anqi S, Xian F, Xiaowei H and Yuzhong Z 2015 *Anal. Biochem.* **469** 71-75
[7] Roy D, Todd K and John M 2009 Gauss View Version 5 *Semichem Inc. Shawnee Mission*
[8] Thalamuthu S Annaraj B and Neelakantan M A 2014 *Spectrochim. Acta, Part A* **118** 120-129

[9] Aljahdali M and Ahmad A E-S 2013 *Inorg. Chim. Acta.* **407** 58-68

[10] Yusoff H. M Hassan H and Ku Bulat K H 2018 *IOP Conf. Series: Mater. Sci. and Eng.* **374** 1-

[11] Salihovic M Huseinovic S Spirtovic-H. S Osmanovic A Dedic A Asimovic Z and Zavrsnik D 2014 *Bull. Chem. Technol.* Bosnia and Herzegovina **42** 31-36

Acknowledgments

Authors wish to thank Universiti Malaysia Terengganu (UMT) and School of Fundamental Science for all the facilities provided also Fundamental Research Grants Scheme (FRGS) vot 59383 for financial support throughout this study. Gratitude to the anonymous reviewers who have spent time and effort, constructive recommendations that enhanced the value of this manuscript.

IOP Conf. Series: Materials Science and Engineering **440** (2018) 012010 doi:10.1088/1757-899X/440/1/012010

Synthesis of Phenylpyridine Iridium(III) Complexes with N-Heterocyclic Carbene as Ancillary Ligands

N H A Bain[1], N M Ali[1], Y Juahir[1], N Hashim[1], I M Isa[1], A Mohamed[1], A Kamari[1] and B M Yamin[2]

[1]Department of Chemistry, Faculty of Sciences and Mathematics, Universiti Pendidikan Sultan Idris, Tanjong Malim, Malaysia

[2]School of Chemical Sciences and Food Technology, Faculty of Science and Technology, Universiti Kebangsaan Malaysia, Bandar Baru Bangi, Malaysia

Email: noorshida@fsmt.upsi.edu.my

Abstract. In this study, [Ir(2,4-F_2ppy)$_2$(2-(1,2,4-triazol-1-yl)pyridine)]PF_6 (**C1**), [Ir(2,4-F_2ppy)$_2$(2-(2-(4-dimethylbenzyl)-1,2,4-triazol-1-yl)pyridine)]PF_6 (**C2**) and [Ir(2,4-F_2ppy)$_2$(2(2-hexyl-1,2,4-triazol-1-yl)pyridine)]PF_6 (**C3**) complexes were successfully synthesised by refluxing the mixture of dichloro-bridged iridium(III) dimer, [Ir(2,4-F_2ppy)$_2$(μ-Cl)]$_2$ and corresponding triazolium salt containing N-heterocyclic carbene (NHC) as ancillary ligands. ^{1}H NMR of all complexes display well-resolved signals between 10.00-5.00 ppm assign to proton at the aromatic region for penylpyridine and pyridyltriazole. Spectra of **C2** and **C3** also display signals at aliphatic region (singlets) between 0.80-1.50 ppm that proved the presence of methylbenzyl and hexyl substituent on the pyridyltriazole ring. In addition, IR analysis reported the presence of C=C and C=N stretching of pyridine at range 1570-1470cm^{-1} and C-H stretching of aromatic pyridine at 2990-3080cm^{-1}. Besides, UV-Vis absorption data showed that the low-energy metal-to-ligand charge-transfer (MLCT) band in **C2** (364 nm) is significantly blue-shifted compared to **C1** (375 nm) and **C3** (381 nm). Overall, this study reported the synthesis and spectroscopic study of phenylpyridine Ir(III) complexes with various types of pyridyltriazole (NHC ancillary ligands).

1. Introduction

Transition metal complexes such as Pt(II) [1-2], Re(I) [3], Ru(II) [4], Au(I) [5] and Ir(III) [6,7] of N-Heterocyclic Carbene (NHC) ligands have been the subject of interest in modern materials research due to their rich of photophysical properties and big potential in photonic applications such as a good emissive dopants in organic light emitting (OLEDs) devices[8-10], solar energy conversion [11], electroluminescence [12] and sensors [11]. However, compared to other transition metals, Ir(III) complexes are particularly highly tuneable in the colour of the iridium emission. For instance, while common Ru(II) and Os(II) polypyridines emit in the red or near-infrared (NIR) region, the emission of Ir(III) complexes can shift from the red to the blue region of the spectrum. Generally, Ir(III) complexes containing 2-phenylpyridine are known to exhibit high triplet quantum yields due to the mixing of the singlet and the triplet excited states via a spin-orbit coupling, leading to high phosphorescence efficiencies [13]. In particular, the emission colour of Ir(III) complexes is particularly sensitive to the substituents on the ligand attached to the central metal [14].

Recent studies have shown that an introduction of neutral pyridine-NHC ligands to the Ir(III) complex can form very strong bonds to the transition metals, which shift the metal-carbene antibonding orbitals to high energy and decreasing or eliminating quenching through the ligand field state. Thus, replacing the pyridyl ring with NHC-based group leads to a significant increase in the Lowest Unoccupied Molecular Orbital (LUMO) energy and increases the emission energy of complexes coordinated to these ligands. Hence, it becomes possible to observe efficient blue or near UV phosphorescence at room temperature from Ir(III) complexes that have cyclometalated NHC based ligands[10]. Although there were many reports on the synthesis and characterisation of cationic Ir(III) complexes, there is no studies on the use of the specific pyridyltriazole derivatives reported. Thus, in this paper we report the synthesis and spectroscopic study of phenylpyridine Ir(III) complexes with various types of pyridyltriazole (NHC ancillary ligands).

2. Experimental section

2.1. Analytical Measurements

^{1}H NMR spectra were recorded at 400 Hz using JEOL, JNM-ECX-500 NMR spectrometer. The FT-IR spectra were measured using Thermo Nicolet Nexus spectrometer in the 4000-600 cm^{-1} regions. UVVis absorption spectra were measured in 1 cm path length quartz cell on Cary 60 Agilent Technologies types of a spectrophotometer. Elemental analysis for carbon, hydrogen and nitrogen was carried out by using a CHNO-S EA 1112 Series Elemental analyser.

2.2. Reagents

All the commercial chemicals and solvents were from reagent quality and were used as received from various suppliers The dimer [Ir(2,4-F_2ppy)$_2$(μ-Cl)]$_2$ was prepared according to the literature methods [10].

2.3. Synthesis

*2.3.1. Synthesis of [Ir(2,4-F_2ppy)$_2$(*2-(1,2,4-triazol-1-yl)pyridine*)]PF_6* ***(C1)***

2-(1,2,4-triazol-1-yl)pyridine (0.075g, 0.5 mmol, 2.1 equiv), silver(I) oxide (0.07g, 0.29 mmol, 1.2 equiv), and [Ir(2,4-F_2ppy)$_2$(μ-Cl)]$_2$ (0.29g, 0.24 mmol, 1.0 equiv) in 12 mL 1,2-dichloroethane were heated to reflux for 18 h. After cooling to ambient temperature, the mixture was filtered through celite to remove silver residues. The filtrate was treated with an aqueous KPF_6 solution (55 mg of KPF_6 in 40 mL of water per cycle). The combined organic layers were washed with water, dried over Na_2SO_4, and concentrated in vacuum. The crude product was purified by column chromatography on silica gel using CH_2Cl_2/acetone (4:1 v/v) as the solvent. After evaporation of the solvent, the pure product obtained was a yellow solid. Yield: 1.27g, 79%. ^{1}H NMR (500 MHz, $CDCl_3$): δ 9.91 (1H, d), 9.71 (1H, s), 8.43 (1H, d), 8.30 (1H, t), 8.19 (1H, d), 7.90 (1H, t), 7.90 (1H, t), 7.70 (4H, t), 7.33 (1H, t), 7.09 (1H, t), 6.45 (1H, t), 6.35 (1H, d), 5.82 (1H, d), 5.60(1H, d). IR: v/cm^{-1} 779 (s, C-H stretch), 1593-1400 (s, C=H, C=N), 3020 (w, C-H from pyridine ring). Anal. Calcd. for $C_{29}H_{20}N_6F_4Ir$: C, 48.33; H, 2.80; N, 11.66%. Found: C, 48.31; H, 2.70; N, 11.61%.

2.3.2. Synthesis of [Ir(2,4-F_2ppy)$_2$(2-(2-(4-dimethylbenzyl)-1,2,4-triazol-1-yl)pyridine)]PF_6 ***(C2)***

2-(2-(4-dimethylbenzyl)-1,2,4-triazol-1-yl)pyridine (0.05g, 0.2 mmol, 2.1 equiv), silver(I) oxide (0.027g, 0.12 mmol, 1.2 equiv), and [Ir(2,4-F_2ppy)$_2$(μ-Cl)]$_2$ (0.12g, 0.1 mmol, 1.0 equiv) in 12 mL 1,2dichloroethane were heated to reflux for 18 h. After cooling to ambient temperature, the mixture was filtered through celite to remove silver residues. The filtrate was treated with an aqueous KPF_6 solution (55 mg of KPF_6 in 40 mL of water per cycle). The combined organic layers were washed with water, dried over Na_2SO_4, and concentrated in vacuum. The crude product was purified by column chromatography on silica gel using CH_2Cl_2/acetone (4:1 v/v) as the solvent. After evaporation of the solvent, the pure product obtained was a yellow solid. Yield: 0.09g, 61%.^{1}H NMR (500 MHz, $CDCl_3$): δ 9.90 (1H, s), 8.28 (3H, t), 7.98 (3H, d), 7.98 (3H, t), 7.25 (1H, t) 7.09 (1H, d), 6.98 (1H, d), 6.96 (1H, d), 6.96 (1H,t), 5.76 (1H, t), 5.50 (1H, t), 2.34 (2H,d), 1.24 (4H, s), 0.92 (3H, t). IR: v/cm^{-1} 824 (s, CH stretch), 1570-1473 (s, C=H, C=N), 2994 (w, C-H from pyridine ring). Anal. Calcd. for $C_{37}H_{30}F_4IrN_6$: C, 53.74; H, 3.66; N, 10.16%. Found: C, 53.68; H, 3.59; N, 10.09%.

*2.3.3. Synthesis of [Ir(2,4-F₂ppy)₂(2-(2-hexyl-1,2,4-triazol-1-yl)pyridine)]PF₆ (**C3**)*

2-(2-hexyl-1,2,4-triazol-1-yl)pyridine (0.08g, 0.35 mmol, 8.4 equiv), silver(I) oxide (0.011g, 0.05 mmol, 1.2 equiv), and [Ir(2,4-F_2ppy)$_2$(μ-Cl)]$_2$ (0.05g, 0.04 mmol, 1.0 equiv) in 12mL 1,2dichloroethane were heated to reflux for 18 h. After cooling to ambient temperature, the mixture was filtered through celite to remove silver residues. The filtrate was treated with an aqueous KPF_6 solution (55 mg of KPF_6 in 40 mL of water per cycle). The combined organic layers were washed with water, dried over Na_2SO_4, and concentrated in vacuum. The crude product was purified by column chromatography on silica gel using CH_2Cl_2/acetone (4:1 v/v) as the solvent. After evaporation of the solvent, the pure product obtained was a yellow solid. Yield: 0.19g, 45%. ^{1}H NMR (500 MHz, $CDCl_3$): δ 9.90 (1H, d), 9.71 (1H, s), 8.37 (1H, d), 8.30 (1H, t), 8.20 (1H, d), 7.80 (4H, t), 7.60 (1H, t), 7.35 (1H, t), 7.34 (t, 1H) ,7.10 (1H, t), 6.43 (1H, t), 6.33 (1H, t), 5.82(1H, d), 5.62(1H, d),1.24 (8H, t), 0.85(5H, d). IR: v/cm^{-1} 782 (s, C-H stretch), 1595-1459 (s, C=H, C=N), 3057 (w, C-H from pyridine ring). Anal. Calcd. for $C_{35}H_{34}F_4IrN_6$: C, 52.10; H, 4.25; N, 10.42%. Found: C, 52.05; H, 4.15; N, 10.36%.

3. Results and Discussion

Complexes of bis(phenylpyridine) iridium(III) pyridyltriazole ligands were successfully synthesised by reacting the dichloro-bridged iridium(III) dimer, [Ir(2,4-F_2ppy)$_2$(μ-Cl)]$_2$ in 1,2-dichloroethane solvent with the presence of silver oxide which gave [Ir(2,4-F_2ppy)$_2$(2-(1,2,4-triazol-1yl)pyridine)]PF_6 (**C1**), [Ir(2,4-F_2ppy)$_2$(2-(2-(4-dimethylbenzyl)-1,2,4-triazol-1-yl)pyridine)]PF_6 (**C2**) and [Ir(2,4-F_2ppy)$_2$(2-(2-hexyl-1,2,4-triazol-1-yl)pyridine)]PF_6 (**C3**). The schematic pathway leading to the phenylpyridine iridium(III) pyridyltriazole complexes was depicted in scheme 1.

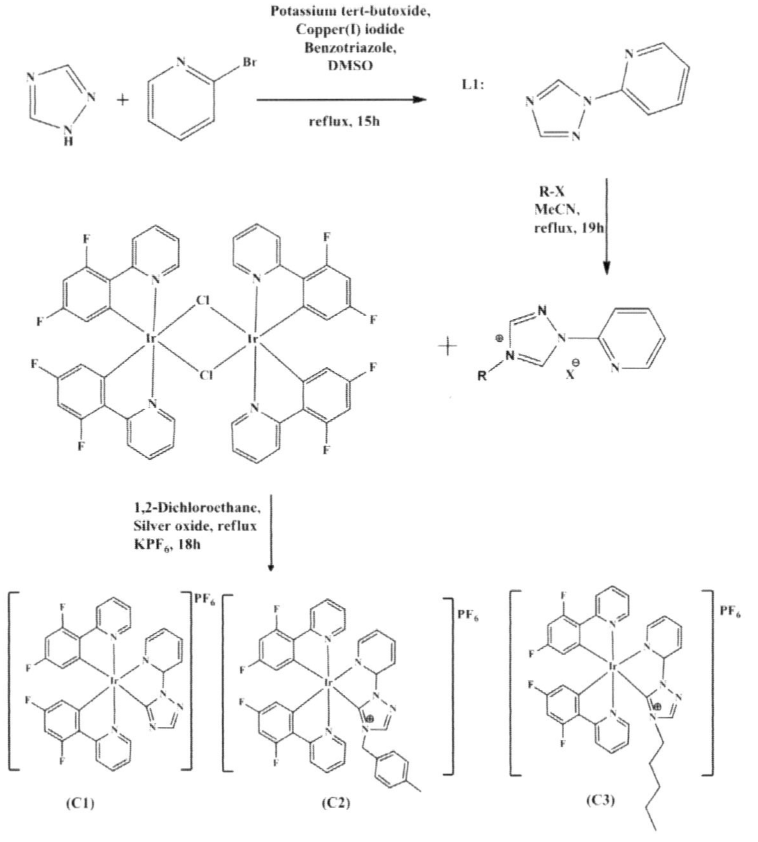

Scheme 1. Synthesis of bis(phenylpyridine) iridium(III) pyridyltriazole complexes **C1-C3**.

In each case, the elemental analysis is consistent with the expected molecular formula. IR spectra exhibit C=N and C=C ring stretching of pyridine for all complexes at 1595-1400 cm^{-1}(**C1**); 1570-1473cm^{-1}(**C2**); 1595-1473 cm^{-1}(**C3**). In addition, the weak C-H stretching of aromatic pyridine shows at 2990-3080cm^{-1} for all complexes and for complex **C3** indicates the presence of C-H from hexyl substituent [18] and this show that introduction of different ancillary ligand to the complex affected the shift of IR wavenumber to the right or to the left. In general, there is no significant change in IR spectral pattern between the starting dichloro-bridged iridium(III) dimer and the complexes **C1C3.** The clear comparison for **C3** shown in Figure 1.

IOP Conf. Series: Materials Science and Engineering **440** (2018) 012010 doi:10.1088/1757-899X/440/1/012010

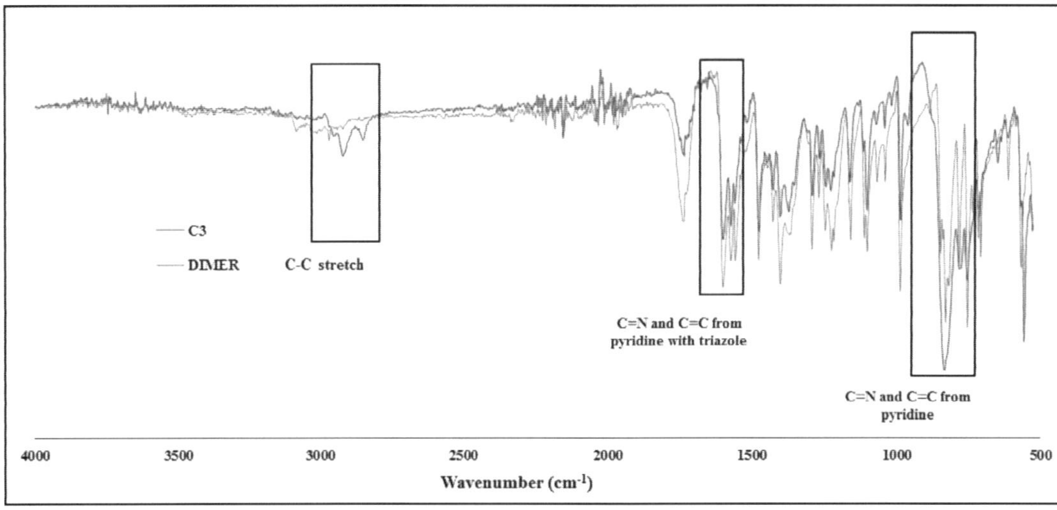

Figure 1. FT-IR Comparison of complex **C3** and the starting dichloro-bridged Ir(III) dimer .

^{1}H NMR spectroscopy used to characterise all three complexes. ^{1}H NMR spectra of **C1-C3** were recorded in deuterated $CDCl_3$ are presented in Table 1 and Table 2, with their labeling scheme in Figure 2. Assignments of proton resonances for **C1-C3** were made by comparison to the ^{1}H NMR spectra of dichloro-bridged iridium(III) dimer and their ligands. All the ^{1}H NMR spectra display wellresolved signals and a similar pattern to each other in the aromatic region between 10.00 and 5.00 ppm that assigned to penylpyridine and pyridyltriazole proton but slightly different pattern at the aliphatic region due to the ancillary ligand modification. The spectrum of **C2** displays 5 well-resolved signals at aliphatic region (singlets) between 1.00-1.50 ppm corresponding to the proton that attached to the aromatic ring in the ancillary ligand, while the spectrum of **C3** displays 13 well-resolved signals at aliphatic region (singlets) assigned to the hexane substituent that attached to triazolium. Additionally, the coupling between hydrogen and fluorine (spin ½) is very strong. Therefore, the proton signals of H7 and H8 are split into a triplet because of coupling with two fluorine nuclei [4].

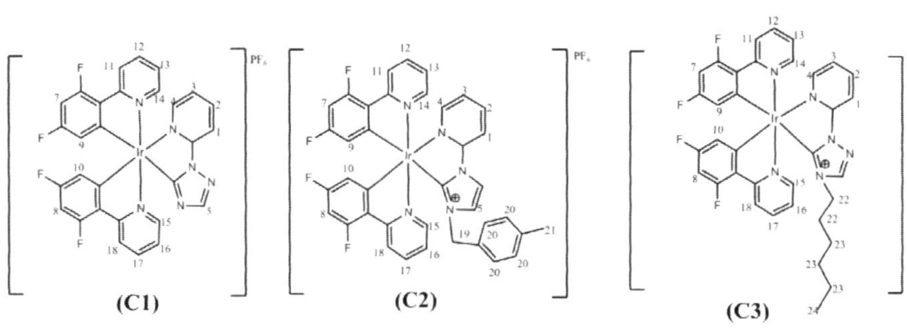

Figure 2. The molecular structure of complexes **C1-C3** with their labeling scheme.

Table 1. ^{1}H NMR for H1-H12 for **C1-C3** and starting dimer in deuterated $CDCl_3$

Complexes	Structural position (ppm), Group											
	H1	H2	H3	H4	H5	H6	H7	H8	H9	H10	H11	H12
C1	8.53 d	8.43 d	8.30 t	9.91 d	9.71 s	-	7.90 t	7.90 t	7.70 t	7.70 t	8.19 d	7.33 t
C2	8.00 d	7.98 d	7.98 t	8.00 s	9.90 s	-	8.28 t	8.28 t	8.28 t	7.98 t	7.25 d	8.28 t
C3	8.57 d	8.37 d	9.30 t	9.90 d	9.71 s	-	7.80 t	7.80 t	7.80 t	7.60 t	8.20 d	7.10 t
Starting dimer							7.90 t	6.90 t	-	-	9.10 d	8.31 t

s = singlet, d = doublet, t = triplet

Table 2. ^{1}H NMR for H13-H24 for C1-C3 and starting dimer in deuterated $CDCl_3$

Complexes	Structural position (ppm), Group											
	H13	H14	H15	H16	H17	H18	H19	H20	H21	H22	H23	H24
C1	7.09 t	5.82 d	5.60 d	6.35 t	6.45 t					-	-	-
C2	7.09 d	6.96 d	6.96 t	5.76 t	5.50 t	6.98 t	2.34 s	7.24 s	0.92 s	-	-	-
C3	7.35 t	5.82 d	5.62 d	6.43 t	6.33 t	-	-	-	-	1.24 s	1.24 s	0.85 s
Starting dimer	6.30 t	5.30 d	-	-	-	-	-	-	-	-	-	-

s = singlet, d = doublet, t = triplet

Table 3 shows the comparison of UV/Vis absorption data for **C1-C3** in dichloromethane solution at 298K. As with most iridium(III) complexes, the absorption spectra can be divided into two regions; the short wavelength π–π* transitions of the ligands in the UV region, and the longer wavelength metal-to-ligand charge-transfer (MLCT) dπ(M)→π*(L-ligand) transitions in the visible region [6,11,14,16].

Table 3. The summary of UV-Vis spectral data for all the complexes.

Compound	**λ_{max}/nm(10^{-3})ε/$M^{-1}cm^{-1}$**
$[Ir(2,4\text{-}F_2ppy)_2(L1)]PF_6$ (**C1**)	262 (17), 375(5.7)
$[Ir(2,4\text{-}F_2ppy)_2(L2)]PF_6$ (**C2**)	262 (12), 364 (6.6)
$[Ir(2,4\text{-}F_2ppy)_2(L3)]PF_6$ (**C3**)	261 (14), 381 (5.6)

Complexes of **C1-C3** show very similar absorption spectra to each other, in which less intense absorptions at lower energy have a mixture of ligand centred (LC) π–π* and metal-to-ligand chargetransfer (MLCT) character (ε between 5500 and 6800 M^{-1} cm^{-1}) in the region 360-385 nm [16], and the expected strong π–π* transitions associated with aromatic ligands occur in the UV region. In particular, the low-energy MLCT band in **C2** (364 nm) is significantly blue-shifted compared to C1 (375 nm) and **C3** (381 nm). The low-lying transitions in the spectral properties of iridium(III) phenylpyridine complexes are categorised as MLCT transitions, though the metal orbitals have a significant mixture of ligand π character, and the high-energy bands above 280 nm are assigned to the intraligand (IL) π–π* transition of 2-phenylpyridine [14]. The absorption spectra of **C1-C3** complexes are shown in Figure 3.

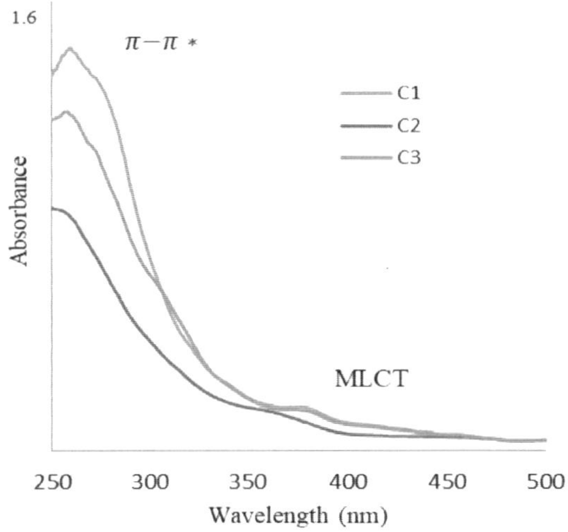

Figure 3. Absorption spectra of complexes **C1-C3** in dichloromethane at 298K

4. Conclusion
A series of phenylpyridines iridium(III) complexes with various types of pyridyltriazole ancillary ligands were successfully synthesised and characterised to gaining a better understanding into reasons behind the varying of electronic properties. This study demonstrated that complexes of **C1-C3** display comparable absorption spectra to each other with a mixture of ligand centred (LC) π–π* and metal-toligand charge-transfer (MLCT) character in the region 364 nm (**C2**), 375 nm (**C1**) and 381 nm (**C3**). Therefore, further photophysical study will be done in order to prove that the luminescence properties of cyclometalated iridium(III) complexes can be tuned by the ligand modification.

5. References
[1] Lee CS, Sabiah S, Wang JC, Wang JC, Hwang WS and Lin IJB 2010 *Organomet. Commun.*, **29** 286–289
[2] Unger Y, Meyer D, Molt O, Schildknecht C, Munster I, Wagenblast G and Strassner T 2010 *Organometallic Communications* **49** 10214–10216
[3] Casson LA, Muzzioli S, Raiteri P, Skelton BW, Stagni S, Massimiliano M and Brown DH 2011 *Dalt. Trans.* **3** 11960-11967
[4] Barbante GJ, Francis PS, Hogan CF, Kheradmand PR, Wilson DJD and Bernard PJ 2013 *Inorg. Chem.* **52** 7448–7459
[5] Pell TP, Wilsin JD, Skelton BW, Dutton JL and Barnard PJ 2016 *Inorg. Chem.* **55** 6882–6891
[6] Monti F, Kessler F, Delgado M, Frey J, Bazzanini F, Accorsi G, Armaroli N,Bolink HJ, Orti E and Scopelliti R 2013 *Inorg. Chem.* **52** 10292–10305
[7] Visbal R and Gimeno MC 2014 *Chem. Soc. Rev.* **43** 3551–74
[8] Che CM, Kwok CC, Lai SW, Rauseh AF, Finkenzeller WJ, Zhu N and Yersin H 2010 *Chem. - A Eur. Journal* **16** 233–247

[9] Lee MT, Liao CH, Tsa CHi and Chen CH 2005 *Advance Material* **17** 2493–2497
[10] Sajoto T, Djurovich PI, Tamayo A, Yousufuddin M, Bau R and Thompson ME, 2005 *Inorg. Chem.* **44** 7992–8003
[11] Nazeeruddin MK, Humphry R, Berner D, Rivier S, Zuppiroli L and Graetzel M 2003 *J. Am. Chem. Soc.* **125** 8790–8797
[12] Zu Y and Bard AJ 2000 *Anal. Chem,* **72** 3223–3232
[13] Barbante GJ, Doeven EH, Francis PS, Stringer BD, Hogan CF, Kheradmand PR, Wilson DJD and Bernard PJ 2015 *Dalt. Trans.* **44** 8564–8576
[14] Ali NM, Ward MD, Hashim N and Daud N 2017 *Mater. Res. Innov.* **89** 17 1–6
[15] G. J. Barbante, P. S. Francis, C. F. Hogan, P. R. Kheradmand, D. J. D. Wilson, and P. J. Barnard 2013 *Inorg. Chem.* **52** 7448–7459
[16] Baranoff E, Fantacci S, Angelis FD, Zhang X, Scopelliti R, Gratzel M and Nazeeruddin MK 2011 *Inorg. Chem.* **50** 451–462
[17] Ali NM, MacLeod VL, Jennison P, Sazanovich IV, Hunter CA, Weinstein JA and Ward MD 2012 *Dalt. Trans.* **41** 2408
[18] Robert MS, Francis XW and David JK 2005 *Spectrometric Identification of Organic Compounds*, (John Wiley & Sons, Inc) p 107-108

Acknowledgment
The authors greatly acknowledge the scientific support from Universiti Pendidikan Sultan Idris and Ministry of Higher Education Malaysia for financial support through Fundamental Research Grant Projects (FRGS) [grant number 2016-0080-101-02].

IOP Conf. Series: Materials Science and Engineering **440** (2018) 012011 doi:10.1088/1757-899X/440/1/012011

Review on cellulose nanocrystals (CNCs) as reinforced agent on electrospun nanofibers: mechanical and thermal properties

S Shafizah[1], A R Saiful Izwan[2, 3,*], F Fatirah [2] and M N Nadirul Hasraf [4]

[1] Faculty of Biosciences and Medical Engineering, Universiti Teknologi Malaysia, 81310 Skudai, Johor Darul Takzim, Malaysia.
[2] Faculty of Biosciences and Medical Engineering, Medical Devices & Technology Group (MEDITEG), Universiti Teknologi Malaysia, 81310, Skudai, Johor, Malaysia
[3] IJN-UTM Cardiovascular Engineering Centre, Universiti Teknologi Malaysia, 81310 Skudai, Johor Darul Takzim, Malaysia
[4] Department of Chemical Engineering Technology, Faculty of Engineering Technology, Universiti Tun Hussein Onn Malaysia, 86400 Batu Pahat, Johor, Malaysia.

Corresponding author email: saifulizwan@utm.my

Abstract. The growing ecological and environmental consciousness has driven efforts for development of new innovative products for various end-use applications. Nanotechnology plays an important role in the development of such products. Electrospun nanofibers which are produced from electrospinning technique are likely to be the next novel nanomaterials designated in the consumer products. Cellulose nanocrystals (CNCs) are nanoscale cellulose produced from renewable resources with strong reinforcing potential when included in a polymer matrix. Nanotechnology plays an important role in the development of such products. Electrospun nanofibers which are produced from electrospinning technique are likely to be the next novel nanomaterials designated in the consumer products. Electrospinning technique is the known prevalent technology for the manufacturing of nano-sized diameter fibers with the advanced functionalities. Various polymers have been successfully electrospun into ultrafine fibers in recent years mostly in solvent solution and some in melt. In this paper, a mini review is presented on the researches and developments related to electrospun polymer nanofibers.

1. Introduction

Nanocellulose, a unique and promising natural material extracted from native cellulose, has gained much attention for its use as biomedical material, because of its remarkable physical properties, special surface chemistry and excellent biological properties (biocompatibility, biodegradability and low toxicity) [1]. Nanocellulose can be define as nano-sectional fibers <100 nm in diameter or micronsized fibers with nano-dimensional cross-sectional structures [2]. For past decade, production of nanocellulose from biomass is gaining attention because of its mechanical and thermal properties, as well as environmental friendliness and cost-effectiveness. Review by Habibi *et al.,* [3] and Samir *et al.,* [4], the development of nanocellulose has attracted significant interest in the last decades due to the unique characteristics, such as high surface area-to-volume ratio, high surface area, high Young's modulus, high tensile strength, and low coefficient of thermal expansion. Eichhorn *et al.,* [5] in his research revealed that the specific modulus of CNCs exceeds those of common engineering materials such as steel, concrete, glass, and aluminium.

Among reported studies, CNCs have received ever-increasing interest, such as reinforced with polylactic acid (PLA) [6-10], poly-vinyl alcohol (PVA) [11-22], poly-vinyl acetate (PVAc) [23], Polyacrylonitrile (PAN) [24], Poly-vinyl pyrolidone (PVP) [25], poly-methyl methacrylate (PMMA) [26], polyethylene terephthalate (PET) [27] and etc. Due to an abundance of hydroxyl groups existed on surface of CNCs, reactive CNCs can be modified with various chemical groups to accomplish expected surface modification, such as esterification, facilitate the incorporation and dispersion of CNCs into different polymer matrices [3] etherification, oxidation, silylation, or polymer grafting, which could successfully functionalize the CNCs and facilitate the incorporation and dispersion of CNCs into different polymer matrices. CNCs is considered as one of the ideal nano-reinforcements for polymer matrices (including water- soluble and water-insoluble polymer systems) and have already been incorporated into many polymer matrices to produce reinforced composites [28].

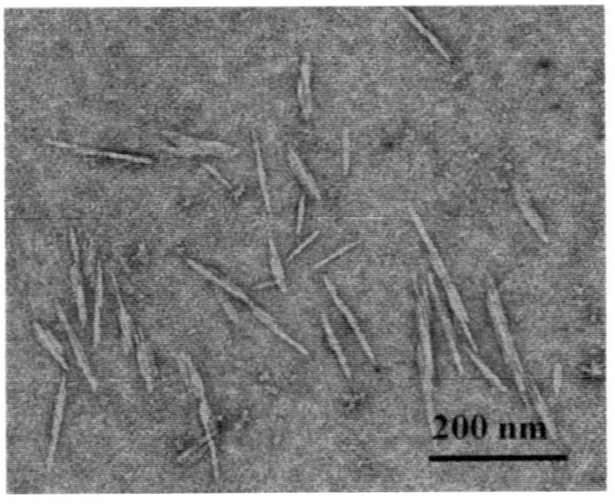

Figure 1. The TEM shows the representative morphology of the CNCs. Based on the TEM images, the CNC diameter is in the range of 6-12 nm and length between 150 and 200 nm [24].

This review is aim at presenting a summary on recent effect of cellulose nanocrystals reinforced in advanced polymer-based nanocomposites using the novel fabrication strategies for targeting electrospun nanofibers.

2. Electrospinning

2.1 Electrospinning technology

Electrospinning technique generates ultrafine electrospun nanofibers through the applied electrical force and then collects those nanofibers to form non-woven or felt-type fabrics [29-30]. The term ''electrospinning'', derived from ''electrostatic spinning'', was used relatively recently (in around 1994), its fundamental idea dates back more than 60 years earlier. A lab-scale electrospinning reactor unit can be straightforwardly completed by using simple laboratory apparatus consists of high voltage power supply, syringe pump and metal collector, as the basic part of the electrospinning setup component. The ultrafine electrospun nanofibers will be formed at the needle tip contained charged spinning solution, once the surface tension of the charged solution has been exceeded. The formation of those electrospun nanofibers will be deposited onto the grounded collector plate, assisted by the electrostatic interaction [31].

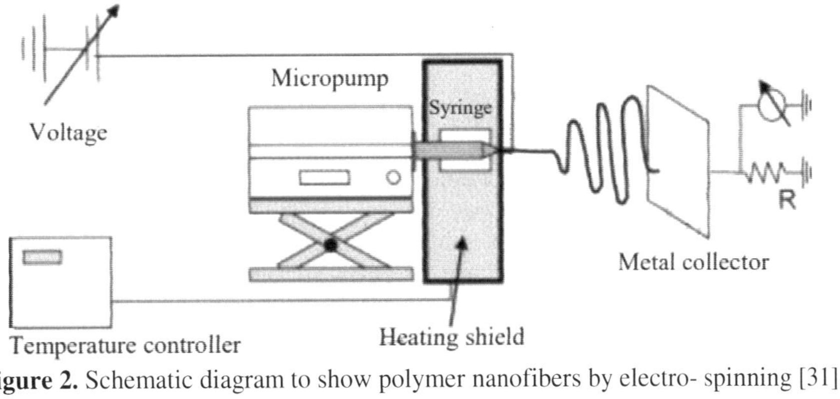

Figure 2. Schematic diagram to show polymer nanofibers by electro- spinning [31].

Polymer solution or melt are able to be electrospun into fibrous structure shaped by electrospinning technique with the appropriate selection of suitable solvents and optimum electrospinning working parameters. Jiang *et al.,* [32] proved that electrospun fibers has provided additional advantages as compared to film and sheet carriers due to their nano- or submicron-scale diameter and consequent very large surface area, and, thus, are ideal as controlled release. The functional additives such as CNCs at one-dimensional (1D) range sizes are also possible to be embedded into the nano-sized diameter fibers during the process of spinning.

2.2 Electrospun nanocomposites fibers
During the past several years, a large number of studies have been conducted to improve mechanical properties of electrospun polymeric nanofibers. Incorporating nanoparticles into polymer matrices is one technique that has been developed and used as one of the most effective methods for reinforcing electrospun nanofibers [33]. CNCs have been successfully used as highly effective reinforcing nanofillers for improving mechanical properties of various electrospun polymer matrices, as summarized below.

2.2.1. Tensile properties
Mechanical properties of nanofibers are important in many practical applications. Table 1 summarized the comparison of maximum tensile stress (σ_{max}), Young's modulus (E), and elongation at break (ε_b) obtained between neat PVA, neat PLA and reinforced CNC-PVA and CNC-PLA.

Table 1. Tensile properties for electrospun nanocomposites

Nanocomposites	σ_{max}(MPa)	E (MPa)	ε_b (%)	References
PLA	82.2 ± 6.6	3455 ± 158.0	5.0 ± 1.1	[7]
PLA/CNC	118.5 ± 16.5	3852 ± 241.0	51.1 ± 11.3	
PLA–PHB	6.0 ± 1.0	210 ± 30.0	50 ± 10.0	[8]
PLA–PHB–CNC1	16.0 ± 2.0	230 ± 10.0	50 ± 5.0	
PVA	2.6 ± 1.1	0.8 ± 0.2	294 ± 101.0	[14]
3CNW/PVA	3.8 ± 0.1	1.1 ± 0.3	343 ± 84.0	
7PVA	4.97 ± 0.12	31.9 ± 0.3	271 ± 13	[15]
7PVA/CNC-20	6.71 ± 0.25	53.9 ± 1.4	86 ± 7	
MPLA	1.6 ± 0.4	7.8 ± 3.1	51.8 ± 3.8	[22]
MPLA/CNC	10.8 ± 1.7	135.1 ± 10.4	44.0 ± 5.1	

Mujiga-Garcia et al., [7] indicated that the improvements are assumed to be due to the alignment of the CNC along to the fiber axis. A very large tensile strength improvement was observed for PLA/CNC compared to neat PLA, which can be considered as an evidence of a flexible interface between PLA/CNC and the PLA matrix and allowed this high stretching which is resulting to higher maximum strength. According to Huan et al., [15] PVA/CNC nanofibers became more uniform and finer with increased CNC content due to the synergetic effect of electric conductivity, interfacial tension, and viscosity of electrospinning solutions. It was also found that thermomechanical and mechanical properties of both types of nanofibrous mats were enhanced as a result of the addition of CNCs to the PVA polymer matrix. This was ascribed to the efficient stress transfer from PVA to CNCs originating from the strong percolation network interactions held by hydrogen bond in the nanofibers and the ordered arrangement of electrospun nanofibers in the nanofibrous mats.

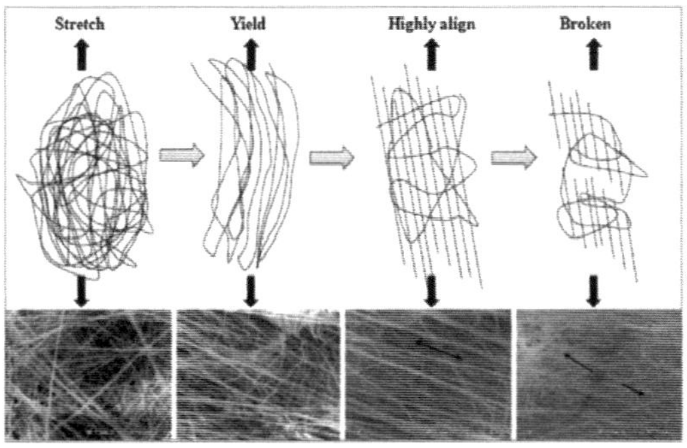

Figure 3. Scheme SEM-based morphologies of fracturing process of electrospun PLA/CNC nanocomposites mats [15].

Figure 3 shows a scheme of the tensile fracturing of electrospun mats based on SEM photographs. At the beginning, the mats were stretched in macroscopic view and most fibers in mats hardly moved attributed to the cohesion between fibers, which determined modulus of mats. When most fibers in mats reached the tightened form, yield of mats appeared, at which the interaction points among fibers were broken. With further increase of tensile strain, the fibers in mats were drawn out to highly align along the tensile direction. At this stage, single fiber has the same tensile behaviour as the cast films as mentioned above. However, the rupture of individual fibers one by one did not influence the tensile properties of the whole mats, resulting in larger elongation at break for electrospun mats compared with the cast films. At the maximum tensile stress level, most fibers were necked and broken, leading to the final rupture of mats [15].

2.2.2. Thermal properties of biocomposites

Thermal properties of electrospun PVA, PLA and reinforced CNC/PVA, CNC/PLA nanofiber are summarized in Table 2. Maximum thermal degradation temperature (T_{max}) is referring to the decomposition temperature corresponding to the maximum weight loss.

Table 2. Thermal properties of electrospun nanocomposites

Nanocomposites	T_{max} ($_{o}C$)	References
PLA	~370.0	[7]
PLA/CNC	~385.0	
PLA–PHB	335.0	[8]
PLA–PHB–ATBC–CNC1	340.0	
7PVA	~220.0	[15]
7PVA/CNC-20	~275.0	
PLA	351.8	[21]
PLA/CNC	357.5	
MPLA	348.0	[22]
MPLA/CNC	353.6	

In general, Table 2 shows that the maximum temperature of nanocomposites increased with increased CNC content. As studied by Shi *et al.,* [21] and Zhou *et al.,* [22] indicated that the incorporation of CNCs in matrix improved the heat resistance of composites. Huan et al., [15] reveals that the char yield of electrospun PLA/CNC nanocomposites mats increased with increased CNC content. The char yield is directly correlated to the potency of flame retardation for the composites. Char yield is usually defined as the percentage (%) of solid amount you obtain at end of pyrolysis. Increased char formation can limit the production of combustible gases, decrease the exothermicity of the pyrolysis reaction, and inhibit the thermal conductivity of the burning materials. Moreover, the increased crystallinity could also improve the T_{max} of the composites [34]. Therefore, the increase of T_{max} of PLA/CNC nanocomposites mats was mostly attributed to the char formation and increased crystallinity in nanocomposites. According to Dong et al., [26], the influence of CNCs on the thermal transitions of PMMA could be explained by interactions between the hydroxyl groups of CNCs and the carbonyl groups of PMMA. Hydrogen bonding interactions between hydroxyl and carbonyl groups have been widely used to facilitate miscibility of polymer blends, in which hydrogen bonding was found to have a significant effect on the thermal properties of polymer blends. The hydrogen bonding interactions between CNCs and PMMA could also facilitate the dispersion of CNCs in PMMA/DMF solutions.

3. Conclusions

Electrospinning is a simple method to produce nano scale fibers both in laboratory and industrially. Different kinds of artificial and natural polymers have been successfully electro spun into small and fine fibers. In the last decade the number of researches about this method and its applications has increased and this demonstrates the importance of electrospinning. In the review, **it is** shown that CNCs have a distinct advantage for improving mechanical properties of both nanocomposite hydrogels and electrospun nanocomposite fibers/mats. The molecular interactions of CNCs and matrix chains could be beneficial to the mechanical and thermal properties of electrospun nanocomposites. This review also indicates that the nanocellulose has a great potential for application in reinforcing polymers. Even though there is numbers of studies related to nanocomposites containing CNC, there are still several major challenges for the further development of CNC-reinforced nanocomposites fibers. These include surface modification and homogeneous dispersion of CNCs, interface and alignment characterization of CNCs within individual electrospun nanocomposite fiber, analytical model for mechanics of single nanocomposite fiber, and assembly and effect of nanocomposite fibers within mats.

4. References

[1] Dufresne A 2012 Berlin/Boston: Walter de Gruyter GmbH
[2] Isogai A, Saito T and Fukuzumi, H 2011 *Nanoscale* **3** 71-85
[3] Habibi Y, Lucia L A and Rojas O J 2010 *Chem. Rev* **110** 3479
[4] Samir M A, Alloin F and Dufresne A 2005 *Biomacromolecules* 612
[5] Eichhorn S, Dufresne A, Aranguren M, Marcovich N, Capadona J, Rowan S, Weder C, Thielemans W, Roman M and Ren- neckar S 2010 *J. Mater Sci* **45** 1–33
[6] Zhang C, Salick M R, Cordie T M, Ellingham T, Dan Y and Turng L S 2015 *Materials Science and Engineering C* **49** 463
[7] Mujica-Garcia A, Hooshmand S, Skrifvars M, Kenny J M, Oksman K and Peponi L 2016 *RSC Advances* **6** 9221
[8] Arrieta M P, López J, López D, Kenny J M and Peponi L 2016 *Industrial Crops and Products* **93** 290
[9] Cacciotti I, Fortunati E, Puglia D, Kenny J M and Nanni F 2014 *Carbohydrate Polymers* **103** 22
[10] Yee Y Y, Ching Y C, Rozali S, Awanis Hashim N and Singh R 2016 *BioResources* **11** 2269
[11] Li W, Wu Q, Zhao X, Huang Z, Cao J, Li J and Liu S 2014 *Carbohydrate Polymers* **113** 403
[12] Costa L M M, de Olyveira G M, Cherian B M, Leão A L, de Souza S F and Ferreira M 2013*Industrial Crops and Products* **41** 198
[13] López de Dicastillo C, Garrido L, Alvarado N, Romero J, Palma J and Galotto M 2017 *Nanomaterials* **7**106
[14] Gonzalez J S, Ludueña L N, Ponce A, and Alvarez V A 2014 *Materials Science and Engineering C* **34**54
[15] Huan S, Bai L, Cheng W and Han G 2016 *Polymer* **92** 25
[16] Rahman M M, Afrin S and Haque P 2014 *Progress in Biomaterials* **3** 23
[17] Qiu K and Netravali A N 2012 *Composites Science and Technology* **72** 1588
[18] Cho M J and Park B D 2011 *Journal of Industrial and Engineering Chemistry* **17**36
[19] Mandal A and Chakrabarty D 2014 *Journal of Industrial and Engineering Chemistry* **20** 462
[20] Ching Y C, Rahman A, Ching K Y, Sukiman N L and Chuah C H 2015 *BioResources* **10** 3364
[21] Shi Q, Zhou C, Yue Y, Guo W, Wu Y and Wu Q 2012 *Carbohydrate Polymers* **90** 301

[22] Zhou C, Shi Q, Guo W, Terrell L, Qureshi A T, Hayes D J and Wu Q 2013 *ACS Applied Materials and Interfaces* **5** 3847
[23] Chaabouni O and Boufi S 2017 *Carbohydrate Polymers* **156** 64
[24] Chang H, Luo J, Liu H C, Bakhtiary Davijani A A, Wang P H, Lolov G S and Kumar S 2017 *Polymer (United Kingdom)* **122** 332
[25] Poonguzhali R, Basha S K and Kumari V S 2017 *International Journal of Biological Macromolecules* **105** 111
[26] Dong H, Strawhecker K E, Snyder J F, Orlicki J A, Reiner R S and Rudie A W 2012 *Carbohydrate Polymers* **87** 2488
[27] de Oliveira Santos R P, Rodrigues B V M, Santos D M, dos Campana-Filho S P, Ruvolo-Filho A C and Frollini E 2017 *Polymer Testing* **60** 422
[28] Cao X D, Habibi Y, Magalhaes W L E, Rojas O J and Lucia L A 2011 *Current Science* **100** 1172
[29] Ragaišienė A, Rukuižienė Ž, Mikučionienė D and Milašius R 2014 *Fibres and Textiles in Eastern Europe* **22** 59
[30] Ali U, Zhou Y, Wang X and Lin T 2011 *Nanofibers–Production, Properties and Functional Applications*153
[31] Agarwal S, Wendorff J H and Greiner A 2008 *Polymer* **49** 5603
[32] Jiang H, Hu Y, Li Y, Zhao P, Zhu K and Chen W 2005 *Journal of Controlled Release* **108**, 237
[33] Huang Z M, Zhang Y Z, Kotaki M and Ramakrishna S 2003 *Composites Science and Technology* **63** 2223
[34] Kumar A P, Depan D, Tomer N S and Singh R P 2009 *Progress in Polymer Science* **34** 479

Acknowledgments
The authors would like to acknowledge Ministry of Higher Education Malaysia for funding through Fundamental Research Grant Scheme

Removal of cadmium using electrospun nanofibers

Nur Shahira Wahida Zulkefeli, Soo Kar Weng and Nurul Syazana Abdul Halim*
Faculty of Earth Science, Universiti Malaysia Kelantan, 17600 Jeli, Kelantan, Malaysia.

*Corresponding author email address: syazana@umk.edu.my

Abstract. The present study reports on the capability of CTA/ Aliquat 336 electrospun fibers in extracting Cd(II) from aqueous phase. The CTA/Aliquat 336 electrospun fibers were produced by electrospun a polymer solution containing CTA and Aliquat 336 in DMC and methanol with a ratio of 8:2 respectively. The effect of different Aliquat 336 concentration on the performance of Cd(II) removal was examined. Extraction experiment was conducted in a batch mode using 5 ppm of Cd(II) in 1 M HCl solution. Results showed that the extraction of Cd(II) increased as Aliquat 336 concentration increased. Morphology structure of the CTA/ Aliquat 336 electrospun fibers were also carried out using scanning electron microscopy (SEM).

Keywords: electrospun fibers, metal ion extraction, electrospinning method, Aliquat 336.

1. Introduction

Heavy metals remain in environment because they are non-biodegradable. Zinc, cadmium and iron are usually used in battery manufacturing (1) and pigment use in paint (2). However, cadmium is one the most toxic heavy metal which can cause kidney damage, renal disorder and also a human carcinogen (3). Conventionally, there are many treatments used for heavy metals removal including membrane filtration (4), adsorbent (5) and solvent-solvent extraction.

Polymer inclusion membranes (PIMs) has received a lot of attention in metal ions extraction due to high stability compared to other liquid membrane processes and more environmental friendly compared to solvent extraction method (6). Basically, PIMs is made from polymer, carrier and/or plasticizer. The polymer is a base layer which provides mechanical strength to PIMs while carrier is acting like a guest specific host which control the selectivity of components that can pass through the film. On the other hand, plasticizer can improves compatibility between polymer and carrier and also enhances flexibility of the membrane. Aliquat 336 is one of the most used carriers in PIMs. Many studies have reported on the successful of metal ions removal using PIMs incorporated with Aliquat 336 (7, 8, 9). However, the amount of Aliquat 336 in PIMs need to reach certain threshold in order to extract metal ions. For example, (10) stated that extraction of metallic ions only occurred when the proportion of Aliquat 336 in PIMs is 30 wt.% or higher. The critical Aliquat 336 content in PIMs was also confirmed by other researcher (11, 12).

Electrospinning method has become popular in producing nanofibers with simple, effective and versatile. This technique is known since 1930's for the interest in application of nanotechnology and biotechnology. In recent years, electrospun fibers have successfully developed for metal ions applications (10, 13, 14). This is due to the higher surface area and less amount of carrier used compare to PIMs. Similar to the polymer solution used in PIMs, the solution is electrospun using electrospinning equipment which consists of high voltage supplier, metal collector and syringe pump. Due to high voltage, the polymer solution was ejected from the syringe needle onto a metal collector producing fibrous mat which can be collected after few hours. (10) reported that the extraction of Cd(II) has improved from 74 % to 98 % of Cd(II) removal when used PVC/Aliquat 336 PIMs containing 40 wt.% of carrier and PVC/Aliquat 336 electrospun fibers containing 25 %wt.

of carrier respectively. (15) Was also reported on the improvement of Cd(II) removal when used PVC electrospun fibers compared to PVC PIM.

In this study, electrsopun fibers made of cellulose triacetate (CTA) incorporated with Aliquat 336 as a carrier was produced. The electrospun fibers were produced at different Aliquat 336 concentration in order to investigate the extraction rate of Cd(II) ions. This study provides insight for further development in CTA/Aliquat 336 electrospun fibers in extracting metal ions.

2. Experimental

2.1 Preparation of polymer solution

For this study, CTA and Aliquat 336 solutions were prepared with different Aliquat 336 concentrations (0, 2, 4, and 6wt%). A total mixture of CTA and Aliquat 336 of 600 mg was dissolved in methanol and dichloromethane in a ratio of 8:2 respectively. The solution was stir vigorously until homogenous. Then, the solution was transferred into media bottle and left until all bubble are removed.

2.2 Preparation of CTA/Aliquat 336 electrospun fibers

CTA/ Aliquat 336 electrospun fibers was prepared by electrospinning method. The polymer solution was placed into a 5 mL plastic syringe that attached with 23-gauge needle tip. The polymer solution was electrospun at 20 kV with 1 mL/h of flow rate for 8 hours. The distance between syringe tip and aluminium collector was 140 mm. Later, the CTA/Aliquat 336 electrospun fibers was collected from the aluminium collector and used in Cd(II) ions extraction.

2.3 Cadmium extraction protocol for CTA / Aliquat 336 electrospun fibers

The extraction study was conducted in batch mode. The CTA/ Aliquat 336 electrospun fibers weighing of 0.05 ± 0.01 g was cut into smaller pieces and placed in a beaker containing 250 mL extraction solution. The extraction solution contained a 5 ppm of Cd(II) in 1 M of hydrochloric acid (HCl). The extraction solution was stirred during the experiment and sample of aliquots was taken at 0, 5, 30, 60, 180, 300 and 1440 minutes for Cd(II) analysis using atomic absorption spectroscopy (AAS). Then, the sample was diluted according to dilution protocol prior AAS analysis. The extraction rate of Cd(II) was calculate by using below equation:

$$Extraction\ efficiency, (\%) = \frac{Ci-Co}{Co} X\ 100\% \tag{A.1}$$

C_i = initial concentration

C_o = final concentration

2.4 Morphology surface analysis

The surface morphology of electrospun fibers at different Aliquat 336 concentration was observed by using scanning electron microscopy (SEM) (JSM-IT100, Jeol USA). Square sections of samples approximately 10 mm by 10 mm were mounted on aluminium stubs. The power used is 10 kV with magnification of times 10, 50 and 100.

3. Results and discussion

3.1 Extraction of cadmium by CTA/ Aliquat 336 electrospun fibers

The removal of Cd(II) ions in 1 M HCl by CTA/ Aliquat 336 elctrospun fibers at different Aliquat 336 concentration (0, 2, 4 and 6wt%) was showed in Figure 1. The extraction rate increased with increasing of Aliquat 336 content. At the highest Aliquat 336 content (6wt%) the Cd(II) removal was 84 % removed. This result showed that the CTA/Aliquat 336 electrospun fibers has a capability to remove Cd(II) ions at lower Aliquat 336 content compared to PIMs using similar carrier. However, CTA electrospun fibers without Aliquat 336 content show insignificant removal of Cd(II) ions.

The transport of Cd(II) ions is based on the ion exchange mechanism which involve the formation of carrier-complex in the aqueous solution. In a chloride solutions, different chloro complexes can be formed depending on metal and chloride concentration. According to (16), trichloro complex is a predominant species form when the chloride concentration is above 0.1 M HCl. Thus, the extraction of Cd(II) can be described as:

$$CdCl_3^- + CH_3(C_8H_{17})_3N^+Cl^- \leftrightarrow CH_3(C_8H_{17})_3N^+CdCl_3^- + Cl^-$$

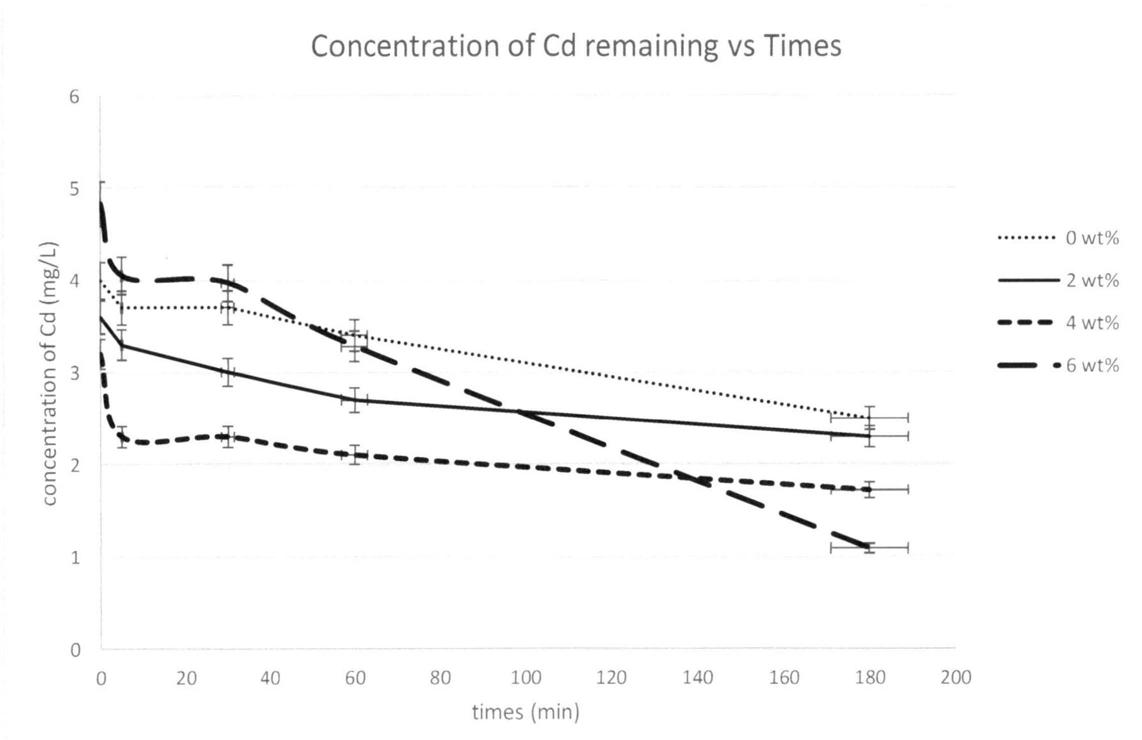

Figure 1: The extraction of Cd(II) against Aliquat 336 content in CTA/Aliquat 336 electrospun fibers.

3.2 Membrane surface morphology

Figure 2, 3, and 4 show the surface morphology of CTA/Aliquat 336 electrospun fibers at Aliquat 336 content of 0, 4 and 6 wt% respectively. A fibrous web like structure could be seen in all samples showing that electrospinning method has successfully produced fibrous mats. The surface morphology of CTA electrospun fibers without Aliquat 336 content revealed a compact and uniform structure of fibers (Figure 2). As Aliquat 336 content increased to 4wt% the diameter of fibers decreased from 1.60 μm (without Aliquat 336) to 0.30 μm (Figure 3). It can be seen that as Aliquat 336 content further increased to 6wt% the fibers diameter increased to 1.06 μm (Figure 4). Besides, the number and size of beads also increased with the increasing of Aliquat 336 concentration. The formation of beads is probably due to either instability in parameter during electrospinning process or during the increasing in concentration of Aliquat 336 (17). However, the structure of nanofibers become dense as the concentration of Aliquat 336 increased.

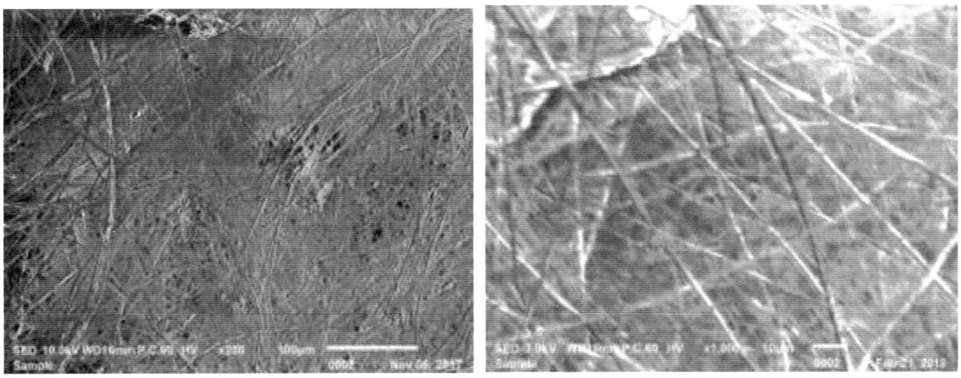

Figure 2: electrospun fiber of 0wt% under x200 and x1000 magnification

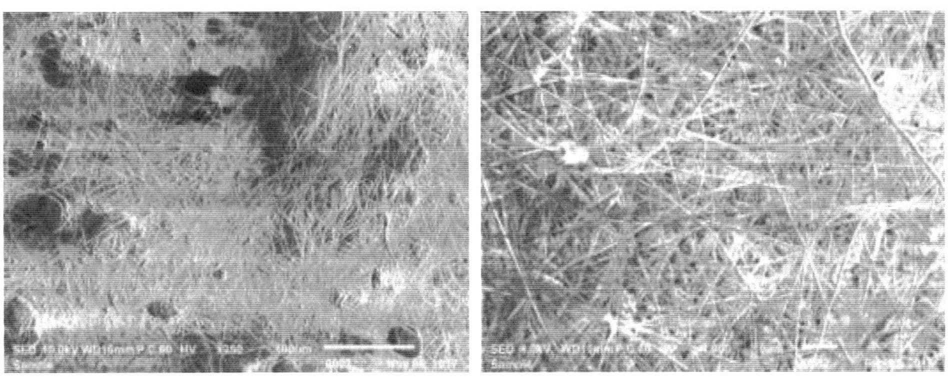

Figure 3: electrospun fiber of 4wt% under x200 and x1000 magnification

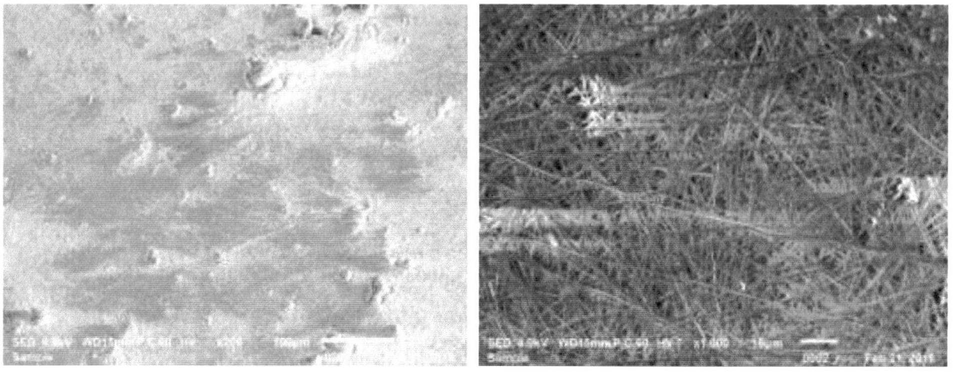

Figure 4: electrospun fiber of 6wt% under x200 and x1000 magnification

4. Conclusion

The CTA/ Aliquat 336 electrospun fibers have been produced in order to removed Cd(II) ions from aqueous solution. The extraction rate of Cd(II) was significantly change after the addition of Aliquat 336. From the result, the CTA/ Aliquat 336 electrospun fibers containing 6 wt% of Aliquat 336 showed the highest removal of Cd(II) ions. It is expected that the percentage removal will increase to more than 90 % if the Aliquat 336 content is further increased. Further work is underway to investigate the optimum of Aliquat 336 concentration, voltage, flow rate and distance from tip distance to metal collector on the performance of CTA/ Aliquat 336 electrospun fibers in extracting different type of metallic ions.

Acknowledgement

This research was funded by the research grant R/FRGS/A08.00/00453A/001/2015/000285.

5. References

1. Mohamed M, Abdul Aziz MS, Abdullah MZ, Ahmed A, Abdul Razab MKA, Khor CY. Experimental Method on Solder Joint of Ball Grid Array Using Reflow Oven. Int J Curr Sci Eng Technol. 2018;487–92.
2. Zulkefeli NSW, Ismail WMIW, Razab MKAA, Masri MN. Effect of Cathode Waste Material Concentration in Conductive Paint Coating. J Trop Resour Sustain Sci. 2016;4(December):78–81.
3. Barakat MA. New trends in removing heavy metals from industrial wastewater. Arab J Chem. 2011;4(4):361–77.

4. Fu F, Wang Q. Removal of heavy metal ions from wastewaters: A review. J Environ Manage. 2011;92(3):407–18.

5. Zhao X, Liu C. Efficient removal of heavy metal ions based on the optimized dissolution-diffusion-flow forward osmosis process. Chem Eng J. 2018;334(November 2017):1128–34.

6. Shahira N, Zulkefeli W, Weng SK, Syazana N, Halim A. Removal of Heavy Metals by Polymer Inclusion Membranes. 2018;

7. Wong CY, Wong WY, Loh KS, Mohamad AB. Study of the plasticising effect on polymer and its development in fuel cell application. Renew Sustain Energy Rev. 2017;79(September 2016):794–805.

8. Manzak A, Yıldız Y, Tutkun O. Characterization of polymer inclusion membrane containing Aliquat 336 as a carrier. Membr Water Treat [Internet]. 2015;6(2):95–102.

9. Vázquez MI, Romero V, Fontàs C, Anticó E, Benavente J. Polymer inclusion membranes (PIMs) with the ionic liquid (IL) Aliquat 336 as extractant: Effect of base polymer and IL concentration on their physical-chemical and elastic characteristics. J Memb Sci. 2014;

10. Abdul-Halim NS, Whitten PG, Nghiem LD. Characteristics and cadmium extraction performance of PVC/Aliquat 336 electrospun fibres in comparison with polymer inclusion membranes. Sep Sci Technol. 2016;6395(April):1–8.

11. Wang L, Paimin R, Cattrall RW, Shen W, Kolev SD. The extraction of cadmium (II) and copper (II) from hydrochloric acid solutions using an Aliquat 336 / PVC membrane. 2000;176:105–11.

12. Xu J, Wang L, Shen W, Paimin R, Wang X. The Influence of the Interior Structure of Aliquat 336/PVC Membranes to their Extraction Behavior. 2010.

13. Huang Y, Miao YE, Liu T. Electrospun fibrous membranes for efficient heavy metal removal. J Appl Polym Sci. 2014;131(19):1–12.

14. Aliabadi M, Irani M, Ismaeili J, Piri H, Parnian MJ. Electrospun nanofiber membrane of PEO/Chitosan for the adsorption of nickel, cadmium, lead and copper ions from aqueous solution. Chem Eng J. 2013;220:237–43.

15. Wong L, Shen W, Truong YB. Investigation of Electrospun and Film-Cast PVC Membranes Incorporated with Aliquat 336 for Efficient Cd Extraction : A Comparative Study. 2011;336.

16. He D, Liu X, Ma M. Transfer of Cd(II) chloride species by a tri-n-octylamine-secondary octyl alcohol-kerosene multimembrane hybrid system. Solvent Extr Ion Exch. 2004;22(3):491–510.

17. Bhardwaj N, Kundu SC. Electrospinning: A fascinating fiber fabrication technique. Biotechnol Adv. 2010;28(3):325–47.

IOP Conf. Series: Materials Science and Engineering **440** (2018) 012013 doi:10.1088/1757-899X/440/1/012013

Evaluation on adsorption isotherms of alizarin red S dye removal by nickel/aluminium layered double hydroxide

M Mamat[1,*], N Roslan[2], K H K Bulat[2], M A A Abdullah[1] and A M Jaafar[3]

[1]Advanced Nanomaterials Research Group, School of Fundamental Science, Universiti Malaysia Terengganu, 21030 Kuala Terengganu, Malaysia
[2] School of Fundamental Science, Universiti Malaysia Terengganu, 21030 Kuala Terengganu, Malaysia
[3]Chemistry Department, Faculty of Science, Universiti Putra Malaysia, 43400 UPM Serdang, Selangor.

Corrsponding author email:

Abstract. The textile industry is one of the swiftest thriving industries contributing to the economic growth of Malaysia, particularly in east coast peninsular state. As the popularity of industry increases, the water usage was found to be elevated and eventually causes high expel rate of wastewater with excessive dye pollutants. In this study, anionic clay of nickel/aluminium layered double hydroxide (NiAL), was chosen to be studied as a potential adsorbent for anionic dye, namely Alizarin Red S (AR). The NiAL was synthesised *via* co-precipitation method and were characterized using powder X-Ray diffraction (PXRD) and Fourier Transform Infrared spectrophotometer (FTIR). NiAL was then used as adsorbent for the removal of AR dye in aqueous solution which was tested at different NiAL dosages. The experimental data were analysed using adsorption isotherm models and the AR adsorption was found to obey the Langmuir model.

1. Introduction

The textile and clothing industries is a vital part of the manufacturing production, trade and employment in numerous evolving countries. It is one of the industry that helps provide economic growth in Malaysia. Handmade textile industry is very popular in the East Coast of Peninsular Malaysia and Sarawak region. This industry is highly commercialized and contributes positively to the economic growth for some of the states, mainly Kelantan and Terengganu. However, this industry has increased water usage and eventually generate exorbitant expel rate of wastewater with excessive amount of pollutants. During textile dyeing and finishing process, dyes residues are released as the main water pollutant [1]. A numerous amount of these dyes have been identified as toxic substances or sometimes even carcinogenic [2]. The existence of dyes in water causes damage to the environment especially aquatic life as it retards photosynthetic activity and causes disturbance in the re-oxygenation capacity because it blocks sunlight, which cause anaerobic condition that restricts the aquatic plants' growth.
Also, there could be contamination in groundwater and surface water that makes the water ill-suited for other uses as it may cause mutagenic effects [3].

There has been multiple treatment techniques such as solvent extraction, chemical precipitation, coagulation, and adsorption being used as metal ions remover from aqueous solutions. Due to economic limitation, cost efficient and clean processes is preferred, for example, electrolysis, solvent extraction, ion exchange, and reverse osmosis. But these methods have a certain number of disadvantages, such as high energy requirement, incomplete removal, and high operational cost. Therefore, it has been proven that the most effective method is adsorption.

Anionic clay known as layered double hydroxides (LDH) attract significant attention from the academia and industries. LDH have also attracted attention on several fields of study, such as biomedical uses, polymer nanocomposites, water treatment, and catalysis [4]. LDH have relatively weak interlayer bonding which allows it to showcase outstanding ability to trap organic and inorganic anions [5]. It also have derived attention because of its great ionic exchange capacities and can be used as an efficient adsorbent for the removal of an assortment of anions from aqueous solution as it has anions in the interlayer that can be exchanged by other anions [6]. Hence, it is possible to use the LDH as an adsorbent for anionic dyes in wastewater. Previous researches reported the utilization of LDH as an adsorbent for dyes including eriochrome black T [4], acid red 27, direct red 23, disperse red 9 and methylene blue [7].

In this study nickel/aluminium layered double hydroxide (NiAL) is investigated as an adsorbent for the removal of Alizarin Red S (AR) dye from aqueous solution. AR, also known as sodium alizarinesulfonate which has a molecular formula of $C_{14}H_7NaO_7S$ (Figure 1) is a type of anthraquinone dye that is commonly used in the textile industries to produce a deep red colouring. It could not be degraded completely *via* general biological and physicochemical processes due to its complex structures of aromatic rings that can sustain high thermal, biological and optical stability [8]. The removal of AR dye from water were previously reported by using activated carbon from *Achyranthes aspera* plant [8], activated carbon/γ-Fe_2O_3 nano-composite [9], alumina [10], and photocatalyst ZnO and TiO_2 [11]. However to best authors knowledge, there is no study on the adsorption of AR by LDH.

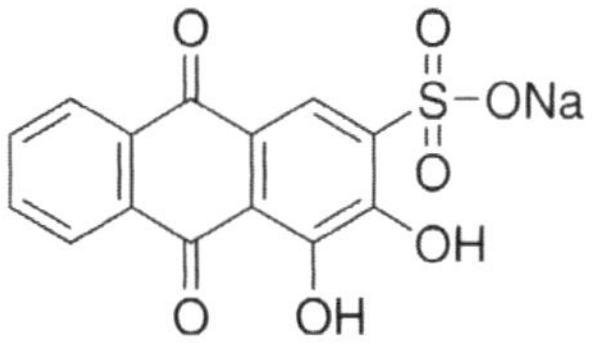

Figure 1. Structural formula of Alizarin Red S

2. Materials and methods

All chemicals were used as purchased without further purification. Nickel(II) nitrate hexahydrate ($Ni(NO_3)_2.6H_2O$, Merck), aluminium nitrate nonahydrate ($Al(NO_3)_3{\cdot}9H_2O$, Merck), sodium hydroxide (NaOH, Merck), alizarin red S ($C_{14}H_7NaO_7S$, R&M).

A direct synthesis method, which is the co-precipitation method, was used to synthesis the NiAL with the Ni/Al molar ratio of 2:1. The synthesis process was conducted under constant flow of nitrogen gas while the pH of the solution was adjusted to 6 by dropwise addition of 2M NaOH solution into the mother liquor. Then the solution was aged in oil bath for 18 hours and the resultant slurry was washed with distilled water before being filtered, and dried in an oven at 70°C. The NiAL synthesized were then grinded and characterized by using Powder X-Ray Diffractometer (PXRD), Rigaku Miniflex II, and Fourier Transform Infra-Red (FTIR) spectrophotometer, Perkin Elmer Precisely, Spectrum 100 in the range of 400 to 4000 cm^{-1}. The adsorption study was conducted at different amount of NiAL (0.05 – 0.5 g) which were placed in contact with 100 mL of 40 ppm AR dye solution for 60 minutes. The remaining dye concentration was determined at $\lambda = 260$ nm using UV-Vis spectrophotometer, Shidmadzu UV1800.

3. Results and Discussion

X-ray diffractogram of NiAL (Figure 2) shows typical peaks of layered double hydroxide with poor crystallinity which is similar to the one reported by previous study [12]. The broad reflections of the planes also indicate that amorphous phase might be present in sample [13]. However the basal spacing value for (003) plane (d_{003}) which twice the basal spacing of (006) plane (d_{006}) showing that the materials comprise of layered structure. The d_{003} of NiAL is 0.90 nm, indicates the thickness of a single layer of the NiAL and the charge balancing interlayer anion [14] which is nitrate.

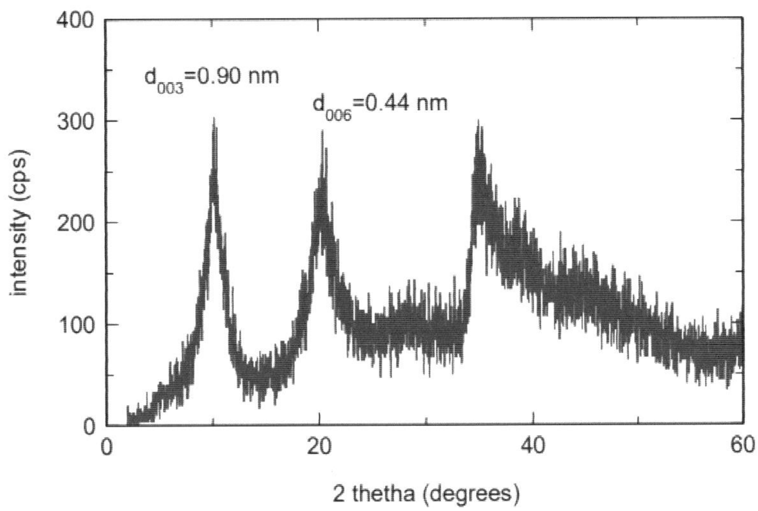

Figure 2. PXRD pattern of NiAL

A broad band was observed in the FTIR spectrum of NiAL (Figure 3) centred at around 3400 cm^{-1} is assigned to OH stretching due to the presence of hydroxyl groups in NiAL layers and interlayer as well as adsorbed water molecules [15]. A peak at approximately 1627 cm^{-1} is attributed to the HOH deformation (δ_{H-O-H}), thus confirming the presence of water molecules in the interlayer region of NiAL [16]. The presence of the nitrate anion located in the interlayer region of NiAL is shown by the appearance of a sharp peak at 1384 cm^{-1}, attributed to the bending vibration of N-O. The peaks within the region lower than 1000 cm^{-1} are assigned to the metal-oxygen vibrations (M-O, O-M-O, M-O-M) within the layers of NiAL.

The ability of NiAL as an adsorbent for the removal of dye was tested using anionic dye of alizarin red S (AR) at different dosages of NiAL. As the mass of NiAL increases it provides greater number of vacant sites, which contributes to the increase of the number of binding site and surface area [17] for dye adsorption hence the removal percentage of AR increases (Figure 4). The removal percentage continues to increase with the increase of NiAL mass but the graph line starts to plateau when it reaches optimum mass at 0.25 g with 73% of AR was removed and no significant changes can be observed beyond this mass. This may be due to the overloading of adsorbent particles hence the binding sites available in NiAL were overlapped [18] and not accessible for dye molecules.

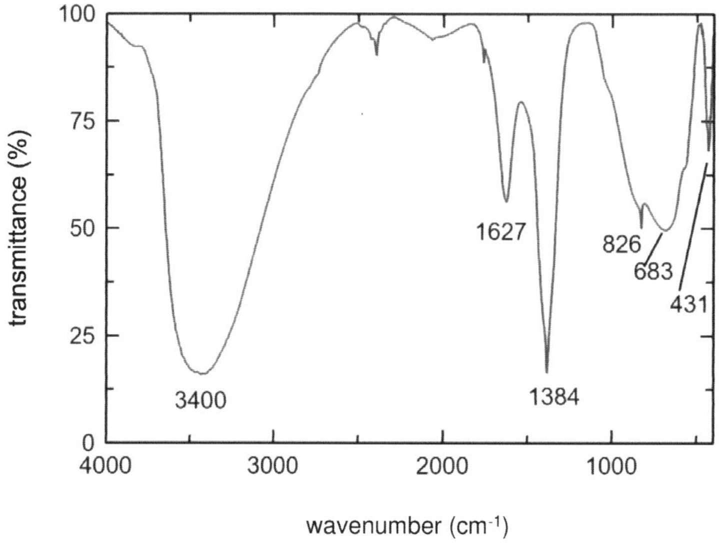

Figure 3. FTIR spectrum of the NiAL

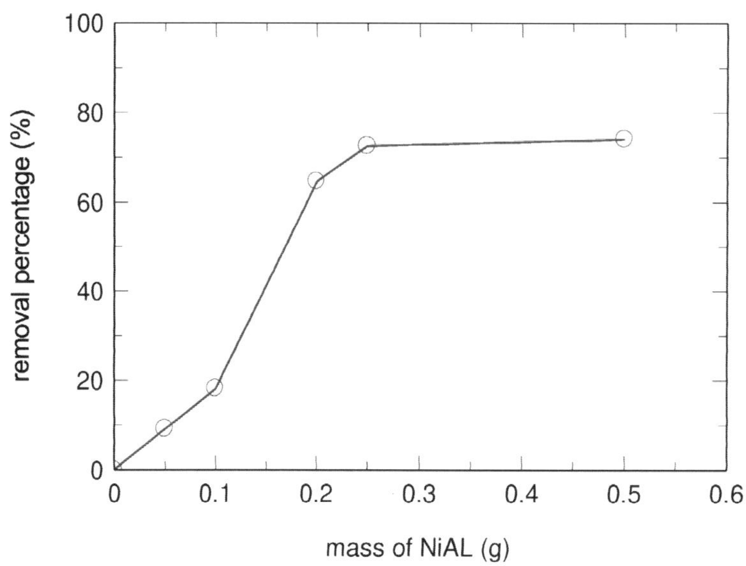

Figure 4. Removal percentage of AR against mass of NiAL

The data were then fitted into two widely used adsorption isotherm models namely Langmuir and Freundlich. The Langmuir isotherm model presumes monolayer adsorption in which the adsorbed layer only one molecule in thickness with no steric hindrance and lateral interaction between the adsorbed molecules [19] and can be presented as shown in Equation 1 [20]:

$$q_e = \frac{q_{mK_L C_e}}{1 + K_L C_e} \tag{1}$$

Where, q_m is the maximal adsorption capacity and K_L is the constant related to the free energy of the adsorption.

Meanwhile, Freundlich isotherm describes the reversible and non-ideal adsorption, which can be used for multilayer adsorption, with non-uniform dispersal of adsorption heat and affinities for the heterogeneous surface [21]. This model can be represented as empirical Equation 2 [20]:

$$q_e = K_F C_e^{1/n} \tag{2}$$

Where, K_F and n are Freundlich constants, respectively are related to the adsorption capacity and the adsorption intensity.

Both Langmuir and Freundlich linearized equations were used in this study to evaluate the suitability of the model in describing the adsorption mechanism. The equations are expressed as equations (3) and (4), for Langmuir [7] and Freundlich [22], respectively.

$$\frac{C_e}{q_e} = \frac{1}{q_m K_L} + \frac{C_e}{q_m} \tag{3}$$

$$\log q_e = \log K_F + \left(\frac{1}{n}\right) logC_e \tag{4}$$

Based on the correlation coefficient (R^2) for the linearized plot of both models (Figure 5), the adsorption of AR onto NiAL can be better explained by the Langmuir model as the R^2 value is higher (0.939) than Freundlich model (0.102). This indicates that the adsorption was made onto a homogenous adsorbent surface and the formation of monolayer coverage of AR dye onto the NiAL. Subsequent, no adsorption site available for other AR dye molecules once the site is occupied with an AR molecule. The adsorption process favours Langmuir model, may due to the affinity of NiAL surface towards AR dye is low thus the it can only capable to hold a layer of dye molecules.

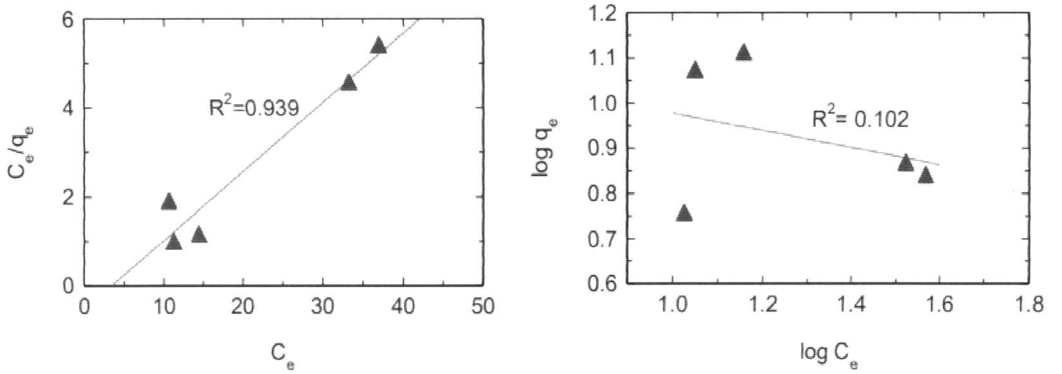

Figure 5. Fitting the data into Langmuir (a) and Freundlich (b) isotherms

4. Conclusion

The NiAL was successfully synthesized via co-precipitation method and was used as an adsorbent for the AR dye in aqueous solution at different amount of NiAL with the optimum mass of 0.25 g. The experimental data is fitted well with Langmuir isotherm model showing that the process is govern by homogeneous adsorption.

5. References

[1] Pang Y L & Abdullah A Z 2013 *Clean - Soil, Air, Water* **41** (8) 751
[2] Dave P N, Kaur S & Khosla E 2011 *Indian Journal of Chemical Technology* **18** 53
[3] Luna M D, Flores E D, Genuino D A, Futalan C M & Wan M 2013 *Journal of the Taiwan Institute of Chemical Engineers* **44** (4) 646
[4] Zubair M, Jarrah N, Manzar M S, Al-Harthi M, Daud M, Mu'Azu N D & Haladu S A 2017 *Journal of Molecular Liquids* **230** 344
[5] Goh K H, Lim T T & Dong Z 2008 *Water Research* **42** (6–7) 1343
[6] Yanming S, Dongbin L, Shifeng L, Lihui F & Shuai C 2017 *Arabian Journal of Chemistry* **10** S2295
[7] Bu R, Chen F, Li J, Li W & Yang F 2016 *Colloids and Surfaces A: Physicochemical and Engineering Aspects* **511** 312
[8] Sushita R & Ravindhranath K 2016 *Der Pharma Chemica,* **8** (9) 63
[9] Fayazi M, Ghanei-Motlagh M & Taher M A 2015 *Materials Science in Semiconductor Processing* **40** 35
[10] Rehman, R, Mahmud T, Anwar J, Salman M, Shafique U, Waheed-Uz-Zaman & Ali F 2011 *Journal Chemical Society of Pakistan* **33** (2) 228
[11] Joshi K M & Shrivastava V S 2011 *International Journal of Environmental Sciences* **2** (1) 8
[12] Mamat, M, Tagg T, Khairul W M, Abdullah M A A, Mohd Tahir N, Jubri Z & As'ari R A 2014 *Applied Mechanics and Materials* **563** 94
[13] Hibino T & Ohya H 2009 *Applied Clay Science* **45** 123
[14] Olfs H-W, Torres-Dorante L-O, Eckelt R and Kosslick H 2009 *Applied Clay Science* **43** 459
[15] Mamat M, Kusrini E, Yahaya A H, Hussein M Z & Zainal Z 2013 *International Journal of Technology* **1** 73
[16] Ferreira O P, de Moraes S G, Durán N, Cornejo L and Alves O L 2006 *Chemosphere* **62** 80
[17] As'ari R, Mamat M, Abdullah M A A & Mohamed Zuki H 2015 *Journal of Applied Sciences Research* **11** (24) 22
[18] Mary P P, Subraman M & Jeevitha 2015 *World Journal of Pharmaceutical Research* **4** (5) 1246
[19] Vijayaraghavan K, Padmesh T V N, Palanivelu K, Velan M 2006 *J. Hazard. Mater. B* **133** 304
[20] Liu Q-S, Zheng T, Wang P, Jiang J-P & Li N 2010 *Chemical Engineering Journal* **157** (2–3) 348
[21] Adamson A W, Gast A P 1997 Physical Chemistry of Surfaces sixth ed. *Wiley Interscience* New York
[22] Mittal A Kaur D & Mittal J 2009 *Journal of Hazardous Materials* **163** 568

Acknowledgements

The authors are grateful to the Malaysian government for the funding (FRGS 59179) and Universiti Malaysia Terengganu for providing research facilities.

IOP Conf. Series: Materials Science and Engineering **440** (2018) 012014 doi:10.1088/1757-899X/440/1/012014

Potential of Polysulfone/Polydimethyl Siloxane Thin Film Composite (PSf-PDMS-TFC) Membrane for CO_2/N_2 Gas Separation

Sunarti Abd Rahman[1*], Nur Atikah Che Embi[1], Abdul Latif Ahmad[2] and Hasrinah Hasbullah[3]

[1]Faculty of Chemical & Natural Resources Engineering, Universiti Malaysia Pahang, 26300 Gambang, Pahang, Malaysia.
[2]School of Chemical Engineering, Engineering Campus, Universiti Sains Malaysia, 14300 Nibong Tebal, Pulau Pinang, Malaysia.
[3]Advanced Membrane Technology Research Centre (AMTEC), Universiti Teknologi Malaysia, 81310 Skudai, Johor Darul Ta'zim, Malaysia.

*Corresponding author email: sunarti@ump.edu.my

Abstract. The capture and storage of carbon dioxide has been identified as one potential solution to greenhouse gas driven climate change. Efficient separation technologies are required for removal of carbon dioxide from flue gas streams to allow this solution to be widely implemented. This study is mainly focusing on the effect of different concentration of polydimethyl siloxane (PDMS) in dip-coating solution on the membrane's performance. The asymmetric thin flat sheet membrane was prepared by dry/ wet phase inversion process consisting 20 %w/v of polysulfone (PSf) as the support layer polymer and 80 %w/v of N- methyl-2-pyrrolidone (NMP) as the solvent. PDMS was coated on the support PSf membrane with the composition of 10, 15 and 20 wt% of PSf in n-hexane respectively. The characterization of morphology of TFC membrane will be conducted by using Scanning Electron Microscopy (SEM) and Fourier Transform Infrared (FTIR). The membrane's performance and the selectivity of CO_2/N_2 separation will be determined by conducting gas permeation test. The result obtained, show that membrane with highest concentration of PDMS in dip-coating solution give a highest performance in selectivity and unfortunately it contributes to lower permeability. It is vice versa from the membrane without PDMS in the top layer which gives highest value of permeability but lowest in selectivity. From the characterization and permeation test of the membrane, hereby the membrane with highest percentage of PDMS should be selected for the future development of membrane due to its highest value of selectivity which contributes to highest efficiency in separating the gas.

1. Introduction

Enhancements of the greenhouse effect in the worldwide have contributed to the study that can recover this problem. The most abundant greenhouse gas CO_2 has risen from preindustrial levels of 280 parts per million (ppm) to present levels of over 365 ppm [1]. Carbon dioxide, CO_2 arise from different emission sources, particularly flue gas from power stations, steel works and chemical industries. According to the study done by Mitsubishi Heavy Industries, Ltd., 98% of CO_2 emissions are derived from fossil fuel combustion, exhausted to atmosphere by flue gas stacks from furnace boilers, gas turbines and engine emission. The content of the flue gases that released at atmospheric pressure usually containing Oxygen, Carbon Dioxide, Nitrogen, NO_x, SO_x and dust particulate impurities.

The separation of gas by thin barrier termed as membranes is a dynamic and rapidly growing field, and it has been proven to be technically and economically superior to other emerging technology. Materials for effective separation of gases can follow one of two overall strategies; increasing the rate of diffusion of gases through the polymeric structure and increasing the solubility of gases through the polymeric structure. Unfortunately, there is frequently a trade-off between selectivity and permeability [2]. In preparation of gas and vapour separating membranes, a polymer should have the following intrinsic properties:(1) High selectivity for the components to be separated, (2) High permeability for the permeating components and (3) High life expectance under operating condition such as good thermal, chemical and mechanical stability. In this particular study, PSf and PDMS were selected as a polymer in developing support membrane and thin film due to their great characteristic such as a tough and rigid polymer, high-temperature properties and chemically inert.

2. Methodology

2.1 Chemicals

PSf pellet purchased from Sigma Aldrich (M) Sdn. Bhd. PSf used is without any modification. Anhydrous NMP supplied from Sigma Aldrich (M) Sdn. Bhd was used as a solvent. Water is used as a coagulation medium. CO_2 and N_2 used in permeation test.

2.2 Membranes Preparation

With the amount of 20% PSf pellet, it is been added gradually into beaker contains 80% of NMP solvent and mixed by mechanical stirrer. The mixing step continues for at least 7 to 8 hours till homogenous and clear solution was formed. The dope solution was prepared within a range of temperature between 60°C to 80°C.

2.3 Dope Solution Casting Process

The dope solution was casted on a glass plate by using a glass rod. It is followed by the immersion in a water bath in accordance of dry/ wet phase inversion process. Casting process was conducted at room temperature (30°C±2). The cast membrane was immersed into water to initiate delamination and slightly peel the membrane film from the glass plate. After that the membrane is transferred and immersed into the coagulation medium with water act as a coagulation medium for 1 hour and further dried in the air at room temperature.

2.4 Thin Film Composite Membrane Preparation

The dip-coating solution is prepared by mixing PDMS with n-hexane solvent with stirring. The dip-coating solution is consisting of three different percentages of PDMS which were 10%, 15% and 20% in n-hexane solvent. The prepared asymmetric PSf membranes were placed on a glass plate. The coating solution is placed on the PSf membrane by dropped the PDMS/n-hexane solution on the skin layer of the membrane. PSf membrane is allowed to dry for 5 minutes and cure process is carried out in oven at 60°C for 1 hour.

2.5 Gas Permeation Test

The thin film composite of PSf/ PDMS membrane is tested on pure gas CO_2 and N_2. The pure gas will be tested at pressure in the range of 0.5 bar to 1.5 bar. The time of the first bubble formed and volume difference of the soap water by each membrane was taken.

The gas permeation rate (P) was calculated by using equation as followed:

$$P = \frac{Vl}{At\Delta p}(cm^3(STP)cm)/(cm^2s.cmHg) \quad (1)$$

The permeance (P/l) was calculated using equation as followed:

$$\frac{P}{l} = \frac{V}{At\Delta p}(cm^3(STP)cm)/(cm^2 s.cmHg) \qquad (2)$$

where, V = the volume (cm^3) displaced in time () A = the effective membrane area (cm^2) Δp=The transmembrane pressure expressed in cmHg.

Meanwhile, Selectivity was calculated using this equation:

$$\alpha = P_i/P_j \qquad (3)$$

P_i = Permeance of one gas component
P_j = Permeance of another gas component

2.6 Membrane Characterization (SEM and FTIR)
The prepared membrane is characterized using two methods which are by using Scanning Electron Microscopy (SEM) and Fourier Transform Infrared Radiation (FTIR). There are several parameters that will be evaluated from this study which were morphological studies and effect of concentration of polymer.

3. Results and Discussion

3.1 Effect of Dip Coating Percentage on the Membrane Morphology
Structure of the surface membranes produced from 0%, 10%, 15% and 20% of PDMS are illustrated in Figure 1 (a), (c), (e) and (g) respectively. As shown in the surface structure of the membrane, as the amount of percentage of PDMS increased, a fewer number of macro-voids in the membrane structure are observed. Figures 1(b) illustrate that membrane without dip coating solution figured as finger-like macro-voids. The finger-like macro-voids ends at the top surface of the skin layer, making the surface defective with pin-holes [4]. The enhancement of the PDMS illustrate in Figure 1 (d), (f) and (h) shows that the sub-layer has a nodular structure, diminishing the chance of pin-hole formation at the membrane surface.
In the sequence of Figure 1 (d), (f) and (h), a new selective layer was formed and covered the entire surface pore in order 10%, 15% and 20% of PDMS percentage respectively. The presence of 10% PDMS in the dip coating solution has no obvious effect on the cross section structure of the membranes. Proceeding in increasing the percentage of PDMS results in higher polymer concentration in the upper layer of the membrane forming films and in a denser coating layer with smaller pores in the final membranes.

IOP Conf. Series: Materials Science and Engineering **440** (2018) 012014 doi:10.1088/1757-899X/440/1/012014

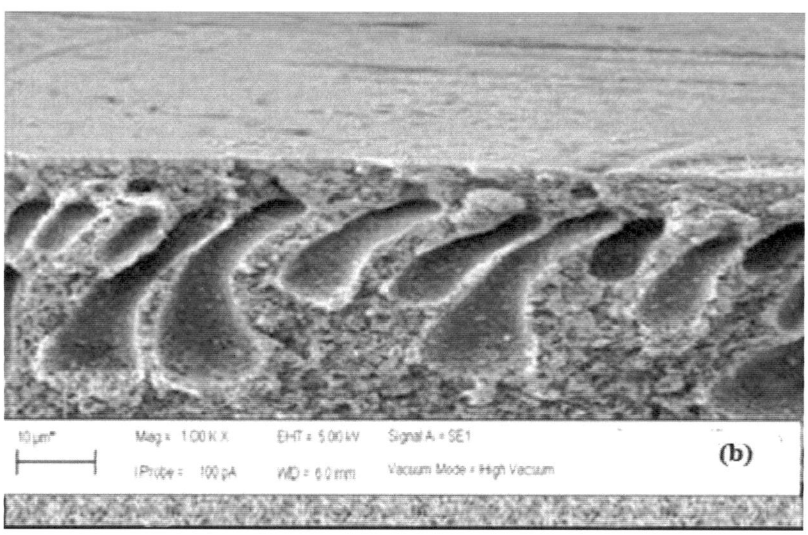

IOP Conf. Series: Materials Science and Engineering **440** (2018) 012014 doi:10.1088/1757-899X/440/1/012014

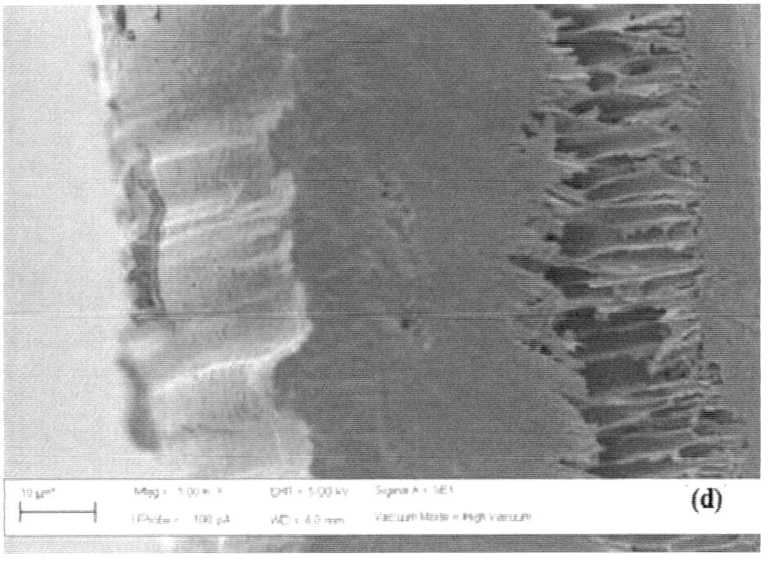

10 µm
(e)

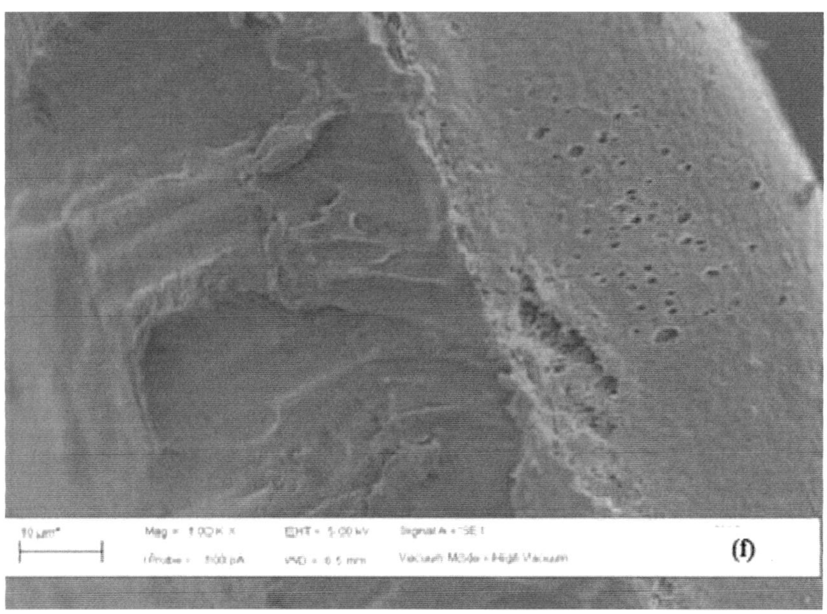
10 µm
(f)

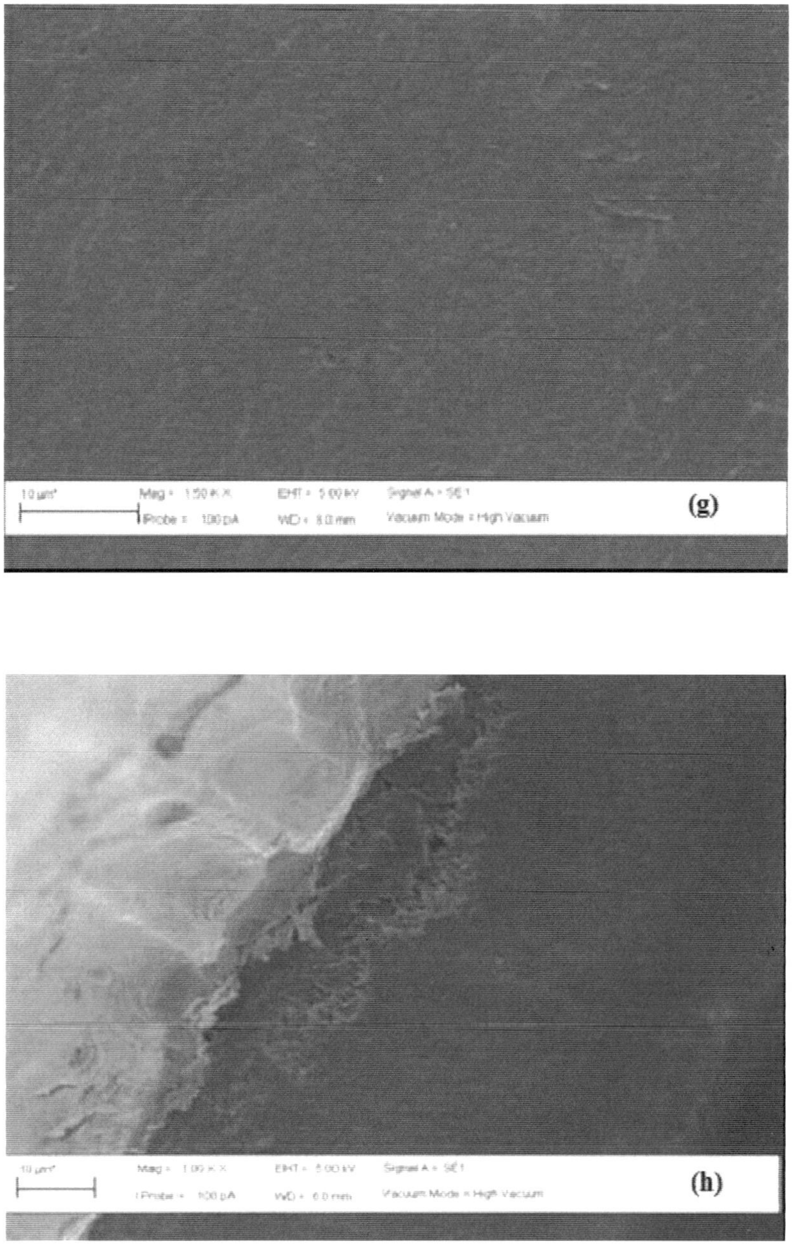

Figure 1. (a) Surface Image of PSf Membrane with 0% PDMS at Magnification 1.50 K X; (b) Cross Section Image of PSf Membrane with 0% PDMS at Magnification 1.00 K X; (c) Surface Image of PSf Membrane with 10% PDMS at Magnification 1.50 K X (d) Cross Section Image of PSf Membrane with 10% PDMS at Magnification 1.00 K X (e) Surface Image of PSf Membrane with 15% PDMS at Magnification 1.50 K X (f) Cross Section Image of PSf Membrane with 15% PDMS at Magnification 1.00 K X (g) Surface Image of PSf Membrane with 20% PDMS at Magnification 1.50 K X (h) Cross Section Image of PSf Membrane with 20% PDMS at Magnification 1.00 K X.

IOP Conf. Series: Materials Science and Engineering **440** (2018) 012014 doi:10.1088/1757-899X/440/1/012014

3.2 Effect of Concentration of Top Layer on the FTIR Analysis

FTIR spectra of 0%, 10%, 15% and 20% of PDMS concentration are shown in Figure 2 (a), (b), (c) and (d) respectively. The spectrum indicates the functional groups that presence in the sample, which are siloxane group (Si-O) stretching vibration at 1000-1100 cm^{-1} and sulfone group (S=O) at 1150-1300 cm^{-1}. The changes in the intensity of the sulfone group can be attributed to the consumption of sulfone group due to increasing siloxane group content. The characteristic IR absorption bands for PSf polymer will decrease in the sequence of 0 % > 10 % > 15 % > 20 % of the PDMS solutions. This is the consequence from the siloxane group is taking place of sulfone group in their reaction. The IR spectrum for the PSfs samples prepared with a 0% of PDMS content shows a peak at 1243.85 cm^{-1} indicating the sulfone groups still in the support polymer sample. For PSf samples prepared with 10, 15 and 20% PDMS content, there are no apparent peaks at 1243.85 cm^{-1}, suggesting the sulfone group is reacted with siloxane groups. The Si-O absorption cannot be clearly detected for the polymer samples prepared at 0% PDMS content. For 10, 15 and 20% PDMS the changes in the intensity of Si-O groups were due to the addition of percentage of PDMS.

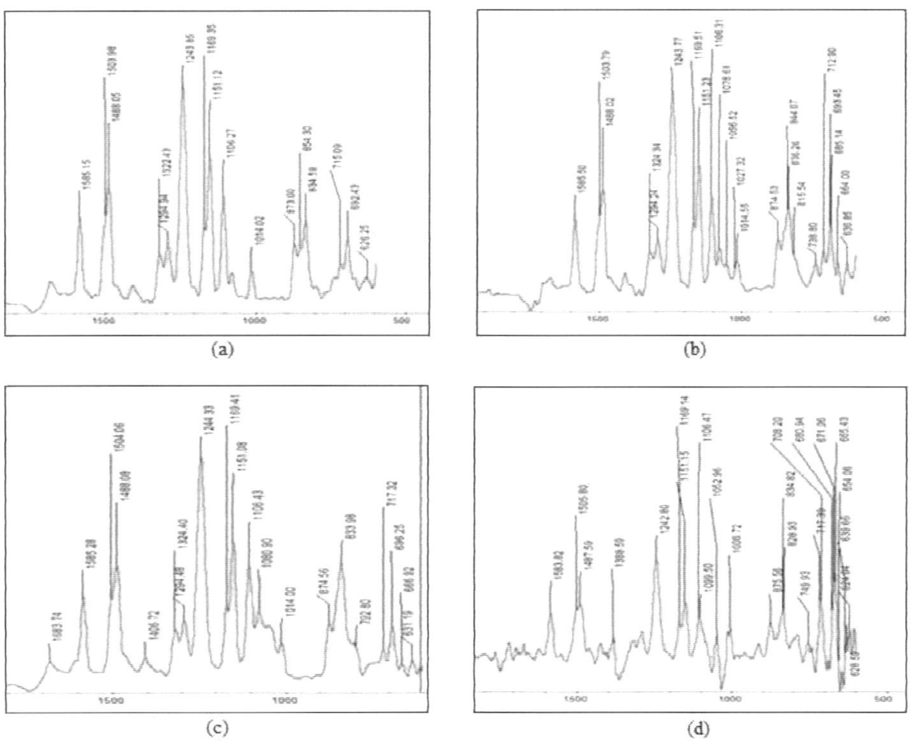

Figure 2. FTIR Absorbance Peak for (a) 0% PDMS of PSf-PDMS Membrane; (b) 10% PDMS of PSf-PDMS Membrane; (c) 15% PDMS of PSf-PDMS Membrane; (d) 20% PDMS of PSf-PDMS Membrane

3.3 Effect of Concentration of Selective Top Layer on the Membrane Performance for Gas Separation

The permeability graph trends obtained in Figure 3 and Figure 4, it is show that the permeability of CO_2 is slightly higher than permeability of N_2 due to kinetic molecular diameter for CO_2 is much smaller which is 3.3 angstroms rather than kinetic diameter of N_2 molecule approximately 3.6 angstroms. CO_2 molecule is easier to pass through the membrane due to its molecular diameter is smaller than N_2 and resulting higher permeability of CO_2 gas [5]. The changes in permeability of single pure gases with change in feed pressure. The decrease in viscosity of polymer for the dip coating solution can allow higher diffusion of gas rate in sub-layer which makes the permeability of certain gas to be higher. As a result, thin film composite membrane with a thin and porous skin layer and open cell sub-layer will result higher permeability value. Meanwhile, N_2 permeability is slightly increased as well as feed pressure is increase. Feed pressure is acting as a driving force of the N_2 molecule to pass through the membrane. Since larger driving forces were applied to N2 molecule, it will contribute to a lot of N_2 particle will pass through the polymeric membranes then resulting to high permeability value.

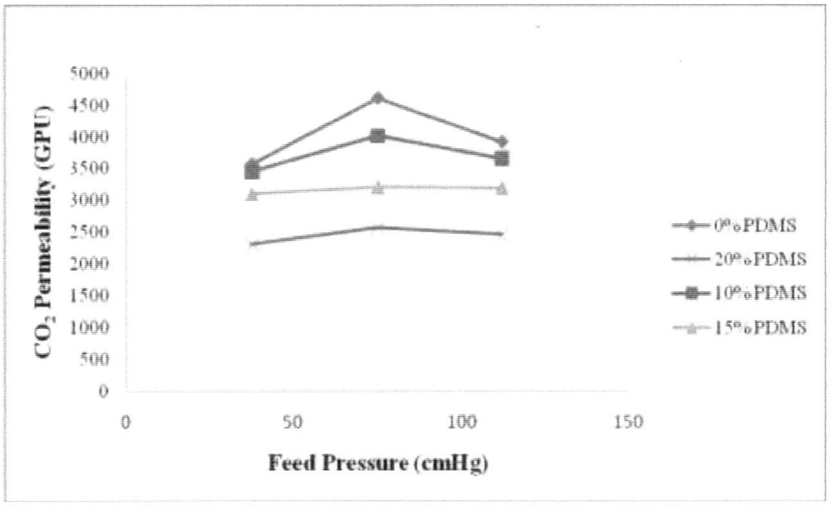

Figure 3. CO_2 Permeability versus Feed Pressure

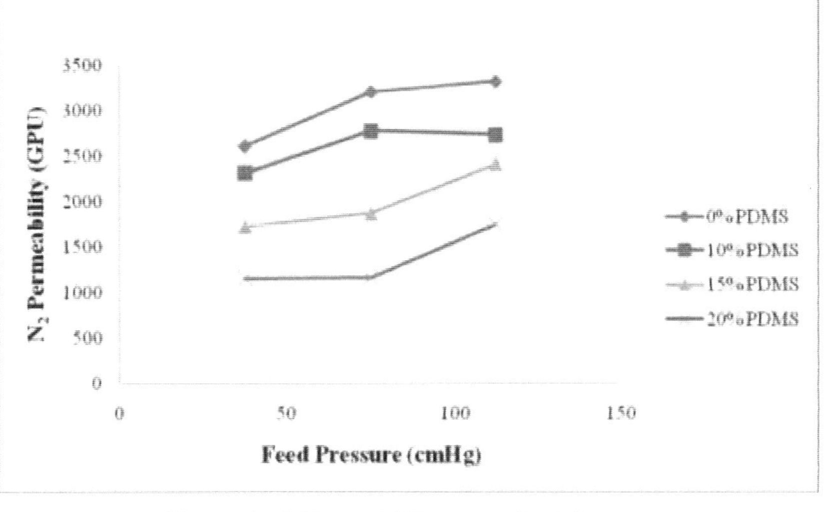

Figure 4. N_2 Permeability versus Feed Pressure

By referring to the Figure 5, the selectivity of the membrane increase with the increase of the percentage of PDMS in the order of 0 % < 10 % < 15 % < 20% of the PDMS concentration. Membrane with concentration 20% PDMS exhibit highest selectivity about 1.57 because increasing the polymer concentration, resulted a denser and thicker skin layer membrane. Furthermore, since the polymer concentration in dip coating solution is increased, it was observed that the skin layer thickness increased while both mean pore size and surface porosity decreased. The uncoated membranes seem to exhibit lowest selectivity probably due to existence of defect (pores) and the enhancement of free volume in the thin skin layers. The relationship between selectivity and the feed pressure show that with increasing the feed pressure, it is resulted lower selectivity. This is suggesting that when there was large driving force that drive the molecule to pass through the membranes, the efficiency of separating the gas is getting lower hence contribute to the bad performance.

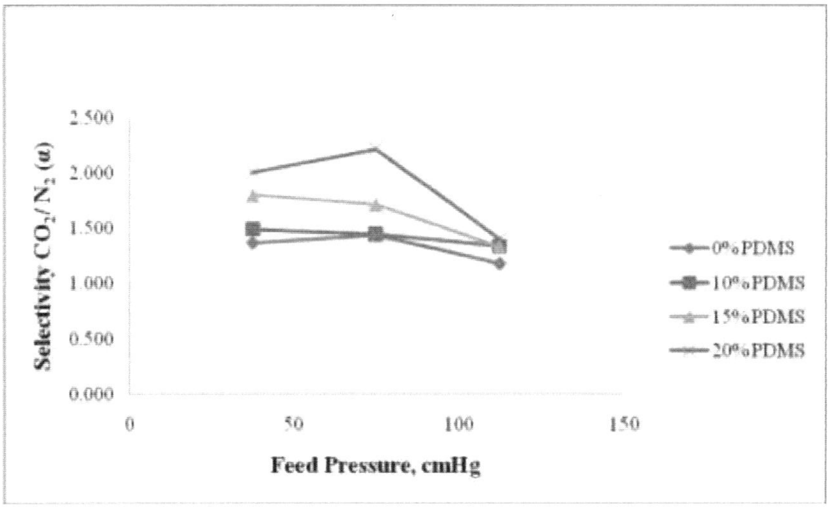

Figure 5. CO_2/ N_2 Selectivity versus Feed Pressure

4. Conclusion

This study successful developed the TFC PVDF/ PEBAX membrane. More sponge-like asymmetric structures were formed across the membrane wall and the macro void formation is suppressed. This is due to the higher viscosity of the casting solution that delayed the phase inversion rate (kinetic effect) which results in a sponge-like structure.

Acknowledgement

The authors thank the sponsor for the grant of RDU1703203 (UMP Grant) for financial support of this research.

5. References

[1] Dantas T L P, Luna F M T, Jr I J S, Azevedo D C S, Grande C A, Rodrigues A E 2011 *Chemical Engineering Journal* **169** 11
[2] Burling I R, Yokelson R J, Griffith D W T, Johnson T J, Veres P and Roberts J M 2010 *Atmospheric Chemistry and Physics* **10** 1115
[3] Pinnau I, and Koros W J 1991 *Journal of Applied Polymer Science* **43** 1491
[4] Aroon M A, Ismail A F, Montazer-Rahmati M M, and Matsuura T 2010 *Separation and Purification Technology* **72** 194
[5] Elashmawi I S 2011 *Materials Chemistry and Physics* **107** 96

IOP Conf. Series: Materials Science and Engineering **440** (2018) 012015 doi:10.1088/1757-899X/440/1/012015

Synthesis and characterization of chromone-based ligand and its fluorescence property towards Zn^{2+}

W M H Arifin[1], Y Juahir[1], N M Ali[1], N A Bakar[1], M S Ahmad[1], F L Supian[2] and U M Osman[3]

[1]Department of Chemistry, Faculty of Science and Mathematics, Universiti Pendidikan Sultan Idris, 35900 Tanjong Malim, Perak, Malaysia
[2]Department of Physic, Faculty of Science and Mathematics, Universiti Pendidikan Sultan Idris, 35900 Tanjong Malim, Perak, Malaysia
[3]School of Fundamental Science, Universiti Malaysia Terengganu, 21030 Kuala Nerus, Terengganu, Malaysia.

Corresponding author emails: yusnita@fsmt.upsi.edu.my

Abstract. Zinc (Zn) is a type of element that are commonly found in the Earth's environment and may be originated either naturally from the environment or produced by human activities. Zn has been identified as a hazardous metal which can give harmful effects to human health and Earth's environment. Therefore, it is very important to develop a simplest method which does not require laborious work in order to detect Zn metal ions. Ligand of AFCMSH was synthesized by the mean of reflux reaction between 2-amino-3-formylchromone (AFC) and 4methylbenzenesulfonylhydrazide (MSH). The molecular structure of the ligand in the ground state was optimized using density functional theory (DFT/B3LYP) method with 6-311G(d,p) basis set. Characterization of ligand was conducted using spectroscopic techniques such as Fourier Transform Infra-Red spectroscopy (FT-IR), Gas Chromatography Mass Spectrometry (GC-MS), Nuclear Magnetic Resonance spectroscopy (NMR) and Ultraviolet-visible spectroscopy (UV-VIS). Meanwhile the fluorescence property of ligand in the presence of metal cations was recorded using Fluorescence Spectrophotometer. Ligand shows almost no fluorescence property in the presence of various metal ions except for Zn^{2+} ion when observed under UV light. Interestingly, ligand exhibits a remarkable fluorescence response by giving yellow-green fluorescence with strong fluorescence intensity in the emission spectrum at 501nm at excitation of 375 nm upon addition of one equivalent of Zn^{2+}.

1. Introduction

Zinc (Zn), the twenty fourth most abundant elements that exists in both natural and anthropogenic sources. The main sources of Zn from human activities can be related to metal production, waste incineration, burning of fossil fuel, phosphate fertilizer and cement production [1]. Excessive concentration of Zn in the environment especially in river might cause toxic response to aquatic animals such as reproductive and behavioral problem [2-4]. Zn exposure towards human may cause bad responses such as skin irritation, immunological effects, abdominal pain, nausea, vomiting, and even cancer if exposed at high dosage and long term [5]. In water, the maximum concentration of Zn have been set at 120ug/L by Water Quality Criteria (WQC) [6], whereas the recommended dietary allowance (RDA) on daily intake of Zn for men and women were set at 11mg/day and 8mg/day respectively [7].

Due to uncontrollable Zn exposure in the environment and their danger towards the ecosystem, a new tool of molecular sensors for detecting Zn^{2+} ions should be designed. The great effort of researchers in this field is directed towards the development of new selective chemosensors based on fluorescence property because they are easy to perform, inexpensive and usually very sensitive (parts per billion/trillion). Although a number of fluorescent chemosensors for zinc ion have been studied, it is still desirable to develop a new potential fluorescent chemosensor with extremely high affinity and good selectivity over other relevant metal ions. In addition, the designing of chemosensor for Zn^{2+} via chelation enhanced fluorescence (CHEF) is still challenging. Reported probes for Zn^{2+} detection includes quinoline-based, rhodamine6G, Tris(3-(2-hydroxyacetophenone)propyl)amine and pyrazoline-based compounds [8-11]. Our developed chemosensor from chromone-based ligand namely as 2-amino-3-formylchromone-toluenesulfonhydrazide (AFCMSH) is thus an important addition to the list of other turn-on fluorescence chemosensors which can recognize Zn^{2+} effectively.

In this study, we report a chromone-based ligand as a turn-on fluorescent chemosensor for the detection of Zn^{2+} (Figure 1). The chemosensor performs remarkably with the addition of Zn^{2+}compared to other metal ions with enhanced fluorescence intensity. The free probe of AFCMSH shows no fluorescence emission, the addition of Zn^{2+} ions causes a significant fluorescence emission with over 15-fold. The molecular structure of AFCMSH in the ground state was optimized using density functional theory (DFT) calculation to give an initial assumption where the most probable sites during the metal chelation.

2. Experimental

2.1. Materials and measurements

All materials and solvents for the synthesis were purchased from Acros and Aldrich and were used without further purification. The ligand AFCMSH was synthesized by method in previous literature [9]. The ligand was characterized by different spectroscopic method. Infrared spectra were obtained from Thermo Nicolet Nexus using KBr pellets. UV-vis spectra were recorded on Agilent, Cary 60 spectrophotometer. Melting point of ligand was obtained from Stuart Scientific Melting Point Apparatus SMP3. The ^{1}H-NMR and ^{13}C-NMR spectra were obtained from Jeol, JNM-ECX-500. Fluorescence spectra were recorded using Agilent, Cary Eclipse. The mass spectra for ligand were collected from Central Laboratory, Universiti Malaysia Terengganu using Gas Chromatograph Mass Spectrometer GCMS-QP2010.

2.2. Synthesis of ligand, AFCMSH

The proposed structure of AFCMSH was given in Figure 1. Mass of 0.189 g (1 mmol) of 2-amino-3formylchromone was dissolved in 15 mL ethanol, and was added gradually to 0.186g (1 mmol) of 4methylbenzenesulfonhydrazide, dissolved in 15 mL ethanol. The reaction mixture was refluxed for 24 h. The resulting yellow precipitate was filtered-off, washed with cold ethanol, and dried in desiccator. Yield, 70%, m.p: 260 °C. FT-IR (KBr, cm^{-1}): v = 3196 (N-H), 1644 (C=O), 1605 (C=N), 1310 ($S{=}O_{as}$), 1158 ($S{=}O_s$). UV-vis: (DMSO): λ_{max}/nm (ε/mol^{-1} Lcm^{-1}) 282, 207. MS (m/z): 357. ^{1}H NMR (DMSO-d_6, 400 MHz): δ 11.04 (1H, s, N-H^{8}), δ 8.25 (3H, s, CH^{17}=N, N-$H^{38, 39}$), δ 7.90 (1H, d, H^{30}), δ 7.89 (1H, d, H^{12}), δ 7.65 (2H, s, $H^{20, 34}$), δ 7.62 (1H, m, H^{40}), δ 7.36 (1H, m, H^{10}), δ 7.32 (1H, m, $H^{22, 36}$), δ 11.04 (3H, s, C-$H^{25, 26, 27}$). ^{13}C NMR (DMSO-d6, 400 MHz): δ 173.8 (C^{32}), 163.0 (C^{14}), 153.1 (C^{13}), 145.5 (C^{23}), 144.2 (C^{16}), 136.0 (C^{18}), 133.9 (C^{9}), 130.3 (C^{29}), 127.6 (C^{21}), 125.6 (C^{31}), 125.5 (C33, C-35), 121.8 (C^{28}), 117.2(C^{19}), 92.6 (C^{15}), 31.2 (C^{11}), 21.5 (C^{24}).

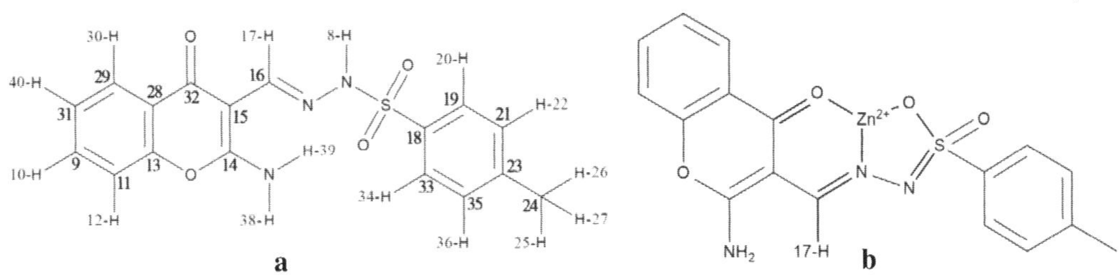

Figure 1. (a) Structure of AFCMSH with atom numbering and (b) Proposed binding mode for interaction of Zn^{2+} with AFCMSH.

2.3. Computational method

The molecular structure of AFCMSH in the ground state was optimized using density functional theory (DFT) by a hybrid functional B3LYP (Becke's three parameter hybrid functional using the Lee, Yang, and Parr correlation functional) methods at 6-311G(d,p) basis set. The calculation was performed on a personal computer using Gauss View 5.0 molecular visualization program [12] and Gaussian 09W program package [13] without specifying any symmetry for the title molecule. In addition, the NBO analysis of the compound was also calculated at the B3LYP/6-311G(d,p) level.

2.4. Analysis of fluorescence property

Stock solutions of metal acetate salts of Zn^{2+}, Cu^{2+}, Co^{2+}, Hg^{2+}, Mn^{2+}, Ni^{2+} and Pb^{2+} (1 mM) in DMSO were prepared. Stock solution of AFCMSH (1 mM) was prepared in DMSO. Test solution for selectivity study was prepared by adding 15 μL of the probe solution into cuvette, later, appropriate aliquot of each metal ions stock solution were added into the cuvette and let to rest for 5 min. The solution was diluted to 3 ml with DMSO solution. Both excitation and emission slit widths for the analysis was set to 5 nm. The excitation wavelength was set to 375 nm.

3. Results and Discussion

3.1. General characterization

The structure of the ligand was explained by IR, UV-Vis spectra, ^{1}H NMR, ^{13}C NMR and mass spectra. The IR spectral of the ligand showed characteristic absorption bands at 3368 and 3196 cm^{-1} assigned to $\nu(NH_2)$ and $\nu(NH)$. The absorption bands at 1644, 1605, 1307 and 1158 cm^{-1} were attributed to $\nu(C{=}O_{ketone})$, $\nu(C{=}N)$, $\nu(S{=}O_{as})$ and $\nu(S{=}O_s)$ respectively. The electronic spectral data of AFCMSH in DMSO were observed for having two bands at 282, and 207 nm. Addition of Zn^{2+} showed a new emission peak at 375 nm. The structure of ligand was further supported by ^{1}H NMR spectral data (δ ppm) dissolved in DMSO-d_6. ^{1}H NMR spectra showed that the ligand was completely synthesized with δ 11.04 signal was assigned for N-H and δ 8.25 signal was assigned for -C=N. ^{13}C NMR spectra obtained showed there are seventeen signals which have the same amount of C in AFCMSH. The mass spectra give out 357 *m/z* similar to the molecular weight of AFCMSH 357 g/mol.

3.2. Optimized structure of AFCMSH

Quantum chemical calculations were carried out using density functional theory (DFT) at the B3LYP/6-311G(d,p) level to obtain the optimized structural parameters and determine the properties of the frontier molecular orbitals. Molecular Electrostatic Potential (MEP) is studied to determine the probability of coordination properties of AFCMSH ligand. The optimized structure of ligand was shown in Figure 2 with atoms numbering. The selected structural parameters of ligand (bond lengths and bond angles) computed from the DFT-B3LYP/6-311G(d,p) basis set level calculation are listed in Table 1.

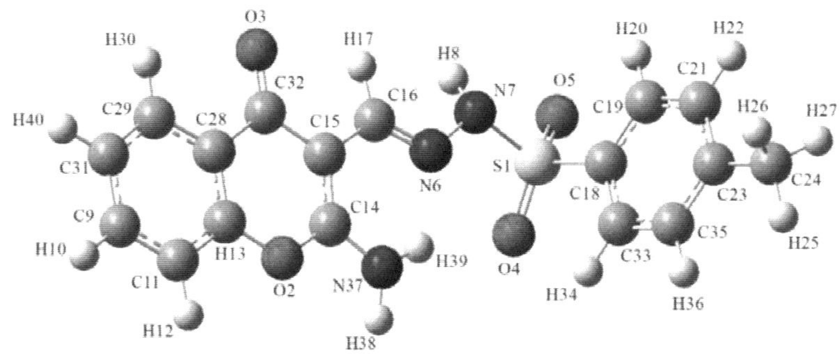

Figure 2. The optimized structure of AFCMSH.

Table 1. Selected bond lengths (Å) and angles (°) determined by B3LYP/6311G(d,p) theoretical calculations.

Bond length	Value (Å)	Bond Angle	Value (°)
S1-O4	1.4239	O4-S1-O5	120.1708
S1-O5	1.4208	O4-S1-N7	109.1677
S1-N7	1.6193	O4-S1-C18	107.8171
S1-C18	1.7525	O5-S1-N7	103.4074
O2-C13	1.3721	O5-S1-C18	108.9114
O2-C14	1.3330	N7-S1-C18	106.5971
O3-C32	1.2318	C13-O2-C14	118.7329
N6-N7	1.3791	N7-N6-C16	113.6930
N6-C16	1.2705	S1-N7-N6	119.7689
N7-H8	0.8594	C11-C9-C31	121.8083
C9-H10	0.9302	C9-C11-C13	118.7535
C9-C11	1.3718	02-C13-C11	116.9712
C9-C31	1.3923	02-C13-C28	121.5257
C14-N37	1.4700	02-C14-C15	125.0063
C15-C16	1.4484	02-C14-N37	117.5023
C16-H17	0.9301	C15-C14-N37	117.4914
C18-C19	1.3699	C14-C15-C16	123.0689
C21-C23	1.3810	N6-C16-C15	124.0056
C23-C24	1.5004	N6-C16-H17	118.0472
C24-C25	0.9604	S1-C18-C19	119.0016
C28-C29	1.4048	C14-N37-H38	109.4712

Most of the calculated bond lengths and bond angles were slightly different from experimentally reported in related ligands. For example the reported experimental bond lengths of S=O, S-N, C=N, NN, C=O and C-O are 1.4240(15)-1.436(15) Å, 1.6336(17)-1.6448(17) Å, 1.274(2)-1.282(3) Å, 1.3969(19)-1.433(2) Å, 1.223(2)-1.233(2) Å, 1.344(2)-1.348 (2) Å respectively [14-17]. Meanwhile the reported experimental bond angles of O=S=O, N-S=O, N-N-S, C=N-N were 119.23(10)119.94(9)°, 103.87(9)-104.12(9)°, 113.01(13)-113.54(12)°, 111.66(16)-115.26(15)° respectively [14, 15]. The differences due to the theoretical calculations were carried out for the gas phase and the experimental results were for the solid phase. Furthermore it could be affected by the existence of a crystal field along with the intermolecular interactions of hydrogen bonds that connect the molecules together in solid phase [18].

3.3. Frontier molecular orbitals analysis

Frontier molecular orbitals analysis provides important information on the electrical and optical properties, as well as in electronic absorption and chemical reactions. The ability of molecule to donate an electron can be explained by the highest occupied molecular orbital (HOMO), while the view of ability to gain an electron can be described from the lowest unoccupied molecular orbital (LUMO) [18, 19]. The HOMO and LUMO for the title compound were calculated at B3LYP/6311G (d,p). Figure 3a showed the view of electron excitation from the ground state HOMO to the excited state LUMO.

In the ground state of HOMO, the charge density are mainly located on the sulfonyl (SO_2), azomethine (C=N), amino group and carbonyl (C=O) functional groups of the ligand. The charge density spread over to the other chromone ring in the case of LUMO. Thus, the HOMO→LUMO electron transition potentially occurs from sulfonohydrazone to chromone moieties with 5.493 eV. The highly charge density of HOMO gives a hypothesis that the oxygen and nitrogen atoms are the potential atoms for intra- or intermolecular interaction and metallic bonding [19].

3.4. Molecular electrostatic potential

Molecular electrostatic potential (MEP) study was carried out to visual more the most probable coordination sites. In the present study, MEP was calculated at the B3LYP/6-311G(d,p) optimized geometry produced 3D plots as illustrated in Figure 3b. The negative region with maximum electron charge value of -0.503 e and -0.481 e is indicated as red colour located at O4 and O5 which belong to sulfonyl oxygen atoms. The other negative region also occurred at O3 that belong to carbonyl oxygen atom in chromone moiety with high electron charge value of -0.0352 e. Both negative regions provide information where the probability of preferred site for electrophilic attack [20]. Meanwhile the probability of preferred site for nucleophilic attack is observed as deep blue colour or maximum positive region occurred around H-N37 (0.278 e) and H-N7 (0.261 e) surface [20]. The probabilities of coordination sites information given by MEP calculation is in good agreement with HOMO-LUMO.

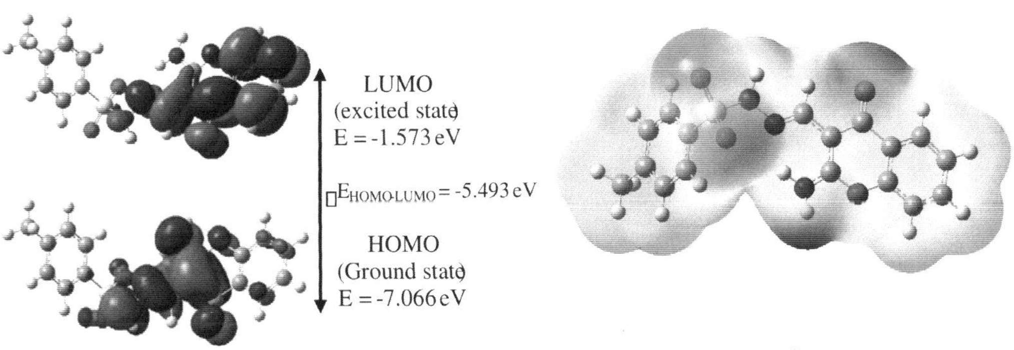

a b

Figure 3. (a) The atomic orbital compositions of the frontier molecular orbital of AFCMSH B3LYP/6-311 G(d,p) level and (b) Molecular electrostatic potential map calculated at B3LYP/6 311G(d,p) level.

3.5. Fluorescence study

The emission changes were investigated in order to obtain information on the fluorescent behavior of AFCMSH. For this purpose, 1.0 eq. ligand was reacted with 1.0 eq. metal ions such as Co^{2+}, Cu^{2+}, Hg^{2+}, Mn^{2+}, Ni^{2+}, Pb^{2+} and Zn^{2+} in DMSO and diluted to 3 ml. The addition of metal ions such Co^{2+}, Cu^{2+}, Hg^{2+}, Mn^{2+}, Ni^{2+} and Zn^{2+} showed colour change from colourless to light yellow except of Pb^{2+} when observed with naked eye (Figure 4a). However, under UV light, the reaction between AFCMSH and Zn^{2+} give out an intense yellow-green fluorescence compared to other ions (Figure 4b). When viewed under UV light, AFCMSH showed no or slight fluorescence property when treated with other metal ions. Fluorescence spectra were obtained by exciting AFCMSH at 375 nm. Ligand showed an interesting behavior towards Zn^{2+} compared to other metal ions. New emission peak was observed at 501nm upon addition of Zn^{2+} (Figure 5a). The addition of other ions showed such as Co^{2+}, Cu^{2+}, Ni^{2+} and Pb^{2+} showed no changes in the fluorescence property of AFCMSH except a slight increase in fluorescence spectra of Mn^{2+} and Ni^{2+}.

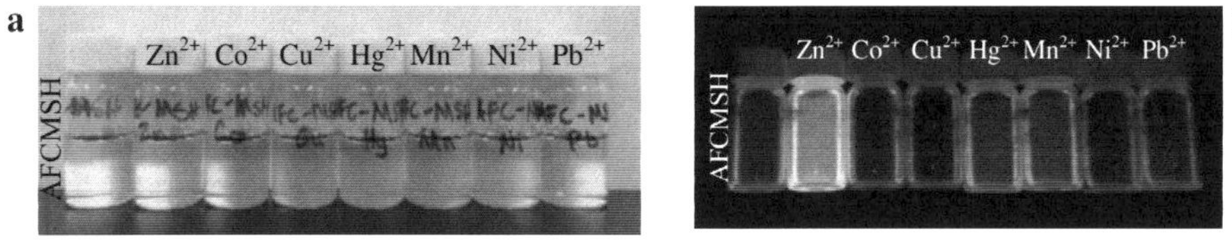

Figure 4. Colour changes of AFCMSH when treated with 1.0 eq. metal ions when observed with naked-eye (a) and under UV light (b).

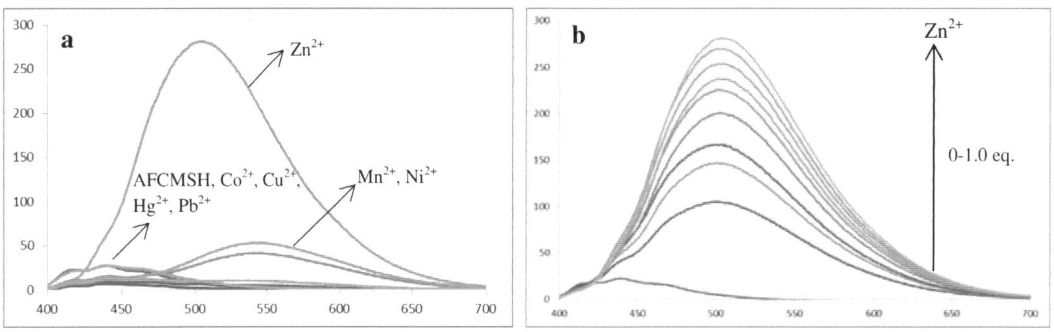

Figure 5. (a) Fluorescence emission spectra of AFCMSH when treated with 1.0 eq. different metal ions and (b) Fluorescence emission spectra when treated with 0-1.0 eq. of Zn^{2+}.

Further study of the fluorescence property of AFCMSH was conducted via fluorescence titration experiment by adding 0-1.0 eq. Zn^{2+} solution to the AFCMSH solution of 25 μl and diluted to 3 ml using DMSO. The addition of Zn^{2+} showed shifting in the emission peak of the ligand from 441 nm to 501 nm (Figure 5b). The fluorescence intensity of the mixed solution also increased by 15-fold which prove the intense yellow-green fluorescence property of AFCMSH-Zn. From the results, AFCMSH can be used as a sensitive chemosensor to detect Zn^{2+}.

3.6. Binding study
From IR spectra, the assigned stretching of carbonyl of chromone 1644 cm^{-1} reduced to lower frequency of 1630 cm^{-1} after addition of Zn^{2+}. The vibration of C=N also shifted from 1605 cm^{-1} to 1600 cm^{-1}. The changes in vibrational stretching suggest that Zn^{2+} coordinated with carbonyl group of chromone and nitrogen from azomethine. The absence of ν(N-H) vibrations after addition of Zn^{2+} suggesting the proton on the amine nitrogen atom is lost upon the complexation [21]. The weaker intensity of $\nu(S{=}O_{as})$ at 1340cm^{-1} together with the appearance of new band at 1086 cm^{-1} assigned to the binding $\nu(SO_2)$ group might give out the possibilities of coordination between oxygen from S=O with Zn^{2+}. The band at 508 cm^{-1} was assigned to ν(Zn-O), which demonstrated formation of a coordinative bond between zinc and oxygen while for ν(Zn-N) was assigned at 421 cm^{-1} for coordination between zinc and nitrogen. The result was in a good agreement with the information given by the Frontier molecular orbitals analysis

and Molecular electrostatic potential that the probabilities coordination are via oxygen and nitrogen atoms.

The absorption titration experiment was also conducted to determine the binding of AFCMSH with Zn^{2+}. Using 0.1-1.0 eq. volume of Zn^{2+}, the solution of metal ion was reacted with 1.0 eq. volume of AFCMSH. The electronic spectrum observed for free ligand AFCMSH exhibited two peaks at 282 and 207 nm at room temperature (Figure 6). The highest energy bands are assigned to π–π* transitions within the aromatic rings while the moderate energy bands can be assigned to n–π* transitions within C=O and C=N groups [22]. Upon treatment with Zn^{2+}, the absorption band at 282 and 207 nm gradually decreased, while a new peak was observed at 375 nm and gradually increased absorption intensity with the addition of Zn^{2+}. The new peak at 375 nm was assigned as charge transfer transition band indicating that the coordination between AFCMSH and Zn^{2+} lead to the fluorescence emission of the ligand [9]. The appearance of isosbestic point at 343 nm and the shifting of absorption bands after addition of Zn^{2+} implies that ligand AFCMSH bonded with Zn^{2+} to form AFCMSH-Zn [23].

The ^{1}H-NMR data of AFCMSH and its treatment with Zn^{2+} were recorded in Table 2. The signal of H8(N-H) and H-17(CH=N) were δ 11.04 and. δ 8.25 respectively. Upon addition of Zn^{2+}, the signal of H-8 disappeared completely while the signal for H-17 remains but shifted to higher field by δ 0.05. The remaining signals for protons in the benzene rings remain unchanged even after the addition of Zn^{2+}. This indicates the involvement of carbonyl of chromone moiety and the nitrogen atom of the –C=N group coordinated with Zn^{2+}.

Table 2. ^{1}H NMR spectral data of AFCMSH ligand.

	-NH(H-8) δ (ppm)	-N=CH(H-17) δ (ppm)
AFCMSH	11.04	8.25
AFCMSH + Zn^{2+}	-	8.30
Change in δ	-	0.05 (up-field)

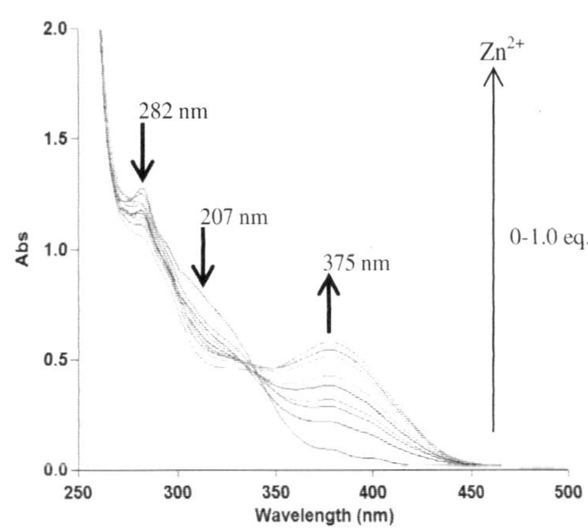

Figure 6. Uv-vis titration of AFCMSHwhen treated with 0-1.0 eq. of Zn^{2+} ion.

Theoretically the very low emission intensity of the ligand was due to quenching by the carbonyl group through a photoinduced electron transfer (PET), induced by the lone pair electrons from the ligand. The coordination AFCMSH to Zn^{2+} through M-N (imine nitrogen atom) and M-O (carbonyl and sulfonyl oxygen atoms) was believed to enhance the fluorescence output via chelation-enhanced fluorescence (CHEF) effect by interrupting the PET process, thus give the "OFF-ON" fluorescence signal in the presence of Zn^{2+} [21].

To the best of our knowledge, there is no report regarding the potential turn-on fluorescence selective towards Zn^{2+} using AFCMSH ligand and our present findings show that it is worth exploring more in the future. Thus, studies of sensitivity against other competing metal ions and the limit of detection should be investigated to confirm the ability of ligand to serve as a practical chemosensor in detecting Zn^{2+}.

4. Conclusion

In summary, we have successfully synthesized chromone-based ligand (AFCMSH) which has potential to be used in detecting Zn^{2+}. The compound exhibited high selectivity for Zn^{2+} over other metal ions with over 15-fold fluorescence enhancement. The finding also showed the compound has a good fluorescence behavior which emits yellow-green fluorescence when viewed under UV light compared to other metal ions. Thus, we believed AFCMSH had the ability to be used as a potential fluorescent chemosensor for detecting Zn^{2+} in environment. In addition, we believed that the oxygen atom of sulfonyl moiety and azomethine nitrogen donor atoms play important role in coordination with Zn^{2+} ion. This study might provide a general insight for designing and producing new sensors useful to detect zinc contaminant in the environments.

5. References

[1] Councell TB, Duckenfield K, R Landa E, and Callender E 2004 *Environ. Sci. Technol* **38** 4206-4214
[2] Lefcort H, Meguire RA, Wilson LH, and Ettinger WF 1998 *Arch. Environ. Contam. and Toxicol* **35** 447-56
[3] Jelmert A and van Leeuwen J 2000 *Water Res.* **34** 1937-40
[4] Itow T, Loveland RE, and Botton ML 1998 *Arch. Environ. Contam. and Toxicol* **35** 33-40
[5] Plum LM, Rink L, and Haase H 2010 *Int. J. Environ. Res. Public Health* **7** 1342-65
[6] Feng C, Wu F, Zhao X, Li H, and Chang H 2012 *Sci. China Earth Sci.* **55** 882-91
[7] Trumbo P, Yates AA, Schlicker S, and Poos M 2001 *J. Am. Diet. Assoc.* **101** 294-301
[8] Zhou X, Yu B, Guo Y, Tang X, Zhang H, and Liu W 2010 *Inorg. Chem.* **49** 4002-4007
[9] Liu Z, Yang Z, Li Y, Wang B, Li Y, Li T, and Ding Y 2013 *J. Fluoresc.* **23** 1239-1245
[10] Hosseini M, Ghafarloo A, Ganjali MR, Faridbod F, Norouzi P, and Niasari MS 2014 *Sens. Actuators B* **198** 411-415
[11] Gong Z, Ge F, and Zhao B 2011 *Sens. Actuators B* **159** 148-153
[12] Frisch M, Trucks G, Schlegel H, Scuseria G, Robb M, Cheeseman J, Scalmani G, Barone V, Mennucci B, Petersson G, and Nakatsuji H 2009 *Gaussian 09, revision a. 02, gaussian. Inc., Wallingford, CT*
[13] Dennington R, Keith T, and Millam J 2009 *Semichem Inc. Shawnee Mission KS, GaussView, version 5*
[14] Ali H, Yusnita J, and Ng S 2007 *Acta Crystallogr., Sect. E: Struct. Rep. Online* **63** o2734
[15] Mehrabi H, and Kia R 2009 *Acta Crystallogr., Sect. E: Struct. Rep. Online* **65** o1056
[16] Ishikawa Y 2015 *Acta Crystallogr., Sect. E: Struct. Rep. Online* **71** o652-o653
[17] Gomes LR, Low JN, Cagide F, Gaspar A,and Borges F 2018 *Acta Crystallogr., Sect. E: Struct. Rep. Online* **71** 1270-1277
[18] İnkaya E, Dinçer M, Ekici Ö, and Cukurovali A 2012 *J. Mol. Struct.* **1026** 117-26
[19] Atiş M, Karipcin F, Sarıboğa B, Taş M, Çelik H 2012 *Spectrochim. Acta Part A* **31** 290-301
[20] Arslan NB, Kazak C,and Aydın F 2012 *Spectrochim. Acta, Part A* **89** 30-38
[21] Fan L, Li T, Wang B, Yang Z, and Liu C 2014 *Spectrochim. Acta, Part A* **118** 760-764
[22] Dueke-Eze C, Fasina T, and Idika 2011 *Afr. J. Pure Appl. Chem.* **5** 13-18
[23] Sohrabi M, Amirnasr M, Farrokhpour H, and Meghdadi S 2017 *Sens. Actuators B* **250** 64758

Acknowledgement

Author wishes to thank Universiti Pendidikan Sultan Idris for facilities and Ministry of Higher Education, Malaysia for the Fundamental Research Grant Scheme (FRGS:2016-0087-102-02). Author thanks to Ku Halim Ku Bulat and Mohd Sukeri Mohd Yusof from Universiti Malaysia Terengganu;

School of Fundamental Science for the sharing wise knowledge of DFT studies during the attachment programme.

Synthesis and characterization of alkoxy substituted p-cyano stilbene schiff bases

H Norhafiefa[1],Yusoff H. M[1] and Rafizah Rahamathullah[1]

[1] School of Fundamental Science, Universiti Malaysia Terengganu, 21030 Kuala Terengganu, Terengganu, Malaysia

Abstract. Schiff bases consist of aromatic have been used in various application including biological, inorganic and analytical chemistry. It also has been applied in optical and electrochemical sensors to enable detection of enhance selectivity and sensitivity of the sensors. These compounds might give excellent prospect in sensor. Schiff bases derived from 4-(4-aminostyryl)benzonitrile and three 4-alkoxybenzaldehyde derivatives are synthesis using ethanol as solvent. All compounds were characterized by Fourier transform-infrared spectroscopy (FT-IR), UV-Vis spectrophotometer and Nuclear Magnetic Resonance (NMR). FTIR showed formation of C=N (imine) stretching vibrations at range of 1604 cm^{-1} to 1609 cm^{-1}. In UV-vis, π-π^* electronic transition of the C=N can be observed at peak range 361 nm. While in the δ_H NMR the peak of CH=N (Imine) group was found at δ 8.32-8.34 ppm and ^{13}C NMR for three Schiff bases were discovered at δ 158.72, 158.73 and 158.71 ppm. In previous study there are several types of linker that have been used in E-DNA, especially alkanethiol. In this study, three new Schiff base derivatives have been synthesized and characterized due to it promising in sensor which prompted us to apply its as future potential linker in electrochemical DNA sensor.

Keywords: Synthesis, Characterization, Schiff base, future potential linker in E-DNA

1. Introduction

The chemistry of compound containing Schiff base have been study from ancient years in various field [1]. The compound was discovered by Hugo Schiff, a German Chemist and a Novel Prize Winner in 1864th [2]. Schiff bases are produced using condensation reaction between ketones or aldehyde with primary amines. The structure of Schiff bases consists of functional group azomethin (-HC=N-) [3]. A Schiff base structurally is a compound that contain nitrogen analogue of an aldehyde or ketone in which an imine or azomethine group replace the carbonyl group (C=O) [4]. Azomethine group is the common structural feature of Schiff base compound with general formula RHC=N-R1, where R and R1 are alkyl, aryl, cyclo alkyl or heterocyclic groups which may be variously substituted. Schiff base compound also known as anils, imines or azomethines. Previous study revealed that the presence

IOP Conf. Series: Materials Science and Engineering **440** (2018) 012016 doi:10.1088/1757-899X/440/1/012016

of a lone pair of electron in sp2 hybridized orbital of nitrogen atom of the azomethine group is chemically significant and biological importance [5].

During the reaction, the water molecules will be eliminated and can be accelerated using acid as catalyst. Basically, the reaction is run via refluxing the mixture of the carbonyl and an amine using Dean Stark apparatus to get rid of water. The removal of water can avoid the conversion of aminal into imine because it can be reversible. The previous study showed, there are several dehydrating agents to remove water molecules including sodium sulphate and molecular sieves. Besides, there are some in situ method that have been reported involving dehydrating solvents such as tetramethyl orthosilicate or trimethyl orthoformate. In addition, there are various acid catalyst that have been reported including mineral acids (H_2SO_4, HCl), organic acids (p-toluene sulphonic acids, pyridium p-toluene sulphonate), acid resin, montmorillonite or Lewis acid ($ZnCl_2$, $TiCl_4$, $SnCl_4$, BF_3Et_2O, $MgSO_4$, $Mg(ClO_4)_2$) [6]. Scheme 1.1 shows the general reaction of Schiff base for imines preparation.

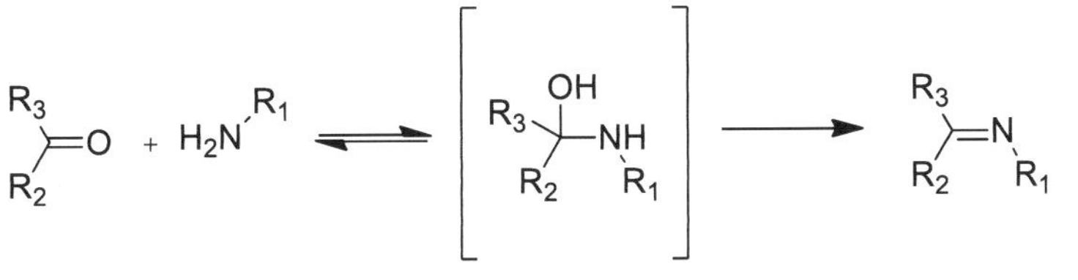

Scheme 1. Schiff base reaction for imine preparation [6].

Schiff base consist of aryl substituents are more stable and readily synthesized compared to those with alkyl substituent which is relatively unstable. Moreover, Schiff base from aliphatic aldehydes are unstable and readily polymerizable compared to those with aromatic aldehyde that have effective conjugation which make it more stable [7]. Nowadays, several of Schiff base compound have been synthesized and their complexation characteristics have been studied due to it great flexibility and diverse structural aspects [8]. Schiff bases comprise of aromatic have been used in many applications in field of biological, inorganic and analytical chemistry. The presence of organic reagents is required in application of many new analytical devices as essential compounds of the measuring system. It is applied in optical and electrochemical sensors, as well in chromatographic method. The advantages of Schiff base in sensor are improved the selectivity and sensitivity of the detection [9]. In addition, aromatic aldehyde Schiff bases display great photoluminescence properties. It might be good to use in sensor, luminescence, electroluminescent devices and biological [10].

Many advance technologies have been developed which provide us tools for cultivate new technique for monitor biorecognition and interaction between solid and in solution over past 15 years [11]. Biosensor is known as an analytical device which covert biological response into the signal [12]. It consists of two main parts which is a bioreceptor and a transducer. Bioreceptor act as a biological compound that will identify the target analyte. Transducer part which also known as physicochemical detector component that will transform the recognition event into signal that can be measured. There are many types of biosensors including; optical biosensor, resonance biosensor, thermal biosensors, electrochemical biosensor, bioluminescence sensors, nuclei acid-based biosensors and nanobiosensors [13]. Electrochemical DNA sensor gained popularity due to their great advantages that offer simplicity, rapidness, low cost and high sensitivity compared to others method for DNA detection [14]. It has become one of the trusted tools for variety applications including diagnosis, reliable forensic analysis, environmental monitoring, and biological research due to its benefits [15]. Nowadays, there are many kinds of working electrode that have been applied in electrochemical DNA sensor including mercury electrode, carbon electrode, gold electrode, indium tin oxide electrode and chemically modified electrode. The properties of electrode can strongly affect the efficiency of DNA hybridization on the surface of electrode, especially in term of material types, surface modification and surface area [16]. Previous study shows the common linkers are used in E-DNA is alkanethiol due to its simplicity and ease of preparation [17]. Most of researcher have modified their electrode using 6-mercapto-1-hexanol (MCH) [18], 3-mercapropionic acid (MPA) [19] and 11-mercaptoudecanoic acid (MUA) [19, 20].

The aim of the study is to synthesize three Schiff bases consist of alkoxy substituted with different length of alkane chain, followed by characterization. The synthetic procedures for all synthesized compounds were showed in scheme 2. In the future work, all three Schiff base derivatives will be applied as potential linker in E-DNA sensor due to its excellent prospect in sensor. These compounds might be excellent linker as it consists of cyano functional group and alkyl chain to chemically binding with gold electrode and phosphate group in DNA. The study about the attachment is still in progress.

2. Experiment

2.1. Materials and Instrumentation

All chemical and solvents were all purchased from Merck, Sigma Aldrich, Acros Organic, Fisher Chemical and HmbG® Chemicals and were used without further purification.

Infrared (IR) spectra of all synthesized compound were recorded on Perkin Elmer 100 Fourier Transform Infrared Spectroscopy using potassium bromide (KBr) pellet in the range 4000 – 450 cm^{-1}. The analysis of Uv spectra were measured by double beam Spectrophotometer Shimadzu UV-1800 in 1 cm^3 cuvette and acetonitrile as solvent. δ_H and ^{13}C nuclear magnetic resonance (NMR) spectra were determined by Bruker Avance II 400 spectrometer in deuterated chloroform ($CDCl_3$-d_1).

2.2. Synthesis of 4-(4-aminostyryl)benzonitrile (1)

4-(4-aminostyryl)benzonitrile **(1)** was synthesized by using Heck-cross coupling method. 4-Vinyl aniline 90% (2.18 mmol) and 4-aminobenzonitrile (3.27 mmol) was dissolved in 4 mL of DMF and 3.5 mL trimethylamine. The solution was refluxed at 70-75 °C for 24 h in the presence of $PdCl_2(PPh_3)_2$ (0.043 mmol) and was monitored by TLC. The product was purified using column chromatography using hexane: ethyl acetate as an eluent. Yellow powder; Yield 257 mg, 58%; IR (KBr, υ cm^{-1}): 3447-3369 v(N-H), 3033 v(C-H) aromatic, 1618, 1439 v(C=C) aromatic, 1594 v(C=C) alkene, 2224 v(C≡N), 1174 v(C-N); δ_H NMR (400 MHz, $CDCl_3$, ppm) δ: 3.76 (2H, NH_2), 6.81- 7.05 (doublet-doublet, 2H, HC=CH), 6.61-7.52 (8H, CH, Ar). ^{13}C-NMR (400 MHz, $CDCl_3$, ppm) δ; 146.34 (Ar-C-NH_2), 118.25 (C≡N), 108.60-146.11 (6 Ar-C), 125.23 (C=C); UV-Vis (Acetonitrile): λmax=252 nm, 360 nm; MS: m/z 240 [M+.].

2.3. Synthesis of 4-alkoxybenzaldehyde (2a, 2b, 2c)

The formation of 4-alkoxylbenzaldehyde was synthesized using general method of Williamson Ether synthesis. This synthesis involved reaction between equimolar mixture of 4-hydroxylbenzaldehyde and bromoalkane in the presence of potassium carbonate, K_2CO_3 and potassium iodide, KI. 4-hydroxylbenzaldehyde (1.5g) and 50 ml of DMF was stirred at temperature 65 °C for 30 min in the presence of K_2CO_3 and potassium iodide, KI. After that, bromoalkane was added into the mixture and was kept stirred under 80-85 ºC for 24 h. The reaction was monitored by TLC. The mixture was transferred into a 500 ml of separatory funnel and 50 ml of diethyl ether was added. The reaction mixture was washed with distilled water repeatedly to remove the K_2CO_3. 25 ml of 5% NaOH was used to removed unreacted 4-hydroxybenzaldehyde by washing the ether layer. Then repeatedly washed the ether layer with distilled water until it became neutral to pH paper and used anhydrous MgSO4 to eliminate water. The diethyl ether was evaporated via rotary evaporator to give transparent pale-yellow liquids [20].

2.3.1 4-propoxybenzaldehyde ***(2a).*** Light Yellow liquid; Yield 55%; IR (KBr, υ cm^{-1}): 3075 v(C-H) aromatic, 2967-2828 v(C-H), 1693 v(C=O), 1598, 1509 v(C=C) aromatic, 1474 (CH_2 bend), 1428,1314 (CH_3 bend), 1254 v(C-O); δ_H NMR (400 MHz, $CDCl_3$, ppm) δ: 0.96 (3H, CH_3), 1.76 (2H, CH_2), 3.91 (2H, CH_2), 6.89-7.74 (4H, CH, Ar), 9.78 (1H, H-C=O). ^{13}C-NMR (400 MHz, $CDCl_3$, ppm) δ; 10.42, 22.41, 69.87 (CH_2-CH_2-CH_3), 114.75, 129.76, 131.97, 164.27 (Ar-C), 190.79 (C=O); UV-Vis (Acetonitrile): λmax=218 nm, 274 nm; MS: m/z 164[M+.].

2.3.2 4-(hexyloxy)benzaldehyde ***(2b).***Light Yellow liquid; Yield 60%; IR (KBr, υ cm-1): 3074 v(C-H) aromatic, 2932-2734 v(C-H), 1691 v(C=O), 1602, 1509 *v*(C=C) aromatic, 1464 (CH_2 bend), 1427,1393 (CH_3 bend), 1255 v(C-O); δ_H NMR (400 MHz, $CDCl_3$, ppm) δ: 0.91 (3H, CH_3), 1.34 (4H, CH_2-CH_2), 1.45 (2H, CH_2), 1.81 (2H, CH_2), 4.02 (2H, CH_2), 6.97-7.82 (4H, CH, Ar), 9.86 (1H, H-C=O). ^{13}C-NMR (400 MHz, $CDCl_3$, ppm) δ; 12.97, 21.54, 24.61, 28.00, 30.49, 67.42 (CH_2-CH_2-CH_2-CH_2-CH_2-CH_3), 113.74, 1128.76, 1130.94, 163.27 (Ar-C), 189.73 (C=O); UV-Vis (Acetonitrile): λmax=219 nm, 275 nm; MS: m/z 206[M+.].

2.3.3 4-(nonyoxy)benzaldehyde ***(2c).*** Light Yellow liquid; Yield 64%; IR (KBr, υ cm-1): 2927-2733 v(C-H), 1692 v(C=O), 1598, 1509 v(C=C) aromatic, 1468 (CH_2 bend), 1428,1393 (CH_3 bend), 1251 v (C-O); δ_H NMR (400 MHz, $CDCl^3$, ppm) δ: 0.80 (3H, CH_3), 1.23 (10H, $5CH_2$), 1.38 (2H, CH_2), 1.72 (2H, CH_2), 3.95 (2H, CH_2), 6.89-7.74 (4H, CH, Ar), 9.79 (1H, H-C=O). ^{13}C-NMR (400 MHz, $CDCl_3$, ppm) δ; 13.07, 21.64, 24.94, 28.05, 28.22, 28.32, 28.48, 30.84, 67.42 ($8CH_2$-CH_3), 113.74, 128.77, 130.94, 163.27 (Ar-C), 189.72 (C=O); UV-Vis (Acetonitrile): λmax=219 nm, 275 nm; MS: m/z 248[M+.].

2.4 Synthesis of 4-(4-((4-xbenzylidene)amino)styryl)benzonitrile ***(3a, 3b, 3c)***
4-(4-aminostyryl)benzonitrile product was dissolved in the ethanol and it will be treated with 4-alkoxylbenzaldehyde. Then, the mixture was stirred under reflux for 2 hours and the solvent was reduced to half by using dean stalk. The reaction was monitored by thin layer chromatography (TLC). Precipitate was formed and recrystallized using acetonitrile.

2.4.1 4-(4-((4-propoxybenzylidene)amino)styryl)benzonitrile anilines, (3a). Yellow solid; Yield 46 %; IR (KBr, υ cm-1):3025 v(C-H) aromatic, 2967-2879 v(C-H), 2224 v(C≡N), 1604 v(C=N), 1584 v(C=C) alkene , 1249 v(C-N), 1164 v(C-O), 843 (C-H out of plane bend) ; δ_H NMR (400 MHz, $CDCl_3$, ppm) δ: 0.98 (3H, CH_3), 1.76, 3.91 (6H, CH_2), 6.90, 7.15, 7.47, 7.50, 7.54, 7.76 (12II, Ar), 6.98, 7.12 (doublet-doublet, 2H, HC=CH), 8.32 (1H, HC=N). ^{13}C-NMR (400 MHz, $CDCl_3$, ppm) δ; 9.47, 21.48, 68.67 ($2CH_2$-CH_3), 109.32, 113.72, 120.49, 126.81, 127.85, 129.61, 130.93, 131.46, 132.68, 140.95, 151.56, 161.04 (Ar-C), 124.44, 125.70 (C=C), 158.72 (C=N), 118.07 (C≡N) ; UV-Vis (Acetonitrile): λmax=288 nm, 385 nm; MS: m/z 386 [M+.].

2.4.2 4-(4-((4-hexyloxybenzylidene)amino)styryl)benzonitrile, (3b). Yellow solid; Yield 48.3%; IR (KBr, υ cm-1): 3025 v(C-H) aromatic, 2935-2871 v(C-H), 2224 v(C≡N), 1605 v(C=N), 1583 v(C=C) alkene , 1257 v(C-N), 1164 v(C-O), 843 (C-H out of plane bend); δ_H NMR (400 MHz, $CDCl_3$, ppm) δ: 0.84 (3H, CH_3), 1.28, 1.41, 1.73, 3.95 (10H, CH_2), 6.90, 7.17, 7.48, 7.52, 7.56, 7.77 (12H, Ar), 6.99, 7.13 (doublet-doublet, 2H, HC=CH), 8.34 (1H, HC=N). ^{13}C NMR (400 MHz, $CDCl_3$, ppm) δ; 13.01, 21.57, 24.66, 28.11, 30.54, 67.22 ($5CH_2$-CH_3), 109.33, 113.73, 120.49, 126.81, 127.84, 129.61, 130.93, 131.46, 132.68, 140.96, 151.57, 161.06 (Ar-C), 124.89, 125.73 (C=C), 158.73 (C=N), 118.07 (C≡N); UV-Vis (Acetonitrile): λmax=286 nm, 387 nm; MS: m/z 428 [M+.].

2.4.3 4-(4-((4-nonyloxybenzylidene)amino)styryl)benzonitrile ***(3c).*** Yellow solid; Yield 66 %; IR (KBr, υ cm-1): 3040 v(C-H) aromatic, 2923-2853 v(C-H), 2222 v(C≡N), 1609 v(C=N), 1585 v(C=C) alkene , 1258 v(C-N), 1162 v(C-O), 846 (C-H out of plane bend; δ_H NMR (400 MHz, $CDCl_3$, ppm) δ:

0.81 (3H, CH_3), 1.21, 1.39, 1.73, 3.94 (16H, $8CH_2$), 6.90, 7.15, 7.47, 7.50, 7.55, 7.76 (12H, Ar), 6.98, 7.12 (doublet-doublet, 2H, HC=CH), 8.32 (1H, HC=N). ^{13}C NMR (400 MHz, $CDCl_3$, ppm) δ; 13.09, 21.65, 24.99, 28.15, 28.23, 28.36, 28.50, 30.85, 67.23 ($8CH_2$-CH_3), 109.34, 113.73, 120.49, 126.81, 127.85, 129.61, 130.94, 131.46, 132.68, 140.97, 151.58, 161.07 (Ar-C), 124.89, 125.73 (C=C), 158.71 (C=N), 118.07 (C≡N) ; UV-Vis (Acetonitrile): λmax=289 nm, 388 nm; MS: m/z 470 [M+.].

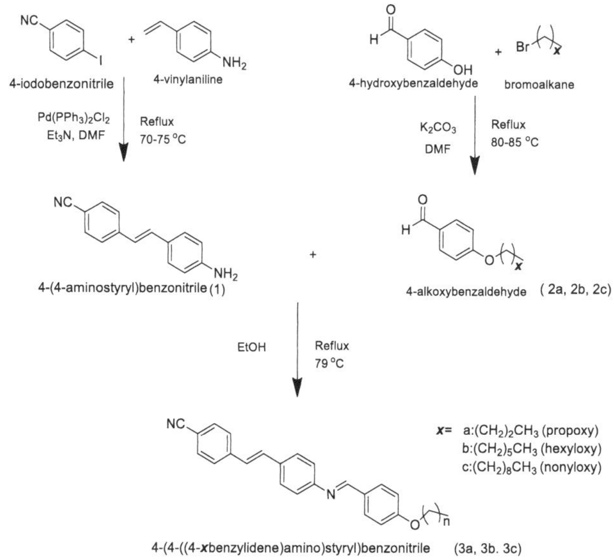

Scheme 2. Synthetic procedures for the 4-(4-((4-xbenzylidene)amino)styryl)benzonitrile

3. Results and Discussion

3.1. Infrared (IR) Spectroscopic Analysis

The FTIR spectra of synthesized compounds 3a, 3b and 3c are shown in Figure 1. The structure of all compound was similar and only differ in length chain of alkoxy. The spectra showed that the band of CH stretch for aromatic in range 3025-3040 cm^{-1}. The peak observed at 2853 – 2967 cm^{-1} show the presence of C-H stretching of alkyl chain in the compound . Absorption band of cyano group (C≡N) was discovered at peak range of 2222-2224 cm^{-1}. The characteristics absorption bands of carbon-nitrogen double band of the imine group (-HC=N-) of compound 3a, 3b and 3c were presence in range 1604, 1605 and 1609 cm^{-1} respectively [21, 22]. Another band around 1583-1585 cm^{-1} assigned to olefinic double bonds (-CH=CH-) [21]. Appearance of band at 1249-1258 cm^{-1} indicate the existence of C-O stretching [20].

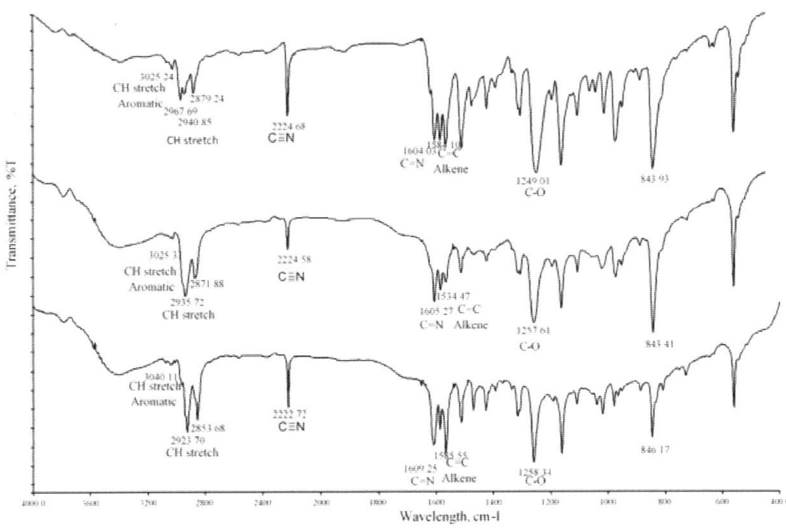

Figure 1. FTIR spectra of compound 3a-3b

3.2. Ultraviolet-visible (UV-vis) spectroscopy

The Ultraviolet-visible (UV-vis) spectroscopy analysis of compound 3a-3b was done by using acetonitrile (1 x 10^{-5} M) as a solvent with cut-off point at 190.0 nm (Figure 2). The ultraviolet-visible electronic spectra of three compound showed the presence π-π* electronic transition of C=C (aromatic) at the wavelength 227 nm to 238 nm. While, band at wavelength 361 nm indicate to the π-π* electronic transition of the C=N group [23].

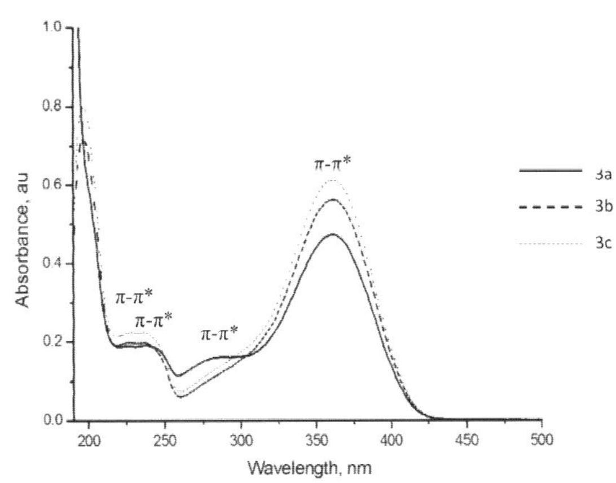

Figure 2. UV-vis spectrum for 3a-3b

3.3. δ_H and ^{13}C Nuclear Magnetic Resonance (NMR)

Table 3.1. represent the δ_H spectra of compound 3a-3b. The spectra display triplet signal moieties of CH_3 of alkyl chain at δ 0.98, 0.84 and 0.81 ppm for compound 3a, 3b and 3c, while the moieties of CH_2 of alkyl chain were observed at δ 1.21 to 3.94 ppm. Aromatic protons of the benzene ring resonated at δ 6.90–7.76 ppm. Doublet-doublet signal of C=C of alkene was discovered at δ 6.98-6.99 ppm and δ

7.12-7.13 ppm for all compound. Finally, the signals at δ 8.34- 8.32 ppm were attributed to the imine proton which show the evidence of the Schiff base.

Table 1 δ_H NMR spectral data of compound **3a-3c**

Compound	CH3	C-H (Alkyl chain)	C-H (Aromatic)	C=C (Alkene)	CH=N
3a	0.98	1.76, 3.91	6.90, 7.15, 7.47, 7.50, 7.54, 7.76	6.98, 7.12	8.32
3b	0.84	1.28, 1.41, 1.73, 3.95	6.90, 7.17, 7.48, 7.52, 7.56, 7.77	6.99, 7.13	8.34
3c	0.81	1.21, 1.39, 1.73, 3.94	6.90, 7.15, 7.47, 7.50, 7.55, 7.76	6.98, 7.12	8.32

The ^{13}C NMR of compound 3a, 3b and 3c in $CDCl_3$ are used as complementary technique for approving the structure of the prepared Schiff bases as shown in Table 3.2. C-C of alkyl chain resonance clearly identified at δ 9.47- 68.67 ppm. The C=C (alkene) were appeared at δ 124.28- 124.29 ppm and δ 125.73 ppm. The resonance of aromatic ring was recorded in the range δ 109.32 - 161.07 ppm. Resonance at δ 118.06-118.07 assign for C≡N in the compound. The resonance of C=N (imine) can been seen at δ 158.72, 158.73 and 158.71 ppm prove that compound 3a-3c were successfully synthesized.

Table 2 ^{13}C NMR spectral data of compound **3a-3c**

Compound	C-C (Alkyl chain)	C-C (Aromatic)	C=C (Alkene)	C=N (Imine)	C≡N
3a	9.47, 21.48, 68.67	109.32, 113.72, 120.49, 126.81, 127.85, 129.61, 130.93, 131.46, 132.68, 140.95, 151.56, 161.04	124.28, 125.73	158.72	118.07
3b	13.01, 21.57, 24.66, 28.11, 30.54, 67.22	109.33, 113.73, 120.49, 126.81, 127.84, 129.61, 130.93, 131.46, 132.68, 140.96, 151.57, 161.06	124.89, 125.73	158.73	118.07
3c	13.09, 21.65, 24.99, 28.15, 28.23, 28.36, 28.50, 30.85, 67.23	109.34, 113.73, 120.49, 126.81, 127.85, 129.61, 130.94, 131.46, 132.68, 140.97, 151.58, 161.07	124.89, 125.73	158.71	118.06

4. Conclusion

In conclusion, three Schiff's bases: 3a, 3b and 3c were successfully synthesized by condensation reaction between primary amine and aldehyde and characterized by UV, FTIR, δ_H-NMR, ^{13}C-NMR spectroscopies and Mass spec. All three Schiff base compounds consist of aromatic substituent which make the compounds more stable. These compounds were believed to give great prospect in sensor which it can enhance selectivity and sensitivity in sensor detection. In future work, all three compounds will be applied as potential linker in E-DNA sensor for cancer detection to improve the sensitivity of the sensor. Besides, we hope the compound can be excellent linker as it consists of cyano functional group and alkyl chain to chemically binding with gold electrode and phosphate group in DNA.

5. Reference

[1] Kumar J, Rai A and Raj V 2017 *Org. Med. Chem. I. J.* **1** (3) 555564

[2] Ay E 2016 *CBU J. of Sci.* **12** (3) 375-392

[3] Silva C M, Silva D L, Modolo L V, Alves R B, Resende M A, Martins C V B and Fatima A 2011 *J. Adv. Res.* **2** 1-8

[4] Sashikala S and Shafi S Syed 2014 *Der Pharma Chem.* **6** (2) 90-97

[5] Ashraf M A, Mahmood K and Wajid A 2011 *Int. Conf. Chem. Chem. Process* **10** 1-7

[6] Qin W, Long S, Panunzio M and Biondi S 2013 *Molecules* **18** 12264-12289

[7] Chigurupati S 2015 *J. Med. Bioeng.* **4**(5) 363-366

[8] Toliwa S, Jadav K, Pavagadhil T, Gupte A and Banu A 2009 *J. Appl. Chem. Res* **11** 7-17

[9] Turki A A 2012 J. *Kerbala Univeristy* **10** (3) 259-266

[10] Wang H, Qian X and Li J 2015 *2nd International Conference on Green Materials and Environmental Engineering* 43-44

[11] Drummond T G, Hill M G and Barton J K 2003 *Nat. Biotechnol.* **21**(10) 1192-1199

[12] Grieshaber D, MacKenzie R, Vörös J and Reimhult E 2008 *Sensors* **8** 1400-1458

[13] Darsanaki R K, Azizzadeh A, Nourbakhsh M, Raeisi G and Aliabadi M A 2013 *J. Biol. Today's World* **2**(1) 53-61

[14] Jin H, Wei M and Wang J 2013 *Chem. Cent. J.* **7** 65

[15] Lee H, Kang O and Choi S 2014 *Int. J. Electrochem. Sci.* **9** 6793 - 6808

[16] Rashid J I A and Yusoff N A 2017 Sens. Bio-Sensing Res. 16 19-31

[17] Han S W and Kim K 2001 J. *Colloid Interface Sci.* 240 492–497

[18] McEwan G, Chen F and Zhou A 2009 Anal. Chim. Acta 643 26–37

[19] Lee H, Kang Y Ok and Choi S 2014 Int. J. Electrochem. Sci. 9 6793 – 6808

[20] Kupčiūnaitė J, Kaušaitė A, Ramanavičius A and Ramanavičienė A 2006 Biologija 3 1-3

[21] Hasan A, Abbas A AND Akhtar M N 2011 *molecules* **16** 7789-7802

[22] Alsaygh A, Al-Humaidi J and Al-Najjar I 2014 *Int. J. Org. Chem.* **4** 116-121

[23] Issa R, Khedr A and Rizk H 2005 *Spectrochimica Acta, Part A: Molecular and Biomolecular Spectroscopy (SAA)* **62** 621–629

Acknowledgments

We would like to thank Universiti Malaysia Terengganu and Ministry of Higher Education (MOHE) for research grant FRGS vot number 59383 for giving us financial support and opportunity to be involved in this research

IOP Conf. Series: Materials Science and Engineering **440** (2018) 012017 doi:10.1088/1757-899X/440/1/012017

Photocatalytic activity of hydrothermally synthesized Al_2O_3graphene nanocomposite

Nur Farihah Mohamad Isa[1], Nik Raihan Nik Yusoff[2], Mohd Hasmizam Razali[3], Mahani Yusoff[1*]

[1]Faculty of Bioengineering and Technology, Universiti Malaysia Kelantan, Jeli Campus Locked Bag 100, 17600 Jeli, Kelantan, Malaysia.
[2]Faculty of Earth Science, Universiti Malaysia Kelantan, Jeli Campus Locked Bag 100, 17600 Jeli, Kelantan, Malaysia.
[3]School of Fundamental Sciences, Universiti Malaysia Terengganu, 21030 Kuala Terengganu, Terengganu, Malaysia

Corresponding author email: mahani@umk.edu.my

Abstract. In this study, Al_2O_3-graphene nanoplatelets (GNP) nanocomposite was evaluated on its properties and performance towards methyl orange (MO) dye. Al_2O_3-GNP nanocomposite was prepared by conventional hydrothermal at 200°C for 24 h. The result showed the crystallite size of Al_2O_3 is decreased and internal strain increased with the increased GNP content. The particle size of nanocomposite becomes larger with the increment GNP amount in the nanocomposite. Nanocomposite with lower graphene contents (20 wt%) and higher Al_2O_3 contents performed (80 wt%) the optimum for the MO absorption with efficiency of 75% in visible light.

1. Introduction

Al_2O_3 is the most widely used ceramic matrix to produce composite due to low price, high hardness and chemical stability [1]. Al_2O_3 also possess high thermal conductivity, melting point and thermal expansion making it possible for high temperature application [2]. However, Al_2O_3 suffers from low fracture toughness [3-4]. Therefore, Al_2O_3 is reinforced with nanoparticles such as carbon nanotube [5], titanium dioxide nanoparticles [6] and silver nanoparticles [7]. Al_2O_3 also has potential to be functionalized in environmental application as it has properties that could reduce the band gap of the materials [8]. However, Al_2O_3 is seldom reinforced with graphene for the used as photocatalytic materials for degradation of dyes.

Reinforcement of the composite is defined as binder that combines the two materials together. Reinforcement usually carries the load of the composite for about 70 to 90% by fibers. The reinforced materials provide structural properties such as stiffness, strength and thermal stability to the composite. The examples of the reinforcement materials used as reinforced phase in Al_2O_3 are silicon nitride (Si_3N_4) [9], carbon nanotube (CNT) [10] and silver nanoparticles [7].

Al_2O_3-based nanocomposite has been found mostly on high temperature application. However, there is limited information regarding the performance of Al_2O_3-based nanocomposite in waste water application. Recently, graphene gains the interest of the most researchers to become reinforcement in the nanocomposite [11]. Graphene with structure of two-dimensional (2D) honeycomb lattice is also an ideal candidate to be reinforced with Al_2O_3. This is because graphene exhibits excellent electrical and mechanical properties [11]. Graphene displays a good compatibility with Al_2O_3 nanocomposite that can enhanced the mechanical properties of nanocomposite [12] - [14] so that it can be used for waste water application. Therefore, this study used Al_2O_3-graphene nanocomposite to degrade the dyes in wastewater treatment. The aim of the research is to evaluate the characteristics of using different content of graphene in Al_2O_3-graphene nanocomposite produced by hydrothermal method. The performance of Al_2O_3-graphene nanocomposite towards the degradation of dyes for environmental application potential will be discussed.

2. Experiment

2.1 Preparation of Al_2O_3-GNP nanocomposite

The raw materials used in this study were graphene powder (>99% purity, average particle size >20 nm) and Al_2O_3 powder (>99.9% purity, average particle size >20 μm) and were purchased by Sigma Aldrich. Al_2O_3 and graphene nanocomposite was prepared according to Table 1. Prior to performing hydrothermal method, the sample was premixed using ball milling for an hour. Then, 100 ml aqueous solution consisted of 1M NaOH and water was added to the mixture. The mixture was stirred for 30 minutes and subjected to hydrothermal treatment at 200°C for 24 h in an autoclave. Then, the sample was washed with 100 ml 0.1 M HCl and distilled water. The sample was dried in an oven at 80°C for 24 h.

Table 1. Nanocomposite composition for hydrothermal process

Nano composite	Al_2O_3 (wt%)	GNP (wt%)
AG80	80	20
AG70	70	30
AG60	60	40
AG50	50	50
A100	100	0

2.2 Characterization

The determination of phase identification of Al_2O_3-graphene nanocomposite was performed by X-Ray Diffraction (XRD) using Bruker D2 Phaser. The step size was 0.02° and the range of 2θ angle between 20° to 90°. Morphology of nanocomposite was studied using scanning electron microscope (SEM) JSM-IT100. Fourier Transformation Infrared Spectrophotometer (FTIR) was used to determine the specific functional groups of Al_2O_3-GNP nanocomposite using Nicolet IS5 Spectrometer. Thermo Scientific OMNIC Specta software was used to analyse the functional group of the composite.

2.3 Photocatalytic performance

The performance of nanocomposite on photocatalytic degradation of methyl orange (MO) was studied. 0.1 g of samples was added into 100 ml of 20 ppm MO dye solution. The aqueous suspension was magnetically stirred throughout the experiment. The suspension was subjected to visible light irradiation start from 1, 2, 3 and 4 h and any discoloration was recorded by UV-vis.

UV-Vis spectrometer HACR DR 5000 with specification wavelength range of 190 to 1100 nm, accuracy of ±1 nm in wavelength range 200 to 900 nm and resolution of 0.1 nm was used to record the absorption spectra. The wavelength was 457 nm and the original absorbance value of MO was taken 1.9. The percentage of the MO degradation was calculated using the following equations:

$$MO\ absorbtion\ (\%) = \frac{A_0 - A_l}{A_0} \quad (1)$$

where A_0 is original absorbance value of MO and A_t is absorbance value of compositions with MO.

3. Results and Discussion

Phase identification of Al_2O_3-GNP nanocomposite was characterized using XRD is shown in Figure 1. Peak pattern of raw Al_2O_3 and after hydrothermal process shows no significant difference between these two patterns. This is because at temperature set up 200°C of hydrothermal was not enough to form new or change structure in Al_2O_3. This is in agreement with Shehata *et al.*, 2011 [15] that even though at thermal treatment of 975°C it still could not change Al_2O_3 form as the Al_2O_3 were uniformly dispersed. In Al_2O_3-graphene nanocomposite, the intensity of Al_2O_3 become sharp with decreasing GNP content indicating this composite is a polycrystalline material. While, peak of GNP was maintained since the amount GNP was higher. In the nanocomposite containing higher Al_2O_3 content, small shift to the right can be detected indicating the presence of graphene affects the Al_2O_3 lattice structure. Moreover, as the Al_2O_3 content increased, the crystal structure of the nanocomposites also changes. The intensity of Al_2O_3 becomes narrower which is due to hexagonal lattice of graphene changes to tetragonal after it is dispersed in Al_2O_3. This indicates that the introduction of graphene to the Al_2O_3 matrix changed the crystal structure of the nanocomposite.

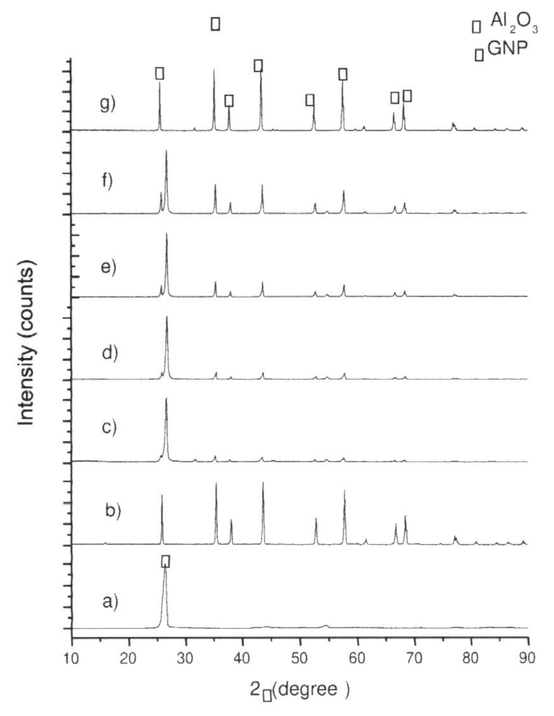

Figure 1. XRD pattern of a) as-received GNP, b) as-received Al_2O_3, c) AG50, d) AG60, e) AG70 and f) AG80 g) A100 nanocomposites

The calculated crystallite size and internal strain of Al_2O_3 is presented in Table 2. It can be observed that as the graphene content decreased, the crystallite size of Al_2O_3 decreased and the internal strain increased. This result also was found similar to the study of Lee and Kale (2008) [16].

Table 2. The crystallite size and internal strain of Al_2O_3-GNP nanocomposite

Nanocomposite	Crystallite size (nm)	Internal strain
AG50	20.3	0.0050
AG60	19.0	0.0096
AG70	14.6	0.0195
AG80	14.3	0.0206

Figure 2 shows the SEM images of different compositions of Al_2O_3 -GNP nanocomposite. In the equal amount of Al_2O_3 and GNP particles, it is obtained that Al_2O_3 particles are distributed in GNP particles. As shown in Figure (b), the Al_2O_3 particles are clearly distributed as a result of increasing Al_2O_3. This implies that hydrothermal with 200°C was enough to homogenously distribute the GNP particles in Al_2O_3 particles.

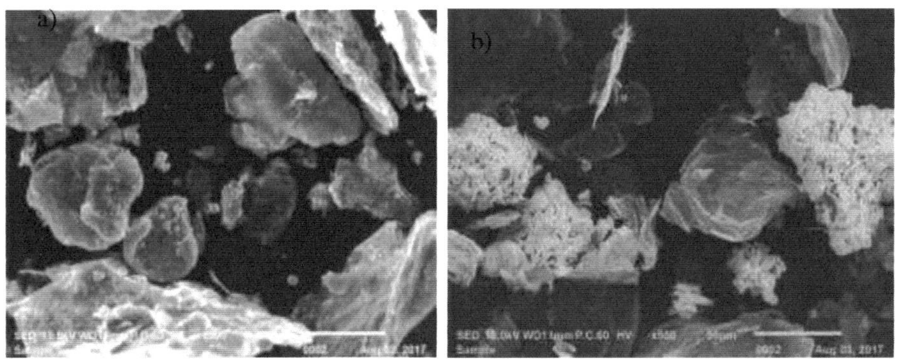

Figure 2. SEM images of (a) AG50 and (b) AG80 nanocomposites

Figure 3 presents the FTIR spectra of as-received GNP, hydrothermally Al_2O_3 (A100) and Al_2O_3-GNP nanocomposites. The significant absorption peak located at 1850 to 2250 cm^{-1} were observed similar absorption at 2104 cm^{-1} to that of as-received GNP which corresponding to the triple bond of carbon stretching vibrations. The percentage transmittance of A100 and as-received GNP was quite far from each other. This means that the intensity of Al_2O_3 is greater than GNP because Al_2O_3 absorbed more amount of radiation than GNP [17]. Moreover, other peak absorptions are observed at 3291 and 3378 cm^{-1} attributed to the OH vibrations of adsorbed water molecules of A100 (OH stretching–Al–OH) [18][19]. The peak located at 1649 cm^{-1} is due to –OH bending. This could be explained that after hydrothermal process, the OH bond was developed inside the Al_2O_3 particles.

The performance of Al_2O_3-GNP nanocomposite on percentage of MO absorption in visible light is given in Figure 4. It can be observed that as the Al_2O_3 content became higher, the absorbance of MO became higher. The highest percentages of MO absorption are AG60 and AG80 with more than 80%. This showed that the photocatalytic efficiencies of Al_2O_3-GNP nanocomposite are influenced by the introduction of GNP as supported by Yan *et al.*, 2013 [20]. It can be seen that AG70 absorbed the lowest MO. AG100 also absorbed MO better with slightly increased percentage in 4 h. This is the same result as reported by Hassena *et al.*, 2008 [21] that Al_2O_3 itself also better in degradation of MO. Ahmad et al., (2017) also proved that nanocomposite shows higher MO adsorption rate [22]. At 3 h, the MO absorption of all the compositions has increased but after that began to decreased gradually. This might due to the surrounding environment factor that could affect the reading such as light intensity and temperature.

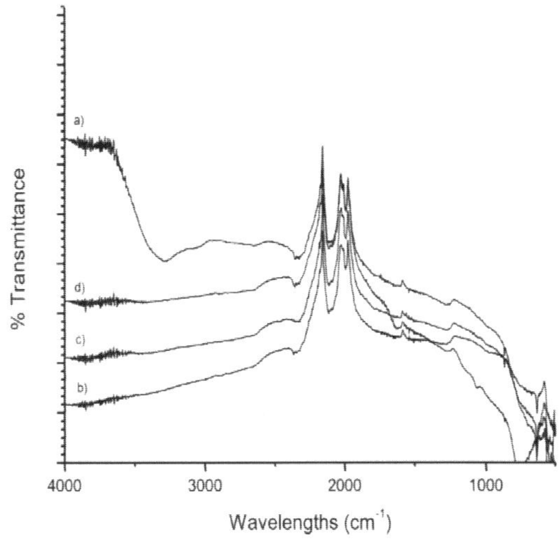

Figure 3. FTIR spectra of a) A100 b) as-received GNP c) AG70 d) AG50 nanocomposites

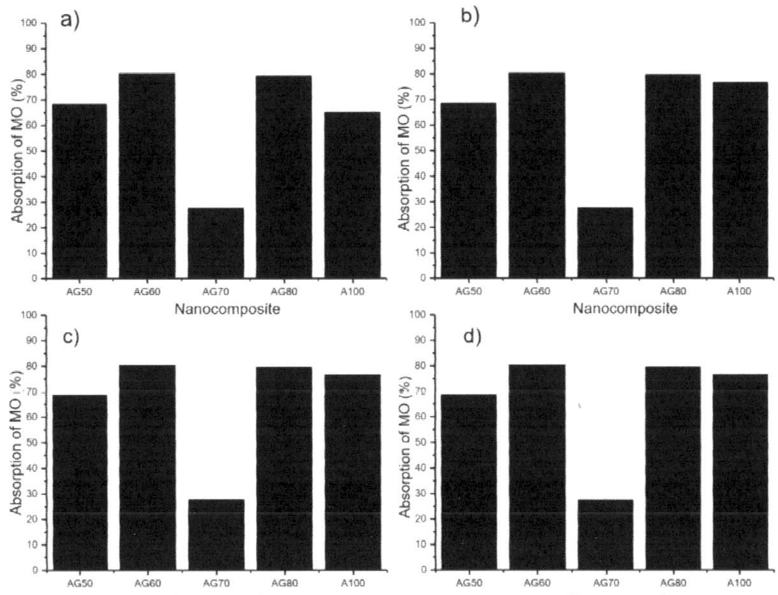

Figure 4. The absorption of MO for a) 1 b) 2 c) 3 and d) 4 h of Al_2O_3-GNP nanocomposites

4. Conclusion

Al_2O_3-GNP nanocomposite was sucessfully prepared by hydrothermal method. The performance of Al_2O_3 with lower graphene contents shows MO absorption with efficiency nearly 75% in visible light. The decreased of crystallite size and internal strain increased with the increased GNP content which also proportionate with the increment of particle size of the nanocomposite. Al_2O_3-GNP nanocomposite can be an alternative of metal oxide photocatalyst with graphene and may carry a great potential in environmental application.

5. Reference

[1] Yazdani B, Xu F, Ahmad I, Hou X, Xia Y and Zhu Y 2015 *Sci. Report.* **5** 11579
[2] Trunec M, Maca K and Chmelik R 2015 *J. Euro. Ceram. Soc.* **35** (3) 1001-1009
[3] Dong Y L, Xu F M, Shi X L, Zhang C, Zhang Z J, Yang J M and Tan Y 2009 *Mater. Sci. Eng. A* **502** (1-2) 49-54
[4] Hanzel O, Sedlacek J and Sajgalik P 2014 *Euro. Ceram. Soc.* **34** (7) 1845-1851
[5] Porwal H, Kasiarova M, Tatarko P, Grasso S, Dusza J and Reece M J 2015 *Adv. Appl. Ceram,* **114** 34-35
[6] Razali M H, Noor A F M and Yusoff, M 2017 *Sci of Adv. Mat.*, ***9*** (6)1032-1041
[7] Chaker H, Ch´erif-Aouali L, Khaoulani S, Bengueddach A and Fourmentin S 2016 *J. Photochem. Photobio. A Chem.* **318** 142-149
[8] Barajas-Ledesma E, García-Benjumea M J, Espitia-Cabrera I, Ortiz-Gutiérrezc M, EspinozaBeltránd F J, Mostaghimi J, Contreras-Garcíaa M E 2010 *Mater. Sci. Eng. B* **174** 71–73
[9] Ramdani N, Wang J, Wang H, Feng T, Derradji M and Wen-bin L 2014 *Compos. Sci. Technol.* **105** 73-79
[10] Long X Bai Y, Algarni M, Choi Y and Chen Q 2015 *Mater. Sci Eng. A.* **645** 347-356
[11] Liu J, Yang Y, Hasssinin H, Jumbu N, Deng S, Zuo Q and Jiang K 2015 *ACS Appl. Mater. Interfaces* **8** (4) 2607–2616
[12] Spanos K N, Georgantzinos S K and Anifantis N K 2015 *Compos. Struct.* **132** 536-544
[13] Simões S, Viana F, Reis M A L, Vieira M F 2015 *Compos. Struct.* **126** 114-122
[14] Drakonakis V M, Aureli M, Doumanidis C C and Seferis JC 2014 *Compos. B* **70** 175-183
[15] Shehata F, Abdelhameed M, Fathy A and Elmahdy M 2011 *Open J. Metal* **01** (02) 25-33
[16] Lee H Y and Kale G M 2008 *Int. J. Appl. Ceram. Technol.* **5** (6) 657–665
[17] Kloprogge J T, Ruan H D and Frost R L 2002 J. Mater. Sci. **37** (6) 1121-1129
[18] Yang G 2015 Int. J. Electrochem. Sci. **10** 9632 - 9640
[19] Khenfouch1 M, Buttner U, Baïtoul M and Maaza M *Graphene* **3** 7-13
[20] Yan J, Wang K, Xu H, Qian J, Liu W, Yang W and Li H 2013 *Chin J. Catal.* **34** (10) 18761882
[21] Hassena H 2008 *Modern. Chem. Appli.* **4** 1 1-5
[22] Ahmad A, Razali M H, Mamat M, Mehamod F S B and Amin K A M 2017 *Chemo.* ***168***474-482

Acknowledgements

The authors would like to thank Universiti Malaysia Kelantan (UMK) for providing the facilities to carry out this project and Malaysia Ministry of Higher Education for the financial support of FRGS 59358.

3D photonic crystals based poly (methyl methacrylate) for active photonic SERS substrates

Tahrin R A A[1] and Kassim S[1*]

[1]Advanced Nano Material Research Group, School of Fundamental Sciences, Universiti Malaysia Terengganu, 21030 Kuala Terengganu, Malaysia

Abstract. Synthesis of 400 nm average size poly (methyl methacrylate) (PMMA) and fabrication onto surface-enhanced Raman scattering (SERS) active substrate were presented.
"Bottom-up" technique was employed to fabricate SERS active substrate while PMMA suspension was obtained via surfactant-free emulsion polymerization. Morphology of fabricated PMMA photonic crystals (PMMAPc) SERS substrate was observed through scanning electron microscopy (SEM) with 20 000 magnification at 10 kV power. Photonic properties measurement was carried out using ocean optic spectrometry at range of 800 nm to 900 nm. Thermal resistivity study was measured through thermogalvanic analysis (TGA) over a range of 0 °C to 500 °C. Subsequently, enhancement percentage assessment by Raman spectroscopy provided evidence that PMMAPc SERS substrate potentiality towards Raman signal by using 4-aminothiolphenol (4-ATP) as probe molecule. Evidence from SEM supported that highly homogenous PMMA particles fabricated onto thin film brings wellordered arrangement in agreement with photonic crystals design. The proposed model was assessed experimentally using ocean optic spectrometry show a wide dip transmission at 899 nm and thermally stable up to 336 °C. Raman spectrum of 4-ATP using normal Raman analysis (without using PMMAPc SERS substrate) and using PMMAPc substrate support the results of the improvement of Raman signal upon 45 %. Low cost and large area fabrication would suggest PMMAPc as convenient yet affordable active SERS substrate for advance and fast detection.

1. Introduction

Optical properties that exist in periodic structure materials are one of excellent candidate towards technology development. This highly ordered material was so called as photonic crystals (Pc) where it has an ability to manipulate light in a very interesting way. The idea of creating these synthetic opals at first adopted from nature for example we can observe on butterflies wings, sea mice and even certain flowers [1-4]. The periodicity of Pc structures give rises to the propagation of electromagnetic (EM) in the material generally possessing photonic bandgap (PBG) where a region of range which light cannot propagate through the structure [5,6]. The potential was discovered and was divided to three types (1D, 2D and 3D) according to its ability in reflecting light. The dimensionality referred to as differing refractive index (ηeff) materials within the Pc which ηeff varies along one, two and three direction respectively. This behavior leads to difference ability in possessing photonic PBG properties [6-8]. Thus 3dimensional (3D) Pc become major interest as it was potential to offer complete photonic band gap properties. Complete PBG is very important in development and advance new technology in telecommunication and imaging [9-11].

The unique properties of Pc attracted considerable interest in creating artificial opal to be fabricated onto thin film especially for surface-enhanced Raman scattering (SERS). There is several technique proposed in fabricating SERS substrate for instance lithographic and dielectric layer stacking of This technique could fabricate a high quality 3D thin film however it required high cost which most company can't afford while it produced non-scalable film [12][13]. In addition, most of SERS substrate was proposed plamonic properties material for the enhanced effect and this might challenge the fabrication process thus specific instrument was used as mentioned before. Therefore it is sensible to introduced Pc as a material for SERS substrate fabricate Pc via "bottom-up" technique in which the particles were let to self-assemble onto thin film. Besides of its inexpensive choice "bottom-up" technique also can be performed in common laboratory practise which no complex procedure and expertise needed. Pc design could be constructed from many kind of material and silica is the commonly used due to its versatility however it could be a problem in etching process if using it as a template. High tendency of stacking fault is another remarkable disadvantage of using silica in construction of Pc because it will distort the sequence of lattice arrangement thus lead in distorted PBG. Yet, another great candidate in replacing is polymer. Among them, Poly (methyl methacrylate) (PMMA) is a correct option due to its short reaction period, flexible in designing, chemically stable, durable and weather resistance [14]. These characteristics of PMMA promising its ability to be incorporate in most photonic devices such as optical lenses in cameras and optical fiber [15].

In this study, 3D photonic crystals based SERS substrate was prepared. PMMA spheres was synthesised at 400 nm average particle size and used as main material in designing Pc. These polymeric materials could be prepared by emulsion polymerization [16], batch polymerization [17] and miniemulsion [18]. However in this project we proposed of using free-surfactant emulsion polymerization due to the enormous bad effect of surfactant to environment especially in causing water pollution [19]. The "bottom-up" technique was used as the fabricating method of the SERS substrates and produced a highly ordered FCC lattice arrangement of PMMA spheres. PMMAPc SERS substrate was experimentally investigated it performance using Raman spectroscopy with probe molecule 4-aminothiolphenol (4-ATP) and successfully enhanced Raman signal up to 45 %. Finally the conclusion are summarised and some finding remarks made.

2. Experimental

This work consists of synthesis of poly (methyl methacrylate) (PMMA) and fabrication of obtained PMMA spheres onto photonic crystals design called as PMMAPc. PMMAPc was investigated its performance as SERS substrate.

2.1. Materials

Methyl methacrylate monomer (MMA, 99%) was purchased from Aldrich (Milwaukee, WI, USA) while, 4-aminothiolphenol (4-ATP), potassium persulphate (KPS), ammonium hydroxide (NH_3OH) and hydrogen peroxide (H_2O_2) were purchased from R&M Marketing (U.K). Deionised water (DI H_2O, resistivity 18.0 MΩ) was used as reaction medium and for the preparation of all solutions. All chemicals and solvent were used as received without further purification.

2.2. Synthesis of homogenous PMMA spheres

PMMA sphere was prepared via surfactant-free emulsion polymerization in line with green-chemistry approach in which toxic solvent and linking agent were also exceptionally used. 64 ml of DI H_2O was heated under nitrogen gas at 90 ºC in tri-neck flask by silicon oil bath. Hotplate used was equipped with stabilizer rod to highly control the temperature throughout the experimental because polymerization was highly dependence on temperature. 15 ml of monomer was added followed with initiator KPS (0.1318M, 16 ml) and the reaction were continued for 45 minutes under stirring at 250 rpm. PMMA suspension was let to cool at room temperature. The filtrates was centrifuged, redispersed and re-agitated several times before fabrication.

2.3. Fabrication onto 3D photonic crystals

2.3.1. Substrate preparation. Microscope glass slide was used as substrate for fabrication of PMMA spheres as 3D photonic crystals. The glass slide was cut and cleaned with acetone by ultrasonication for 30 minutes and further treated with a mixture of DI H_2O, NH_3OH and H_2O_2 with ratio of 3:2:2 for at least an hour. This treatment is to create hydrophilic surface on the glass slide and ensure the attachment of PMMA spheres feasibly during fabrication.

2.3.2. PMMAPc growing film. 400 µl of PMMA suspension was dispersed in 9.6 ml of DI H_2O in a glass vial. Treated glass slide was immersed in the dispersed suspension vertically at approximately 45° settlement. PMMAPc thin film was let grow in an oven at 60 °C for 2 days. Fabrication required highly control evaporation rate thus the temperature was kept constant during the process without any disturbance.

2.4. Characterizations

Average size of PMMA spheres was measured using particle size anayzer (PSA, Malvern, UK) in wet mode by adding several drops of PMMA suspension onto 300 ml distilled water with turbine speed of 4000 rpm upon analysis. The analysis was taken 3 times and the average was recorded. Morphology of PMMAPc was observed through scanning electron microscopy (SEM, JEOL model 6360F) at magnification of 20 000. The PMMAPc substrate was sputter coat with gold before viewing to minimize charging effects. Thermal resistivity study was performed using Mettler-Toledo thermo galvanic analysis (TGA) with a heating temperature over a range of 0 °C – 500 °C with argon gas flowrate at 10 ml per minute. Optical property was investigated using a halogen white light source. The transmitted light was collected with a detector and recorded via Ocean Optics spectrometer.

2.5. PMMAPc as SERS substrate

4-aminothiolphenol (4-ATP) was used as SERS probe molecule and was prepared at concentration of 1.0×10^{-7} M. 50 µl of prepared 4-ATP was drip onto fabricated PMMAPc and let dried in ambient condition for several minutes. The analysis was performed using microconfocal Raman spectroscopy (InVia Renishaw) at 514 nm with laser power of 200 mW, 20x lenses magnification and 2 s accumulation time. The analysis was taken 7 times to ensure the stability of PMMAPc as SERS substrate.

3. Results and discussion

3.1. Physical characterizations

A size distribution plot of PMMA measured by PSA is shown in figure 1. In view of the results obtained narrow distribution was observed from the analysis graph and the average size was obtained at 400 nm. Accordingly, the uniformity index was recorded at 0.244 showing narrow distribution properties. These statistics provide information of highly homogenous PMMA spheres was successfully synthesised even with exceptionally used of surfactant. In this work the homogeneity of PMMA sphere was an important parameter in construction to photonic crystals. This was advance checked by SEM as shown in figure 2 (a) and the average size with uniformity was found to be no significant different with those measured using PSA.

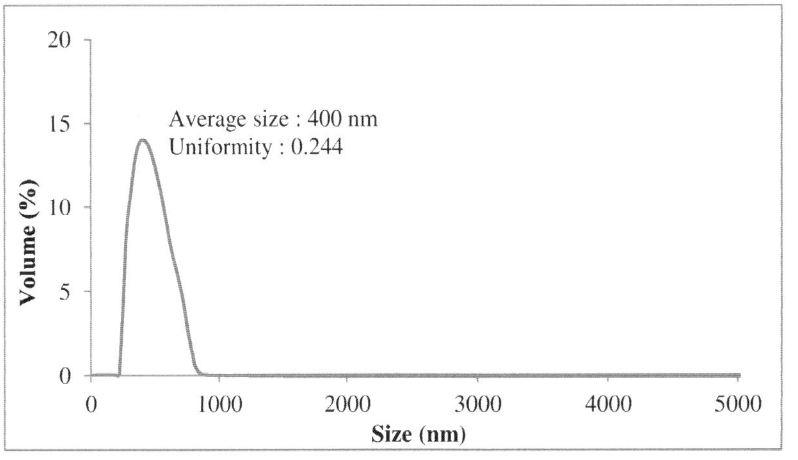

Figure 1. PSA analysis of PMMA suspension

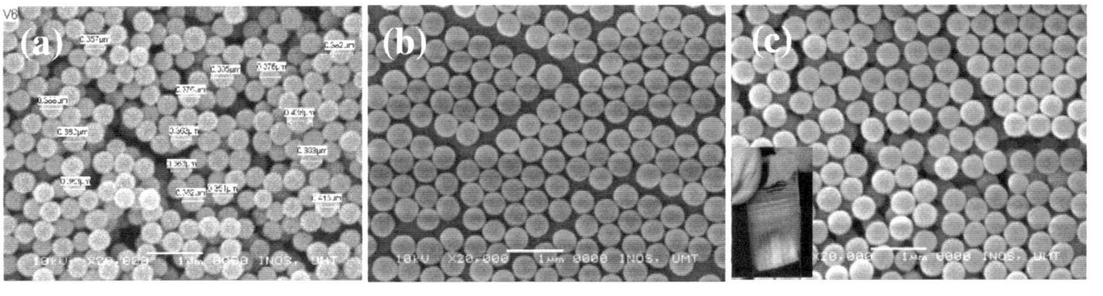

Figure 2. SEM image of PMMA spheres (a) in dispersion (b) monolayer PMMA film and (c) double layer PMMA film

Figure 2 shows SEM images of PMMA sphere in dispersion and after the fabrication. As earlier mentioned the average size of PMMA obtained have nearly similar as measured using PSA, however slight shrinkage might be observed from SEM compared to PSA might due to the shrinkage of PMMA spheres from heating proccess during sample preparation for SEM viewing. In figure 2 (b), the image of fabricated PMMA or PMMAPc was display a highly homogenous and well-ordered in face-centred cubic (FCC) lattice structure. From the image, the self-arrengemnt of PMMA colloidal sphere onto FCC lattice structure would give rise to photonic band gap (PBG) properties hence beneficial in manipulating Raman signal [20, 21]. Accordingly, homogenous spheres play an important role in succesfully creating FCC lattice strucure thus exhibit PBG properties. Figure 2 (c) shows SEM image of PMMAPc film that consist of two layer of particles. The size and parameter is equal however the volume of PMMA suspension was double during the fabrication proccess. It can be oberved clearly the existance of second layer beneath. This indicate that layer of PMMAPc was tunable by changing the volume of uspension during preparation as mentioned in section 2.3.2. The supplementary insert photograph image displays the fabricated PMMA thin film that appear opalescence on naked eyes when interact with light. Simple and cost-effective SERS substrate could be developed by highlighting the uniqe yet powerful photonic crystals properties in manipulating Raman signal. Furthermore, "bottom-up" technique could be practised upon fabrication of photonic crystals film which promising high quality large area substrate. The overall performance of PMMAPc substrate will be discussed in the next section.

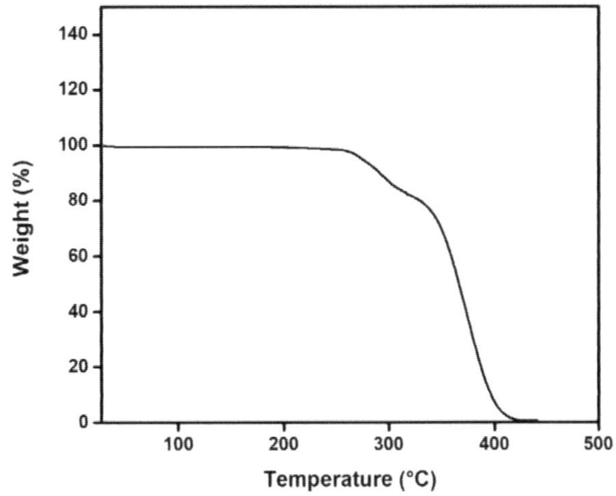

Figure 3. TGA curve of PMMA spheres

Figure 3 shows percentage loss of PMMA under thermal stress over a range of 0 to 500 °C using TGA at 10 °C/minute in argon atmosphere with a purge rate of 10 ml/minute. In view of the result obtained, a little weight loss (8.06 %) occurred up to a temperature of 297 °C. However, immediately beyond this temperature, a major decrease in weight was observed, and at 441 °C, almost complete weight loss was observed. This major weight lost is recorded to 68.05 % corresponding to degradation of random chain scission proccess of PMMA polymer [22][23].

3.2 Optical properties

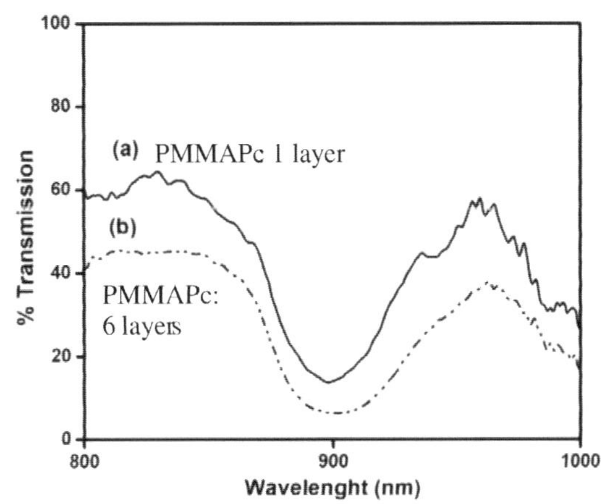

Figure 4. Transmission spectra of PMMAPc film in corellation to layer of the film (a) 1 layer and (b) 6 layers.

Figure 4 shows transmission spectra obtained from ocean optic spectrometer for a layer and 6 layers of PMMAPc according to curve (a) and (b) respectively. White light was strikes at 0° and the spectra were collected from the detector. As can be seen wavelength dependence for transmission which their periodicity and amplitude decrease when increasing the number of layers. Wide dip in the transmission spectra at 899 nm corresponding to the Bragg reflection peak, when propagation through the opal is nearly forbidden.

3.3 PMMAPc as SERS substrate

Final investigation towards PMMAPc ability in enhancing Raman signal was perfomed using Raman spectrsocopy with 4-ATP as SERS probe molecule. 4-ATP was analysed without using PMMAPc SERS substrate and by using the susbtrate and the obtain spectrum was summarized in figure 5.

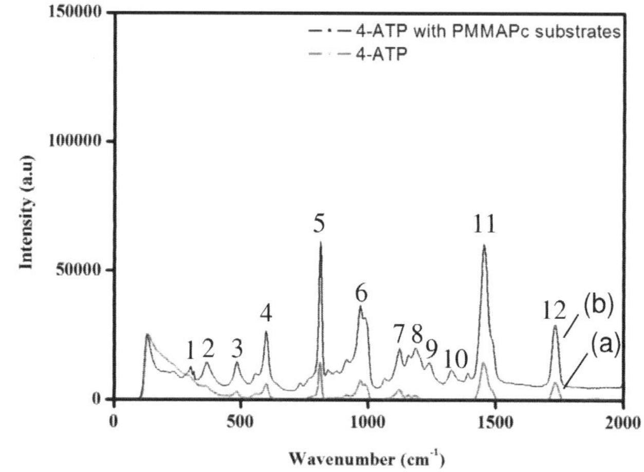

Figure 5. Raman spectra of 4-ATP (a) without PMMAPc substrate and (b) with PMMAPc substrate

Figure 5 shows graph obtained from Raman spectroscopy analysis of 4-ATP by normal Raman and by using PMMAPc substrate. From the graph the red line (line (a)) indicating the curve of 4-ATP from normal Raman while black line (line (b)) is corresponding to the curve of 4-ATP using PMMAPc substrate. It can be seen clearly that the obtained curved of 4-ATP using PMMAPc substrate was found to be higher than the one who does not using the susbtrate. Several absence peak of 4-ATP were also remark from line (a) compared to line (b) indicated that the sensitivity of detection was increased by using PMMAPc SERS substrate. Graph of enhancement for each peak was summarized in figure 6 to visual a clearer enhancement for each peak accordingly.

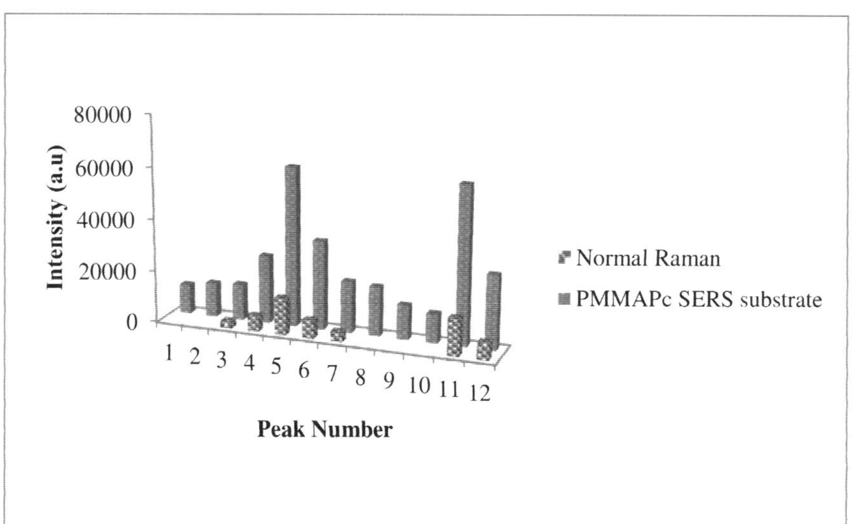

Figure 6. Enhancement of intensity for each corresponding peak in figure 5

Figure 6 shows intensity for the enhancement of each peak accordingly (refer figure 5). Peak 1, 2, 8, 9 and 10 does not show up in normal Raman analysis however enhancement up to 11595, 13320, 1194, 1250 and 1333 a.u was obtained from SERS analysis. To be noted, the analysis was performed on a very low concentration of 4-ATP which is 1.0×10^{-7} M. This reveals the sensitivity of PMMAPc SERS substrate that would beneficial in detection of limited sample for example in forensic and cancer studies. The other major enhancement were also spotted from peak 3, 4, 5, 6, 7, 11 and 12 up from 20 to 45 % in average. An excellent performance of PMMAPc SERS substrate was succesfully proved from the analysis thus the proposed design of using Pc as one of developed materials in SERS substrate was reasonable.

4. Conclusion

This work was devoted to assess the ability of PMMAPc to enhanced Raman analysis. This was performed by investigating the enhancement peak for normal Raman of 4-ATP with a 4ATP/PMMAPc SERS substrate. The results of PSA shows a homogenous PMMA sphere was successfully obtained from surfactant-free emulsion polymerization. From SEM observation PMMAPc substrate have well-ordered FCC lattice structure that would exhibit PBG properties as explained by ocean optic analysis. TGA analysis shows PMMAPc have experienced polymer degradation at temperature 336 °C this properties supported the uses of PMMAPc as SERS substrate as it could stand at high temperature during sample preparation or analysis process. An experiment was carried out to assess the proposed SERS substrate and comparison to normal Raman analysis using 4-ATP probe molecule. Sensitive PMMAPc SERS substrate able to reveals several peak that was absence during normal Raman analysis and also increase the enhancement of Raman spectra up to 45 %. It is considered that the PMMAPc SERS substrate can provide a low cost, robust, large area and sensitive detection over Raman analysis. All the obtained results suggest that photonic crystals have a bright potentiality in enhanced Raman analysis thus the combination of this material with metallic nanoparticles into such metallodielectric photonic crystals (MDPCs). The MDPCs structure was expected to enhance Raman signal thus serve as ultrasensitive SERS substrate.

5. References

[1] White T P 2005 *Thesis of PhD*, University of Sydney
[2] Bockstaller M, Kolb R, and Thomas E L 2001 *NSLS activity, science highlight* **13** 23
[3] David J N, Erin J A, Linli M, Ruth H and Scriven L E 2004 *J. of adv. mater* **16** 16
[4] Kassim S, Padmanabhan S C, Salaun M, and Pemble M 2011 AIP Proceeding **1391** 1
[5] Kassim S 2014 *Thesis of PhD*, University college cork
[6] Sibu C, Padmanabhan S C, Keith L, Shane O, Kassim S, Hugh D, Ian M P, Michael S and Pemble M 2014 *J. of mat. Chem. C* **2** 1675
[7] Zuocheng Z and Zhao X S 2005 *Langmuir* **21** 10
[8] Johnson S and Joannopoulos J D 2003
[9] Brewer D, Allen J, Miller M, Juan M, Kumar S, Norris D, Tsapatsis M and Scriven L E 2008 *Langmuir* **24** 23
[10] Waterhouse G I N and Waterland M R 2007 *Polyhedron* **26** 2
[11] Schroden R, Mohammed A, Blanford C and Stein A 2002 *J. of chem. Mat.* **14** 2
[12] Tang B, Wu C, Lin T and Zhang S 2013 *Dyes and Pigments* **99**
[13] Hattona B, Mishchenkoa L, Davis S, Sandhagec K and Aizenberga J, 2010 *J. of app. phy. Sci.* **107** 23
[14] Tanrisever, Taner, okay O and Sonmezoclu I C 1996 *J. of App. Poly. Sci.* **61**
[15] Arora P, Jain R, Mathur K, Sharma A and Gupta A 2010 *African J. of Pure and App. Chem.* **4** 8
[16] Sathish S and Shekar Chandar S 2013 *Indian J. of Pure & App. Phy.* **52**
[17] Harun N A 2013 *Thesis of PhD*, Newcastle University, UK
[18] Swisher R D 1966 *The Soap and Detergent Assoc.* Report No. 4, Madison Avenue, New York
[19] Wong S, Kitaev V and Ozin G A 2003 *J. of American chem. Soc.* **125** 20

[20] Escuti M J, Qi J, and Crawford G P 2003 *Appl. Phys. Lett.* **83** 1331
[21] Escuti M J, Qi J and Crawford G P 2003 *Appl. Phys. Lett.* **28** 7
[22] Cao X T, Bach L G, Islam Md R and Lim K T 2015 *Molecular Crystals and Liquid Crystals*, **618** 1
[23] Elshereksi N W, Mohamed S H, Arifin A and Ishak Z A M 2014 *J. of Phy. Sci.* **25** 2

Acknowledgments
We would like to express our acknowledgement to Ministry of Higher Education Malaysia (MOHE) for research grant FRGS 59391, Universiti Malaysia Terengganu for research supports and facilities.

IOP Conf. Series: Materials Science and Engineering **440** (2018) 012019 doi:10.1088/1757-899X/440/1/012019

Synthesis and characterization of Al-Fe-Cu tri-doped TiO_2 by in-situ hydrothermal for degradation of methylene blue dyes

R C Zulkifli[1, a], M H Razali[1, b],F Azaman,[2, c] A Ali[2, d] and M A A M Nor[1, e*]

[1]School of Fundamental Science, Universiti Malaysia Terengganu, 21030 Kuala Terengganu, Malaysia
[2]School of Ocean Engineering, Universiti Malaysia Terengganu, 21030 Kuala Terengganu, Malaysia

Abstract. Pure TiO_2 and a series of Al-Fe-Cu tri-doped TiO_2 were synthesized by using in-situ hydrothermal method. The synthesized materials were characterized using X-ray diffraction (XRD), Fourier Transform Infrared (FTIR) and Scanning Electron Microscope (SEM). The photocatalytic activity of the samples was evaluated for degradation of methylene blue (5mgL^{-1}) in aqueous solution under uv-visible light (λ= <420 nm). The XRD pattern showed that all samples exhibited anatase phase after calcined at temperature 500°C with crystallite size in the range of 10.00-13.23 nm. The FTIR spectra showed three important vibrational modes for νOH, δOH and νTiO that were assigned at range 3411.75-3399.27 cm^{-1}, 1626.911630.53 cm^{-1} and 503.30- 530.93 cm^{-1}, respectively. The degradation rate of $Al_{(1mmol)}$-Fe-Cu tri-doped TiO_2 showed the highest photocatalytic degradation with 93% compared to pure TiO_2, $Al_{(2\ mmol)}$-Fe-Cu tri-doped TiO_2 and $Al_{(3mmol)}$-Fe-Cu tri-doped TiO_2 were calculated as 76%, 85% and 80%, respectively. Hence, it can be seen that, the amount of Al ion-doping have influence on the photocatalytic degradation.

1. Introduction

Nowadays, waste water is the major pollution to the environmental issues. On a daily industrial and basis activities, huge amount of organic and inorganic dyes uncontrolled released in our rivers and seas. Effect from the continual exposure to this dyes can threatened humankind health such as irritations, toxic effects, and limitation to the safe drinking water [1]. Thus, science was interested in searching any low cost alternatives and environmentally friendly methods for filtering up waste water contamination. Recently, tremendous efforts have been devoted to minimize percentage issues of waste water contamination by using a filtration process from semiconductor materials. Among the studies of semiconductor materials, TiO_2 have received much attention due to their good performance such as high efficient photo catalytic activity, high refractive index, high chemical stability, nontoxicity and low cost [2]. However, TiO_2 suffer from some limitations, due to its large band gap, which around 3.0 eV and 3.2 eV for anatase and rutile respectively, occupy only 3-5% of total solar spectrum [3]. On the other hand, performance of photocatalytic activity might be decreased because of the fast recombination of photogenerated electron-hole pairs [4-5].

Therefore many studies have been reported to enhance the photocatalytic activity performance such as widen the absorption scope of TiO_2 to visible region and reduce the recombination of electron-holes. Previous study has reported that addition of dopant into TiO_2 lattice is commonly be used to increase the photocatalytic activities under visible light [6]. Selection of doping TiO_2 from metal such as Ce, La, Bi, and Mg and non-metal ions such as N, S and C have been reported to prolong the absorption edge and could enhance the photocatalytic activities under visible light [7]. Xing *et al.*, (2010) reported that, tri-dopant is better than co-dopant and single-dopant because of metal ions from tridopant can trap the photogenerated electrons and holes, which increasing photocatalytic activity [8]. However tri-dopant TiO_2 nanostructures for both nanoparticles and nanotubes are still in limited study and rarely reported. Hence, our study is to examine the effect of metals ion used as tri-dopant towards photocatalytic activity. In this work, $Al_{(1mmol)}$-Fe-Cu tri-doped TiO_2, $Al_{(2mmol)}$-Fe-Cu tri-doped TiO_2 and $Al_{(3mmol)}$-Fe-Cu tridoped TiO_2 were synthesized by in-situ hydrothermal method. In-situ hydrothermal process was chosen due to easy preparation method and high capability to form large surface area of nanostructure materials. In this case study, aluminium (Al) ion, iron (Fe) ion or copper (Cu) ion has been selected as dopants since they have highly abundance on the earth's crust, low cost, can reducing the electrons and holes recombination rates, most stable in the presence of sulfuric acid and owned smaller ionic radii compared to Ti^{4+} [9-12].

In-situ hydrothermal method involved of two stages including stirring and sonication to increase concentration of TiO_2 with NaOH solution as well as ended with washing process with HCl-dionized water to promote ionic exchange from Na^+ in TiO_2 and H^+ in HCl and deionized water. The effects of various amount of aluminium (III) nitrate nanohydrate were used to optimize the tri-dopant towards degradation study.

2. Experimental

2.1 Preparation

In this study all compounds, pure TiO_2, $Al_{(1mmol)}$- Fe-Cu tri-doped TiO_2, $Al_{(2mmol)}$-Fe-Cu tri-doped TiO_2 and $Al_{(3mmol)}$-Fe-Cu tri-doped TiO_2 were synthesized by in-situ hydrothermal method. 2 g of titanium dioxide powder, aluminium (III) nitrate nanohydrate (1 mmol, 2 mmol and 3 mmol), 1 mmol of iron (III) nitrate nanohydrate and 1 mmol of copper (II) nitrate trihydrate were dispersed in an aqueous solution of 10M NaOH (100 mL) and placed in a stainless-steel with Teflon lining reactor. Then, the reactor was transferred into an autoclave. The autoclave was heated at 150°C for 24 hours. After the treatment, the resultant precipitate was separated by centrifugation and washed with diluted HCl solution until pH 7. The sample was dried in an oven at 80°C for 24 hours. After drying, the obtaining powder was calcined at 500°C. Pure TiO_2 also was synthesized by using similar procedure, but without the presence of tri-dopant, aluminium (III) nitrate nanohydrate, iron (III) nitrate nanohydrate and copper (II) nitrate trihydrate.

2.2 Characterization

The X-Ray Diffraction (XRD) pattern was collected by using Rigaku Miniflex with Cu-Kα (0.1541 nm) at scan rate from 10°- 80° (2θ). Crystallite size was calculated by using Scherrer's Equation as stated below [13]

$$Scherrer's\ equation\ (D) = \frac{k\lambda}{\beta cos\theta} \quad (1)$$

Where (D =crystallite size in nm);
K= 0.9 constant
λ= 0.1541 nm (wavelength of the X-ray radiation)
β= FWHM (full width at half maximum)
θ= the diffraction angle

Infrared analysis was used to determine and identify functional groups present in pure TiO_2 and Al-FeCu tri-doped TiO_2 series. The model of FTIR used is Perkin Elmer Spectrum 100 with recorded in range of 4000-400 cm^{-1} utilizing potassium bromide (KBr) pellet. SEM micrographs were captured using JSM 6360 Joey Scanning Electron Microscopy with accelerated voltage 20kV and magnification from 10000X.

2.3 Photocatalytic degradation test

The photocatalytic degradation test was examined by adding 5mg of samples into 100ml of 5 ppm methylene blue dye solution [13-14]. The solution was directed to UV lamp for 1 hour and continues stirring. At every 10 minutes of interval time, 5ml of the solution was taken out by using syringe and filtered through 0.45 um milipore syringe filter to separate the powder from the supernatant. The absorption was recorded via UV-vis spectrophotometer (Spectrophotometer Shimadzu UV-1800).

3. Results and discussion

Figure 1 shows the XRD pattern of the prepared samples. It can be concluded that the diffraction peaks of all studied samples were represented to anatase TiO_2 phase as at 2θ = 25.26°, 37.78°, 48.00°, 53.93°, 55.03°, 62.78°, 70.19° and 74.98°. Anatase peak was formed by treating with 10M NaOH aqueous solution at 150°C for 24 hours and calcined at 500°C. Washing part has been introduced to eliminate Na^+ from $Na_2Ti_3O_7$ (titanate) (1). $Na_2Ti_3O_7$ was treated by using 0.1M of HCl and distilled water to form $H_2Ti_3O_7$ (3). At this stage, H^+ from HCl promoted ion exchange of Na^+. Generally, formation of TiO_2 anatase (5) reaction was occurred as stated below;

$$3TiO_2\,(commercial) + 2NaOH \rightarrow Na_2Ti_3O_7 + H_2O \quad (2)$$
$$Na_2Ti_3O_7 + 2HCl \rightarrow H_2Ti_3O_7 + 2NaCl \quad (3)$$
$$H_2Ti_3O_7 \rightarrow TiO_2 + H_2O \quad (4)$$
$$H_2Ti_3O_7 \xrightarrow{Calcinated\ at\ 500°C} TiO_{2(anatase)} \quad (5)$$

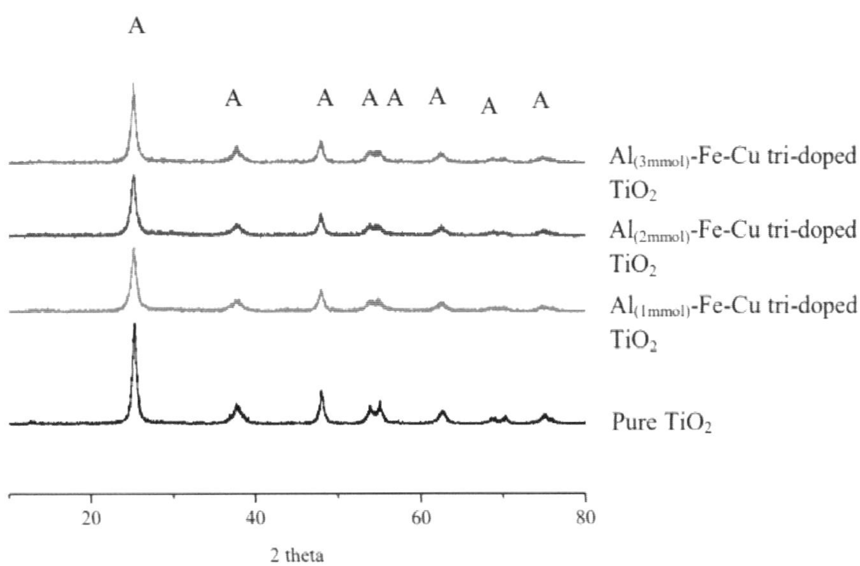

Figure 1. XRD patterns of pure TiO_2, $Al_{(1mmol)}$-Fe-Cu tri-doped TiO_2, $Al_{(2mmol)}$-Fe-Cu tri-doped TiO_2 and $Al_{(3mmol)}$-Fe-Cu tri-doped TiO_2

The XRD pattern also shows that there has no phase change in TiO_2 during addition of tri-dopant process and calcination at 500 °C. Furthermore, peak for Al^{3+}, Fe^{3+} and/or Cu^{2+} did not observed at XRD patterns. This is might be due to lower amount of aluminium (III) nitrate nanohydrate, iron (III) nitrate nanohydrate and copper (II) nitrate trihydrate used in the synthesis. In other hand, this is proving that, all the dopants successfully incorporated into the lattice site of TiO_2, since ionic radii of Al^{3+}(0.39Å), Fe^{3+}(0.58Å) and Cu^{2+} (0.57Å) are smaller than ionic radii of Ti^{4+} (0.65Å) [15]. Table 1 showed that, the crystallite size of $Al_{(1mmol)}$-Fe-Cu tri-doped TiO_2, $Al_{(2mmol)}$-Fe-Cu tri-doped TiO_2 and $Al_{(3mmol)}$-Fe-Cu tri-doped TiO_2 were found to be 10.00 nm, 10.57 nm and 11.29 nm, respectively. The crystallite sizes of tri-dopant were much smaller compared to pure TiO_2. Therefore, the smaller crystallite size contributes to the large surface area and high volume of pores resulting to the rapid adsorption-oxidation reaction occur [15].

Table 1. The average crystallite size

Sample	Crystallite size (nm)
Pure TiO_2	13.23
$Al_{(1mmol)}$-Fe-Cu tri-doped TiO_2	10.00
$Al_{(2mmol)}$-Fe-Cu tri-doped TiO_2	10.57
$Al_{(3mmol)}$-Fe-Cu tri-doped TiO_2	11.29

FTIR spectra of pure TiO_2, $Al_{(1mmol)}$-Fe-Cu tri-doped TiO_2, $Al_{(2mmol)}$-Fe-Cu tri-doped TiO_2 and $Al_{(3mmol)}$-Fe-Cu tri-doped TiO_2, respectively as shown in Figure 2. The FTIR spectra showed that three important vibrational modes for νOH, δOH and νTiO were assigned at range 3411.75-3399.27 cm^{-1}, 1626.91-1630.53 cm^{-1} and 503.30- 530.93 cm^{-1}, respectively. The νOH clearly observed in the spectra due to the physisorbed water molecule that present in the surface of TiO_2.

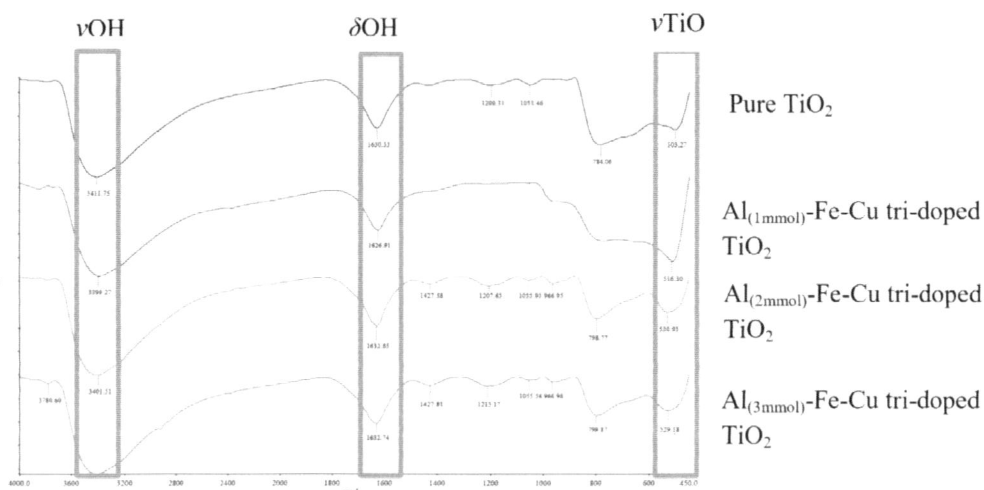

Figure 2. FTIR spectra of pure TiO_2 and Al-Fe-Cu tri-doped TiO_2 series

Based on the IR spectra, the peak Ti-O-Ti stretching of pure TiO_2 showed the lowest frequency which is 503.30 cm^{-1}. However, after the addition of dopants into TiO_2, the vibration of Ti-O-Ti stretching started to change to the higher frequency. This is might be due to the distribution of dopants in the molecules that shifted the Ti-O-Ti stretching into the high frequency.

SEM images of the pure TiO_2 and Al-Fe-Cu tri-doped TiO_2 series powder were represented in Figure 3. All of samples revealed that spherical structures aggregated and have several long fibres were found assembled together. This finding is in agreement with the results as reported by Razali *et al.*, 2014 [15]. From visual inspection of SEM morphology, nanotubes were expected successfully obtained. In contrast, the synthesized pure TiO_2 has larger particle size than tri-dopant as reliable with the crystal size estimation from XRD analysis (Table 1). However, to confirm the obtaining nanotubes, TEM analysis should be done in future.

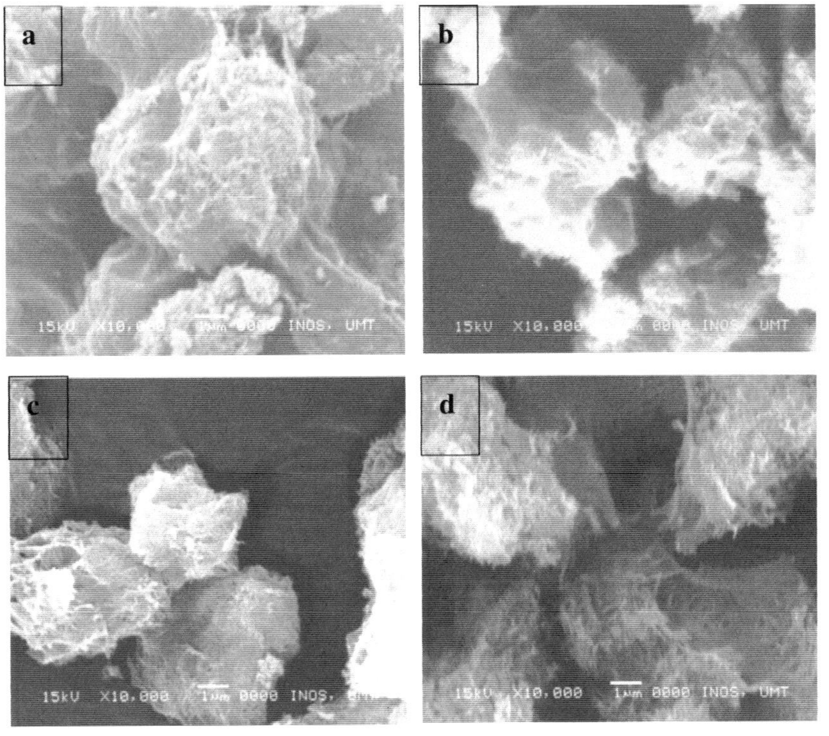

Figure 3. SEM images of (a) pure TiO_2, (b) $Al_{(1mmol)}$-Fe-Cu tri-doped TiO_2, (c) $Al_{(2mmol)}$-Fe-Cu tridoped TiO_2 and (d) $Al_{(3mmol)}$-Fe-Cu tri-doped TiO_2 viewed under 10000x magnification

The degradation photocatalytic was study by using different interval time towards irradiation. 5ppm of methylene blue solution was used to evaluate the performance of all samples upon discoloration. The degradation rates for all samples were determined by using formula as stated below:

$$degradation\ (\%) = \frac{C_o - C_1}{C_o} \times 100\% \tag{6}$$

C_o= initial abs
C_l=time abs

Figure 4 shows the degradation rate of $Al_{(1mmol)}$-Fe-Cu tri-doped TiO_2 sample exhibited the highest photocatalytic degradation, with 93% compared to the pure TiO_2, $Al_{(2mmol)}$-Fe-Cu tri-doped TiO_2 and $Al_{(3mmol)}$-Fe-Cu tri-doped TiO_2 were calculated as 76%, 85% and 80%, respectively. As expected, $Al_{(1mmol)}$-Fe-Cu tri-doped TiO_2 showed the highest photocatalytic degradation due to smaller crystallite size (10.00 nm) that fasten the degradation process. The principle of photocatalytic study is, when a distortion occurs because of the addition of any dopants into the lattice site of TiO_2, the TiO_2 crystallization maybe increased [16]. The surface area would be larger thus increasing the photocatalytic degradation [9]. Based on this fact, it can be seen that the presence of a small amount of Al^{3+} dopant in TiO_2 can enhance the photocatalytic degradation of methylene blue. Thus, 1mmol of aluminium (III) nitrate nanohydrate that used in this study perhaps is the best amount to the optimum level.

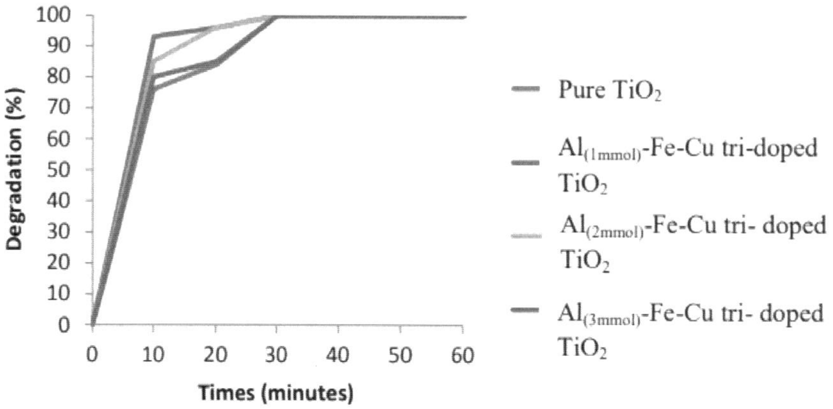

Figure 4. Photocatalytic degradation of Methylene Blue towards pure TiO_2 and Al-Fe-Cu tri-doped TiO_2 series

The mechanism throughout this case study may be summarized as previous study by equation below (6-16) [9,12,15].

$TiO_2 + h\nu \rightarrow h + e$ (6)
$Ti_{4+} + e \rightarrow Ti_{3+}$ (7)
$O_{2-} + h_{+} \rightarrow O_{-}$ (8)
$Cu^{2+} + Ti^{3+} \rightarrow Cu^{+} + Ti^{4+}$ electron trap (9)
$Fe^{3+} + O^{-} \rightarrow Fe^{4+} + O^{2-}$ hole trap (10)
$Fe^{4+} + OH \rightarrow Fe^{3+} + \cdot OH$ hole release (11)
$Cu^{+} + O_2 \rightarrow Cu^{2+} + \cdot O_2^{-}$ electron release (12)
$Al_{3+} + O_{-} \rightarrow Al_{4+} + O_{2-}$ hole trap (13)
$Al^{3+} + Ti^{3+} \rightarrow Al^{2+} + Ti^{4+}$ electron trap (14)
$Al^{4+} + OH \rightarrow Al^{3+} + \cdot OH$ hole release (15)
$Al^{2+} + O_2 \rightarrow Al^{3+} + O^{2-}$ electron release (16)
$Dye + O_2^{-}/\cdot OH \rightarrow$ The degraded product

This study also revealed that, these three metal ions used, Al^{3+}, Fe^{3+} and Cu^{2+} were proven gave the excellent performance towards photocatalytic degradation of methylene blue by reducing the recombination. As comparison, pure TiO_2 degraded only 78% of methylene blue after 10 minutes reaction due to the electrons and holes were recombined and affected the degradation process to the lowest rate [15]. However, Al^{3+} dopant may give the major contribution due to have smallest ionic radii. The insertion of Al^{3+} ion easily enter into lattice site of Ti^{4+} managed to create a trap site for electrons and holes thus, contributed to the highest photocatalytic activity.

4. Conclusion

Pure TiO_2 and a series of Al-Fe-Cu tri-doped TiO_2 were successfully synthesized by using in-situ hydrothermal method. All samples exhibited anatase phase after calcined at temperature 500°C with crystallite size in the range of 10.00-13.23 nm. The FTIR spectra showed three important vibrational modes for νOH, δOH and νTiO that were assigned at range 3411.75-3399.27 cm-1, 1626.91-1630.53 cm^{-1} and 503.30- 530.93 cm^{-1}, respectively. This experimental result revealed that Al^{3+}, Fe^{3+} and Cu^{2+} supressed the recombination rate of photogenerated electrons and holes, hence gave high photocatalytic degradation in methylene blue compared to pure TiO_2. However, $Al_{(1mmol)}$-Fe-Cu tridoped TiO_2 solution showed the highest photocatalytic degradation.

5. References

[1] Gnanasekaran L, Hemamalini R, Saravanan R, Ravichandran K, Gracia F and Gupta V K 2016 *Journal of Molecular Liquids* **223** 652

[2] Ge M, Li Q, Cao C, Huang J, Li S, Zhang S, Chen Z, Zhang K, Al_Deyab SS and Lai Y 2017 *Advanced Science* **1** 4

[3] Cowan AJ, Tang J, Leng W, Durrant JR and Klug DR. 2010 *The Journal of Physical Chemistry C* **114** 4208

[4] Kim W, Tachikawa T, Kim H, Lakshminarasimhan N, Murugan P, Park H, Majima T and Choi W 2014 *Applied Catalysis B: Environmental* **147** 642

[5] Wang H, Zhang L, Chen Z, Hu J, Li S, Wang Z, Liu J and Wang X 2014 *Chemical Society Reviews* **43** 5234

[6] Rafieh A I, Ekanayake P, Tan A L and Lim C M 2017 *Solar Energy* **141** 249

[7] Drăgan N, Crişan M, Răileanu M, Crişan D, Ianculescu A, Oancea P, Şomăcescu S, Todan L, Stănică N and Vasile B 2014 *Ceramics International* **40** 12273

[8] Xing M, Wu Y, Zhang J and Chen F 2010 *Nanoscale* **2** 1233

[9] Li F, Yin X, Yao M M and Li J 2011 *Journal of Nanoparticle Research* **13** 4839

[10] Shao G N, Jeon S J, Haider M S, Abbass N and Kim H T 2016 *Journal of colloid and interface science* **15** 179

[11] Hernández J V, Coste S, Murillo A G, Romo F C and Kassiba A 2017 *Journal of Alloys and Compounds* 710 355

[12] Zhou M, Yu J, Cheng B and Yu H 2005 *Materials Chemistry and Physics* **93** 159

[13] Sabry R S, Al-Haidarie Y K and Kudhier M A 2016 *Journal of Sol-Gel Science and Technology* **78** 299

[14] Ke S, Cheng X, Wang Q, Wang Y and Pan Z 2014 *Ceramics International* **40** 8891

[15] Razali M H, Ahmad-Fauzi M N, Mohamed A R and Sreekantan S 2014 *In Advanced Materials Research* **911**126

[16] Razali M H, Ahmad-Fauzi M N, Mohamed A R and Sreekantan S 2013 *International journal of material, mechanics, and manufacturing* **1** 314

Acknowledgement

The authors would like to thank The Ministry of Higher Education, Malaysia for the FRGS (Grant No: 59450) for financial support.

Synthesis and Characterization of B-S Co-doped TiO_2 Photocatalyst with Variable Boron Concentration

N A I Adnen[1,a], M A A Abdullah[1,b] and M A A M Nor[1,c*]

[1]School of Fundamental Science, Universiti Malaysia Terengganu, 21030, Kuala Terengganu, Malaysia.

[a]aslizaadnen93@yahoo.com, [b]aidil@umt.edu.my, [c]al_amin@umt.edu.my

Abstract. This study aimed to enhance photocatalytic activity of TiO_2 photocatalyst in methylene blue (MB) photodegradation. Boron (B)-sulphur (S) codoped TiO_2 photocatalysts were synthesized using sol-gel method with addition of TTIP as precursor. The reaction was carried out using two dopants (0.05 wt. % of S and 0-7 wt. % of B), stirred for 4 hours followed by washing until pH 7. The samples were dried in oven at 100 °C for 12 hours and calcined at 560 °C with heating rate 5 °C/min. Photodegradation of MB were carried out under ultraviolet (UV) irradiation. The undoped and codoped TiO_2 photocatalysts were characterized using FTIR, TGA, XRD and UV-Vis. The presence of TiO_2 peaks (687-719 cm^{-1}) were confirmed using FTIR analysis and thermal analysis of samples has been evaluated using TGA instrument. XRD analysis revealed the presence of anatase phase at 560 °C and the crystallite size (12-22 nm). Undoped and codoped TiO_2 were tested in removing MB from the solution and the results possessed codoped TiO_2 with 3 wt. % of B was the best photocatalyst for this study by degrading 59% and 100% of MB after 10 and 35 minutes, respectively.

1. Introduction

Recently, many manufacturing industries have discarded their unwanted by-products into surroundings and lead to the expansion of environmental pollution problems. There are various of catalytic techniques have been carried out using solar light and artificial light irradiation in order to solve these problems [1]. Despite of other methods, photocatalysis technique has become favourable because of highly promising and efficient method for environmental cleaning. According to Fujishima et al., (2005), it had been three years since the scientific studies on photocatalysis were carried out [2]. Photocatalysis is engaged with the presence of photocatalyst and energy from sunlight.

Photocatalyst such as titanium dioxide (TiO_2), zinc oxide (ZnO), tin (IV) oxide (SnO_2) and cerium (IV) oxide (CeO_2) has been widely used in photocatalytic applications [3, 4, 5, 6]. However the most used photocatalyst is TiO_2 because it possesses an excellent characteristic such as robust, higher photo-activity, eco-friendly, high stability and non-toxicity [7,8]. It also had been extensively used in solar energy conversion and gas sensing field [9]. However, TiO_2 only worked efficiently under UV irradiation and it was able to utilize less than 6 % of solar light because it had band gap energy of 3.2 eV for anatase phase and 3.0 eV for rutile phase, in which the absorption thresholds corresponded to 380 and 410 nm, respectively [10].

The morphology of TiO_2 can be improved by adding double dopants. As mentioned by Livraghi (2009), codoping technique generated synergistic effect which enhanced the properties of final TiO_2 products [11]. An optimum codoping amount can improve the properties of TiO_2 such as larger specific surface area, smaller crystallite size, shifting the absorption edge from UV to visible light region thus preventing the recombination process to occur [12]. Addition of sulphur (S) can improve the photocatalytic efficiency under visible light reaction and boron (B) helps the modification of TiO_2

by inhibiting the crystallite size, increasing the surface area as well as inducing the crystalline process [13, 14]. In this study, sol gel method was used because it is reproducible and can lead to nanoparticles, meso to macroporous materials, high purity and high surface area [15]. Samples were characterized using Fourier-transform infrared spectroscopy (FTIR), thermogravimetric analysis (TGA), X-ray diffraction (XRD) and ultraviolet-visible spectroscopy (UV-Vis).

1.1 Photocatalytic Mechanism of TiO_2

Generation of electron-hole pairs in photocatalytic reaction

$$\mathrm{TiO_2} + h_v \rightarrow e_{cv}{}^- + h_{vb}{}^+ \quad (1)$$

Formation of superoxide radical anion ($O_2 \bullet^-$)

$$e_{cb}{}^- + \mathrm{O_2} \rightarrow \mathrm{O_2} \bullet^- \quad (2)$$

Formation of hydroxyl radical (H^+) and surface adsorbed hydroxyl radical ($\bullet OH_{ad}$)

$$h_{vb}{}^+ + \mathrm{H_2O} \rightarrow \bullet \mathrm{OH} + \mathrm{H^+} \quad (3)$$

$$h_{vb}{}^+ + \mathrm{OH^-} \rightarrow \bullet \mathrm{OH_{ad}} \quad (4)$$

Recombination of electron and positive hole due to the abundance of electron occurs in conduction band

$$e_{cb}{}^- + h_{vb}{}^+ \rightarrow \mathrm{TiO_2} \quad (5)$$

Complete photocatalytic oxidation process has oxidize organic pollutant into salts, carbon dioxide (CO_2) and water (H_2O)

$$\bullet \mathrm{OH} + \text{pollutant} + \mathrm{O_2} \rightarrow \text{Products (salts}, \mathrm{CO_2}, \mathrm{H_2O}) \quad (6)$$

2. Methodology

Boron oxide (B_2O_3) and thiourea (CH_4N_2S) were used as the source of B and S respectively. 200 mL of cold distilled water was added into a beaker. While stirring, 18.5 mL of titanium (IV) isopropoxide (TTIP) was added into the beaker and stirred for 30 minutes. Under continuous stirring, 1 wt. % of B_2O_3 and 0.05 wt. % of CH_4N_2S were successively added into the solution, followed by adding glacial acetic acid (CH_3COOH). Stirring was continued for 3 hours. Then, 50 mL of 10 M sodium hydroxide (NaOH) was poured into the mixture and stirred for 30 minutes. Mixture was washed using distilled water until pH 7. The precipitate was dried in oven at 100 °C for 12 hours. Dried powder was crushed into finer powder and calcined at 560 °C. All steps were repeated for 0, 3, 5, 7 wt.% B. Samples were labelled as pure TiO_2, TiO_2.S0.05.B1, TiO_2.S0.05.B3, TiO_2.S0.05.B5 and TiO_2.S0.05.B7. All samples were characterized using FTIR, TGA, XRD and UV-Vis.

3. Results and Discussion

3.1 FTIR

In this study, IR spectra was analysed using potassium bromide (KBr) pellet technique and the IR spectrum of synthesized pure and co-doped TiO_2 photocatalysts were in the range of 4000-450 cm^{-1} wavenumber.

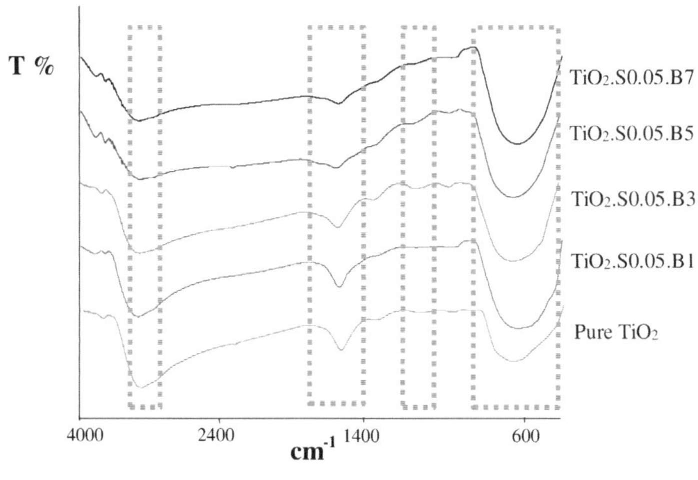

Figure 1. FTIR spectra of pure and co-doped TiO_2

The obtained FTIR spectra in Figure 1 showed the following main patterns. The strong absorption band at 3367-3396 cm^{-1} and 1624-1635 cm^{-1} are attributed to O-H stretching and bending vibration which comes from water absorbed on the TiO_2 surfaces and hydrogen-bonded hydroxyl groups [16,17]. It is in agreement with the statement from Mahalingam (2017), which stated that the broad band which corresponds to –OH stretching was recorded in the range of 3000–3500 cm^{-1} [18]. The absorption range around 1630 cm^{-1} may be related to hydroxyl (bending) groups of molecular water [19]. Peaks appeared at 1220 belonged to the B−O bond in the B_2O_3 crystal as reported by Chen (2006) [17]. Functional group of S was not identified due to its small amount used in this study. The strong absorption observed at 687-719 cm^{-1} is attributed to lattice vibrations of TiO_2. Lu *et al.* (2008) mentioned that the broad absorption band appeared at 500-1000 cm^{-1} was ascribed to the stretching vibration absorption of the Ti-O-Ti linkages in TiO_2 nanoparticles. The calcination of TiO_2 powders at 560 °C lead to significant sharpening of absorption bands in the region of 400-700 cm^{-1} and clearly indicates the formation of anatase phase [20]. There is no peak appeared at 2900 cm^{-1} in all spectrum which indicates that all organic compounds (C-H stretching band) decomposed from the samples after calcination [21]. The results revealed that addition of dopants will narrow the band gap of TiO_2 leading to improvement in the performance of photocatalytic activity.

3.2Thermo-gravimetric Analysis

Thermal properties of synthesized pure and co-doped TiO_2 are shown in Figure 2. All samples were examined from room temperature to 900 °C with heating rate 10 °C/min. The trend showed decreasing in weight loss and followed with plateau as the temperature increased to 900 °C. The weight loss of samples decreased with the addition of dopant. Among all curves, pure TiO_2 showed a larger overall weight loss. The significant weight lost was approximately 10.40 to 16.23% which occurred before 250 °C. As the temperature increased up to 400 °C, the decomposition of 3 to 4 % organic compounds (non-reactive titanium precursor) occurs along with the elimination of OH groups attached on the surface of synthesized TiO_2 nanoparticles [22, 23, 24]. Doped samples results confirmed a denser microstructure and lower surfaces area has caused less water absorption compared to undoped TiO_2 [25]. After 400 °C, the thermo-gravimetric curves showed plateau characteristic.

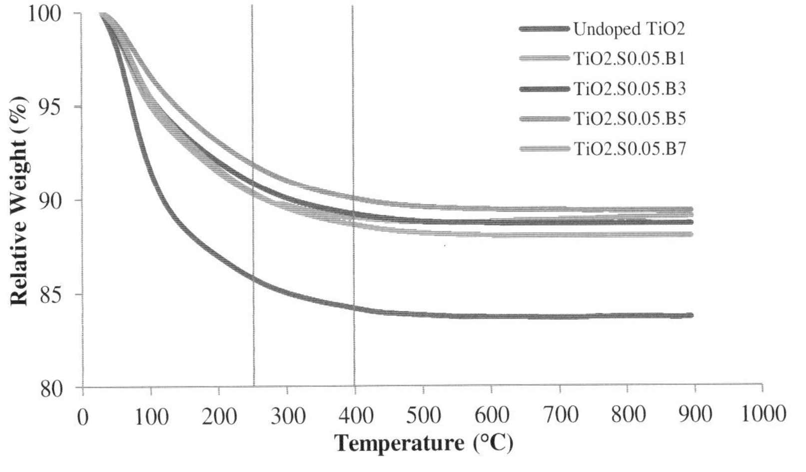

Figure 2. TGA traces of the synthesized TiO_2 powders measured at a heating rate of 10 °C/min

3.3 X-ray Diffraction

XRD analysis was carried out to determine the crystalline phase and to estimate the crystallite size of synthesized TiO_2 photocatalyst. Figure 3 displays the XRD diffractograms of pure and codoped TiO_2 photocatalyst calcined at 560 °C.

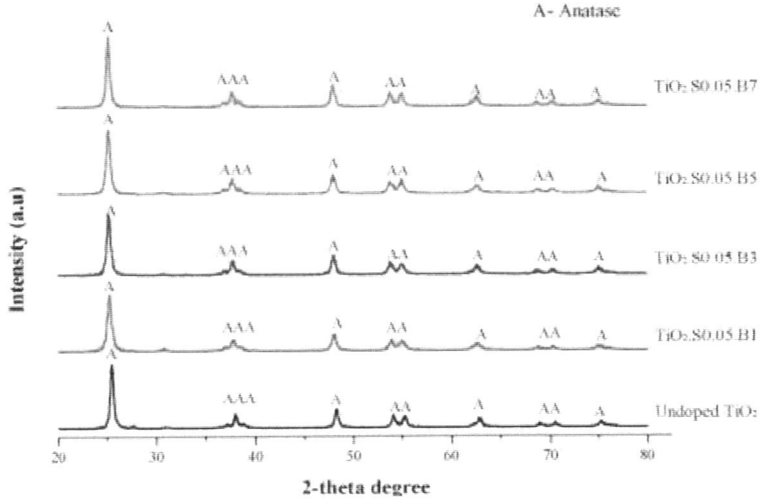

Figure 3. Crystalline anatase phases for synthesized photocatalysts

All samples were confirmed the presence of anatase phase of titania with the corresponding (101) plane of high intensity for both pure and codoped TiO_2 (JCPDS No. 21-1272). The peaks centered at 25.37, 37.99, 48.12, 54.26, 57.23, 62.79, 68.85, 70.44 and 75.15 ° are corresponding to anatase (101), (004), (200), (105), (211), (204), (116), (220) and (301) crystal planes, respectively. It can be observed that the increasing in B weight does not alter the polymorph of TiO_2. The peak of B and S was not detected, and this may due to good incorporation and dispersion of dopant into TiO_2 lattice.

IOP Conf. Series: Materials Science and Engineering **440** (2018) 012020 doi:10.1088/1757-899X/440/1/012020

The results revealed the presence of cystalline phase along with pure anatase phase when the samples were calcined at 560 °C. The crystalline sizes were calculated using Sherrer's equation in Equation 1 [20]:

$$D = \frac{k\lambda}{\beta \cos\theta} \quad (7)$$

Where D is the crystallite size in nanometer (nm), k is a constant (0.9), with the assumption of spherical particles), λ is the wavelength of the X-ray radiation (CuKα = 0.1541 nm), β is the full width at half maximum (FWHM) of selected peak and θ is the diffraction angle for the peak (Bragg's angle).

Table 1 shows the crystallite sizes decreased as the amount of boron increased to 3 wt.% and increased as the amount of boron exceeded 3 wt.%. This is because the addition of dopants will suppress the crystallite size of photocatalyst due to the insertion of dopant into the lattice structure of TiO_2 octahedral thus altering its properties [12].

Table 1. Crystallite sizes of synthesized photocatalyst

Photocatalyst	Crystallite size (nm)	Peak assignment
Undoped TiO_2	22	Anatase TiO_2
TiO_2.S0.05.B1	12	Anatase TiO_2
TiO_2.S0.05.B3	12	Anatase TiO_2
TiO_2.S0.05.B5	15	Anatase TiO_2
TiO_2.S0.05.B7	15	Anatase TiO_2

3.4 Photogedradation of MB

The findings showed that the degradation rate was directly proportional to increasing of time. As depicted in Figure 4, the graph showed a similar trend. However, by comparing codoped TiO_2 and pure TiO_2, the findings portrayed that codoped TiO_2 showed a better performance compared to pure TiO_2 [26, 27]. Addition of dopant into TiO_2 may alter the properties of photocatalyst by narrowing the band gap, lessening crystallite size and increasing the surface area, leading to higher absorption of dye molecules [28].

All samples were successfully degrading MB at different irradiation time. It reached a maximum activity at 3 and 5 wt.% of boron dopant and the performance decreased with increasing B concentration. The performance of 7 wt.% B doping was slower compared to 3 wt.% B doping but faster than pure TiO_2. Addition of 3 wt.% of B improved the performance of photocatalyst by degrading 100% MB at 35 min of UV irradiation as compared to 65 minutes of pure photocatalyst. However, as the doping concentration increased to 7 wt.%, it showed a decrease in degradation rate and it was fully absorbed dye molecules at 45 min of UV irradiation.

These results indicated that the performance of TiO_2 photocatalyst can be improved by adding appropriate amount of dopant and its performance can be deteriorated when excessive amount of dopant was added into TiO_2 photocatalyst [12]. The deterioration may due to the recombination by reducing the opportunity of oxygen photo-adsorbed on the surface of TiO_2 to react with free electrons to form O^{2-} ion or the corresponding hole to form •OH radical [12]. Formation of diboron trioxide phase after exceeding the optimal doping also inhibits the photocatalytic activity of photocatalyst [29]. Addition of S and B as codoping agent helps to promote the photocatalytic performance by suppressing the growth of TiO_2 nanoparticles as well as increasing the surface area thus enhancing the photocatalytic performance [13, 14, 30]. Hence, the photocatalytic activity was optimal when the doping amount of boron was 3 wt. %.

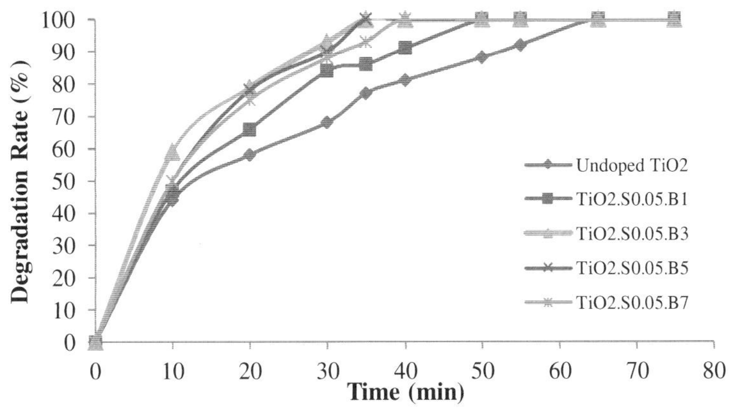

Figure 4. Effect of different composition of boron towards degradation rate of MB

4. Conclusion

Pure and codoped TiO_2 photocatalysts were synthesized via sol gel method and TTIP as a precursor. Formation of TiO_2 and the presence of boron were confirmed by FTIR. TGA results also demonstrated that there was significant mass loss in pure TiO_2. All samples showed a pure anatase peak with crystalline size ranging from 12 to 22 nm and incorporation of dopants were examined using XRD analysis. The codoped TiO_2 exhibited a higher photodegradation rate for MB under UV irradiation compared to pure TiO_2. Exceeding optimum amount of dopant deteriorated the photocatalytic performance. 3 wt. % was identified as the optimal composition for the best photodegradation in this research.

5. References

[1] Seyghali B and Zanjanchi M A 2015 Spectrochimica Acta Part A: Molecular and Biomolecular Spectroscopy **151** 104
[2] Fujishima A and Zhang X 2015 C. R. Chimie 9 750
[3] Chen X, Wu Z, Liu D and Gao Z 2017 Preparation of ZnO Nanoscale *Res. Lett.* **12** 143.
[4] Kim S P, Choi M Y and Choi H C 2016 Mater. Res. Bull. **74** 85
[5] Milenova K, Zaharieva K, Stambolova I, Blaskov V, Eliyas A and Dimitrov L 2017 *J. Chem. Tech. Metallurgy* **52** (1) 13
[6] Sabarinathan M, Harish S, Archana J, Navaneethan M, Ikeda H and Hayakawa Y 2017 RSC Advances **7** 24754
[7] Sabry R S, Al-Haidarie Y K and Kudhier M A 2016 *J. Sol-Gel Sci. Tech.* **78** 299
[8] Atout H, Álvarez M G, Chebli D, Bouguettoucha A, Tichit D, Llorce J and Medina F 2017 *Mater. Res. Bull.* **95** 578
[9] Rao A R and Dutta V 2007 *Sol. Energy Mater. Sol. Cells* **91** 1075
[10] Whang T J, Huang H Y, Hsieh M T and Chen J J 2009 *Int. J. Mol. Sci.* **10** (11) 4707
[11] Livraghi S, Elghniji K, Czoska A M, Paganini M C, Giamello E and Ksibi M 2009 *J. Photochem. Photobiol.* A **205** (2-3) 93
[12] Tan Y N, Wong C L and Mohamed A R 2011
[13] Ho W K, Yu J C and Lee S C 2006 *J. Solid State Chem.* **179** (4) 1171

[14] Xiao J R, Peng T Y, Li R, Peng Z H and Yan C H 2006 *J. Solid State Chem.* **179** (4) 1161
[15] Parra R, Góes M S, Castro M S, Longo E, Bueno P R and Varela J A 2008 *Chem. Mater*. **20** 143
[16] Zhang W, Yang B and Chen J 2012 *Int. J. Photoenergy*
[17] Chen D, Yang D, Wang Q and Jiang Z 2006 *Ind. Eng. Chem. Res.* **45** (12) 4110
[18] Mahalingam T, Selvakumar C, Ranjith Kumar E and Venkatachalam T 2017 *Phys. Lett.* A **381** 1815
[19] Zhao Y, Li C, Liu X, Gu F, Jiang H, Shao W, Zhang L and He Y 2007 *Mater. Lett.* **61** 79
[20] Haider A J, Jameel Z N and Taha S Y 2015 *J. Eng. Tech*. **33** 5
[21] Bagheri S, Shameli K and Abd Hamid S B 2012 J. Chem.
[22] DeBenedetti B, Vallauri D, Deorsola F and Garcia M M 2006 *J. Electroceram.* **17** 37
[23] Luca V 2009 J. of Phys. Chem. C **113** 6367
[24] Muniz E C, Góes M S, Silva J J, Varela J A, Joanni E, Parra R and Bueno P R 2011 **37**(3) 1017
[25] Caratto V, Setti L, Campodonico S, Carnasciali M. M, Botter R and Ferretti M 2012
[26] Chao-hai W, Xin-hu T, Jie-rong L and Shu-ying T 2007 *J. Environ. Sci*. **19** 90
[27] Yu C, Cai D, Yanga K, Yu J C, Zhou Y and Fan C 2010 J. of Phys. Chem. Sol. **71** (9) 1337-1343.
[28] Mesheshaa D S, Ravi Chandra Matangi R C, Siva RaoTirukkovalluri S R and Bojja S 2017 J. Asian Ceram. Soc. **5** (2) 136-143
[29] Zaleska A, SPBCZAK j w, Grabowska E and Hupka J 2008 App. Catal. B. **78** 92-100
[30] Hamadanian M, Reisi-Vanani A and Majedi A 2009 Mater. Chem. Phys. **116** (2–3) 376-382

Acknowledgements
The authors acknowledge the financial support by Ministry of Education Malaysia under Niche research grant scheme 53131/27 and Universiti Malaysia Terengganu for their support in carrying out this research work regarding FT-IR, TGA, XRD and UV-Vis analysis.

Effect of Milling Time on Alumina-Titania-Carbon Nanotube by Powder Metallurgy Method

M N Masri[1], M Yusoff[1] and N N Nordin[1]

[1]Faculty of Bioengineering and Technology, Universiti Malaysia Kelantan, Jeli Campus Locked Bag 100, 17600 Jeli, Kelantan, Malaysia.

Abstract. The Recent work discovers the evaluation on alumina-titania-carbon nanotube (Al_2O_3–TiO_2–CNT) composite prepared by powder metallurgy method. The properties of Al_2O_3–TiO_2–CNT composite have high hardness, chemical and thermal resistance, wear and corrosion resistance but lacking to low fracture toughness which lead to brittle properties. Hence, CNTwas added to the system due to attractive mechanical properties, low density and high fracture toughness. Elemental powders of Al_2O_3–TiO_2–CNT were milling in low energy ball milling using different milling time at 15, 30, 45 and 60 hours. The peak XRD of TiO_2 and CNT become diminished with increasing milling time. The powder that milled at 60 hour possible to has lowest crystallite size and highest internal strain due to finer and homogenous particle size deform during milling.

Keywords- Alumina, titania, carbon nanotube, powder metallurgy, compaction and milling

1. Introduction

Composites will be defined as one or more discontinuous phases embedded in a continuous phase on macro scale that can improve the properties of material [1]. The discontinuous phase are called reinforcement is usually harder and stronger than the continuous phase that are called as matrix. Composite will be divided into three classifications which are metal matrix composite (MMC), polymer matrix composite (PMC) and ceramic matrix composite (CMC).

CMC is material that promises as potential candidates in view of excellent physical and mechanical properties. This is due to their properties like high temperature stability, chemical inertness, strength and corrosion resistance that appropriate usage in industry [2-3]. In previous years, many researchers have been carried out to produce CMC in advance material that used in various industries such as electrical, medical application, aerospace and automotive part [4].

Alumina (Al_2O_3) ceramic material is mostly used as matrix for CMC application because of their superb properties such as high hardness, high compressive strength (0.5-5.0 wt.%) and good chemical resistance. This was proved according to many of CMC that applied of Al_2O_3 based ceramic as matrix material. For example, Al_2O_3 based silver (Al_2O_3 Ag) that improve fracture toughness in range 3.2–$MPa{\cdot}m^{1/2}$ and Al_2O_3 based silicon carbide (Al_2O_3–SiC) [5].

Based on research that was reported [6], alumina–titania (Al_2O_3-TiO_2) is one of CMC that already fascinating study by others due to good ceramic material for many applications such as machinery, textile and printing industries. Al_2O_3-TiO_2 also has tremendous properties due to their high hardness, excellent wear corrosion, chemical and thermal resistance. However, Al_2O_3-TiO_2 composite have limitation leading to poorer fracture toughness and high porosity because of not fully homogeneous of the grain size.

In order to improve the properties of Al_2O_3-TiO_2 composite, carbon nanotube (CNT) will be added to become Al_2O_3-TiO_2 systems prepared by powder metallurgy (PM) technique. CNT play an important role in order to improve the Al_2O_3-TiO_2 composite because of extent excellent properties such as high tensile strength in range of 10–100 times greater than steel [7], low density (~2.1 g/cm^3), light weight, promising stiffness, Young's modulus (973 GPa) and high strength (up to ~100 GPa) [8-9].

2. Methodology

The material used in this experiment were Al_2O_3 (> 99.9% purity, average particle size > 20 µm), TiO_2 (> 99.5% purity, average particle size > 21 nm) and CNT (99.9% purity, average particle size > 20 µm). The composition was used in this study to prepare Al_2O_3–TiO_2–CNT composite and has been purchased from Sigma Aldrich Company.

3. Results and Discussion

The characteristic of pure powders were analysed using XRD. Figure 1 shows the XRD pattern of Al_2O_3, TiO_2 and CNT. Based on XRD result, the pure powders were analysed to have crystalline structure. The Al_2O_3 powder was observed that have four strong peaks at 25.62°, 35.21°, 43.43° and 57.61. While, the° analysis shows the TiO_2 powder also have three sharp peaks at 25.31°, 36.92° and 48.04°. However, the other peak appeared show no strong because of short and broadening.

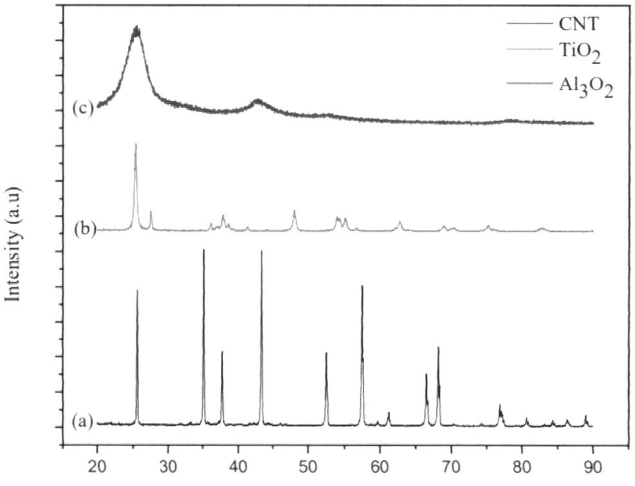

Figure 1: XRD patterns of (a) Al_2O_3 (b) TiO_2 and (c) CNT powder

The Al_2O_3 revealed as corundum phase while TiO_2 showed well defined diffraction peaks corresponding to anatase phase, that form stable rather than rutile and brookite. This similar with the previous research [7]. For both of Al_2O_3 and TiO_2 peak exhibit the strongest peak are at 35.03° and 35.15° that also present same to previous study that revealed in Figure 2.10. In Figure 4.1 the crystalline peak of CNT powder, highest intensity is obtained at position 24.66° and short peak at 44.67. Both crystalline peak of CNT powder was show not strong peak, this is due to high solid solubility of CNT and peak found in amorphous structure [10].

Figure 2 shows the result of XRD pattern of Al_2O_3-TiO_2–CNT powder mixture at different milling time (15, 30, 45 and 60 h) with 10 mm alumina ball. Al_2O_3 and TiO_2 phases are observed having

IOP Conf. Series: Materials Science and Engineering **440** (2018) 012021 doi:10.1088/1757-899X/440/1/012021

crystalline structure while CNT phase has very weak peak obtained after milled. The crystalline phases were obtained with different milling time provide no sign of amorphous phases have seen in diffraction peak for all milling time. The first crystalline peak of Al_2O_3, TiO_2 and CNT indicate was overlapped each other, due to their peak are seen much closed at 15 and 30 h of milling time. The peak crystallite structure of Al_2O_3 was seen at 25.57° for both 15 h and 30 h. The peak TiO_2 phase was presented at 25.30° for 15 h and small shifted to the left at 25.48° and no difference of 2θ for CNT peak at 25.63° for both 15 h and 30 h.

Based on Figure 2, the four strong peaks of Al_2O_3 has appeared at hkl $(\bar{1}0\bar{1})$, $(\bar{1}14)$, $(\bar{2}1\bar{3})$, and $(\bar{2}1\bar{6})$ become shorten and diminished with increasing milling time. Besides that, the peak of Al_2O_3 also reduce size of particle and become more wider according to the grain size become refining with increasing of milling time [11] - [12]. Peak of TiO_2 also appeared at hkl $(\bar{1}01)$, $(00\bar{4})$, and $(\bar{1}0\bar{5})$ and slightly diminished due to small size with increasing the milling time. Apart from that, TiO_2 phase was easy to diffuse into matrix phase of Al_2O_3. CNT has shortened peak and not observed at all milling times, presumably as a result of its faster diffusion into the Al_2O_3 matrix than TiO_2. That is, solid solubility for CNT in Al_2O_3 is far higher than that of TiO_2, and can be later diffused into a solid Al_2O_3 solution. Moreover, the evidence the changes of Al_2O_3 lattice can be described by shifted Al_2O_3 peaks to the left with prolonged milling time.

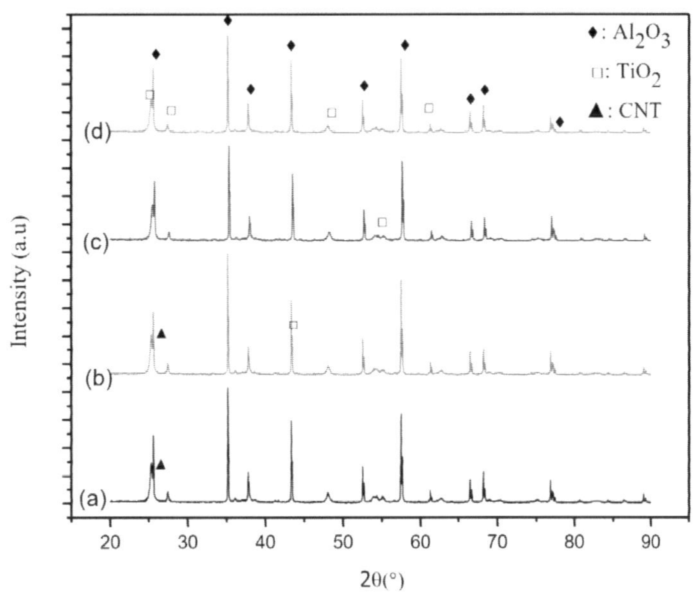

Figure 2. XRD patterns of milled powder at (a) 15 (b) 30 (c) 45 and (d) 60 h

Brittle phase powder basically dissolves into ductile matrix phase during milling. Two methods was used to describe the solubility of powder, there are peak disappearance and peak shift. In this experiment, the methods were used to discuss the solubility of Al_2O_3, TiO_2, and CNT in Al_2O_3 again milling time. The formation of second element into matrix phase of Al_2O_3 was discovered by peak disappearance in XRD pattern. The peak shifts either to the right or left of 2θ angle axis is also another method to determine dissolution of second phase into matrix lattice.

In this experiment, no visible peak shifting is observed at early milling time. This due to low energy ball milling was used at only 200 rpm and shorten milling time that not enough to form new phase of solution. This can explain by limitation amount of Al_2O_3 solution that not contributes to the change of Al_2O_3 lattice. There was no transformation of TiO_2 and CNT into Al_2O_3 phase during the milling process. Thus, it can discovered that peaks of Al_2O_3 and TiO_2 are show broadened.

Besides, the wider peak occurs due to internal stress accumulation induced by formation and the presence of tiny particle of second phase that introduce defect. The crystallite size and internal strain of powder mill not easy to change by milling time 15, 30, 45 and 60 h due to little shift peak to left was formed. Based on result, the 60 h of milling time provide strong formation of second phase of $Al_2O_3TiO_2$ –CNT powder.

4. Conclusion
As a conclusion, based on XRD pattern no visible new phase was formed when increasing milling time with only the changes of lattice parameter and crystallite size. The lowest crystallite size of and highest internal strain Al_2O_3−TiO_2−CNT are at 60 h.

5. References
[1] Banga H, Singh V K and Choudhary S K 2015 *Innov. Sys. Des. Eng.* **6** (1) 84–99
[2] Zahedi A M, Javadpour J, Rezaie H R and Mazaheri M 2015 *Ceram. Int.* **42** (1) 1653-1659
[3] Zhao J, 2014 *The use of ceramic matrix composites for metal cutting applications: Advances in Ceramic Matrix Composites* Woodhead Publishing
[4] Kanthavel K, Sumesh K R and Saravanakumar P 2016 *Alexandria Eng. J.* **5** (1) 13–17
[5] Erdemir F, Canakci A and Varol T 2015 *Transac. Nonferr. Met. Soc. China* **25** (11) 3569– 3577
[6] Bian H, Yang Y, Wang Y and Tian W 2012 *Powder Technol.* **219** 257–263
[7] Vijayanand P, Amitesh Kumar K R, Kumar V, Hussain N, Kumaran P and Arungalai Vendan S 2017 *Rasayan, J. Chem.* **10** (2) 652 - 664
[8] Chen B, Li S, Imai H, Jia L, Umeda J and Takahashi M 2015 *J. Alloys Compd.* **651** 608– 615
[9] Esawi A M K, Morsi K, Sayed A, Taher M and Lanka S 2010 *Compos. Sci. Technol.* **70** (16) 2237–2241.
[10] Mahani Y and Zuhailawati H 2013 *Adv. Mater. Res.* **795** 343–346
[11] Mahani Y, Radzali O, Zuhailawati H 2011 *Mater. Des.* **32** (6) 3293-3298
[12] Liu J, Lina L, Li J, Liu J, Yuana Y, Ivanovc M, Chen M, Liu B, Ge L, Xie T, Koub H, Shib Y, Pan Y and Guo J 2014 *Ceram. Int.* **40** 9841-9851

Acknowledgement
This work was supported by short term research grant scheme (SGJP) code (R/SGJP/A13.00/01136A/002/2018/000455).

IOP Conf. Series: Materials Science and Engineering **440** (2018) 012022 doi:10.1088/1757-899X/440/1/012022

Synthesis and crystal structure of dinuclear based coordination polymer derived from *N*-6-[(4-pyridylmethylamino)carbonyl]-2-pyridinecarboxylic acid methyl ester

M A Kadir [1,2], S M T Mansor[2], F S Mehamod[1,2] and M S M Yusof[2*]
[1]Advanced Nano Materials Research Group (ANoMa), School of Fundamental Science, Universiti Malaysia Terengganu, 21030 Kuala Nerus, Terengganu, Malaysia.
[2]School of Fundamental Science, Universiti Malaysia Terengganu, 21030 Kuala Nerus, Terengganu, Malaysia.

Corresponding author emails: mohdsukeri@umt.edu.my ; maisara@umt.edu.my

Abstract. This study describes the facile synthesis and crystal structure of one dimensional coordination polymer which has been successfully obtained from reaction between semi rigid monoamide ligand namely *N*-6-[(4-pyridylmethylamino)carbonyl]-2-pyridinecarboxylic acid methyl ester (L2) with copper perchlorate. The synthesized coordination polymer, with formula given by elemental analysis and X-ray crystallography as $\{[Cu_3(L2\text{-}CH_3)_3(L2)_2(H_2O)_2(CH_3OH)](ClO_4)_3\}_n$ was crystallized in monoclinic crystal system and space group C2/c. The asymmetric unit of this complex consists of three copper atoms, three anionic carboxylate forms of ligand L2, two neutral forms of ligand L2 and four partially-occupied non-coordinated perchlorate anions. This asymmetric unit acts as the repeating fragment to form one-dimensional networks which was stabilized by intramolecular hydrogen bonds.

1. Introduction

The derivatives of aminopyridine and aminomethylpyridine have been widely used as organic linker in the synthesis of coordination polymers [1-2]. Rigid linkers such as bipyridine or aminopyridine normally provide limited access to complex supramolecular architectures, while in reverse, flexible ligands such as aminomethylpyridine can be readily used to construct interesting and complex supramolecular structures [3-6]. The flexibility conferred by the ligands would allow for the formation of larger metallo- supramolecular assemblies. We in particular have deep interest in the use of aminomethylpyridine and aminopyridine derivatives as organic linker due to their propensity to form various motifs of coordination polymers and discrete metallomacrocycles. It was evidenced from our previous studies that the ligand derived from 3-aminomethylpyridine prefer to self-assemble with copper(II) and formed dinuclear ($[M_2L_2]$) metallo-macrocyclic complexes.

Meanwhile, the rigid ligand derived from 3-aminopyridine provides formation of planar macrocyles [6-7]. Upon reaction with transition metals, the ester of this ligand was hydrolyzed and formed a tridentate 2,6-pyridine dicarboxylate (ONO) donor moieties and the pyridine N donor remains as monodentate binding site (Figure 1). Hence, to continue our interest in the study of aminomethylpyridine derivatives as organic linker, we have synthesized *N*-6-[(4-pyridylmethylamino)carbonyl]-2- pyridinecarboxylic acid methyl ester (L2) from combination of 2,6-pyridinedicarboxamide with 4-

aminomethylpyridine. The obtaining ligand was reacted with copper salts to identify its tendency to form either dinuclear ($[M_2L_2]$) metallo-macrocyclic complexes or coordination polymers. We have confidence that new motifs of coordination will be obtained by L2, thus this study is pursued to investigate the coordination chemistry of L2 with copper salts. Thus, in this account, the synthesis, characterization and crystal structure of the complex are discussed.

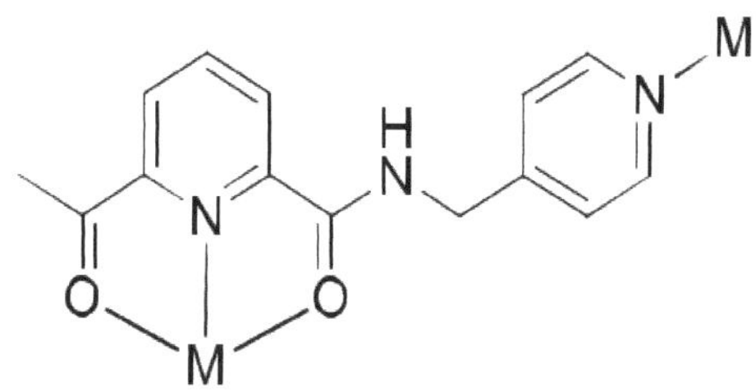

Figure 1. The structure of L2 with ONO tridentate donor monodentate N donor (where M= metal ion)

2. Materials and Methods

Elemental analysis was performed by the Campbell Microanalytical Laboratory at the Universiti Malaysia Terengganu. Infrared spectra were collected on a Perkin Elmer Spectrum BX Infrared spectrometer as KBr disk in the spectral range of 4000 – 400 cm^{-1}. All chemicals were obtained from commercial sources and used as received. The surface morphology of the composite was observed using Scanning Electron Microscope (SEM; JEOL JSM-6360LA). Crystals were mounted under paratone-N oil on a plastic loop. X-ray diffraction datasets were collected with Mo-Kα radiation (0.71073 Å) using Bruker-AXS Single Crystal Diffraction System fitted with an Apex II CCD detector at 110(2) K.

2.1 Synthesis of N-6-[(4-Pyridylmethylamino)carbonyl]-2-pyridinecarboxylic acid methyl ester (L2)

Ligand L2 was prepared following the method described in the literature [8]. A suspension of dimethyl pyridine-2,6-dicarboxylate (3.92 g, 20.0 mmol) and 4-aminomethylpyridine (2.00 mL, 20.0 mmol) was refluxed in toluene (80 mL) for 36 h. The toluene was removed using rotaevaporator under reduced pressure. The residue was redissolved in dichloromethane (100 mL), washed with dilute hydrochloric acid (0.2 M, 2×100 mL) and the dichloromethane layer was discarded. The aqueous extract was neutralized with sodium bicarbonate until pH 7-8. The neutral solution was extracted with dichloromethane (3×100 mL), dried over sodium sulfate and the solvent removed using rotaevaporator under reduced pressure. The resulting residue was purified by column chromatography on silica gel eluting with 1:9 methanol–dichloromethane. The product was obtained as oil from column. The oil was solidified on standing in two days to give L2 as yellow precipitate (2.31 g, 85 %).

2.2 Synthesis of $\{[Cu_3(L2\text{-}CH_3)_3(L2)_2(H_2O)_2(CH_3OH)](ClO_4)_3\}_n$

$Cu(ClO_4)_2 \cdot 6H_2O$ (0.034 g, 0.12 mmol) was dissolved in methanol (5 mL) to give a pale blue solution. This solution was heated for a few minutes before being added dropwise to a solution of L2 (0.063 g, 0.23 mmol) which was also dissolved in hot methanol (15 mL). The addition of copper perchlorate solution has given to a light blue solution. The mixture was heated for 45 minutes and left to evaporate at room temperature. After a week, the blue solution afforded turquoise blue crystals as product (0.095 g, 45%).

3. Results and Discussion

Reaction of L2 with copper salts led to formation of oily product except for copper perchlorate. Due to its polarity in methanol, this complex crystallized as blue crystals in 45% yield of products. Elemental analysis suggested that this compound has general formula of $Cu_3(L2)_3$, incorporated with methanol or water to meet the experimental value (Table 1).

Table 1. Elemental analysis data for copper complex

Compound	C(%)		H(%)		N(%)	
	Calcd	Found	Calcd	Found	Calcd	Found
Complex	42.43	42.27	3.55	3.09	10.56	10.51

FTIR spectrum show several important stretching bands for ν(N-H), ν(C=O), ν(C-N), ν(Cl-O) and ν(Cu-O) at 3409, 1708, 1523, 1082 and 513 cm^{-1}, respectively (Figure 2). The stretching band of ν(N-H) has shifted to higher frequency compared to its ligand due to the additional electron withdrawing group near to the NH moiety. The strong absorption bands around 1690-1708 cm^{-1} are attributed to the stretching of ν(C=O). The increasing in C=O frequencies are affected by the increasing of electron densities from aminopyridine pendant arms and central pyridine. This phenomenon has also resulted to intramolecular hydrogen bonding at amide moieties which at end given rise to a new peak at 3202 cm^{-1}. A strong stretching vibration for ν(C=N) was indicated at 1523 cm^{-1}. The medium intensity band around at 513 and 618 cm^{-1} were assigned for Cu-O and Cu-N, respectively [9-10]. The data obtained from FTIR shows promising formation of a copper complex as expected. Involvement of solvents such water molecule and methanol were indicated 3558 and 3000 cm^{-1}, standing for OH stretching, respectively.

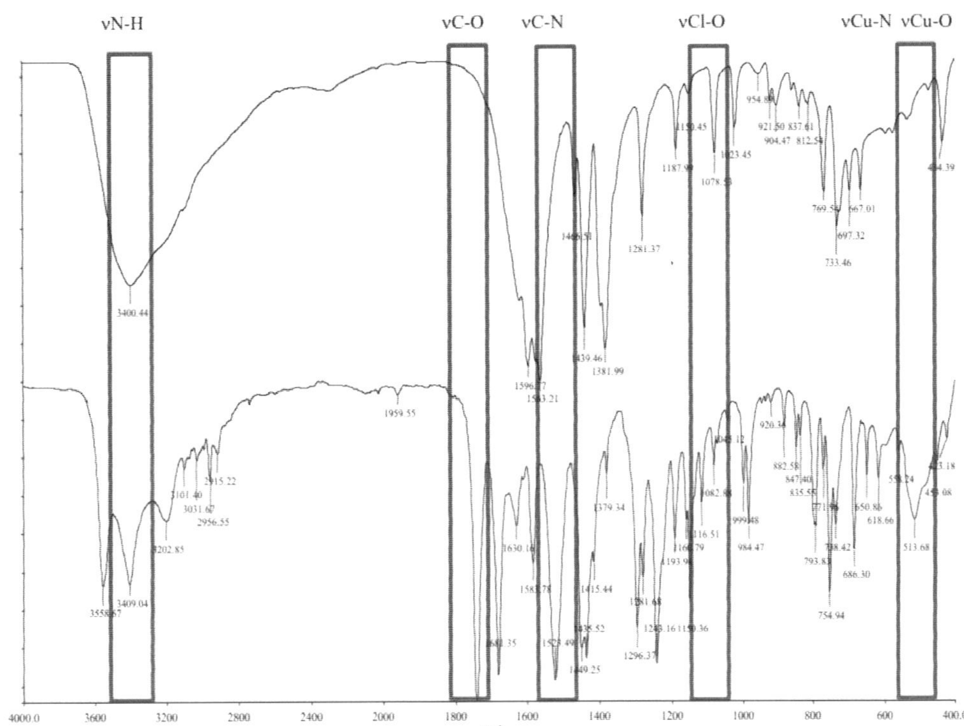

Figure 2. FTIR spectrum of the ligand(a) compared with copper complex (b)

4. X-ray crystallography

Due to the limited solubility of this complex, the only method that is suitable to determine the crystal structure of the blue crystals is single crystal X-ray crystallography techniques. These blue crystals were grown from the reaction solutions of ligand L2 with copper perchlorate in methanol and obtained as irregular shape crystals. The crystals crystallized in monoclinic space group C2/c. The asymmetric unit contains a repeating unit of the coordination polymer, with three copper atoms, three anionic carboxylate forms of the ligand L2, two neutral forms of ligand L2 and four partially-occupied non-coordinated perchlorate anions (Figure 3). An extended view of this structure reveals that it is a 1-D coordination polymer in which M_2L_2 units are linked by the carboxylate donors (Figure 4). The copper ion exhibits six coordination number and displayed distorted octahedral geometry as observed in Figure 4. The bond lengths about the copper environment are in the range of 2.361-2.423 Å. The present of weak hydrogen bonds between the amide moieties (N-H···N = 2.205Å) and perchlorate molecules (N-H···Cl= 2.172 Å) were defined to stabilize the crystal structure (Figure 5). It was witnessed from the x-ray crystallography that the copper ion was coordinated by ONO tridentate donor and the repeating unit was linked by N monodentate from the pyridine pendant arm. This also has supported the formation of *in situ* ligand when reacted with metal salts as reported by us [6-7]. The detailed on the crystal data is shown in Table 2.

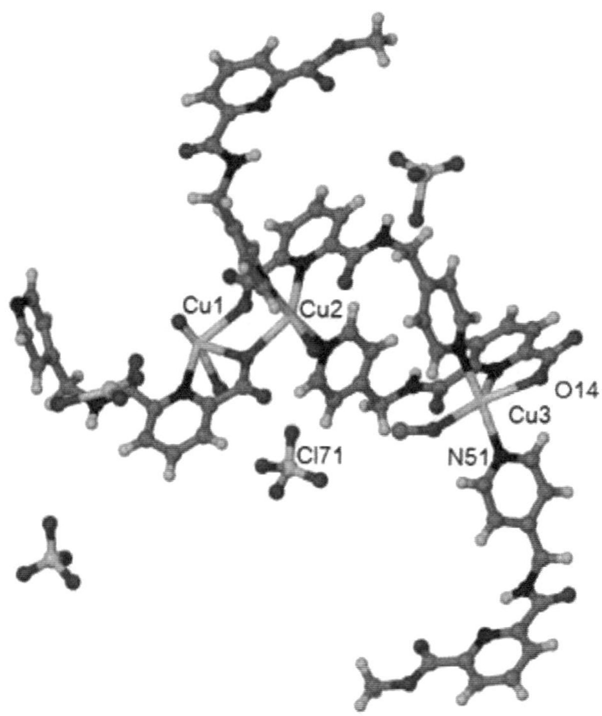

Figure 3. The asymmetric unit of $\{[Cu_3(L2\text{-}H_3)_3(L2)_2(H_2O)_2(CH_3OH)](ClO_4)_3\}_n$

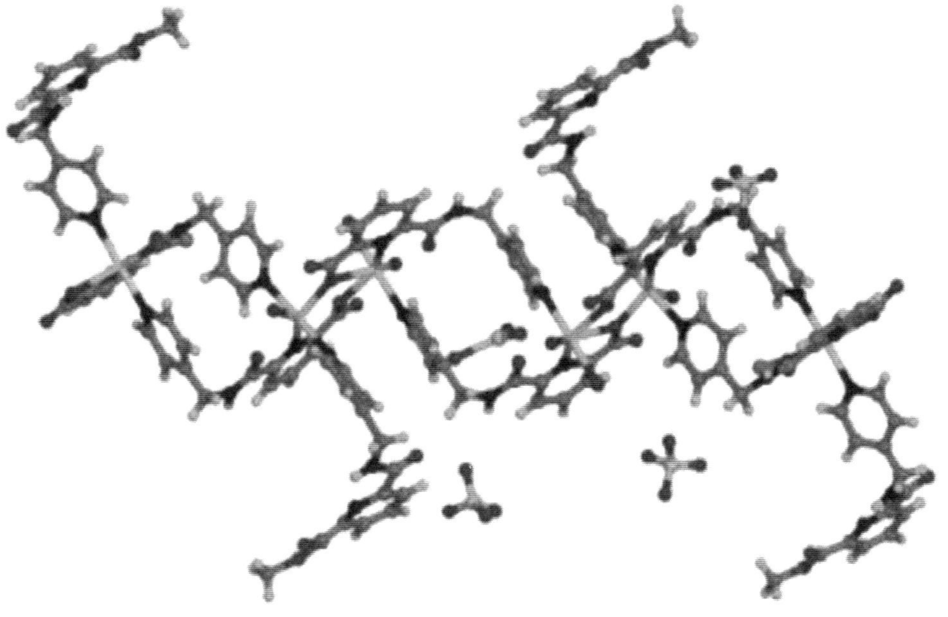

Figure 4. Extended structure of one dimensional coordination polymer with copper ions displayed distorted octahedral environment (turquoise colour)

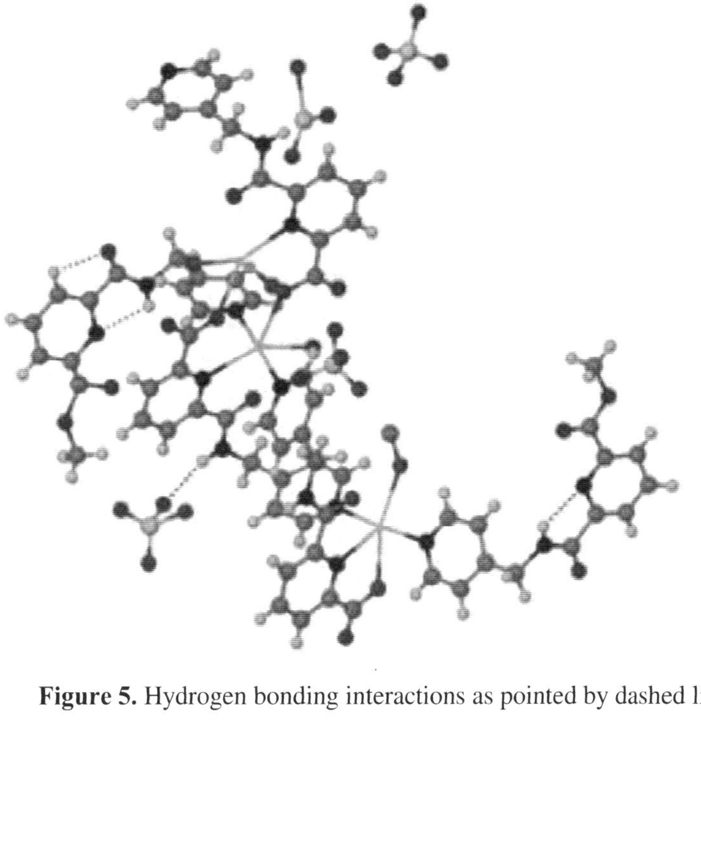

Figure 5. Hydrogen bonding interactions as pointed by dashed line

IOP Conf. Series: Materials Science and Engineering **440** (2018) 012022 doi:10.1088/1757-899X/440/1/012022

Table 2. Crystal data of the one-dimensional coordination polymer

Crystal data	Data
Empirical formula	C68.50H68Cl3Cu3N15O33.50
Formula weight	1934.35
Crystal system	Monoclinic
Space group	C2/c
a (Å)	49.155(10)
b (Å)	10.849(2)
c (Å)	33.877(7)
(°)	90
(°)	112.48(3)
(°)	90
Volume (Å3)	16693(6)
Z	8
Density (calculated) (Mg/m3)	1.539
Absorption coefficient (mm-1)	0.948
F(000)	7920
Crystal size (mm3)	0.40x0.25x0.06
Theta range for data (°)	0.90 – 24.20
Reflections collected	12557
Observed reflections [I>2s(I)]	10953
Data/restraints/parameters	12557/6/922
Goodness-of-fit on F2	2.146
R1 [I>2s(I)]	0.1438
wR2 (all data)	0.4275
Largest diff. peak and hole (e.Å-3)	2.020 and -2.042

Figure 6 (a) depicts the micrographs of surface morphology for the ligand L2. The result shows that this ligand has irregular shaped flakes with different sizes. The size of each flake is measured as approximately 20-100µm. In contrast, the SEM results for the coordination polymer, has better surface morphology compared to the ligand. This material distributed homogeneously throughout the surface with irregular morphology and reasonable uniform sizes within 1-5µm (Figure 6(b)). This analysis has proved that the metal complex was formed in smaller sizes and homogeneous compared to the ligand.

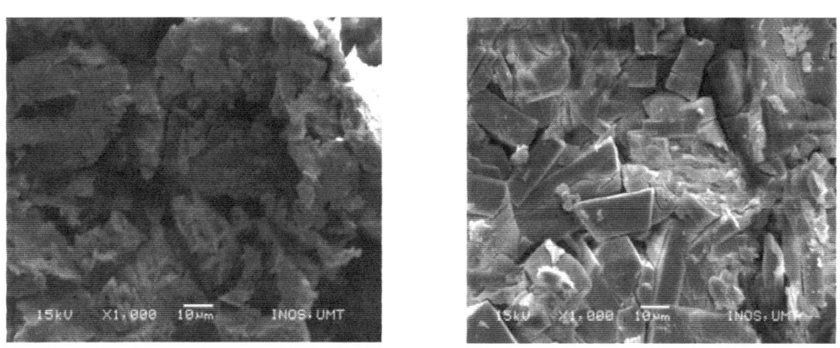

(a) (b)

Figure 6. Surface morphology of (a) ligand b) $\{[Cu_3(L2\text{-}CH_3)_3(L2)_2(H_2O)_2(CH_3OH)](ClO_4)_3\}_n$

5. Conclusion

In conclusion, a one-dimensional coordination polymer has been successfully synthesized using L2. Ligand *N*-6-[(4-pyridylmethylamino)carbonyl]-2-pyridinecarboxylic acid methyl ester (L2) formed in situ ONO tridentate ligand when reacted with copper perchlorate and arranged in dinuclear based, one dimensional coordination polymer. Incorporation of copper ion has affected the particle size of the product.

6. References

[1] Adarsh N N, Kumar D K, Sresh E and Dastidar P 2010 *Inorg. Chim. Acta.* **363** 1367.
[2] Dursun K, Gunes S, Okan Z Y I and Murat T 2014 *J. Mol. Struct.* **1074** 339.
[3] Kadir M A, Mansor N, Yusof M S M and Sumby C J 2014 *Complex Metals.* **32** (40) 4497
[4] Kadir M A, Mansor N, Khairul W M, Yusof M S M and Ismail M 2016 *Int. J. Appl. Chem.* **12** 257
[5] Kadir M A, Ramli R, Kassim N, Mansor N and Sumby C J 2015 *Malaysian Journal of Analytical Sciences.* **3**(19) 520
[6] Abdul-Kadir M, Lyall R H and Sumby C J 2012 *Dalton Trans.* **41** (40) 4497
[7] Abdul-Kadir M, Hanton, L R and Sumby, C J 2011 *Dalton Trans* **40** (45)1237
[8] Sumby C J and Hanton, L R 2008 *Tetrahedron.* **65**: 4681
[9] Arthur E L and David J P 1983 *Inorg. Chim. Acta.* **74** 43
[10] Koleva, B B, Trendafilova, E N, Arnaudov M G, William S S and Heike M-F 2006 *Transition Met Chem* **31** 866

Acknowledgement

The authors greatly acknowledge the scientific support from School of Fundamental Science, Universiti Malaysia Terengganu.

Preparation and Characterisation of Gellan Gum Hydrogel containing Curcumin and Limonene

A M Jaafar[1,2,*] and V Thatchinamoorthi[2]

[1]Chemistry Unit, Centre of Foundation Studies for Agricultural Science
[2]Chemistry Department, Faculty of Science, Universiti Putra Malaysia, 43400 UPM Serdang, Selangor.

Abstract. Success in wound treatment requires careful assessment on the wound and suitable wound dressing. In this project a novel incorporation of both curcumin and limonene into gellan gum hydrogel have been done. This system is believed to have a better permeation of curcumin to the target area. The samples prepared are denoted as GGTF for gellan gum thin film, Cur-GGTF for curcumin gellan gum thin film. Samples of Cur-GGTF incorporated with limonene with three different concentrations of 0.0005 M, 0.0006 M and 0.0007 M are known as CurLim-GGTF05, CurLim-GGTF06 and CurLim-GGTF07, respectively. The hydrogels were crosslinked by calcium ion. From the FTIR analysis, all of the expected functional groups from the gellan gum, curcumin and limonene were found present in the hydrogel samples. The TGA-DTG analysis has proven that the incorporation of limonene into hydrogels has decreased the thermal stability of the hydrogels. The addition of limonene has caused a hike in the gel fractions of the hydrogels. The water vapour transmission rate (WVTR) and swelling test showed that the introductions of high concentrations of limonene into the hydrogels have hampered the absorption of water molecules thus causing the swelling degree to decrease.

1. Introduction

Hydrogel is a polymer in three dimensional networks which is hydrophilic and capable of assimilating vast amount of biological fluids and water [1]. Due to their adjustable physical and chemical characteristics, the hydrogels have evolved as crucial substance for wound healing and management. The hydrogel adhesives act as a boundary against bacterial infection by adhering to the tissues and staying intact by acquiring plentiful mechanical strength. In addition, the wound exudates are being absorbed by the hydrogels and subsequently the moisture of the wound are kept for an accelerated cure. These cater the wound an outright protection and healing process [2]. The utilisation of hydrogel minimises the wound area with the increasing duration of wound healing [3].

Gellan gum is a polysaccharide produced by Pseudomonas elodea, a Gram-negative bacterium which is non-pathogenic [4]. High acyl and low acyl forms of gellan gum are obtainable commercially. The low acyl gellan gum latterly has been the choice of interest in wound care and tissue engineering fields. This is due to the resemblance of the low acyl gellan gum as the natural extracellular matrix and it is biologically inert [5]. High acyl gellan gum is less stable than low acyl gellan gum due to the presence of bulky acyl group which produce a weaker gel by disrupting the gelling process. Due to the high stability and optimum stiffness properties of low acyl gellan gum, it makes a promising material for wound dressing application [6].

Curcumin is a phytochemical agent in the spice, turmeric that is responsible for the distinctive taste and yellow colour of Indian curry. Traditional Indian medicine also vastly used curcumin. The antiinflammatory and antioxidant properties of curcumin have been the reason for it being used as a medium of wound healing and treatment of various diseases for millennia [7]. Curcumin is a primary curcuminoid, which is a group of phenolic compounds isolated from the rhizome of Curcuma longa L, turmeric from the Zingiberaceae family [8]. Curcuminoid is referred as a group of compounds such as curcumin, demethoxycurcumin, bis-demethoxycurcumin and cyclic curcumin. Curcumin is the major component and cyclic curcumin is the minor component out of these compounds [9]. Chemically, curcumin is termed as diferuloylmethane and the IUPAC name is 1,7-bis(4-hydroxy-3-methoxyphenyl)1,6 heptadiene-3,5-dione [10]. Curcumin provides protection to skin by decreasing any inflammation and extinguishing free radicals off the skin. In addition, it is found that curcumin elevates fibroblast and vascular density in wounds, enhance collagen deposition and consequently, minimize the time taken for wound healing. Besides, curcumin is a powerful non-toxic agent for curing skin disorders. The proangiogenic property of curcumin induces accumulation and angiogenesis of extracellular matrix, which continues through the remodeling phase of wound healing [11].

Limonene (4-isopropenyl-1-methylcyclohexene) is a monoterpene which can be found in lemons, grapefruits, oranges, limes and other citrus fruits. Terpene is a large group of unsaturated hydrocarbons found in essential oils of plants especially in citrus trees [12]. There are two isomers of limonene which are D-limonene and L-limonene. Yet, D-limonene is the prime active form of limonene [13]. Dlimonene is the main flavour compound in citrus oil. It has been utilised numerously as food flavouring and tumor treating agent [14]. Besides, it has been discovered to be used in chemotherapy as an antimicrobial agent and antioxidant against cancer [15]. By certain cellular mechanisms, the growth of tumor can be inhibited by limonene. The prevention of the development of new blood vessels by angiogenesis process is one of them which aids in the reduction of inflammation [16, 17]. Above all, the tissue-repair properties of D-limoenene have been proved significantly due to the remarkable antiinflammatory effects in wound-healing [17]. Nonetheless, the skin absorption of inadequately permeable drugs can be escalated by incorporating limonene, which is an effective enhancer into the system [18]. On top of that, burn wound antimicrobial treatment could be enriched by limonene via the amplification of drugs' permeation [19].

2. Materials and methods

2.1 Material

Low-acyl gellan gum (Gelzan™ CM), glycerin, calcium chloride, $CaCl_2$ were obtained from Sigma Aldrich. Curcumin and Limonene were purchased from Acros Organic. All materials were used as received. No further purification was required.

2.2 Preparation of Samples

2.2.1 Preparation of Gellan Gum Thin Film (GGTF)

GGTF was prepared by dissolving 0.5 g of gellan gum powder in 30 mL deionized water (18 MΩ cm^{-1}). Then, the solution was mixed with 2.5 mL of glycerin which acts as a plasticizer. Next, 5 mL of 0.03 M of calcium chloride, $CaCl_2$ was added dropwise to establish the physical crosslinking process. The solution was then stirred using magnetic stirrer on hot plate set at temperature of 80 °C for a total mixing of 2 hours. After 2 hours, the solution was placed in a petri dish and left in the oven drying at 72 hours and 35 ° C. Finally, the GGTF was stored in the desiccator at room temperature for further characterizations.

2.2.2 Preparation of Curcumin Gellan Gum Thin Film (Cur-GGTF)

The same method to prepare the GGTF was adapted here as mentioned in 2.2.1, only that 2.1 g of curcumin was added to form 0.06 M Cur-GGTF.

2.2.3 Preparation of Curcumin–Limonene Gellan Gum Thin Film (CurLim-GGTF)
The same method to prepare the Cur-GGTF was adapted here as mentioned in 2.2.2, only that 5 mL of limonene solution of 0.0005 M, 0.0006 M and 0.0007 M were added into the mixture solution.

2.3 Characterization
The prepared GGTF, Cur-GGTF and CurLim-GGTF were characterized by using few analytical techniques. The techniques involved were Fourier Transform Infrared (FTIR) spectroscopy and Thermogravimetric Analysis-Differential Thermogravimetric Analysis (TGA-DTG).

2.3.1 Fourier Transform Infrared (FTIR) Spectroscopy
Fourier Transform Infrared (FTIR) spectroscopy was used to determine the functional groups which are present in the GGTF, Cur-GGTF, CurLim-GGTF, curcumin and limonene. The instrument used was Perkin Elmer Fourier Transform Infrared (FTIR) spectrometer model 100 Series which used the Universal Attenuated Total Reflection (UATR) accessory (single-bounce beam path, 45 ° incident angle, 16 scans, and 4 cm-1 resolution). All samples were tested in the transmission range of 400 to 4000 cm1. Infrared spectrum obtained from the analysis is used as a verification to determine the incorporation of gellan gum, curcumin and limonene.

2.3.2 Thermal Analysis (TGA-DTG)
Thermal analysis (TGA-DTG) for GGTF, Cur-GGTF, CurLim-GGTF, curcumin and limonene was carried out at 10°C/min heating rate from 50 °C to 1000 °C under nitrogen flow using a Q5000 (TA Instruments) thermo balance. Based on the thermograms obtained, the decomposition and mass change over time of GGTF, Cur-GGTF, CurLim-GGTF, curcumin and limonene were determined.

2.3.3 Gel Fraction
The thin film samples were cut into 20mm x 20mm square piece, sealed with aluminium foil and parafilm in a beaker, and dried at room temperature for 24 hours. The samples were weighed and immersed in 10 mL of deionized water at room temperature for 24 hours. Then, the samples were removed from the solution and oven dried at 50 ºC for 24 hours. The samples were then weighed again and the gel fraction was calculated as shown below:

$$Gel\ fraction\ (\%) = \left(\frac{W_f}{W_i}\right) \times 100 \tag{1}$$

Where,
W_i = initial weight of samples
W_f = final weight of samples

2.3.4 Water Vapour Transmission Rate (WVTR)
The WVTR procedure was carried out using ASTM E96 international standard method. The thin film samples were cut into 30 mm x 30 mm and put as a cap on a glass vial with diameter of 23 mm that contains 10 mL of deionized water. Then, the samples were weighed and left in the desiccators for 24 hours. The silica gel was oven dried for 24 hours before they were used. After 24 hours, the samples were weighed again. The value of water vapour transmission rate (WVTR) was calculated as shown below:

$$WVTR\ (gm^{-2}d^{-1}) = \Delta m / A \cdot t \tag{2}$$

Where,
Δm = the change in weight of sample system
A = the area of glass vial opening
t = the time in days

2.3.5 Swelling Test
The dried samples with a dimension of 2cm x 2cm were immersed in a properly sealed 100 mL beaker that contains 20 mL of pseudo-extracellular fluid (PECF) buffer solution of pH 5.5 that resembles human skin condition at a controlled temperature of 37 °C in an oil bath shaker with a medium shaking frequency for the duration of 24, 48 and 72 hours. The wet samples were weighed after drying them using a tissue paper. The swelling data for GGTF, Cur-GGTF and CurLim-GGTF of different concentrations were recorded. The swelling percentage for each sample was calculated using the equation as shown below:

$$Swelling\ Percentage\ (\%) = [(M_f - M_i)/M_i\ ] \times 100 \quad (3)$$

Where,
M_f = weight of sample after swelling
M_i = initial weight of the sample

3. Results and Discussion

3.1 Fourier Transform Infrared (FTIR) Spectroscopy

The FTIR spectrum obtained for GGTF, curcumin, limonene, Cur-GGTF, CurLim-GGTF05, CurLimGGTF06 and CurLim-GGTF07 were shown in Figure 1. Based on the GGTF spectrum analysis, a broad absorption band at 3317 cm^{-1} is observed indicating the presence of O-H, hydroxyl group. At wavenumber of 2943 cm^{-1}, C-H bending group has shown absorption. The absorption band at 2887 cm^{-1} attributed to the carboxyl, COOH group is witnessed. Then, carbonyl, C-O group from the glycosidic bond showed strong absorption band at 1639 cm^{-1}. The peak at 1037 cm^{-1} demonstrates the presence of ester, C-O-C group [20, 21]. For curcumin, the absorption peak at 3510 cm^{-1} consisting of aromatic hydroxyl, O-H vibration group could be observed. Then, C-H alkane group showed a peak of absorption at 2918 cm^{-1}. An absorption band is noticed at 1625 cm^{-1} which indicates the carbonyl, C=O stretching group.

The strong and sharp peak of the carbonyl group is due to the large dipole of the bond. At wavenumber of 1600 cm^{-1}, alkene C=C group has been identified [22]. Subsequently, the symmetric aromatic benzene group, C_6H_5 is noted at the absorption peak of 1427 cm^{-1}. The resonance effect in the benzene ring lowers the stretching frequency. An alkoxy group, which is methoxy $O\text{-}CH_3$,

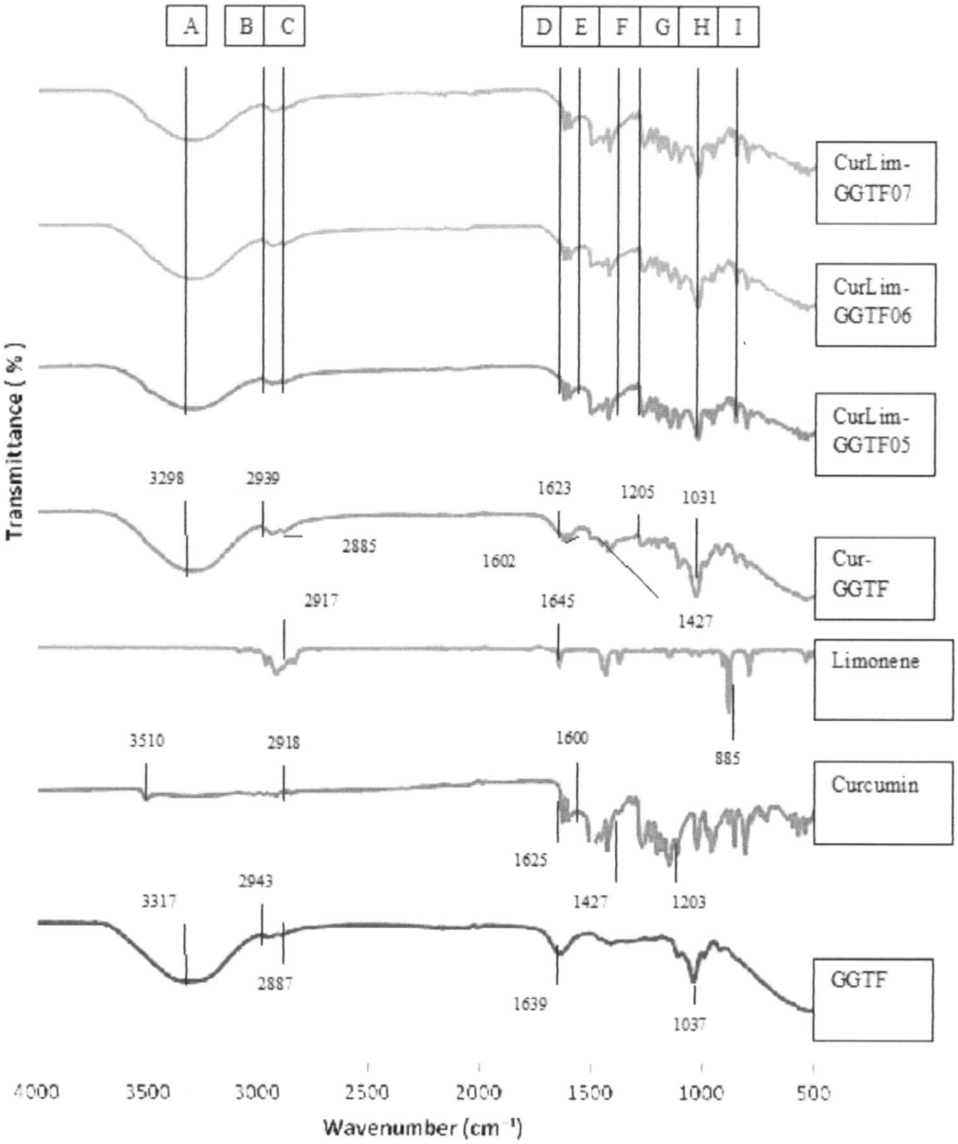

Figure 1. FTIR spectra for GGTF, curcumin, limonene, Cur-GGTF, CurLim- GGTF05, CurLimGGTF06 and CurLim-GGTF07

exhibited an absorption peak at 1203 cm^{-1} [23]. In the limonene, alkane C-H group is denoted at absorption peak of 2917 cm^{-1}. Moreover, at 1645 cm^{-1} absorption band, the appearance of alkene C=C group is marked. Besides, an alkenyl $C=CH_2$ group is displayed in a strong absorption peak manner at 885 cm^{-1} [24]. In the Cur-GGTF, the hydroxyl O-H group has been detected at the absorption peak of 3298 cm^{-1}. The shift in this peak compared to that of in GGTF and curcumin shows that the incorporation of these compounds were successful. In addition, the absorption peak at 2939 cm^{-1} and 1602 cm^{-1} were discovered for C-H alkane and C=C alkene groups respectively. There is a shift in absorption band of

COOH carboxyl group from that of in the GGTF which is from 2887 cm^{-1} to 2885 cm^{-1} [20]. Then, carbonyl C=O group showed up at wavenumber of 1623 cm^{-1}. 1427 cm^{-1} and 1205 cm^{-1} which are referred to aromatic benzene ring, C_6 H_5 and methoxy group, O-CH_3, respectively with some alteration in the peaks due to the assimilation of the compounds [23]. Ester C-O-C group has shown a shift to the peak from that of GGTF at 1037 cm^{-1} to 1031 cm^{-1} after the integration of both GGTF and curcumin. On top of that, the absorption peaks and functional groups present in the samples with the incorporation of GGTF, curcumin and limonene which are CurLim-GGTF05, CurLim-GGTF06 and CurLim-GGTF07 are listed in Table 1. All of the functional groups existed in the GGTF, curcumin and limonene were found in the prepared samples with some perceived shifts in the absorption peaks. This has proven that there are interactions between the compounds and the incorporation of curcumin and limonene into GGTF were successful [20].

Table 1. List of functional groups and absorption peaks for CurLim-GGTF05, CurLim-GGTF06 and CurLim-GGTF07.

Symbol	Functional Group	Samples		
		CurLim-GGTF05(cm^{-1})	CurLim-GGTF06(cm^{-1})	CurLim-GGTF07(cm^{-1})
A	Broad hydroxyl, O-H	3309	3305	3304
B	Alkane, C-H	2941	2940	2940
C	Carboxyl, COOH	2884	2885	2887
D	Carbonyl, C-O	1625	1625	1624
E	Alkene, C=C	1600	1602	1601
F	Benzene, C_6 H_5	1427	1426	1425
G	Alkoxy (methoxy), O-CH_3	1203	1203	1203
H	Ester, C-O-C	1025	1026	1024
I	Alkenyl, C=CH_2	885	884	882

3.2 Thermal Analysis

Figure 2 shows the DTG for GGTF, curcumin, limonene, Cur-GGTF, CurLim-GGTF05, CurLimGGTF06 and CurLim-GGTF07. The DTG provide the percentage and peak temperature of sample mass change. As given in Table 2, there are two stages of samples decomposition, part A which occurred at 50 -200 °C and part B which occurred at 130-450 °C. Part A is the first stage which represents the loss of plasticizers, low molecular weight and volatile liquid components in the samples. Part B is the second stage indicates the loss of polymer chains in the samples [25]. The introduction of limonene definitely has an effect on the thermal behavior of the hydrogels. The weight loss decreases as the concentration of limonene increases in the hydrogels indicating there is an interaction between the limonene molecule, curcumin and gellan gum. Consequently, the weight loss in part B increases as the concentration of limonene in the hydrogel increases due to the successful incorporation of limonene onto the polymer chains. The weight loss in part A in the limonene sample is the highest due to its nature of being in a liquid state. This causes the liquid to be evaporated easily with the presence of heat. Moreover, there is no weight loss in part A for curcumin as there is no volatile liquid exists in the sample.

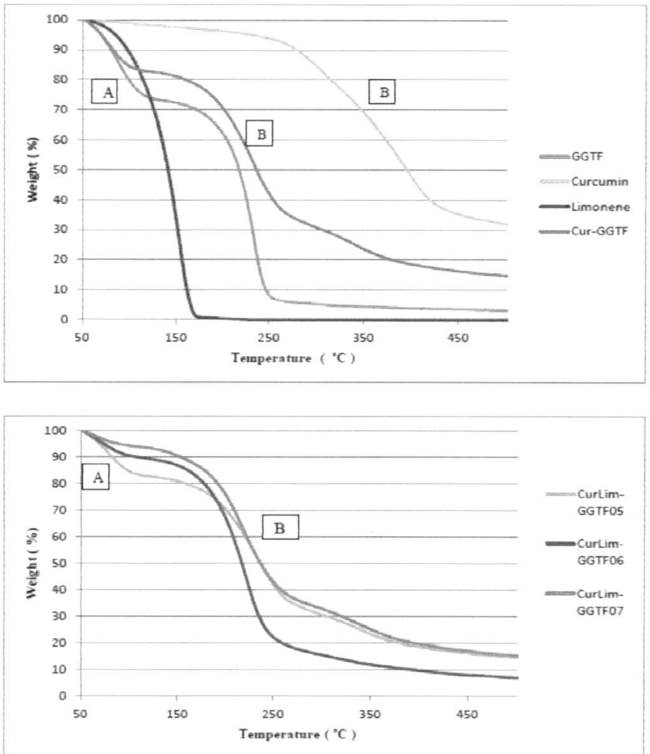

Figure 2. DTG curves for GGTF, curcumin, limonene, Cur-GGTF, CurLim-GGTF05, CurLimGGTF06 and CurLim-GGTF07

Table 2. The data obtained from DTG for all the samples GGTF, curcumin, limonene, Cur-GGTF, CurLim-GGTF05, CurLim-GGTF06 and CurLim-GGTF07.

Sample	A			B		
	Temperature range (°C)	Weight loss (mg)	Percentage of weight loss (%)	Temperature range (°C)	Weight loss (mg)	Percentage of weight loss (%)
GGTF	58-136	4.08	25.59	136-380	11.01	69.05
Curcumin	-	-	-	136-451	6.30	75.60
Limonene	57-200	39.51	99.33	-	-	-
Cur-GGTF	52-125	2.86	17.56	125-540	11.22	68.78
CurLim-GGTF05	54-117	0.91	10.49	118-363	6.74	78.10
CurLim-GGTF06	53.6-109.8	0.70	6.11	110-544	9.12	79.65
CurLim-GGTF07	53-103	0.46	2.13	102-545	9.87	80.16

Cur-GGTF exhibited decomposition temperature slightly higher than that of GGTF but lower than curcumin. Hence, after the curcumin was incorporated into the hydrogel, it exhibits the physical property of the curcumin. This has verified the interaction has occurred between the curcumin and gellan gum.

In addition, for the samples of CurLim-GGTF05, CurLim-GGTF06 and CurLim-GGTF07, the decomposition temperature decreases respectively as the limonene concentration gets higher in the hydrogel. This means that the thermal stability of the limonene incorporated hydrogels decline accordingly with the surge of limonene concentration in the hydrogels [26]. CurLim-GGTF07 can withstand maximum temperature of 220.95 °C before it started to decompose. According from the previous research, lowering of the decomposition temperature of a delivery system can cause significant escalation in the transdermal permeation of the compounds or drugs from the system to the skin [27]. Therefore, even though the thermal stability of the CurLim-GGTF incorporated with limonene were lowered than that of without limonene, better permeation of drugs and chemical compounds to the wound area could be expected. Thus, limonene has been verified to increase the penetration of drugs for better wound treatment. The decomposition temperatures for the all of the samples are given in Table 3.

Table 3: The decomposition temperatures for all of the samples, GGTF, curcumin, limonene, CurGGTF, CurLim-GGTF05, CurLim-GGTF06 and CurLim-GGTF07

Sample	Decomposition Temperature (°C)
GGTF	233.83
Curcumin	392.96
Limonene	149.66
Cur-GGTF	239.25
CurLim-GGTF05	228.63
CurLim-GGTF06	222.13
CurLim-GGTF07	220.95

3.3 Gel Fraction

Gel fraction is a measure of crosslinking degree in a polymer which indicates the crosslinking between the gellan gum, curcumin and limonene. Figure 3 demonstrate the gel fraction percentage obtained for all the samples. The gel fraction for the samples of CurLim-GGTF05, CurLim-GGTF06 and CurLimGGTF07 increases as the concentration of limonene increases. This is due to the intensified compactness of the hydrogel structure in which resulted to denser packing of the incorporated molecules, curcumin and limonene into the gellan gum hydrogel. Furthermore, Cur-GGTF has higher gel fraction than GGTF but lower than that of in the CurLim-GGTF05, CurLim-GGTF06 and CurLim-GGTF07. This is due to the presence of gellan gum and curcumin molecules which makes the packing is less tight when compared to hydrogel containing both curcumin and limonene. The existence of curcumin molecules in Cur-GGTF has occupied the space in the hydrogel structure causing it to have a higher gel fraction than GGTF. Moreover, GGTF has the least gel fraction percentage due to the unoccupied space of other molecules which makes it to have a less dense structure thus having the lowest degree of crosslinking. In a nutshell, higher number of molecules present in a hydrogel causes the crosslinking degree to be amplified. This leads to an increment in the gel fraction percentage in the hydrogel [28].

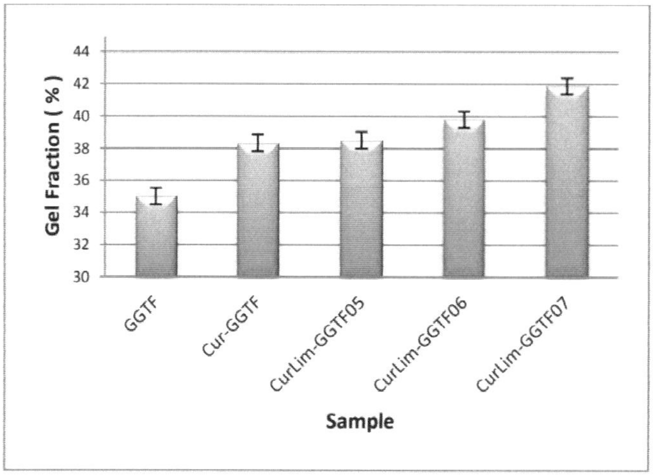

Figure 3. Gel fraction percentage obtained for GGTF, Cur-GGTF, CurLim-GGTF05, CurLim-GGTF06 and CurLim-GGTF07

3.4 Water Vapour Transmission Rate

Water vapour transmission rate (WVTR) is crucial in wound dressing application. WVTR controls the depletion of body fluid caused by the exudation and evaporation. The loss of body fluid causes reduction in the body temperature which results in the escalation of metabolism rate. This will enhance the wound healing process. However, the WVTR value less than 76 g m^{-2} d^{-1} is not suitable for wound dressing as the accumulation of exudates can occur which leads to bacterial growth [29]. This will decelerate the wound healing process. Figure 4 show the WVTR for all of the samples. The WVTR for GGTF is the highest because of the hydrophilicity property of gellan gum that allows more water molecules to be transmitted through the thin film. The presence of hydrophobic compound, curcumin in Cur-GGTF has caused the WVTR value to be slightly less than that of GGTF. For the samples of CurLim-GGTF05, CurLim-GGTF06 and CurLim-GGTF07, the values of the WVTR decreased as the concentration of limonene in the thin film increases. This is due to the higher number of limonene molecule per unit volume that present in the thin film which has interrupted the diffusion of water molecules through the thin film. Thus, the tight packing of limonene which has increased as the concentration of limonene in the thin film increases causes the rate of water vapour transmission across the film to decline accordingly. Subsequently, higher concentration of limonene causes a hike in the gel fraction which affects the WVTR values to decrease accordingly. The higher gel fraction which indicates a stronger structure of the hydrogel results in a reduction of the WVTRs. The optimum WVTR for wound dressing is in the range of 76 to 9360 g m^{-2} d^{-1} [29]. Hence, the WVTR values obtained for all the samples are suitable for wound dressing purpose.

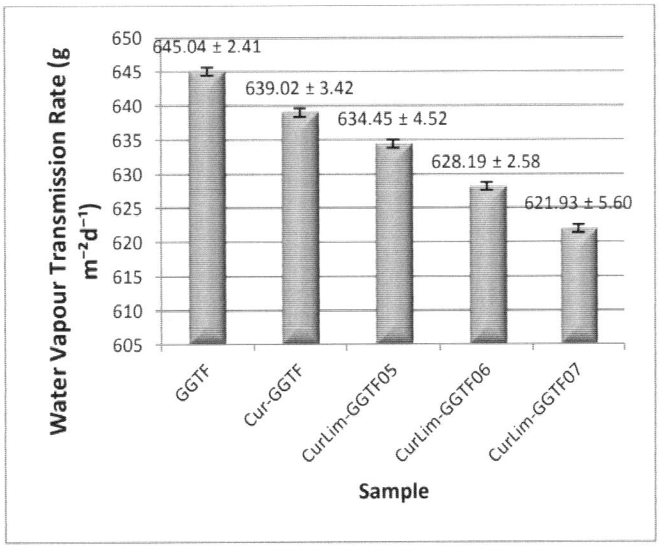

Figure 4. Water Vapour Transmission Rate for GGTF, Cur-GGTF, CurLim-GGTF05, CurLimGGTF06 and CurLim-GGTF07

3.5 Swelling Test

Swelling degree is the measure of water uptake by the hydrogel. This is an essential in vivo study for the wound management. Figure 5 show the swelling pattern and percentage of swelling at 24, 48 and 72 hours for GGTF, Cur-GGTF, CurLim-GGTF05, CurLim-GGTF06 and CurLim-GGTF07. The percentage of swelling for all of the samples, GGTF, Cur-GGTF, CurLim-GGTF05, CurLim-GGTF06 and CurLim-GGTF07 increases with the time frame of 24, 48 and 72 hours. The swelling percentage of GGTF is the highest due to the hydrophilic behaviour of the gellan gum hydrogel. Hence, it can absorb and conserve a large amount of water. This has caused the swelling degree of GGTF to be the most highest. The percentage of swelling decreases accordingly from GGTF to CurLim-GGTF07 due to the increase in the crosslinking magnitude of the hydrophobic molecules which are curcumin and limonene in the thin films. This affects the water absorption by the hydrogels which caused the swelling percentage to decrease with the presence of curcumin in the hydrogel. Moreover, the decreased in the swelling degree from CurLim-GGTF05, CurLim-GGTF06 to CurLim-GGTF07 is due to the concentration of limonene that gets amplified, respectively. High concentration of limonene denotes a large quantity of limonene molecule present per volume of hydrogel. Consequently, the crosslinking of the hydrogel is enhanced and the assimilation of water molecule into the hydrogel is diminished. This crosslinking effect has minimised the mobility of the polymer chains in the hydrogels which eventually lessen the swelling degree [1]. Hence, the addition of higher concentration of limonene causes higher gel fraction that restricts the swelling potential of the hydrogels. Thereafter, an immense amount of water absorption by the hydrogels can make them to be highly swollen. A highly swollen hydrogel can be fractured easily and disintegrate before the healing of the damaged tissues or wound on the body. Therefore, the incorporation of limonene into curcumin gellan gum thin film will make a better wound dressing material.

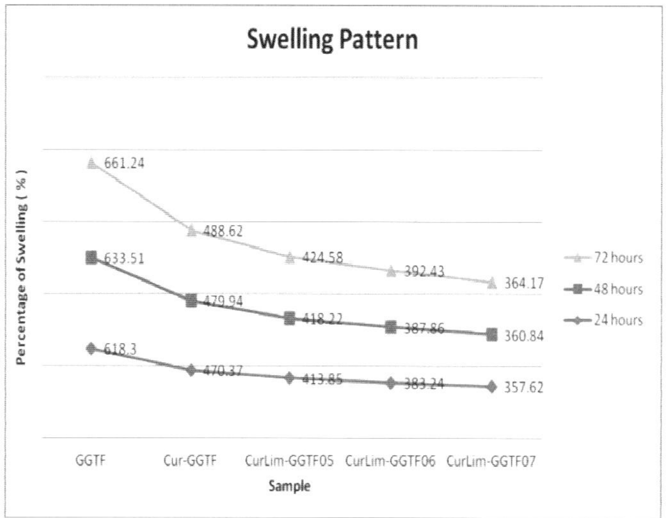

Figure 5. Swelling pattern of GGTF, Cur-GGTF, CurLim-GGTF05, CurLim-GGTF06 and CurLim-GGTF07

4. Conclusion

GGTF, Cur-GGTF containing only 0.06 M curcumin and Cur-GGTF containing 0.06 M curcumin with three different concentrations of limonene of 0.0005 M, 0.0006 M and 0.0007 M has been successfully prepared. From all of the characterizations that have been done, it has been proven that the incorporation of curcumin and limonene into gellan gum hydrogel was successful. Subsequently, the lowering of the decomposition temperature after the incorporation of the curcumin and limonene could lead to a pathway for a higher permeation of curcumin. This is expected to enhance the wound healing process. Moreover, WVTR study has verified the suitability of all the prepared hydrogels for wound dressing purpose. The swelling test has resulted that the hydrogels incorporated with limonene have lesser absorption of moisture in which makes it a better option for wound dressing as it is less likely to be broken and tear apart during body movement. The incorporation of curcumin and limonene into gellan gum hydrogel has shown high capability to be applied as good wound dressing material.

5. References

[1] Peppas N A, Bures P, Leobandung W & Ichikawa H 2000 *Journal of Pharmaceutics and Biopharmaceutics* **50** (1) 27
[2] Ghobril C & Grinstaff M W 2015 *Chemical Society Reviews* **44** (7) 1820
[3] Gupta B, Agarwal R & Alam M S 2010 *Indian Journal of Fibre & Textile Research* **35** 174187
[4] Antony P J, & Sanghavi N M, 1997 Drug development and industrial pharmacy **23** (4) 417418
[5] Kirchmajer D M, Steinhoff B, Warren H, Clark R & in het Panhuis M 2014 Enhanced gelation properties of purified gellan gum *Carbohydrate research* **388** 125-129
[6] Lee H, Fisher S, Kallos M S & Hunter C J 2011 *Journal of Biomedical Materials Research Part B: Applied Biomaterials* **98** (2) 238
[7] Singh S 2007 *Cell* **130** (5) 765-768
[8] Joe B, Vijaykumar M & Lokesh B R 2004 *Critical reviews in food science and nutrition* **44** (2) 97
[9] Priyadarsini K I 2014 *Molecules* **19** (12) 20091
[10] Wathoni N, Motoyama K, Higashi T, Okajima M, Kaneko T & Arima H 2017 *International Journal of Biological Macromolecules* **98** 268

[11] Thangapazham R L, Sharma A & Maheshwari R K 2007 Beneficial role of curcumin in skin diseases. In The Molecular Targets and Therapeutic Uses of Curcumin in Health and Disease (pp. 343-357) Springer US
[12] Umagiliyage A L, Becerra-Mora N, Kohli P, Fisher D J & Choudhary R 2017 *Postharvest Biology and Technology* **128** 130
[13] Arruda D C, Miguel D C, Yokoyama-Yasunaka J K, Katzin A M & Uliana S R 2009 *Biomedicine & Pharmacotherapy* **63** (9) 643
[14] Lević S, Lijaković I P, Đorđević V, Rac V, Rakić V, Knudsen T Š & Nedović V 2015 *Food Hydrocolloids* **45** 111
[15] Vigushin D M, Poon G K, Boddy A, English J, Halbert G W, Pagonis C 1998 *Cancer chemotherapy and pharmacology* **42** (2) 111
[16] Kim K J, Li B, Winer J, Armanini M, Gillett N, Phillips H S & Ferrara N 1993 *Nature* **362** (6423) 841
[17] A d'Alessio P, Mirshahi M, Bisson J F & Bene C M 2014 *Anti-Inflammatory & Anti-Allergy Agents in Medicinal Chemistry* **13** (1) 29
[18] Okabe H, Suzuki E, Saitoh T, Takayama K & Nagai T 1994 *Journal of controlled release* **32** (3) 243
[19] Moghimi H R, Makhmalzadeh B S & Manafi A 2009 *Burns* **35** (8) 1165
[20] Prakash S J, Santhiagu A & Jasemine S 2014 *J Pharm BioSci* **2** 63
[21] Anghel I, Holban A M, Grumezescu A M, Andronescu E, Ficai A, Anghel A G, Maganu M & Chifiriuc M C 2012 *Nanoscale research letters* **7** (1) 690
[22] Modasiya M K & Patel V M 2012 *International Journal of Pharmacy & Life Sciences* **3** (3) 1490
[23] Venkatasubbu G D & Anusuya T 2017 *International Journal of Biological Macromolecules* **98** 366
[24] Arrieta M P López J, Ferrándiz S & Peltzer M A 2013 *Polymer Testing* **32** (4) 760
[25] Dutta J 2012 *Synthesis* **2** (2) 6
[26] Zhao L, Mitomo H, Zhai M, Yoshii F, Nagasawa N & Kume T 2003 *Carbohydrate Polymers* **53** (4) 439
[27] Stott P W, Williams A C & Barry B W 1998 *Journal of controlled release* **50** (1) 297
[28] Budianto E, Muthoharoh S P & Nizardo N M 2015 *Asian Journal of Applied Sciences* **3** (05) 581
[29] Wu P, Fisher A C, Foo P P, Queen D & Gaylor J D S 1995 *Biomaterials* **16** (3) 171

IOP Conf. Series: Materials Science and Engineering **440** (2018) 012024 doi:10.1088/1757-899X/440/1/012024

Influence of Manuka Honey on Mechanical Performance and Swelling Behaviour of Alginate Hydrogel Film

Nur Aisyah Nasuha Mohd Azam[1] and Khairul Anuar Mat Amin[1]

[1]School of Fundamental Science, Universiti Malaysia Terengganu, 21030 Kuala Nerus, Terengganu, Malaysia.

Corresponding author emails: kerol@umt.edu.my

Abstract. In this study, sodium alginate (SA) hydrogel film incorporated with Manuka honey was successfully developed. Swelling properties and gel fraction of hydrogel films decreased upon addition of Manuka honey in SA film. SA hydrogel film has the highest swelling percentage at 512 ± 21% and decreased to 197 ± 9% in SA hydrogel film containing 10% (v/v) of Manuka honey (SA-H10). In contrast, the SA-H10 hydrogel film has the lowest gel fraction at 12 ± 1% compared to SA hydrogel film at 45 ± 3%. The Young's Modulus of SA hydrogel films with Manuka honey were increased as the concentrations of Manuka honey increased. The hydrogel films with higher concentration of Manuka honey (SA-H10) were soft with tensile strength, tensile strain and Young's Modulus at 1.06 ± 3 MPa, 15 ±1% and 34 ± 3 MPa, respectively. The water vapour transmission rates (WVTRs) of SA-H hydrogel films were in the range of 1051 ± 48 – 1428 ± 84 g m^{-2} d^{-1}, which comparable to WVTRs values with the commercial wound dressing products. The SA-H hydrogel films can be a potential material to be used as a wound dressing product.

1. Introduction

The biodegradable films based on biopolymers have been widely used in medical application due to its biocompatibility, biodegradability and properties that are similar to the human tissues [1]. Alginate is one of the examples of natural biopolymers that are commonly used in wound healing, drug delivery, soft tissue engineering, cell delivery, and in pharmaceutical industries [2, 3]. Alginates polymers primarily consist of blocks containing two uronic acids consist of two chain-forming heteropolysaccharides made up of blocks of ß-(1,4)-linked D-mannuronic (M) and a-(1,4)-linked Lguluronic (G) acids. The latter forms complex with divalent cations to form gels, and their structure varies in dependence on the monomer position in the chain, forming either homopolymeric (MM or GG) or heteropolymeric (MG or GM) blocks [4]. Physical properties of alginates as well as formation of gels depend on the relative proportion of these blocks. The ability of alginates to form gels in the presence of divalent calcium ions is one of their main bio-functional properties and has a great industrial significance [5].

An ideal wound dressing should be able to keep moisture and promote faster healing. Film dressings are designed to stick firmly to the skin surrounding a wound without sticking to the wound itself. Alginates have special properties to offer as an ideal property of dressing materials, which are gelling properties, haemostatic and "moist" dressings [6]. Sodium alginates interact with the wound exudate to form a moist and non-adherent gel. The strength of the materials increased when soaked with exudate or blood. The contact with exuding wound will exchange the calcium ions in the dressing with sodium ions in serum or wound fluid and resulting in swelling of fibre and partially dissolves a gel-like mass. Alginate structural match to extracellular matrices of living tissues which are widely used applications in wound healing such as leg ulcers, pressure sores and infected surgical wounds [7].

On other hand, Manuka honey is a natural product that has been widely used for its therapeutic effects [8]. It has been used over thousand years ago in traditional medicine and commonly used to treat wounds [9]. Commonly, the healing or other properties of honey are studied in its pure state, i.e. in liquid form. Numerous studies have been reported the advantages of Manuka honey such as the anti-bacterial, anti-inflammatory, anti-viral and anti-oxidant effects [10-12]. Besides numerous studies of using pure honey have been used to treat wound and burns, limited studies have been reported to incorporated honey as a composite with bio-polymer, i.e. transforming honey with biopolymers to a film or hydrogels. A few studies have been reported to use chitosan and gelatin incorporated with Sunflower honey as a sheet [13], and some using polyvinylpyrrolidone (PVP) entrapped with Gelam honey to produce hydrogels [14]. This study fabricated and characterized the sodium alginate hydrogel films incorporated manuka honey. The mechanical characteristics, water vapor transmission rates, swelling and gel fraction of the hydrogel films were investigated.

2. Materials and method

2.1 Materials

Sodium alginate was obtained from Sigma Aldrich, St Louis, MO, USA. Glycerine and anhydrous calcium chloride, $CaCl_2$ were obtained from Sigma Aldrich, St Louis, MO, USA and Manuka honey was obtained from Nature's Way, New Zealandas. All materials were used as received without any further purification.

2.2 Preparation of sodium alginate hydrogel films

2% of sodium alginate is dissolved in 70ml of deionised water at 80°C at constant stirring. Then 35% of glycerol is dissolved in 30mL of deionised in another beaker and is added into sodium alginate solution. The 5mM $CaCl_2$ solution is added dropwise and solution is heated and stirred for 1 hours and 45 minutes. Then, Manuka honey is added and the solution is stirred for another 15 minutes. These sodium alginate solutions containing 2%, 4%, 6%, 8% and 10% (w/v) Manuka honey are known as SA-H2, SA-H4, SA-H6, SA-H8 and SA-H10 respectively. After 2 hours, sodium alginate solution is poured into 5mm thickness of petri dish and dried in oven at 50°C, 50% fan for 24 hours. Then, the film is cut into 2mm x 2mm and let it immersed into 7mM $CaCl_2$ for 3 minutes. Then after immersion into $CaCl_2$ solution, the film is slightly dried with tissue paper before used for further characterization.

2.3 Characterization of SA-H films

2.3.1 FTIR Characterization.

ATR-FTIR spectra were determined by using Perkin Elmer Spectrum 100 FTIR spectrophotometer with PIKE Miracle ATR accessory with single-bounce beam path, 45° incident angle, and 4 cm^{-1} resolutions. Then all spectra were corrected by using Perkin Elmer Spectrum 100 software. The resulting spectra and functional peaks were recorded.

2.3.2 Swelling Percentage.

The swelling percentage test was carried out by immersing SA-H (20 mm x 20 mm) in phosphate buffer solution at pH 7.2 in water bath (37 ± 0.5 °C). The samples were removed after 24 hours and lightly wipe with wet filter paper to expel surface solution. The swelling degree was determined from equilibrium swelling ratio as in Equation 2.1:

$$Swelling\ percentage\ (\%) = \frac{M_f - M_i}{M_i} \times 100 \quad (1)$$

Where, M_f = final weight of sample and M_i = initial weight of sample.

2.3.3 Gel fraction.
The film sample was cut into 20 mm x 20 mm and let dried for 24 hours. Then the film was weighed (W_i) and the film was immersed in 10 ml deionised water at room temperature for 24 hours. After removing film from solution, the film was dried in oven at 50 °C for 24 hours and weighed again (W_f) and gel content was calculated based on Equation 2.2:

$$Gel\,fraction\,(\%) = \frac{W_f}{W_i} \times 100 \qquad (2)$$

Where, W_f = final weight of sample and W_i = initial weight of sample.

2.3.4 Water Vapour Transmission Rate.
The film was cut into 30 mm x 30 mm and put as a cap on a vial with diameter of 16 mm that contain 10 ml deionised water. Then it was weight (W_i) and left in the desiccator for 24 hours. After 24 hours, it was weighed again (W_f). The value of WVTR (g m^{-2} d^{-1}) was calculated based on Equation 2.3:

$$WVTR = (W_i - W_f)/A \qquad (3)$$

Where, W_i = initial weight, W_f = final weight and A= area of vial opening

2.3.5 Tensile Strength.
The tensile strength measurement was obtained by using Instron universal testing machine, model 3366 with cross-speed set at 10 mm/min. All films were cut into 20 mm x 60 mm for tensile stress-strain measurement. The tests were repeated triplicates per sample for defined ratio.

3. Result and discussion

3.1 FTIR Characterization
The ATR-FTIR spectra of SA films incorporated with Manuka honey were shown in Figure 1. There are a few functional peaks were appeared in SA-H films compared to blank SA film. Shifting and appearance of peak indicates that there was interaction occur between sodium alginate and honey [15]. Blank SA film shows a few main peaks at 3363.38 cm^{-1} was due to stretching vibration of O-H group [16]. A peak was shown at wavelength 2873.44 cm^{-1} which is saturated C-H stretch due to sp3 hybridization of C-H branched alkane, and a few peaks at 1609.88 cm^{-1} and 1412.60 cm^{-1} due to C=O bond stretching, bending or stretching mode of -COOH of sodium alginate. The peak shown around 1230.76 cm^{-1} could be attributed to the C-OH stretching vibration, while 1032.10 cm^{-1} to the O-H bending vibration of alginate [17].

Inclusion of Manuka honey in SA films show shifting vibration in SA film at 3363.38 cm^{-1} to SAH10 at 3354.21 cm^{-1}. This indicates there is a O-H vibration of H-bonded between sodium alginate and Manuka honey. The C=O bond stretching of blank SA at 1609.88 cm^{-1} and 1412.60 cm^{-1} were shifted to 1608.43 cm^{-1} and 1415.83 cm^{-1} in SA-H10 films which indicates the stretching band of carbonyl groups C=O and C=C stretching related to phenolic molecules present in Manuka honey. This shifting indicates that there was interaction occurred between sodium alginate films and Manuka honey [18].

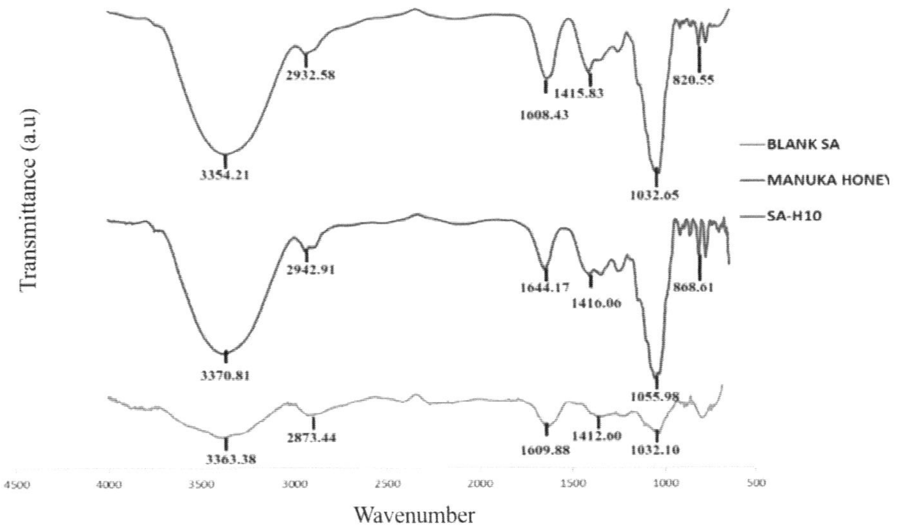

Figure 1. ATR-FTIR spectra of blank SA, Manuka Honey and SA-H10 hydrogel film.

3.2 Swelling percentage

The swelling percentages of SA-H films were recorded in Table 1. Swelling percentage of SA films decreases with addition of Manuka due to the formation of rigid structure of SA films. Blank SA absorbed 512 ± 21% of phosphate-buffered saline (PBS) solution (pH 7.4) and decreased to 430 ± 10% in SA-H2 film and further decreased to 197 ± 9% in SA-H10 film. This might due to hydrogen bond that occurred between SA and Manuka honey limits the availability of free-hydroxyl group (OH) to bind with water molecules. The higher of Manuka honey content incorporated in SA, the higher water resistance was exhibited.

Table 1. Swelling percentage, gel fraction and water vapour transmission rate (WVTR) of sodium alginate (SA) SA-H films

Sample	Swelling percentage (%)	Gel fraction (%)	WVTR ($g\ m^{-2}\ d^{-1}$)
Blank SA	512 ± 21	45 ± 3	1428 ± 84
SA-H2	430 ± 10	41 ± 1	1345 ± 61
SA-H4	345 ± 16	38 ± 2	1297 ± 46
SA-H6	316 ± 8	29 ± 1	1204 ± 35
SA-H8	257 ± 13	18 ± 2	1139 ± 25
SA-H10	197 ± 9	12 ± 1	1051 ± 48

3.3 Gel Fraction and Water Vapour Transmission Rates

Gel fraction is measured to examine the crosslinking degree of polymer chains within the SA films which can indicates the crosslinking behaviour of films. Table 1 shows the gel fraction of SA films depending to the Manuka honey contents. Higher percentage of Manuka honey incorporated in SA films caused the gel fraction values were decreased. Blank SA has the highest gel fraction at 46 ± 3%, and SA-H10 film has the lowest value at 13 ± 1%. It shows that increases the Manuka honey content strengthen the crosslinking with sodium alginate. This result is in agreement with swelling percentage of SA-H films, in which strong crosslinking behaviour decreased the swelling values of the films. Similar observations have been reported by Sirousazar et al. for swelling behaviour and structural characteristics of polyvinyl alcohol/montmorillonite nanocomposite hydrogels[18].

Water vapor transmission rates (WVTRs) is an important test to be determine in examine the loss of body fluid due to the evaporation process. The loss of huge quantity of exudates could causes the decrease in body temperature, disrupted the process of evaporation and could build up the pressure around wound and thus give pain to the patient [19]. Because of that, the WVTRs values are crucial to be examined and confirm the ability of the films to allow the transmission of body fluid. Water vapour transmission rates (WVTRs) of blank SA and SA-H films were in the range of 1428-1051 g $m^{-2} d^{-1}$ (Table 1). SA-H10 film has the lowest WVTR values which 1051 ± 49 g $m^{-2} d^{-1}$ that has highest incorporation of Manuka honey. The higher amount of Manuka honey in SA-H10 resist the water uptake in SA film thus resulting in lowest WVTR values. As compared to other, SA-H2 with low concentration of Manuka honey has the higher WVTR values at 1345 ± 61 g $m^{-2} d^{-1}$ than SA-H10. The WVTR values of SA films is comparable to commercial wound dressing product such as OpSite at 792 g $m^{-2} d^{-1}$ and Metalline at 1272 g $m^{-2} d^{-1}$ that are suitable and safe as wound dressing materials [20].

3.4 Tensile test

The tensile test is conducted to study the mechanical and to measure the strength and flexibility of the polymers films. The characterization of tensile is important for wound dressing materials to be examined and make sure it can function properly and protect wounds. Figure 2 shows the stress-strain curves of the blank SA and SA-H films at different loadings of Manuka honey and the tensile stress (σ), tensile strain (ε) and Young's modulus (YM) were summarized in Table 2. In general, addition of Manuka honey into SA films increased the YM values of the hydrogel films. YM of the SA-H10 was observed at 34 ± 3 MPa compared to SA-H2 at 28 ± 3 MPa.

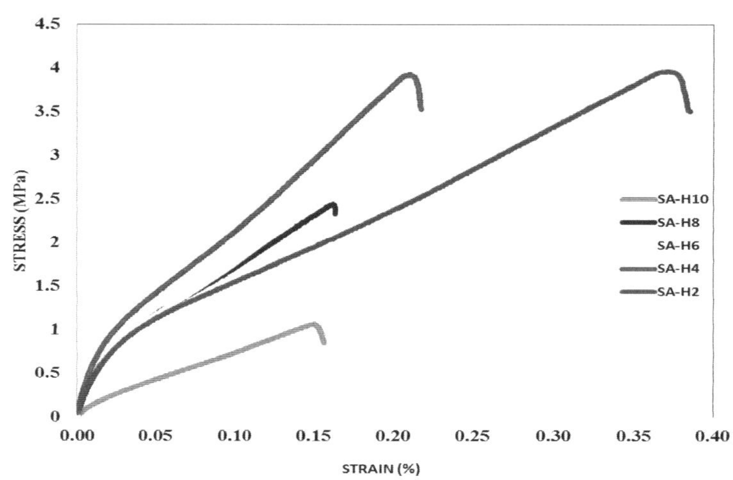

Figure 2. Tensile stress-strain curve of blank SA and SA-H films.

In contrast, the tensile stress and tensile strain of the films were decreased upon addition of Manuka honey. SA-H10 film has the lowest tensile strength and tensile strain at 1.0 ± 3 MPa and 15 ± 1%, respectively, meanwhile SA-H2 film exhibited the highest tensile strength and strain at 3.9 ± 4 MPa and 37 ± 2%, respectively. The behaviour of honey, i.e. soft and stickiness contributed to the decreased of tensile stress and tensile strain of the films at higher concentration of honey. The slight improvement of the Young's Modulus of the hydrogel films is a result of an increased in hydrogen bonding interaction between biopolymer-honey. Even though the film has an improved hydrogen bond, higher content of honey caused the stickiness to the films and reflected to low tensile stress and strain values. Without honey, the interaction occurs between SA-SA and therefore contribute to the brittle property of the hydrogel films and could not be measured [21]. The addition of honey was successfully improved the YM of the hydrogel films and make it possible to be used on different contours of our body as dressing materials.

Table 2. Tensile strength (σ), Young's Modulus (YM) and tensile strain (ε) of sodium alginate films incorporated with Manuka honey.

Sample	σ (MPa)	YM (MPa	ε (%)
SA-H2	3.9 ± 4	28 ± 3	37 ± 2
SA-H4	3.9 ± 3	21 ± 2	21 ± 1
SA-H6	2.9 ± 9	26 ± 2	18 ± 3
SA-H8	2.4 ± 5	29 ± 5	16 ± 5
SA-H10	1.0 ± 3	34 ± 3	15 ± 1

4. Conclusion

The sodium alginate (SA) film incorporated with Manuka honey (SA-H) was successfully prepared. The incorporation of Manuka honey in sodium alginate films (SA-H) caused the swelling percentage and gel fraction were decreased. The water vapour transmission rate of SA-H films was in the range of commercial dressings such as OpSite and Metalline, meanwhile tensile test of SA-H film shown improved Young's Modulus with insertion of Manuka honey. The SA-H film exhibit good properties to be used as a wound dressing application.

5. References

[1] Pereira R, Mendes A and Bártolo P 2013 Alginate/Aloe vera hydrogel films for biomedical applications *Procedia CIRP* **5** 210-5
[2] Hay I D, Wang Y, Moradali M F, Rehman Z U and Rehm B H 2014 Genetics and regulation of bacterial alginate production *Environmental microbiology* **16** 2997-3011
[3] Soni M L, Kumar M and Namdeo K 2010 Sodium alginate microspheres for extending drug release: formulation and in vitro evaluation *International Journal of Drug Delivery* **2**
[4] Castro L S E P W, de Sousa Pinheiro T, Castro A J G, Santos M d S N, Soriano E M and Leite E L 2015 Potential anti-angiogenic, antiproliferative, antioxidant, and anticoagulant activity of anionic polysaccharides, fucans, extracted from brown algae Lobophora variegata *Journal of applied phycology* **27** 1315-25
[5] Ruvinov E and Cohen S 2016 Alginate biomaterial for the treatment of myocardial infarction: progress, translational strategies, and clinical outlook: from ocean algae to patient bedside *Advanced drug delivery reviews* **96** 54-76
[6] Pawar S N and Edgar K J 2012 Alginate derivatization: a review of chemistry, properties and applications *Biomaterials* **33** 3279-305
[7] Lee K Y and Mooney D J 2012 Alginate: properties and biomedical applications *Progress in polymer science* **37** 106-26
[8] Robson V, Dodd S and Thomas S 2009 Standardized antibacterial honey (Medihoney™) with standard therapy in wound care: randomized clinical trial *Journal of advanced nursing* **65** 565-75
[9] Vandamme L, Heyneman A, Hoeksema H, Verbelen J and Monstrey S 2013 Honey in modern wound care: a systematic review *Burns* **39** 1514-25

[10] Lee H, Churey J J and Worobo R W 2008 Antimicrobial activity of bacterial isolates from different floral sources of honey *Int. J. Food Microbiol.* **126** 240-4
[11] Tan H T, Rahman R A, Gan S H, Halim A S, Hassan S A, Sulaiman S A and Kirnpal-Kaur B 2009 The antibacterial properties of Malaysian tualang honey against wound and enteric microorganisms in comparison to manuka honey *BMC complementary and alternative medicine* **9** 34
[12] Yaghoobi R and Kazerouni A 2013 Evidence for clinical use of honey in wound healing as an anti-bacterial, anti-inflammatory anti-oxidant and anti-viral agent: A review *Jundishapur journal of natural pharmaceutical products* **8** 100-4
[13] Wang T, Zhu X-K, Xue X-T and Wu D-Y 2012 Hydrogel sheets of chitosan, honey and gelatin as burn wound dressings *Carbohydrate polymers* **88** 75-83
[14] Yusof N, Hafiza A A, Zohdi R M and Bakar M Z A 2007 Development of honey hydrogel dressing for enhanced wound healing *Radiation Physics and Chemistry* **76** 1767-70
[15] Azam N A N M and Amin K A M 2017 The Physical and Mechanical Properties of Gellan Gum Films Incorporated Manuka Honey as Wound Dressing Materials *IOP Conference Series: Materials Science and Engineering* **209** 012027
[16] Shalumon K, Anulekha K, Nair S V, Nair S, Chennazhi K and Jayakumar R 2011 Sodium alginate/poly (vinyl alcohol)/nano ZnO composite nanofibers for antibacterial wound dressings *Int. J. Biol. Macromol.* **49** 247-54
[17] Xiao Q, Gu X and Tan S 2014 Drying process of sodium alginate films studied by twodimensional correlation ATR-FTIR spectroscopy *Food Chem.* **164** 179-84
[18] Sirousazar M, Kokabi M and Hassan Z 2012 Swelling behavior and structural characteristics of polyvinyl alcohol/montmorillonite nanocomposite hydrogels *Journal of Applied Polymer Science* **123** 50-8
[19] Augustine R, Kalarikkal N and Thomas S 2014 Role of wound dressings in the management of chronic and acute diabetic wounds *Diabetes Mellit Hum Health Care Holist Approach Diagn Treat* 273-314
[20] Wu P, Fisher A, Foo P, Queen D and Gaylor J 1995 In vitro assessment of water vapour transmission of synthetic wound dressings *Biomaterials* **16** 171-5
[21] Amin M, Anuar K, Gilmore K J, Matic J, Poon S, Walker M J and Wilson M R 2012 Polyelectrolyte Complex Materials Consisting of Antibacterial and Cell-Supporting Layers *Macromolecular bioscience* **12** 374-82

Acknowledgments

The authors wish to thanks the Ministry of Higher Education, Malaysia for financial assistance under the Fundamental Research Grand Scheme (FRGS– vote no. 59449) and School of Fundamental Science and Central Laboratory, Universiti Malaysia Terengganu for providing the facilities for undertaking this work.

Review of Metal Supported on Mesoporous MCM41 as Catalyst for CNTs Growth

Amirah Ahmad, Mohd Hasmizam Razali, Khairul Anuar Mat Amin*

School of Fundamental Science, Universiti Malaysia Terengganu, 21030 Kuala Nerus, Terengganu, Malaysia

*Corresponding author email: kerol@umt.edu.my

Abstract. Carbon nanotubes (CNTs) is one of the most unique nanomaterials in the field of nanotechnology because of their high strength, stiffness and electrical conductivity. Growth CNTs through chemical vapour deposition (CVD) method offers controlled synthesis of ordered and aligned CNTs. Catalytic growth also can be designed CNTs in variety forms either straight filament, helically wound tubes and bent. Recently, transition metal incorporated Mobil Composition Matter (known as MCM41) molecular sieves as growth catalyst has been used for the production of CNTs. The invention of mesoporous molecular sieves by Mobil researchers in 1992 has given a new direction to the field of porous materials. MCM41 exhibit a good hexagonal arrangement, uniform channel structure, tuneable pore size (15-100 Å) and large surface area (≈1200 m^2g^{-1}). Transition metal catalysts can be incorporated into the pore walls of the mesoporous molecular sieves and stabilizing the dispersed catalytic sites. The mesoporosity and the well-defined pore structure in combination with high surface area make MCM41 materials as promising candidates for the synthesis of CNTs. In this article, we review and discuss the role of catalyst and their catalyst support using MCM mesoporous material.

1. Introduction

Intensive research to produce high quality and high yield of CNTs have been attracted many researchers due to their unique properties. CNTs is a graphene layers that rolled up into cylinder and have special properties such as high surface area, high electrical and thermal conductivity, high mechanical properties and flexible [1]. The CVD method is currently used for the growth of CNTs due to simple method and can control the growth of CNTs during the growth process. Most of the CNTs synthesis techniques via CVD method require the introduction of catalyst either in the form of gas particulates or as a solid support. CVD method is also allowing the diameter of CNTs is controlled at certain range by modifying the size of catalyst nanoparticles [2].

The quality and yield of the growth of CNTs obtained depends on different parameters such as hydrocarbon sources, catalyst, and synthesis condition such as temperature, time and flow rate. By selecting proper conditions, the physical properties like length, shape, diameter and chemical properties

IOP Conf. Series: Materials Science and Engineering **440** (2018) 012025 doi:10.1088/1757-899X/440/1/012025

of CNTs can be controlled and designed. In this paper, we highlight about catalyst and it can be describe as a substance that increases the rate of reaction without itself being consumed [3]. But the main role of catalyst in CNTs growth is to decompose hydrocarbon molecules [4]. The catalyst materials may affect the CNTs growth in term of morphology, types of CNTs either single-walled carbon nanotubes (SWNT) or multi-walled carbon nanotubes (MWNTs), diameter and growth mechanism. Thus, catalyst material is very important for CNTs growth.

One of the most effective way to control the selectivity of CNTs with high yield is by using catalytic from mesoporous material. Mesoporous silica materials that have uniform pores and high specific areas have been used in numerous applications including catalysts, separation and adsorbent for large molecules [5]. These materials were first discovered by researchers at Mobil Research and Development Corporation in 1992 and known as the M41S family [6]. M41S family consist three members depends on their structure which has a hexagonal phase (MCM41), cubic phase (MCM48) and lamellar phase (MCM50). MCM41 has received most attention because of its simple structure, ease synthesis and good surface properties. The unique feature of MCM41 is the large uniform pore structure, large pore volume (> 0.7 cm3/g) and the pore can be tailored by several different strategies [7]. The unique feature of MCM41 is the large uniform pore structure and hence offered promising properties for both adsorption and catalysis [7].

However, purely siliceous MCM41 mesoporous materials have limited application, due to its lack of intrinsic acidity and ion exchange capacity. In order to utilize this material as efficient catalyst, it is necessary to generate the appropriate catalytically active sites to the silicate framework of MCM41. In previous study, different metal catalysts such as Cu, La, Gd, Ni, Co, Fe, Pd, Cr, Zn, Mn, Ti, Cd and Zr have been used to endow the properties of mesoporous materials [8]. It is because the structures and morphology can be influenced by particle size, metal content and the hollow pore structures of MCM41. But the most important property of the metals with regard to CNTs formation was their ability to catalytically decompose gaseous carbon-containing molecules [9]. The good thermal stability and high productivity of the MCM41 mesoporous molecular sieves could be a catalytic template for synthesizing CNTs. Most of the CNTs growth that we have performed in this paper so far used MCM41 supported with metal particles as growth catalyst. Due to MCM41 has high surface area that leads to high catalyst efficiency and high yield of CNTs. Therefore, we rely on the results obtained from previous study and discuss the performance of metal-MCM41 as catalyst to growth CNTs.

2. Catalyst

Catalyst is described as a substance that increases the rate of chemical reaction until approaches equilibrium without itself being involved in the reaction. Catalyst are used widely in the industry and played a major role in the publishing the economic strength of the chemical industry. According to Clark [10], the major goals of "Green Chemistry" and the use of efficient solid catalysts are to achieve these goals; a) to increase process selectivity; b) to maximize the use of starting materials; c) to replace stoichiometric reagents with catalyst; d) to facilitate easy separation of final reaction mixture and e) efficient recovery of the catalyst.

In CNTs growth, the most important property of catalyst regard to CNTs formation is their ability to catalytically decompose gaseous carbon-containing molecule. The catalyst used for the production of CNTs are usually transition metals supported on silica [11], mesoporous silica [12], zeolites [13], magnesium oxide and calcium carbonate. Therefore, the selection of metal and support material may affect the morphology and growth of CNTs. The most important function of support material is to provide high surface area and porosity for the active component. The interaction between catalyst and support material affects the dispersion and morphology of catalyst. Huh et al [14] reported that controlled growth of carbon nanotubes (CNTs) has been achieved by CVD of acetylene gas over nanometer-sized cobalt particles. They found CNTs with well-aligned, uniform diameter and high purity is successfully synthesized over cobalt. Amama et al [15] studied the synthesis of SWNTs over Fe-MCM41 using acetylene as carbon source. They found SWNTs successfully synthesized and claimed the reduction of Fe by the hydrogen produced during acetylene decomposition was responsible for the formation of metallic Fe, which provided the growth of CNTs.

Previous study reported that large amounts of CNTs can be formed by CVD method using acetylene as carbon source and cobalt and iron catalyst supported on silica or zeolite as catalyst [16]. However, the CNTs were proposed to grow over the external surface of the support materials. Thus, the growth directions are still random and the pore sizes are not narrowly distributed. To solve this problem, the mesoporous molecular sieves such as M41S materials are a kind of new catalyst supports with many interesting properties such as high surface area and periodic arrays of uniform pores. It suggested that MCM41 can be used as synthesis template for CNTs with regular pore structure.

3. MCM41 as Support for Metal-MCM41 Catalyst

The most important function of support material is to provide high surface area and porosity for the active component. The interaction between catalyst and support material directly affect the dispersion and morphology of catalyst. In CNTs growth, the catalyst needs an appropriate support material for selective controlling the morphology and yield of CNTs. It has been found that a single metal and mixture of metals supported on oxides, clays or porous materials like zeolite and MCM41 have great contribution in growth CNTs [17]. Chai et al., (2006) studied the effects of different support materials on CoO catalyst. They investigated titania, silica, alumina, magnesia, zeolite and calcium oxide as support and methane as carbon precursor at 700°C. They found that the carbon capacity decreased in order of alumina > zeolite > silica > titania > calcium oxide > magnesia [18].

Another approach was synthesized CNTs on different support materials which are quartz, porous alumina, conductive glass and nickel plates, using Co and Fe as catalyst and ethanol as the carbon sources. The SWNTs successful synthesized on porous alumina and nickel plates and MWNTs on conducting glass [19]. Su et al., (2000) studied the synthesized CNTs using aerogel support and Fe/Mo catalyst. They found this method improved the CVD method for SWNT preparation and achieved high quality SWNTs [20].

Besides, the CNTs were proposed to grow over the external surface of the support materials. Thus the growth directions are still random and the pore sizes are not narrowly distributed. The limited number of applications is related to the lack of mesopores in the zeolite that only contain micropores with diameter less than 2 nm. Therefore, only a fraction of the active sites in the zeolites is effectively used for catalytic conversions. To solve this problem, the mesoporous molecular sieves such as M41S materials are a kind of new catalyst supports with many interesting properties. It has unique properties such as high surface area and periodic arrays of uniform pores and also can be used as synthesis template for CNTs with regular pore structure. Recently, there are several reports of production of CNTs using mesoporous MCM-41 materials [21-23]. Hernadi et al [24] reported the comparison performance of Fe or Co supported on zeolite and MCM-41as catalyst for growth CNTs using acetylene as carbon source

IOP Conf. Series: Materials Science and Engineering **440** (2018) 012025 doi:10.1088/1757-899X/440/1/012025

at 700°C in 1 hour reaction by ion-exchange and impregnation method. From results, they found only a tiny CNTs formation was observed with ion-exchanged Co-MCM41.

However, pure siliceous mesoporous molecular sieves possess a neutral framework, which limits their applications. In order to provide molecular sieves with potential catalytic applications, it is possible to modify the nature of the framework by introduction of heteroatoms by hydrothermal methods. Besides, other elements can also be incorporated on the surface of the materials by grafting or impregnation. Recently, there are several reports of formation of carbon nanotubes over these mesoporous MCM41 materials [22, 23]. Table 1 shows the results obtained in in the previously reported work on various metal-MCM41 catalysts for CNTs production.

Table 1 Previously reported metal supported on mesoporous as growth catalyst for CNTs production

Catalyst	Reaction condition	Type of CNTs	CNTs size	References
Fe_2O_3-mesoporous	Carbon source : acetylene, CVD, 800°C	MWCNTs	5-15 nm	[25]
Ni-SBA15	Carbon source : acetylene, CVD, 850°C	Carbon sphres and filaments	0.5-0.8 μm	[26]
Ni-APTES-MCM41	Carbon source : acetylene, CVD, 600-800°C	MWCNTs	3-8 nm	[27]
Fe-Co-MCM41	Carbon source : acetylene, CVD, 750°C	SWCNTs	0.78-1.35nm	[22]
Fe/Co-Zeolite	Carbon source : ethane, pyrolysis, 850°C	MWCNTs	0.3-1.5nm	[28]

MWCNTs = multi-walled carbon nanotubes, SWCNTs = single-walled carbon nanotubes

Numerous studies had been reported that the presence of SWNTs is evidenced via Raman spectroscopy by observing the radial breathing modes (RBM) peak in the range 100 cm^{-1} to 350 cm^{-1} [29]. Meanwhile, the absence of RBM peaks indicates the CNTs produced is multi-walled type [30]. The two peaks were detected at ~1340 cm^{-1} (known as D band) and ~1580 cm^{-1} (known as G band) corresponding to the vibration of sp2 bonded carbon atoms in a 2D hexagonal lattice, and the vibration of carbon atom with dangling bond in the plane, respectively [31]. Figure 1 shows the Raman spectra of single-walled and multi-walled, respectively [32]. Chen et al [33] found the SWNTs with low defects and uniform diameter was obtained by using Co-MCM41 as growth catalyst.

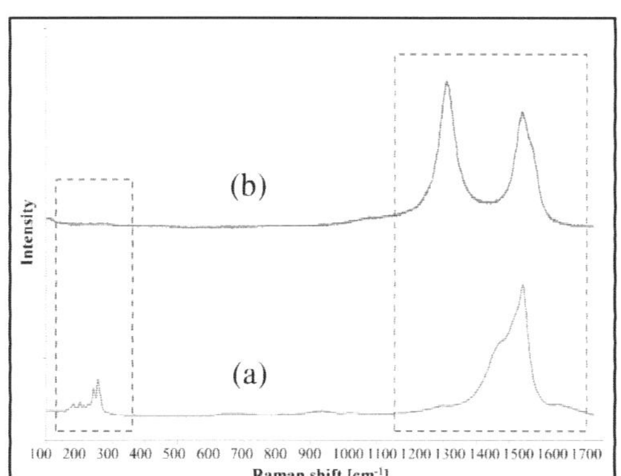

Figure 1. Raman spectra of (a) SWCNTs and (b) MWCNTs.

Urban et al [34] have reported TEM results show the diameter distribution of CNT formed was very narrow in between 5 to 6 nm in diameter and 200 nm in length. The quality of CNTs proved to be high, since their length exceeds 200 nm and had a narrow diameter distribution. Ahmad et al. [27, 32] reported that SEM and TEM images show the CNTs grown on Ni-MCM41 appearing looks like as a hair-like structure with the diameter in between 2 to 5 nm and a few hundred nanometres in lengths, as shown in Figure 2 [35]. Duxiao et al [36] reported the catalytic formation of CNTs on Fe-loading hexagonal mesoporous and microporous molecular sieves. The results confirmed that CNTs were grown inside the channel of mesoporous molecular sieves and on the external surface of microporous molecular sieves. The diameter of SWNTs formed was in the range 1 to 3 nm.

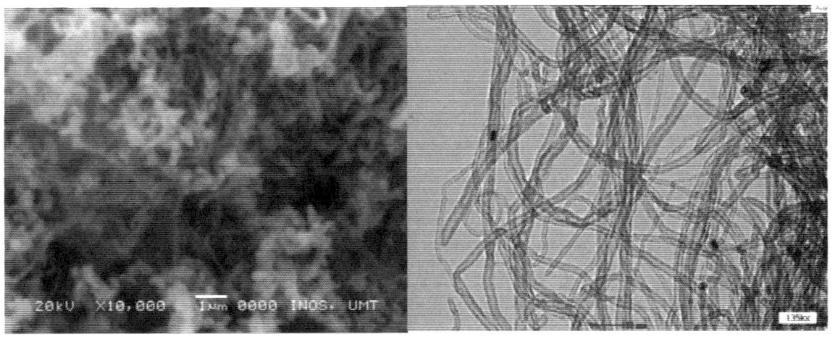

Figure 2. SEM and TEM images of MWCNTs grown on Ni-MCM41 catalyst

4. Conclusions

For commercial applications, the production of CNTs in high yield at low cost is importance and CVD method using growth catalyst has all the advantages in comparison with the other known method. Mesoporous MCM41 materials are considered as convenient candidates for the formation of carbon nanotubes. It is confirmed that MCM41 material with the incorporation of transition metals are considered to form better quality CNTs with high yield.

References

[1] Wildgoose G G, Banks C E, Leventis H C and Compton R G 2006 Chemically modified carbon nanotubes for use in electroanalysis *Microchimica Acta* **152** 187-214
[2] Liu W, Cai W, Yao L, Li X and Yao Z 2003 Effects of methane partial pressure on synthesis of single-walled carbon nanotubes by chemical vapor deposition *Journal of materials science* **38** 3051-4
[3] Richardson J T 1970 Principles of Catalyst Development, 1989. New York: Plenum Press)
[4] Dupuis A-C 2005 The catalyst in the CCVD of carbon nanotubes—a review *Progress in Materials Science* **50** 929-61
[5] Cho W, Park J-w and Ha C-S 2004 Direct synthesis of titanium-modified hybrid periodic mesoporous organosilicas from 1, 2-bis (triethoxysilyl) ethane and titanium isopropoxide *Materials Letters* **58** 3551-4
[6] Beck J, Vartuli J, Roth W J, Leonowicz M, Kresge C, Schmitt K, Chu C, Olson D H, Sheppard E and McCullen S 1992 A new family of mesoporous molecular sieves prepared with liquid crystal templates *Journal of the American Chemical Society* **114** 10834-43
[7] Chen C-Y, Li H-X and Davis M E 1993 Studies on mesoporous materials: I. Synthesis and characterization of MCM-41 *Microporous materials* **2** 17-26

[8] Taguchi A and Schüth F 2005 Ordered mesoporous materials in catalysis *Microporous and mesoporous materials* **77** 1-45
[9] Moisala A, Nasibulin A G and Kauppinen E I 2003 The role of metal nanoparticles in the catalytic production of single-walled carbon nanotubes—a review *Journal of Physics: condensed matter* **15** S3011
[10] Clark J H 2012 *Chemistry of waste minimization*: Springer Science & Business Media)
[11] Huang Z, Wang D, Wen J, Sennett M, Gibson H and Ren Z 2002 Effect of nickel, iron and cobalt on growth of aligned carbon nanotubes *Applied Physics A* **74** 387-91
[12] Hoyos-Palacio L, García A, Pérez-Robles J, González J and Martínez-Tejada H 2014 Catalytic effect of Fe, Ni, Co and Mo on the CNTs production. In: *IOP conference series: materials science and engineering*: IOP Publishing) p 012005
[13] Yu G, Gong J, Zhu D, He S, Cao J and Zhu Z 2006 Efficient synthesis of carbon nanotubes over rare earth zeolites by thermal chemical vapor deposition at low temperature *Diamond and related materials* **15** 1261-5
[14] Huh Y, Lee J Y, Cheon J, Hong Y K, Koo J Y, Lee T J and Lee C J 2003 Controlled growth of carbon nanotubes over cobalt nanoparticles by thermal chemical vapor deposition *Journal of materials chemistry* **13** 2297-300
[15] Amama P B, Lim S, Ciuparu D, Yang Y, Pfefferle L and Haller G L 2005 Synthesis, characterization, and stability of Fe– MCM-41 for production of carbon nanotubes by acetylene pyrolysis *The Journal of Physical Chemistry B* **109** 2645-56
[16] Ciambelli P, Sannino D, Sarno M, Fonseca A and Nagy J 2005 Selective formation of carbon nanotubes over Co-modified beta zeolite by CCVD *Carbon* **43** 631-40
[17] Rakov E G 2001 The chemistry and application of carbon nanotubes *Russian Chemical Reviews* **70** 827-63
[18] Chai S-P, Zein S H S and Mohamed A R 2006 Preparation of carbon nanotubes over cobalt-containing catalysts via catalytic decomposition of methane *Chemical Physics Letters* **426** 345-50
[19] Ortega-Cervantez G, Rueda-Morales G and Ortiz-Lopez J 2005 Catalytic CVD production of carbon nanotubes using ethanol *Microelectronics Journal* **36** 495-8
[20] Su M, Zheng B and Liu J 2000 A scalable CVD method for the synthesis of single-walled carbon nanotubes with high catalyst productivity *Chemical Physics Letters* **322** 321-6
[21] Somanathan T, Pandurangan A, Subashini D and Sathiyamoorthy D 2006 Growth of single walled carbon nanotubes over Co-MCM-41 mesoporous molecular sieves *Indian Journal of Pure and Applied Physics* **44** 173
[22] Somanathan T and Pandurangan A 2008 Catalytic activity of Fe, Co and Fe-Co-MCM-41 for the growth of carbon nanotubes by chemical vapour deposition method *Applied Surface Science* **254** 5643-7
[23] Ramesh P, Kishi N, Sugai T and Shinohara H 2006 High-yield synthesis of single-wall carbon nanotubes on MCM41 using catalytic chemical vapor deposition of acetylene *The Journal of Physical Chemistry B* **110** 130-5
[24] Hernádi K, Kónya Z, Siska A, Kiss J, Oszko A, Nagy J and Kiricsi I 2003 On the role of catalyst, catalyst support and their interaction in synthesis of carbon nanotubes by CCVD *Materials Chemistry and Physics* **77** 536-41
[25] Barreca D, Blau W, Croke G, Deeney F, Dillon F, Holmes J, Kufazvinei C, Morris M, Spalding T and Tondello E 2007 Iron oxide nanoparticle impregnated mesoporous silicas as platforms for the growth of carbon nanotubes *Microporous and mesoporous Materials* **103** 142-9
[26] Chiang H-L, Wu T-N, Ho Y-S and Zeng L-X 2014 Carbon material formation on SBA-15 and Ni-SBA-15 and residue constituents during acetylene decomposition *Journal of hazardous materials* **276** 43-51

[27] Ahmad A, Razali M H, Mamat M, Mehamod F S B and Amin K A M 2017 Adsorption of methyl orange by synthesized and functionalized-CNTs with 3-aminopropyltriethoxysilane loaded TiO 2 nanocomposites *Chemosphere* **168** 474-82
[28] Khatri I, Kishi N, Zhang J, Soga T, Jimbo T, Adhikari S, Aryal H R and Umeno M 2010 Synthesis and characterization of carbon nanotubes via ultrasonic spray pyrolysis method on zeolite *Thin Solid Films* **518** 6756-60
[29] Dehouche Z, Lafi L, Grimard N, Goyette J and Chahine R 2005 The catalytic effect of single-wall carbon nanotubes on the hydrogen sorption properties of sodium alanates *Nanotechnology* **16** 402
[30] Masoumi M, Mehrnia M R, Montazer-Rahmati M and Rashidi A M 2010 Templated growth of carbon nanotubes on nickel loaded mesoporous MCM-41 and MCM-48 molecular sieves *International Journal of Nanoscience and Nanotechnology* **6** 88-96
[31] Yadav Y, Kunduru V and Prasad S 2008 *Nanopackaging*: Springer) pp 325-44
[32] Ahmad A, Razali M H And Amin K A M Growth And Characterization Of Multi-Walled Carbon Nanotubes Using Ni-MCM41 Mesoporous Catalyst And Their Application As Adsorbent Journal of Sustainability Science and Management Special Issue **No.2** 2017 16-26
[33] Chen Y, Ciuparu D, Lim S, Haller G L and Pfefferle L D 2006 The effect of the cobalt loading on the growth of single wall carbon nanotubes by CO disproportionation on Co-MCM-41 catalysts *Carbon* **44** 67-78
[34] Urbán M, Méhn D, Kónya Z and Kiricsi I 2002 Production of carbon nanotubes inside the pores of mesoporous silicates *Chemical physics letters* **359** 95-100
[35] Ahmad A, Razali M H, Kassim K and Amin K A M 2017 Synthesis of multiwalled carbon nanotubes supported on M/MCM-41 (M= Ni, Co and Fe) mesoporous catalyst by chemical vapour deposition method *Journal of Porous Materials* **25** 433-441
[36] Duxiao J, Nongyue H, Yuanying Z, Chunxiang X, Chunwei Y and Zuhong L 2001 Catalytic growth of carbon nanotubes from the internal surface of Fe-loading mesoporous molecular sieves materials *Materials chemistry and physics* **69** 246-51

Acknowledgments
The authors wish to thanks the Ministry of Higher Education, Malaysia for financial assistance under the Fundamental Research Grand Scheme (FRGS– vote no. 59294) and School of Fundamental Science and Central Laboratory, Universiti Malaysia Terengganu for providing the facilities for undertaking this work.

IOP Conf. Series: Materials Science and Engineering **440** (2018) 012026 doi:10.1088/1757-899X/440/1/012026

The Effect of Acetaminophen on Physical, Compression Strength and Thermal Behaviours of Kelcogel Hydrogel Films

N F K Ramli[1], S I A Razak[2] and K A M Amin[1,3]*

[1]School of Fundamental Science, Universiti Malaysia Terengganu (UMT), 21030 Kuala Nerus, Terengganu, Malaysia.

[2]Medical Devices and Technology Group, Faculty of Biosciences and Medical Engineering, 54702 Universiti Teknologi Malaysia, Skudai, Johor, Malaysia

[3]Advanced Nano-materials Research Group, School of Fundamental Science, Universiti Malaysia Terengganu (UMT), 21030 Kuala Nerus, Terengganu, Malaysia.

Corresponding authors email: fatenkartiniramli@gmail.com; kerol@umt.edu.my*

Abstract. This study investigates the physical properties, compression strength and thermal behaviors of the Kelcogel hydrogel films. Kelcogel hydrogel film at 2 mm thickness (KEL-2) shows the optimum swelling ratio and compression strength at 99 ± 10% and 155 ± 33 kPa, respectively compared to other samples. The KEL-2 hydrogel films were further incorporated with acetaminophen and the properties were examined. The swelling ratio and compression strength of KEL-2 hydrogel films contained 30% (w/w) of acetaminophen (KEL-A30) show optimum values at 108 ± 10% and 423 ± 42 kPa, respectively. The WVTRs of the KEL hydrogel films incorporated acetaminophen were in the range of 2329 - 4699 g $m^{-2} d^{-1}$. Thermal stability of KEL hydrogel films also increased depending to the amount of the acetaminophen added. This study shows that the addition of acetaminophen improved the physical properties, compressive strength and thermal behavior of the Kelcogel hydrogel film and promising to be applied as dressing materials.

1. Introduction

Hydrogel films are an appealing material due to its structure is similar to the extracellular matrix of many tissues, and often be processed under relatively mild conditions. Hydrogel films have been utilized as dressing materials, scaffold materials for drug and growth factor delivery, engineering tissue replacements, and a variety of other applications [1, 2]. Nowadays, biopolymers are the primary choice to be used in biomedical application than synthetic petrochemical based polymers. Several biopolymers are available such as chitosan, carrageenan, xanthan gum and gellan gum with different properties. In this study, gellan gum, i.e. a linear anionic polysaccharide produced by *Pseudomonas elodea* is chosen because it is approved by the United States Food and Drug Administration (US FDA) and the European Union (EU) to be used in the food industry [3]. Not limited to that, it is an emerging scaffold material for tissue engineering application [4, 5]. Various gellan gum brands are available in commercial market such as Kelcogel and Gelzan. Kelcogel is preferred for pharmaceutical, oral care and personal care, and Gelzan for microbiological media and plant tissue culture. Acetaminophen (AP) or also known as Paracetamol is mostly used as an analgesic and antipyretic [5]. It is considered as the safest analgesics available with 200 over products containing acetaminophen as an active ingredient. AP is also used as a pain-killer to treat headache, back pain, acute pain resulting from soft tissue injury and post-operative pain [6]. Up to our knowledge, no study has been examined the effect of acetaminophen in kelcogel hydrogel films. This study examined the physical properties, compression strength and thermal behaviours of the hydrogel films specifically to be used as dressing materials.

2. Materials and Method

2.1 Materials

Low-acyl gellan gum (Kelcogel, batch number: 5C1574A) were obtained from CPKelco, Chigaco, IL, USA. Glycerin (product number-G2289), anhydrous calcium chloride, $CaCl_2$ (product numberC5670), and acetaminophen (product number-A7085) were obtained from Sigma Aldrich, St Louis, MO, USA. All materials were used as received without any purification.

2.2 Preparation of hydrogel film at different thickness (KEL)

Kelcogel films were prepared using evaporative casting method. Hydrogels solution were prepared by dissolving 1% (w/v) of kelcogel in distilled water under continuous stirring with the temperature maintain at 80°C. Then, 50% (w/w) of glycerol was added into the hot solution, followed by addition of 0.1 M calcium chloride solution. Later, the solution was deposited onto petri (90 mm × 15 mm) up to 8 mm in thickness and dried in the oven at 50°C for 4 hours for 6 mm, 8 hours for 4 mm and 24 hours for 2 mm.

2.3 Preparation of hydrogel film incorporated acetaminophen (KEL-AP)

1% (w/v) of kelcogel was dissolve in distilled water with the temperature at 80°C under continuous stirring to produce hydrogel solution. After the kelcogel was completely dissolved, 50% (w/w) of glycerol was added into the solution. 10% (w/w) of acetaminophen by weight relative to hydrogel was prepared by dissolving using methanol to produce acetaminophen solution. Then the solution of acetaminophen was added drop wise into the hydrogel solution, and hereafter known as KEL-AP10. Lastly, 0.1 M $CaCl_2$ solution was added drop wise into the hydrogel solution under continuous stirring for 2 hours. After 2 hours of reaction, the hydrogel solution was pour into the petri dish at 8 mm thickness and dried it in the oven at 50 °C for 24 hours. The steps were repeated for 20% (w/w) and 30% (w/w) of acetaminophen, and the samples were known as KEL-AP20 and KEL-AP30 hydrogel films, respectively.

2.4 Characterization of hydrogel

2.4.1 Swelling Test.

Dried sample of 2 cm × 2 cm hydrogel were cut and immersed in 20 mL phosphate buffer solution (PBS) of pH 7.2 at 37 ± 0.5 °C for 24 hours. After 24 hours, the samples were removed and the excess water on the surface was wiped using tissue. The weight of hydrogel before (W_i) and after (W_f) immersed in PBS were recorded and the swelling percentage was calculated using the equation 2.1:

$$\text{Swelling percentage } (\%) = (W_f - W_i) / W_i \times 100\% \qquad \text{Eqn. 2.1}$$

where W_i and W_f are initial and final weight, respectively.

2.4.2 Water Vapour Transmission Rate (WVTR).

The WVTR was carried out by cutting 3 cm × 3 cm hydrogel and placed them on the circular opening (test area) of cylindrical glass vial containing 10 mL water. Then the glass vials were left in the desiccator for 24 hours. The weight before and after the glass vials left in the desiccator were recorded and the WVTRs was calculated using the equation 2.2:

$$\text{WVTR} = (W_i - W_f) / A \times 24 \ \frac{g}{m^2}/\text{day} \qquad \text{Eqn. 2.2}$$

where W_i and W_f are initial and final weight of glass vial and A is the test area of the sample in m^2.

2.4.3 Compression Test.
The compression test was carried out using Instron Universal Mechanical machine (model 3366) at a cross speed set at 10 mm/min. All hydrogels were cut to 2 cm × 2 cm dimension. A minimum of three independent measurements will be obtained per sample of a defined ratio.

2.4.4 Thermogravimetric Analysis (TGA).
Thermogravimetric analyses were performed using Thermogravimetric Analyzer model Mettler Toledo. Films were cut into small pieces and about 10 mg were used. Then the films were heated from 30°C to 700°C at a heating rate of 10°C/min under nitrogen atmosphere with a nominal gas flow rate of 30 mL/min.

2.4.5 Differential Scanning Calorimetry (DSC).
Differential scanning calorimetry was performed using Pyris 6, Perkin-Elmer-TGA7. The samples were weighted about 4 mg. The heating rate was 10°C/min and the nitrogen gas was purged at a flow of 50 mL/min. The heating range of the sample was from room temperature to 200°C.

3. Result and Discussion

3.1 Swelling Test
Swelling test was conducted to measure the maximum amount of fluid absorbed and retained by the film. The swelling percentage of all hydrogel films were presented in Table 1. KEL-8 shows the lowest percentage of water uptake due to the thickest hydrogel (8 mm) at 2.2 ± 0.8% and has higher water content compare to other sample which resulting to the smallest absorption of water. Besides that, it is typical for a thick film (or known as hydrogel) of KEL-8 to absorb minimum water due to low water holding capacity. In contrast, KEL-2 film absorbs optimum water at 99 ± 10% due to its high holding capacity. This optimum swelling value is needed for wound dressing materials to absorb exudate from wound and promote moist environment to the wound bed. The swelling percentage of the KEL hydrogel films incorporated acetaminophen (KEL-AP) increased as the concentration of the latter increased. The enhancement of swelling values of KEL-AP film might be due to the present of higher number of the hydroxyl group (OH) in KEL-AP which able to bind with the water. The swelling percentage of KEL-AP30 film increased to 108 ± 10%. It is reported that an ideal wound dressing should be able to absorb water in the range of 100-900% [6]. Thus, the addition of acetaminophen enhanced the swelling properties of the hydrogel films.

Table 1. Swelling percentage and water vapour transmission rates (WVTRs) of the kelcogel film (KEL) and kelcogel film with acetaminophen (KEL-AP).

Sample	Swelling Percentage (%)	WVTR (g m^{-2} d^{-1})
KEL-8	2 ± 0.2	N/A
KEL-6	18 ± 1	N/A
KEL-4	55 ± 18	2690 ± 77
KEL-2	99 ± 10	2385 ± 76
KEL-AP10	82 ± 7	2329 ± 55
KEL-AP20	83 ± 8	4361 ± 154
KEL-AP30	108 ± 10	4699 ± 596

*N/A =not applicable

3.2 Water Vapour Transmission Rate (WVTR)

Water vapour transmission rates (WVTRs) is used to measure the amount of water loss across the film over defined times. Film wound dressing should prevent or reduce the body liquid lost by controlling absorption and transmission as well as maintaining the humidity in the wound area, in order to accelerate the formation of granule and epithelisation process [7]. Therefore, WVTR is an important test to examine the potential of dressing materials in controlling the body liquid or wound exudate. The WVTRs values of all films were shown in Table 1. The WVTR values decreased as the thickness of the film decreased from 2690 ± 77 g m^{-2} d^{-1} of KEL-4 to 2385 ± 76 g m^{-2} g^{-1} of KEL-2. The WVTR values of KEL-8 and KEL-6 films were not carried out due to tendency of the samples to shrink within 24 hours of measurement and resulted in inaccurate results. The addition of acetaminophen in KEL films increased the WVTRs value to 4699 ± 596 g m^{-2} d^{-1}. All the WVTRs values obtained were in the range of acceptable WVTRs values and comparable with commercial wound dressing products, i.e. in the range of 90-2893 g m^{-2} d^{-1} [8].

3.3 Compression Test

Compressive test was used to determine the mechanical strength of films and the effect of addition of acetaminophen into the films. The mechanical property of wound dressing materials is very important because certain strength was needed for handling and replacement of dressing materials at difference contours of human body [9]. The compressive stress (kPa), compressive strain (%) and Young's Modulus (kPa) of the hydrogel films were summarized in Table 2. The compressive strength of the hydrogel films increased as the thickness of the hydrogel film decreased. KEL-2 yielded the highest compressive stress and Young's Modulus at 155 ± 33 kPa and 2165 ± 470 kPa respectively. Inclusion of acetaminophen increased the compressive stress to 160 ± 10 for KEL-AP10 hydrogel film. The Young's Modulus also increased to 2275 ± 284 kPa compare to free-standing KEL-2 film. The inclusion of higher concentration of acetaminophen further increased the stress to 423 ± 42 kPa for KEL-AP30, in which an increment of almost 3-fold compared to KEL-2 film. Meanwhile, the Young's Modulus value also increased to 4-fold at 8773 ± 342 kPa compared to KEL-2. The increased of compressive strength and Young's modulus of the KEL-AP films could be due to the improvement of cross-linking behaviour among the film network which creates the strong hydrogen bond interactions between kelcogel and acetaminophen. In contrast, the strain value for KEL-AP10 was decreased to almost 1-fold after the addition of acetaminophen due to limited stretching of the network in the film structure (hydrogen bond). KEL-AP30 recorded the lowest strain value at 9 ± 1 %. The addition of acetaminophen was successfully improved the strength of the hydrogel films and make it possible to be used on different part of our body as dressing materials.

Table 2. Compressive stress (kPa), compressive strain (%) and Young's Modulus (kPa) of the kelcogel film (KEL) and kelcogel film with acetaminophen (KEL-AP).

Sample	Compressive Stress (kPa)	Compressive Strain (%)	Young's Modulus (kPa)
KEL-8	50 ±7	20 ±4	486±7
KEL-6	55 ±3	15 ±2	590±26
KEL-4	72 ±6	19 ±1	670±77
KEL-2	155 ±33	14±2	2165±170
KEL-AP10	160 ±10	12±1	2275±184
KEL-AP20	360 ±75	16±5	5477±113
KEL-AP30	423 ±42	9±1	8773±242

IOP Conf. Series: Materials Science and Engineering **440** (2018) 012026 doi:10.1088/1757-899X/440/1/012026

3.4 Thermogravimetric Analysis (TGA)

The thermogram and derivative thermogram of pure hydrogel film (KEL-2) and KEL-AP films were presented in Figure 1 and summarized in Table 3. The TGA/DTG result shows two main region observed, at (i) 30 °C to 140 °C, which resulting from the removal of free water from the polysaccharide [10], and (ii) 180 °C to 280 °C, which related to the decomposition of the aromatic bond of acetaminophen (Fig 1).

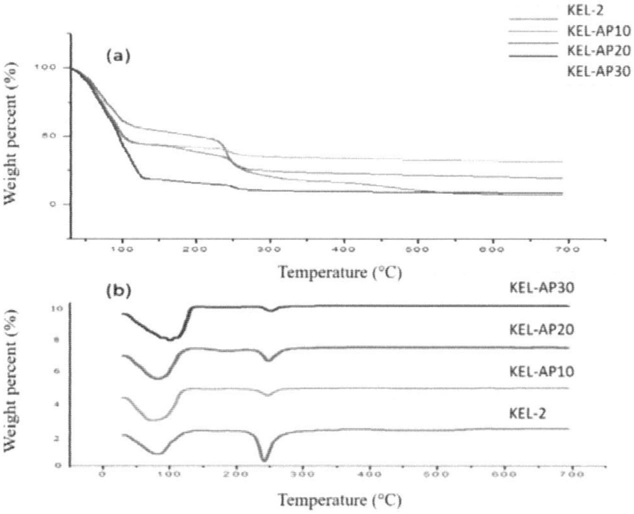

Figure 1. (a) Thermogravimetric thermograms (TGA) and (b) derivative thermograms (DTG) of the kelcogel film (KEL) and kelcogel film with acetaminophen (KEL-AP).

Table 3. Temperature onset *(T_o)*, temperature completion (T_c) and weight loss (%) of KEL-2 and kelcogel film with acetaminophen (KEL-AP).

Sample	Temperature onset, T_o (°C)	Temperature Completion, T_c (°C)	Weight loss (%)
KEL-2	38	600	93
KEL-AP10	38	69	69
KEL-AP20	39	80	80
KEL-AP30	39	92	92

The degradation of KEL-2 film started approximately at temperature onset, T_o=38°C and the temperature completion, T_c at 600°C (Table 3). Inclusion of acetaminophen increased the temperature onset (T_o = 39°C), and further increased the temperature completion, T_c (T_c= 580°C- 670°C) depending to the concentration of acetaminophen. KEL-AP20 films yield the highest temperature completion at T_c=670°C.

3.5 Differential Scanning Calorimetry (DSC) analysis

DSC result shows that the inclusion of acetaminophen affect the thermal stability of the hydrogel film (Figure 2 and Table 4). The melting point, T_m is increased depending to the concentration of acetaminophen, in the range of 106 °C to 147 °C. KEL-AP30 yields the highest melting point (T_m=147 °C) than KEL-2 hydrogels (T_m= 104 °C). Broad exothermic peaks were detected for KEL-A hydrogel films. Higher content of acetaminophen increased the T_c to 278 (KEL-AP30) and confirming the improved thermal stability of the films with addition of acetaminophen.

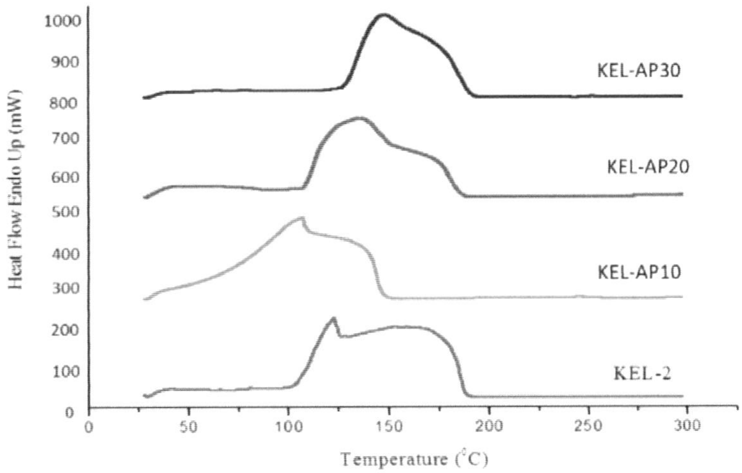

Figure 2. Differential scanning calorimetry thermograms of kelcogel film and kelcogel film with acetaminophen (KEL-AP) at different concentrations.

Table 4. Differential scanning calorimetry properties of kelcogel film and kelcogel film with acetaminophen (KEL-AP) at different concentrations.

Sample	Temperature (°C)				Range (°C)
	Glass Transition, T_g	Temperature Onset, T_o	Temperature Melting, T_m	Temperature Completion, T_c	T_c-T_o
KEL-2	29	104	123	275	171
KEL-AP10	28	63	106	278	215
KEL-AP20	28	108	134	285	177
KEL-AP30	29	128	147	278	150

4. Conclusion

Kelcogel film incorporated with acetaminophen (KEL-AP) was successfully examined for their physical properties, compression and thermal behaviours. It shows that the inclusion of acetaminophen improved all the properties particularly the swelling and compression strength of the films. The water vapour transmission rates of the KEL-AP film were comparable to the commercial dressings. In conclusion, the KEL-AP films shows promising properties to be used as wound dressing materials.

References

[1] Syazwani Mohd S, Abdullah M A A, Amin M and Anuar K 2016 Compression Strength of Gellan Gum Hydrogel Incorporated with Organo-Montmorillonite and Cloisite 15A *Materials Science Forum* **840** 236-9

[2] Stevens L, Gilmore K J and Wallace G G 2016 Tissue engineering with gellan gum *Biomaterials science* **4** 1276-90

[3] Amin M, Anuar K, Gilmore K J, Matic J, Poon S, Walker M J and Wilson M R 2012 Polyelectrolyte Complex Materials Consisting of Antibacterial and Cell_Supporting Layers *Macromolecular bioscience* **12** 374-82

[4] Oliveira J T, Martins L, Picciochi R, Malafaya P, Sousa R, Neves N, Mano J and Reis R 2010 Gellan gum: a new biomaterial for cartilage tissue engineering applications *Journal of Biomedical Materials Research Part A* **93** 852-63

[5] Sebria N J M and Amin K A M 2016 Gellan Gum/Ibuprofen Hydrogel for Dressing Application: Mechanical Properties, Release Activity and Biocompatibility Studies *International Journal of Applied Chemistry* **12** 483-98

[6] Morgado P I, Aguiar-Ricardo A and Correia I J 2015 Asymmetric membranes as ideal wound dressings: An overview on production methods, structure, properties and performance relationship *Journal of Membrane Science* **490** 139-51

[7] Morgado P I, Miguel S P, Correia I J and Aguiar-Ricardo A 2017 Ibuprofen loaded PVA/chitosan membranes: A highly efficient strategy towards an improved skin wound healing *Carbohydrate polymers* **159** 136-45

[8] Wu P, Fisher A, Foo P, Queen D and Gaylor J 1995 In vitro assessment of water vapour transmission of synthetic wound dressings *Biomaterials* **16** 171-5

[9] Zhou Y, Zhao Y, Wang L, Xu L, Zhai M and Wei S 2012 Radiation synthesis and characterization of nanosilver/gelatin/carboxymethyl chitosan hydrogel *Radiation Physics and Chemistry* **81** 553-60

[10] Capanema N S, Mansur A A, de Jesus A C, Carvalho S M, de Oliveira L C and Mansur H S 2018 Superabsorbent crosslinked carboxymethyl cellulose-PEG hydrogels for potential wound dressing applications *International Journal Biological Macromolecules* **106** 1218-34

Acknowledgments
The authors wish to thanks the Ministry of Education, Malaysia for financial assistance under the Fundamental Research Grand Scheme (FRGS– vote no. 59449), School of Fundamental Science and Central Laboratory, Universiti Malaysia Terengganu for providing the facilities for undertaking this work.

Physiochemical and Mechanical Properties Study of Gellan Gum Incorporating Norfloxacin Biofilm

Nur Arifah Ismail, Khairul Anuar Mat Amin, and Mohd Hasmizam Razali*

School of Fundamental Sciences, Universiti Malaysia Terengganu, 21030 Kuala Terengganu, Terengganu Darul Iman, Malaysia.

*Corresponding author email: mdhasmizam@umt.edu.my

Abstract. The aim of this work was to investigate the physiochemical and mechanical properties of gellan gum incorporating norfloxacin (GG-NOR) biofilm at different concentration of NOR. Four different concentration of NOR (0.01%, 0.1%, 0.5%, and 1% (w/w)) were used to produce GG-NOR biofilms. The biofilms was successfully produced with good shape and were transparent at ~95%. The addition of norfloxacin was found to increase the toughness and elongation-at-break values of biofilm to 0.52 J g^{-1} and 3.8 % at higher concentration of NOR. On top of that, the thermal stability of biofilm also was increased when the concentration of norfloxacin was increased attributed to the interaction between gellan gum and norfloxacin as proved by FTIR analysis.

1. Introduction

The ideal biodegradable materials are obtained from renewable biobased resources, usually called biopolymers, with excellent mechanical and barrier properties and biodegradable at the end of their life. Biopolymers have been considered as a potential environmentally-friendly substitute for the use of non-biodegradable and non-renewable plastic packaging materials. Biofilm materials may also serve as gas and solute barriers and complement other types of packaging by improving the quality and extending the shelf life of foods. Furthermore, biodegradable film packaging materials are excellent vehicles for incorporating a wide variety of additives such as antioxidants, antifungal agents, antimicrobials, colors, and other nutrients [1,2].

However, several concerns must be addressed before the commercial use of biobased primary film materials. These concerns include degradation rates under various conditions, changes in mechanical properties during storage, the potential for microbial growth, and release of harmful compounds into packaged foods. In reality, relatively poor mechanical and high hydrophilic properties coupled with the poor process ability of biopolymer-based packaging materials are a major limitation on their industrial use [3]. Thus, in this study norfloxacin has been incorporated into gellan gum (GG) biopolymer in order enhanced the mechanical properties of the film. Previously GG biopolymer widely used in biomedical applications due its biocompatibility and biodegradability properties [4-6]. The selection of norfloxacin because it offer an antibacterial activity due to the presence of fluoroquinolone group in the structure [7, 8]. Many studies have been reported the effectiveness of the norfloxacin in combating the Gram-positive and Gram-negative bacteria [9-16]. At the same time norfloxacin is more active and less cytotoxicity compared to other fluoroquinolone groups such as levofloxacin and ofloxacin [17-18].

2. Experiments

2.1. Materials

Low-acyl gellan gum (Gelzan™ CM, Mw ≈ 2-3 x 10^5 Da, product number-G1910, lot number SLBB0374V) was obtained from Sigma-Aldrich, Malaysia. The commercial antibiotic, Norfloxacin (product number-N9890) was obtained from Fluka, USA. All materials were used as initially received.

2.2 Preparation of Biofilm

A stock solution of gellan gum (GG) was made by dissolving 1% (w/v) of GG in 100 mL deionized water (18.2 MΩ) with continuous stirring for 2 h at 70 °C. GG-NOR solutions were prepared by adding 0.01%, 0.1%, 0.5% and 1% (w/w) norfloxacin (NOR) into the stock solution of GG. The GG and GG-NOR solutions were next deposited onto petri dishes (90 mm x 15 mm) and placed inside Venticell oven at 30 °C for at least 24 h. All biofilms were pre-conditioned in a desiccator (24 °C, 50 % relative humidity (RH)) for at least 2 days prior to testing [19].

2.3 Characterization of Biofilm

Ultraviolet (UV-vis) transmission spectra of biofilms were performed using a spectrophotometer (Varian, Cary 50) with data interval = 5 nm, scan rate = 24 000 nm/min and wavelength range 200-800 nm. Quartz cells with path length 10 mm were used. The UV-vis transmittance was conducted by measured the attached biofilms on the cuvette's surface. ATR-FTIR spectra were collected using a Perkin Elmer Spectrum 100 FT-IR spectrophotometer with PIKE Miracle ATR accessory (single-bounce beam path, 45 ° incident angle, 16 scans, 4 cm^{-1} resolution) and all spectra were corrected by the Perkin Elmer spectrum 100 software. Stress-strain measurements were obtained using an Instron Universal Testing machine (model 3366) with ± 10 kN grips and the cross-speed set at 20 mm/min. Biofilm thickness (2.0 cm x 6.0 cm) was measured by a hand-held micrometer (Mitutoyo). Young's modulus (E), tensile strength (TS), and toughness (T) were calculated from the slope of the linear part of the stress-strain curve, maximum stress, and through integration of the area under the curve, respectively. Elongation-at-break (EAB) was also recorded. Thermogravimetric analyses were performed on a Pyris 6, Perkin-Elmer-TGA6. Biofilm samples were analysed in platinum pans at a heating rate of 10 °C/min to 900 °C in an atmosphere of N2 flow at 50 ml/min. Sample used was approximately 10 mg.

3. Results and Discussion

The addition of norfloxacin in gellan gum biofilm does not change the physical appearance of the biofilm compared to pure GG biofilm (Figure 1 (a)-(e)). All the biofilms, namely GG-NOR001, GG-NOR01, GG-NOR05, and GG-NOR1 which containing 0.01 %, 0.1 %, 0.5 %, and 1 % (w/w) of norfloxacin, respectively, were transparent with the transmittance at ≈95 % at 700 nm (Figure 1(f)). The transparent property of this biofilm can be an advantage for the packaging materials *i.e.*, to increase the initial perception of food freshness, colors and texture of the products as well as to highlight the quality of food ingredient.

The chemical interaction of gellan gum and norfloxacin was further investigated by using ATR-FTIR spectroscopy as shown in Figure 2. Norfloxacin shows eight prominent peaks, *i.e.*, at 3393 cm^{-1} corresponding to the *v*(N-H) stretching vibration of the imino-moiety of piperazinyl groups [20], 2987 cm^{-1} referred to the presence of ethyl group and 1716 cm^{-1} represented the carbonyl *v*(C=O) stretching of carboxylic acid. Apart from that, the peaks at 1619 cm^{-1} represented ν(N-H) bending vibration of quinolones group, 1489 cm^{-1} due to ν(O-C-O) of acids while 1029 cm^{-1} represented the ν(C-F) group. Besides that, the peak at 921 cm^{-1} suggested the ν(N-H) bending vibration of amines. Last but not least, for pure norfloxacin absorption peak was observed at 704 cm^{-1} which was assigned due to the meta distribution of the aromatic protons [21-24].

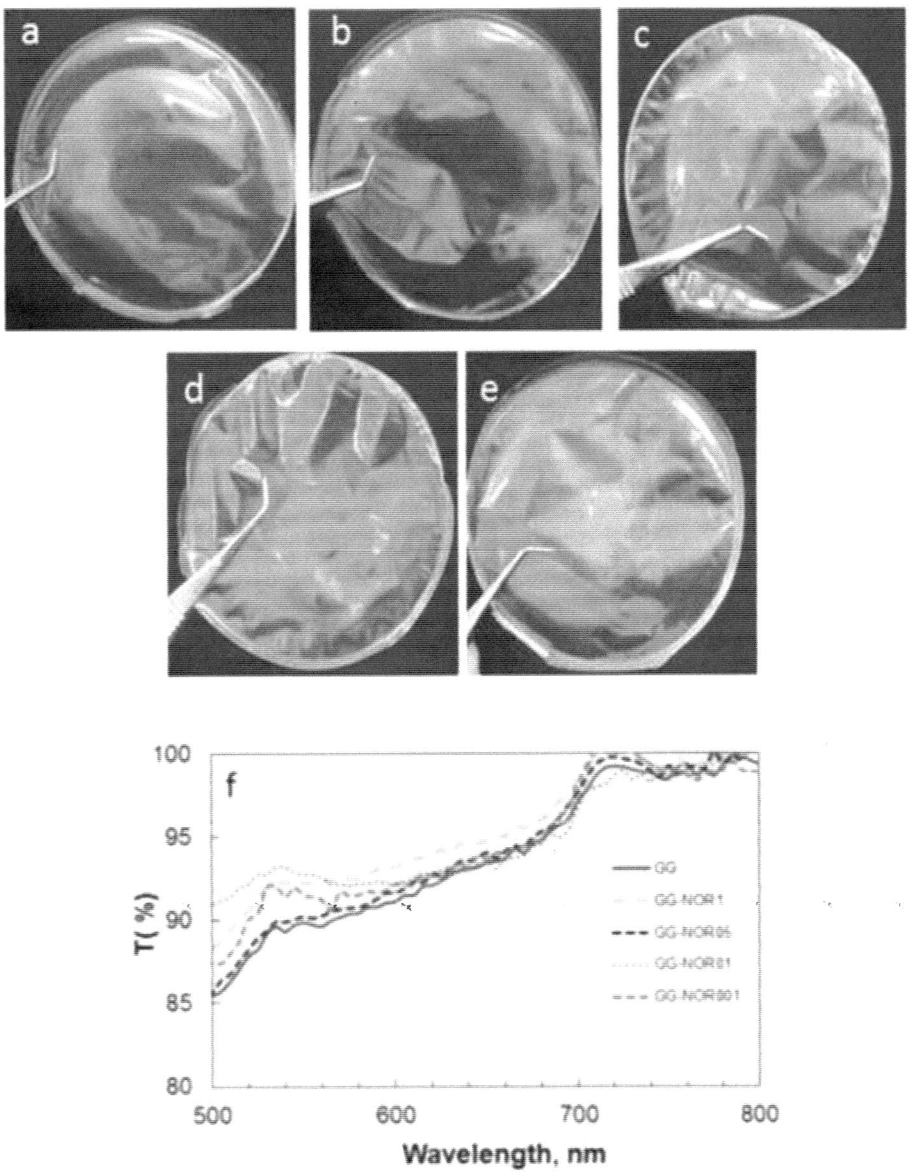

Figure 1. The appearance of biofilms (a) GG (b) GG-NOR001, (c) GG-NOR01, (d) GG-NOR05, (e) GG-NOR1 biofilms, and (f) transmittance of pure GG and GG-NOR biofilms. Scale bar represents 2 cm

The band at 3463 cm^{-1} and 1748 cm^{-1} in gellan gum spectrum was detected due to the O-H and C=O group stretching [25]. Further, the peaks at 1736-1748 cm^{-1} were detected in GG-NOR composite biofilms. All of these peaks are detected with strong intensity compared in pure norfloxacin as mentioned earlier. The intra-molecular hydrogen formation might be occurred between the atoms inside the gellan gum molecule due to the sharped of absorption peak were detected [20]. Furthermore, the inclusion of norfloxacin into gellan gum produced the signal present at range 1622-1675 cm^{-1}. All of these peaks were approved that those two materials have chemical interaction due to the ν(N-H) bending, which is absent in free standing gellan gum structure but appeared in blends. The conformation interactions between gellan gum and norfloxacin also was proved by the presence of the fluoroquinolone group in the blends at range 1108-1122 cm^{-1} which is prominent in norfloxacin for antibacterial properties [26]. Further, the fourth distinctive peaks that appeared in the blends are the region 950-900 cm^{-1} which is due to the ν(N-H) bending vibration of amines [20]. The ν(N-H) bending vibrations were observed broadly appeared in GG-NOR05 and GG-NOR1 at 912 cm^{-1} and 906 cm^{-1}, respectively, while almost absent in GG-NOR001 and GG-NOR01 due to the lesser interaction of norfloxcin added. Apart from that, the aromatic protons were detected at 720 cm^{-1}, 751 cm^{-1}, 706 cm^{-1}, and 735 cm^{-1}; for GG-NOR001, GG-NOR01, GG-NOR05, and GG-NOR1, respectively. All of these peaks were shifted to the higher wavelength, which is proved the interaction between the blends.

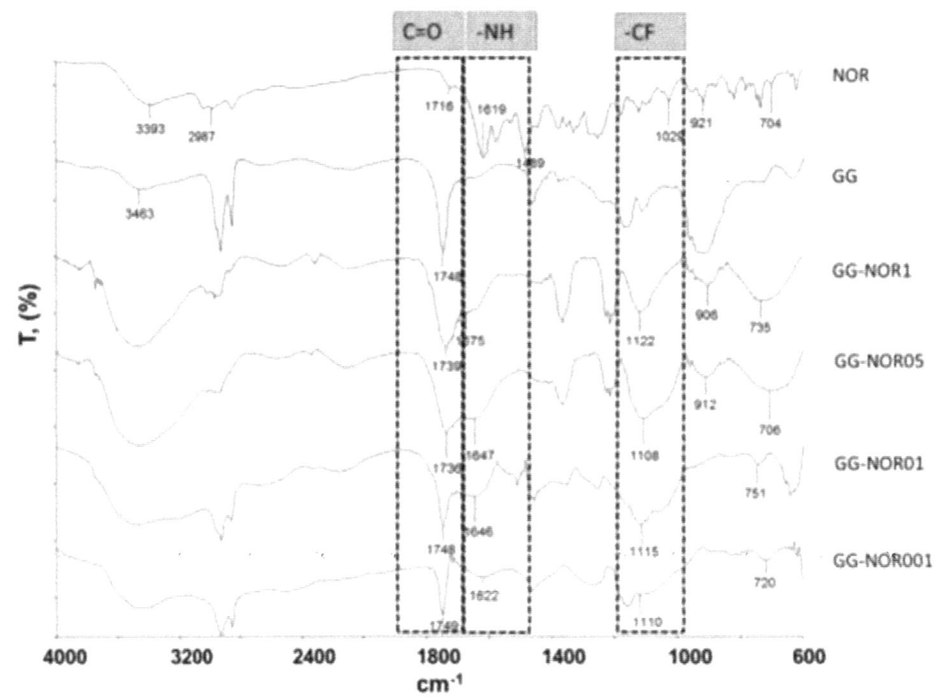

Figure 2. FTIR spectra of pure GG, NOR and GG-NOR biofilm

Figure 3 shows the TGA thermogram and derivative thermogram of GG, GG-NOR001, GG-NOR01, GG-NOR05, and GG-NOR1 composite biofilms. The initial and final degradation of the samples were observed from the onset and offset temperature. From the Figure 3, it can be summarized that GG and GG-NOR001 biofilms was the most stable in thermal based on its highest onset temperature at 246 °C (Table 1). However, GG-NOR05 biofilms shows the longest time to complete the degradation process due to the offset temperature recorded at 747 °C followed by GG-NOR1, GG-NOR001, and GG-NOR01 biofilms at 698 °C, 580 °C, and 579 °C, respectively. A consequent mass loss occurred due to the degradation of polymer matrix of all GG-NOR biofilms [27-28]. Addition of norfloxacin exhibits better outcome in increasing thermal stability due to the certain degree of interaction between gellan gum and norfloxacin as demonstrated by FTIR spectroscopy.

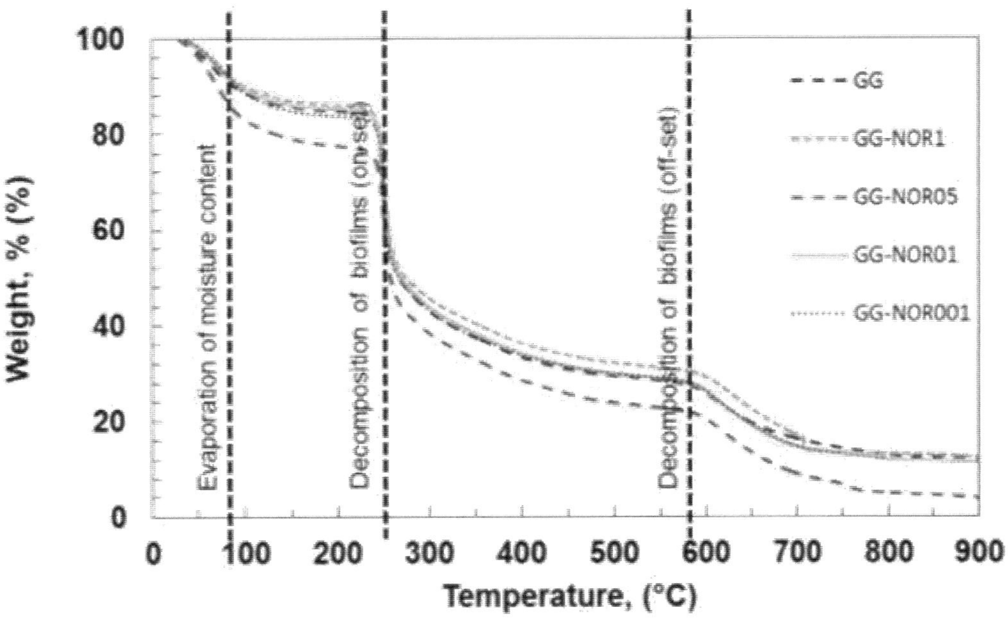

Figure 3. (a) Thermogravimetric thermograms of GG, GG-NOR001, GG-NOR01, GG-NOR05 and GG-NOR1 biofilms

Table 1. Thermal gravimetric properties of GG and GG-NOR001, GG-NOR01, GG-NOR05, and GG-NOR1 biofilms

Sample	**Onset temperature** (°C)	**Offset temperature** (°C)	**Weight loss** (%)
GG	246	570	71
GG-NOR001	246	580	71
GG-NOR01	245	579	71
GG-NOR05	244	747	92
GG-NOR1	245	698	77

The strain curve of GG-NOR composite biofilms are shown in Figure 4. It can be seen that the stress and strain curve of GG-NOR composite biofilms increased with increased of norfloxacin concentration up to 1 % (w/w); *i.e.*, GG-NOR01. The addition of norfloxacin in gellan gum biofilms increased the toughness (T) and elongation-at-break (EAB) values proportionally with addition of more norfloxacin. However, it decreased the tensile strength (TS) and Young's modulus (E) of biofilms as summarized in Table 2.

The inclusion of norfloxacin at 0.01 % (w/w) in gellan gum biofilms insignificant increase the toughness to 0.28 J g^{-1} compared to gellan gum biofilm at 0.26 J g^{-1} which equivalent to 1-fold increase. However, the TS and E values of the GG-NOR001 biofilms were decreased to 59±11 MPa and 3671±639 MPa, respectively, compared to gellan gum biofilm, but it values are highest compared to the other GG-NOR blends. This is due to the modification imposed on the composite biofilms lead to weakening the TS of the composite biofilms. This is may be due to the contribution of norfloxacin into the GG biofilm which is changed the chemical structure in the blends. Our result show that the reason as to the increase of toughness and strain-at-break of GG-NOR biofilm was due to the formation of hydrogen bonds by reaction of carbonyl group of norfloxacin with gellan gum as proved by ATR spectroscopy.

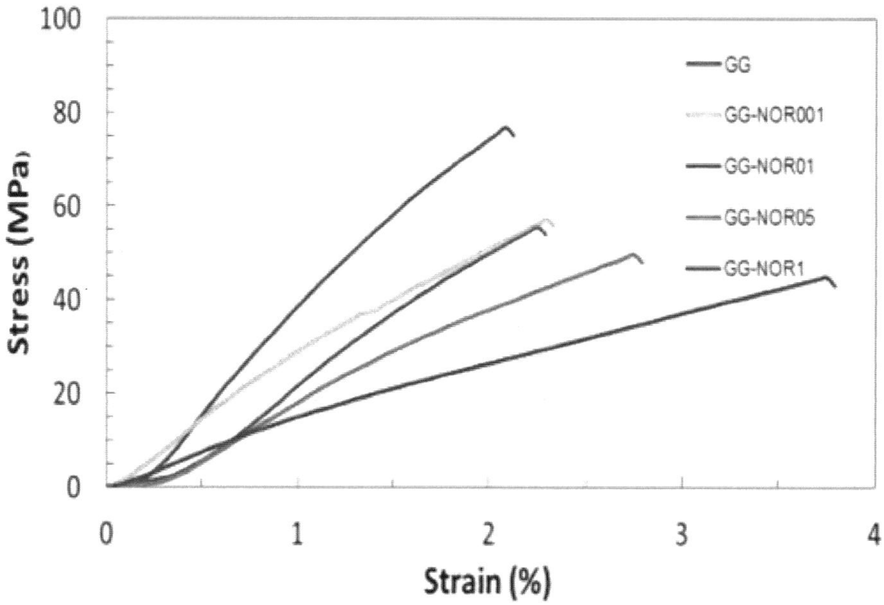

Figure 4. Typical stress-strain curves of pure GG and GG-NOR biofilms

Table 2. The mechanical properties through GG, GG-NOR001, GG-NOR01, GG-NOR05, and GG-NOR1 biofilms (mean±SD) (n=3)

	Thick (μm)	TS (MPa)	E (MPa)	T (J g^{-1})	EAB (%)
GG	30±0.010	80±12	4600±330	0.26±0.09	2.1±0.2
GG-NOR001	27±0.010	59±11	3671±639	0.28±0.06	2.1±0.5
GG-NOR01	27±0.004	55±10	3375±935	0.32±0.03	2.2±0.5
GG-NOR05	34±0.001	53±6	3292±380	0.39±0.09	2.4±0.4
GG-NOR1	36±0.001	47±9	2349±617	0.52±0.18	3.8±0.4

4. Conclusion

The chemical, physical and thermal properties of GG-NOR biofilms were investigated. Inclusion of different concentrations of norfloxacin to the gellan gum caused the toughness property increased with increasing the amount of norfloxacin added. Spectroscopy analysis proved that the hydrogen bonding involved in the bonds formation. The GG-NOR biofilms were thermally stable.

5. References

[1] Clarinval A M, and Halleux J 2005 Classification of biodegradable polymers In Smith, R. (Ed.), *Biodegradable Polymers for Industrial Applications* Woodhead Publishing Ltd Cambridge UK 3–31

[2] Imran H, Revol-Junelles A M, Martyn A, Tehrany E A, Jacquot M, Linder M, Desobry S 2010 *Crit. Rev. Food Sci. Nutr.* **50** 799

[3] Cabedo L, Feijoo J L, Villanueva M P, Lagarón J M, Giménez E 2006 *Macromol. Symp.* **233** 191

[4] Ismail N A, Mat Amin K A, and Razali M H 2018 *Materials Letters* **228** 116
[5] Ismail N A, Razali M H, and Mat Amin K A 2017 *AIP Conference Proceedings* **1885** 020045
[6] Ismail N A, Mohamad S F, Ibrahim M A, and Mat Amin K A 2014 *Advances in Biomaterials* **2014** 1
[7] El-Megharbel S M, Refat, Moamen S 2012 *Life Science Journal* **9** 1234
[8] Turel I 2002 *Coordination Chemistry Reviews* **232** 27
[9] Karp J E, Merz W G, Hendricksen C, Laughon B, Redden T, Bamberger B J, Bartlett J G, Saral R, Burke P 1987 *J. Annals of Internal Medicine* **106** 1
[10] Ginés P, Rimola A, Planas R, Vargas V, Marco F, Almela M, Forne M, Miranda M L, Llach J, Salmeron J M 1990 *Hepatology* **12** 716
[11] Gay J, DeYoung D, Roberts G 1984 *Antimicrobial Agents and Chemotherapy* **26** 94
[12] Moon A, Kamble V 2012 *Asiatic Journal Biotech Research* **3** 1447
[13] Khan M, Gruninger R, Nelson S, Klicker R 1982 *Antimicrobial Agents and Chemotherapy* **21** 848
[14] Goldstein E J, Citron D M, Bendon L, Vagvolgyi A E, Trousdale M D, Applemen M D 1987 *Archives of Ophthalmology* **105** 991
[15] Oliveira P R, Mendes C, Klein L, Sangoi M S, Bernardi L S, Silva M A, Segato N 2013 *BioMed Research International* **2013**, 1
[16] Shungu D, Tutlane V, Weinberg E, Gadebusch H 1985 *Chemotherapy* **31** 112
[17] Yamamoto T, Tsurumaki Y, Takei M, Hosaka M, Oomori Y 2001 *Toxicology in Vitro* **15** 721
[18] Alvarez-Lorenzo C, Yañez F, Barreiro-Iglesias R, Concheiro A 2006 *Journal of Controlled Release* **113** 236
[19] Mat Amin K A and Panhuis M i h 2012 *Polymers* **4** (1) 590
[20] Sahoo S, Chakraborti C K, Naik S 2011 *International Journal of Pharmaceutical Sciences and Nanotechnology* **4** 1505
[21] Al-Mustafa 2002 *J. Acta Chim. Slov.* **49** 457
[22] Dani V 1995 *Organic Spectroscopy*, New Delhi, Tata McGraw-Hill Publishing Company Limited
[23] Webster F X, Silverstein R M, Kiemle D 2002 *Spectrometric Identification of Organic Compounds* New York: John Wiley and Sons
[24] Shivanyboregowda S, Rajamma A J, Shekar H S, Mutahar R K M, Jayanthi A 2010 *Asian Journal Pharmaceutical Sciences* **5** 175
[25] Xu X, Li B, Kennedy J, Xie B, Huang M 2007 *Carbohydrate Polymers* **70** 192
[26] Sarkozy G 2001 *Veterinary Medicine-Czech* **46** 257
[27] Annapurna B, Meenakshi S, Bhat S D, Seshadri S 2013 *Chemical Engineering* **231** 373
[28] Noor I M, Majid S R, Arof A K, Djurado D, Claro Neto S, Pawlicka A 2012 *Solid State Ionics* **225** 649

Acknowledgments
The authors are grateful to Universiti Malaysia Terengganu (UMT) for financial assistance under internal research grant (TPM, Grant no. 68006/2012/46) and for providing the facilities for undertaking this work.

IOP Conf. Series: Materials Science and Engineering **440** (2018) 012028 doi:10.1088/1757-899X/440/1/012028

Comparison of Antibacterial Activity in Ethanol Extract and Essential Oil of *Citrus sinensis* (L.) Peels Obtained by Sohxlet and Distillation Methods

E Kusrini[*1], DP Mawarni[2], M Mamat[3], EA Prasetyanto[4], A Usman[5]

[1]Department of Chemical Engineering, Faculty of Engineering, Universitas Indonesia, Depok, 16424, Indonesia
[2]Faculty of Medical, Veteran National Development University of Jakarta, 16514, Indonesia
[3]Advanced Nano Materials Group, School of Fundamental Science, Universiti Malaysia Terengganu, 21030 Kuala Nerus, Terengganu, Malaysia
[4]Faculty of Medicine, Atma Jaya Catholic University of Indonesia, Jakarta, 14440, Indonesia
[5]Department of Chemistry, Faculty of Science, Universiti Brunei Darussalam, Jalan Tungku Link, Gadong BE1410, Negara Brunei Darussalam

Corresponding author email: ekusrini@che.ui.ac.id

Abstract. Ethanol extract and essential oil of *Citrus sinensis* (L.) peels have been characterized and compared for its antibacterial activity. Ethanol extract of *C. sinensis* (L.) peels was obtained by Soxhlet extraction method in ethanol with purity of 96% and the essential oil was obtained by distillation method for 5 hours at 100°C. The antibacterial activities of the extracts were investigated against *Staphylococcus aureus*. The experimental results showed that the ethanol extract yielded 29.50% (w/w), with total flavonoid compounds of 4.74 mg/g and polyphenolic of 20.54 mg/g, while the water extract (essential oil) yielded 6.80% (w/w) with total polyphenolic compounds of 0.65 mg/g. The flavonoid compound was not observed in the essential oil obtained from distillation method. The bioactive compounds in the ethanol extract obtained by Soxhlet extraction are larger than those found in the essential oil obtained by distillation method. The ethanol extract showed stronger growth inhibitory effect of *Staphylococcus aureus* with zone inhibition diameter of 18 mm, whereas the essential oil and the commercial d-limonene extract had similar zone inhibition diameter of 15 mm.

1. Introduction

The amount of dust and pollution in Jakarta, Indonesia, causes some health problems to human such as, skin infections, respiratory tract disorders, and other diseases. One of the most common skin infection is acne. Acne has been known to affect all gender, men and women, as well as attacking various ages after puberty. Hormonal changes after puberty can cause the occurrence of acne, in addition to that, acne can also arise because of skin pore clogging by dust and pollution, worsen by bacteria infection. One of the culprit that causes acne is bacteria *Staphylococcus aureus* [1]. In addition to causing acne, bacteria from the genus *Staphylococcus* also causes carbuncle and furuncle on the skin [2].

Although there are many medicines used for acne, there are some of them that do not suitable for certain people. Side effects and allergies can occur by applying these medicines. Acne medicines usually contain antibacterial agent that inhibit the formation and/or destroy of pathogenic bacterial walls. Thus, it encourages us to search natural antibacterial agent that is safer for acne medicine application.

Previously, the extract of *Citrus limon* peel has been reported as antimicrobial and antioxidant that applied for skin diseases [2]. Another reported that extract of myrtle lemon had an active antibacterial agent, but myrtle anise essential oil did not have any inhibitory activity against bacteria [3]. The application of natural resource as antibacterial agent has the advantages of being save and generally accepted by the consumer [4]. In Indonesia, production of *Citrus sinensis* (L.) is about 2.6 million tons in a year. One fruit of *Citrus sinensis* (L.) can yield about 10 - 20% (w/w) of peels, thus we assumed that the availability of *C. sinensis* (L.) peel is about 0.25 million tons in a year. Citrus peel is also a potential source for vitamin C, alkaloids compounds, and essential oil which is used as ingredients of perfume, refresher [5], and applied for skin rash treatments. By utilizing citrus peels, we could reduce the disruptive effects of organic waste to the environment.

Based on the literature study, *Citrus sinensis* (L.) peels contained flavonoid and polyphenolic compounds. These compounds are used in topical medicines for skin diseases. *Citrus limon* (Rutaceae), well-known as lemon, was reported to have anticancer and antibacterial activities [6].

In this paper, we extracted the peels of *C. sinensis* (L.) using Soxhlet extraction and distillation method to obtain the antimicrobial agent and characterized its chemical and biological properties. The aim of this study is to investigate and compare the antibacterial activity of these extracts. This study will provide important data for the utilization of *C. sinensis* peels as the source of bioactive ingredients and natural compounds for health care products. This study is also to show which extraction method is more feasible for production of antibacterial agent from *C. sinensis* peels in real application based on the antibacterial activity of each extract.

2. Materials and Method

2.1. Preparation of ethanol extract of C. sinensis (L) by Soxhlet extraction

Fresh orange fruits (*C. sinensis* (L.)) (1 kg) was obtained from the local supermarket in Depok, Indonesia. The peel of *C. sinensis* (L.) fruit (30.01 g) were cut and dried in oven at 70°C for 6 hours. The dried peels were grinded to size of 0.1 μm. The dried peels were put into conical flasks containing 200 mL ethanol (96%) for 4 hours. The suspension was separated and evaporated to obtain the filtrate. The yield of ethanol extract of *C. sinensis* (L.) was 29.5% (w/w) with yellowish brown and smells of citrus.

2.2. Preparation of essential oil by Distillation method

The dried peels of *C. sinensis* (L.) (30.01 g) were put in the conical flasks containing aquoues 300 mL and evaporated for 5 hours at 100°C in a heating mantle. The essential oil was separated from water layer using a drop pipette. The colourless essential oil with yield of 6.8% (w/w) was obtained.

2.3. Antibacterial test

This assay was done according to the procedure of [7]. The growth of bacterial in each sample were analysed by microbiological test using agar well-diffusion method against *Staphylococcus aureus*. The pure cultures of bacteria were sub-cultured on nutrient agar (NA) medium. After incubation at 37°C for 24 hours, the difference of zone inhibition diameter of bacteria was observed

2.4. Total polyphenolic content

The total polyphenolic content of the ethanol extract and essential oil were determined by the Folin–Ciocalteu method. This method was according to the method that reported by Baba and Malik [8]. The total phenolic content was calculated from the calibration curve. This result was represented as mg of gallic acid equivalent per g dry weight.

IOP Conf. Series: Materials Science and Engineering **440** (2018) 012028 doi:10.1088/1757-899X/440/1/012028

2.5. Total flavonoid content

The total flavonoid content of ethanol extract and essential oil were determined by the aluminium chloride colorimetric method. This method was according to the method that reported by Baba and Malik [8]. The result was revealed as mg rutin equivalent per g dry weight. Rutin was used as calibration standard.

3. Results and Discussion

Both preparations of ethanol extract and essential oil are shown in Figures 1 and 2. The ethanol extract of *C. sinensis* (L.) from Soxhlet method in 96% ethanol had total flavonoid compounds of 4.74 mg/g and polyphenolic compound of 20.54 mg/g. The flavonoid content observed in this study was smaller when compared to the reported ethanol extract of *Citri reticulatae* in 80% ethanol solvent with flavonoid of 233.4 ± 4.1mg/g and polyphenol of 79.5 ± 7.7 mg gallic acid equivalents/g [9]. The yield of essential oil obtained from distillation method is 6.80% (w/w) and had total polyphenolic compounds of 0.65 mg/g, less than the ethanol extract. Flavonoid compound was not observed in the essential oil extract of *C. sinensis*.

Step 1. *C. sinensis* (L) peels was dried in oven at 70°C

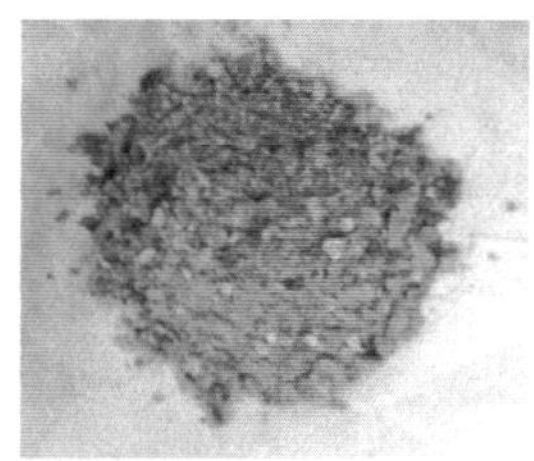

Step 2. The *C. sinensis* (L) powder

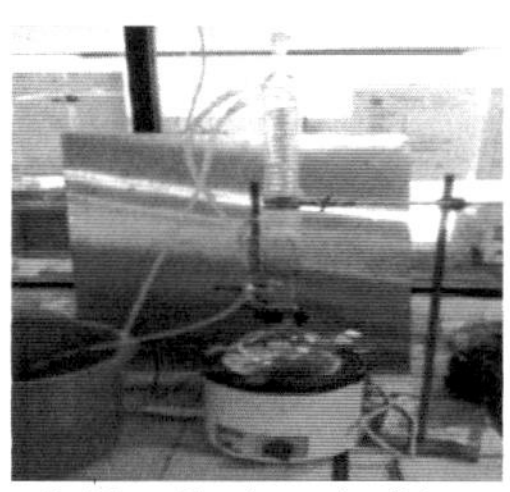

Step 3. The *C. sinensis* (L) powder was mixed with 200 mL ethanol by Soxhlet extraction

Step 4. The ethanol extract of C. *sinensis* (L.)

Figure 1. Schematic representation process for Soxhlet extraction

The inhibition zone of ethanol extract and essential oil of *C. sinensis* (L.) against *Staphylococcus aureus* is shown in the Table 1. Higher antibacterial activity against *Staphylococcus aureus* was observed from the ethanol extract with zone inhibition of 18 mm (see Table 1). On the other hand, the essential oil showed the zone inhibition of 15 mm, less effective than the ethanol extract. The essential oil of C. *sinensis* (L.) as antifungal activity was also reported by Viuda-Martos and co-workers [10].

The antimicrobial agent contained in citrus extract killed *S. aureus* due to the change in the *S. aureus* cell wall and membrane structure. The changes were caused by the interaction between bioactive compounds and polyphenolic compounds which contain a lot of aromatic and hydroxyl groups.

Zone inhibition of ethanol extract and essential oil of *C. sinensis* (L.) are presented in Figure 3A and B. Compared to the ethanol extracts and essential oil of *C. sinensis* (L.), the essential oil had only polyphenolics compounds (see Table 1). The growth inhibition on *Staphylococcus aureus* was clearly associated with the content of the total phenolic.

Table 1. Comparison of antibacterial activities of *Citrus sinensis* (L.) extracts againts *Staphylococcus aureus*

Sample	Total of polyphenolics (mg/g)	Total of flavonoids (mg/g)	Zone of inhibition (mm)
Ethanol extract	20.54	4.74	18
Essential oil	0.65	0 (not observed)	15
Commercial d-limonene	0.14	0 (not observed)	15

Step 1. *C. sinensis* (L.) peels was dried in oven at 70°C

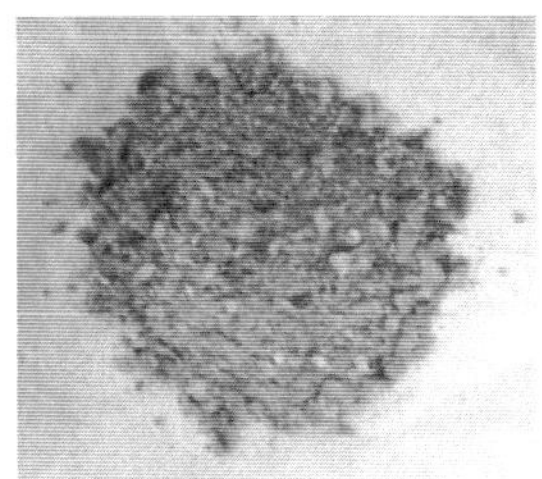

Step 2. The *C. sinensis* (L.)powder

Step 3. The *C. sinensis* (L.) was mixed with 300 mL aquadest

Step 4. The distillation of *C. sinensis* (L.) for 5 hours

Step 5. Separation between the oil layer and water layer

Step 6. The essential oil

Figure 2. Schematic representation process for distillation method

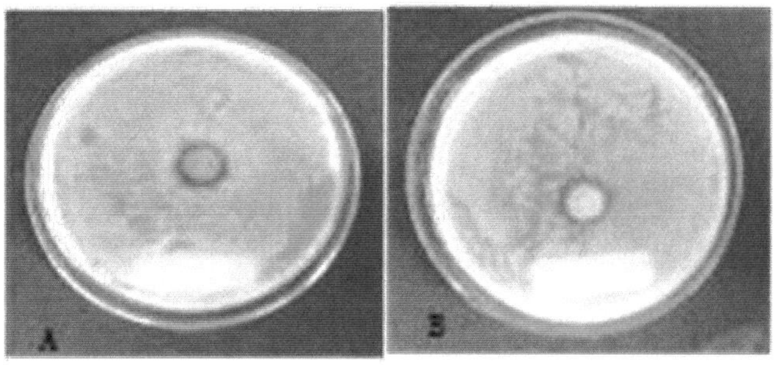

Figure 3. Zone inhibition of ethanol extract of *C. sinensis* (L.) from Soxhlet extraction (A) and the essential oil from distillation method (B)

For comparison, the antibacterial activity of *Citrus paradise* extract from ethyl acetate, methanol, and water solvents against *Staphylococcus aureus* was reported to have zone inhibition of 8 ± 0.3, 10 ± 0.7 and 12 ± 0.5 mm [4]. Other reported that the antibacterial activity of silver nanoparticles synthesized using *Citrus sinensis* (L.) extract showed zone inhibition of 7.8 mm at 25°C and 9.2 mm at 60°C [11]. However, the zone of inhibition against *S. aureus* using *C. sinensis* (L.) in this study are smaller than those found in the acetone extract of *Citrus limon* with zone of inhibition against *E. faecalis* and *B. subtilis* was 23 and 20 mm, respectively [2]. In this study, we also investigated the antibacterial activity of the commercial d-limonene extract of *Citrus terpene* against *S. aureus*. The zone inhibition of 15 mm was observed for d-limonene. It is similar value with the prepared essential oil of *C. sinensis* (L.). The d-limonene has polyphenolic compounds of 0.14 mg/g. The phenolic compounds are well-knowns as antimicrobial activity [4]. As we know that limonene is the major component (96.1% w/w) of the volatile fraction from the orange essential oil [12].

4. Conclusion

We evaluated the extracts of *Citrus sinensis* (L.) peels obtained by Soxhlet extraction and distillation methods. Both extracts showed antibacterial activities against *Staphylococcus aureus* contributed by its polyphenolic compound. Ethanol extract has more polyphenolic compound than the essential oil, thus showed better antibacterial activity compared to that of the essential oil. Based on this study, Soxhlet extraction method would be more feasible for production of antibacterial agent from *C. sinensis* peels in real application.

5. References

[1] Dunlavey E S 2000 *Current Problems in Dermatology* **12** 216-221
[2] Otang W M and Afolayan A J 2016 *South African Journal of Botany* **102** 46–49
[3] Nirmal N P, Mereddy R, Li L and Sultanbawa Y 2018 *Food Chemistry* **254** 1-7
[4] Sayari N, Sila A, Balti R, Abid E, Hajlaoui K, Nasri M and Bougatef A 2015 *Biocatalysis and Agricultural Biotechnology* **4** 616–623
[5] Nekvapil F, Brezestean I, Barchewitz D, Glamuzina B, Chiş V and Pinzaru S C 2018 *Food Chemistry* **242** 560-567
[6] Kawaii S, Tomono Y, Katase E, Ogawa K, Yano M, Koizume M, Ito C and Furukawa H 2000 *Journal of Agriculture and Food Chemistry* **48** 3865–3871
[7] Kusrini E, Usman A, Wisakanti C D, Arbianti R, and Nasution D A 2015 *Makara Journal of Technology* **19** 148-152
[8] Baba S A, and Malik S A 2015 *Journal of Taibah University for Science* **9** 449–454
[9] Yi Z B, Yu Y, Liang Y Z, and Zeng B 2008 *LWT* **41** 597–603

[10] Viuda-Martos M, Ruiz-Navajas Y, Ferna´ndez-Lo´pez J and Pe´rez-A´ lvarez J 2008 *Food Control* **19** 1130–1138

[11] Kaviya S, Santhanalakshmi J, Viswanathan B, Muthumary J and Srinivasan K 2011 *Spectrochimica Acta Part A: Molecular and Biomolecular Spectroscopy* **79** 594–598

[12] Gonçalves D, Costa P, Rodrigues C EC and Rodrigues AE 2018 *J. Chem. Thermodynamics* **116** 166–17

Acknowledgments

This work was funded by the IbM grant No. 602/H2.R12/PPM.01/2015. The authors are grateful to Mr Diki Firdaus for his technical help.

Selective Adsorption of Textile Dyes using Pre-treated Graphite Waste and Graphite/Fe_3O_4 Composites

E Kusrini[1*], O Sakadewa[1], G Pasca[1], A Usman[3], Y Yulizar[4]

[1]Department of Chemical Engineering, Faculty of Engineering, Universitas Indonesia, Depok 16424, Indonesia
[2]Advanced Nano Materials Group, School of Fundamental Science, Universiti Malaysia Terengganu, 21030 Kuala Nerus, Terengganu, Malaysia 3
[3]Department of Chemistry, Faculty of Science, Universiti Brunei Darussalam, Jalan Tungku Link, Gadong BE1410, Negara Brunei Darussalam 4
[4]Department of Chemistry, Faculty of Mathematics and Natural Sciences, Universitas Indonesia, Kampus Baru UI Depok, 16424, Indonesia

*Corresponding Author: Tel. 62-21-7863516 Ext. 204; Fax: 62-21-7863515; E-mail: ekusrini@che.ui.ac.id

Abstract. In the present study, we describe the application pre-treated graphite waste (PTG) and graphite waste/Fe_3O_4 composites as adsorbents to remove different textile dyes from water. Selective adsorption of methylene blue (MB), methyl orange (MO), methyl violet (MV), and rhodamine B (RB) textile dyes on PTG and graphite waste/Fe_3O_4 composites were studied. The removal of dyes in water was assessed to evaluate the selectivity of PTG and graphite waste/Fe_3O_4 composites. The graphite/Fe_3O_4 (1:0.5), graphite/Fe_3O_4 (1:1), and graphite/Fe_3O_4 (1:2) composites based PTG is abbreviated GF1, GF2, and GF3, respectively, were prepared with different weights of Fe_3O_4 magnetic nanoparticles with ratio of 1:0.5, 1:1, and 1:2 (w/w%). The adsorption experiment was carried out in batch reactor with adsorbent dosage of 0.1 g and dye concentration of 10 mg/L. The results showed that the optimum adsorption capability of the PTG was 96.3, 89.2, 100, and 75.3 for MB, MO, MV, and RB, respectively. Under comparable experimental condition, the GF2 (2M HCl) showed the optimum removal of dyes of 81.6% (MB), 40.9% (MO), 76.0% (MV) and 20.9% (RB). The adsorption capacity of the dyes is governed by electrostatic interaction between the dyes and the functional groups on the adsorbent. The PTG and the GF2 (2M HCl) composite are promising candidates as low-cost adsorbents for removal of pollutive dyes from waste water.

Keywords: Composites, Fe_3O_4 magnetic nanoparticles, Graphite waste, Removal of dye, Selective adsorption

1. Introduction

Most of textile dyes are colorful, toxic, and non-biodegradable, thus they can pollute in the ground waters as well as in the surface water. Removal of textile dyes from wastewaters is not easy due to their chemical properties, complex structure, and synthetic origins. For removal of textile dyes from aqueous solutions, several methods including physical, chemical and/or combine both physicohemical methods have been reported [1]. The adsorption method is the most effective and has many advantages such as high efficiency, less time and energy consumption, feasibility, easy operation, low cost, low-concentration and with minimum unwanted by-products [1-3].

Recently, modification of graphite surface using magnetic nanoparticles is one of the best way to increase the surface area of materials [4]. By surface modification with chemical treatment using Fe_3O_4 magnetic nanoparticles, the surface area of pre-treated graphite waste was increase up to 35.52 m^2/g [4].

Magnetic nanoparticles are versatile and advanced materials, and they can be used as adsorbent for removal of pollutants from aqueous solution becuase they have strong magnetic properties, high surface area, and easy to be recovered by external magnet [2]. In particular, Fe_3O_4 nanoparticles are suitable for removal of dye because they are easy to be re-collected and separated from water [5] as well as easy to be regenerated adsorbent [6].

In this work, pre-treated graphite waste (PTG) and graphite waste/Fe_3O_4 composites have been successfully synthesized and utilized as adsorbents for removal of textile dyes from aqueous solution. We have selected four different textile dyes, namely methylene blue (MB), methyl orange (MO), methyl violet (MV), and rhodamine B (RB), as representative model of the textile dyes. The removal of dyes in water was assessed to evaluate the selectivity of PTG and graphite waste/Fe_3O_4 composites. The aim of this study was to use PTG and graphite waste/Fe_3O_4 composites and evaluate it to remove textile dyes in water.

2. Experimental

2.1. Materials

Graphite waste was obtained from PT. X (Indonesia). Graphite cathode waste contained 31.4% carbon, 1.4% moisture, and other compounds. MB, MO, MV, RB, $FeCl_2.4H_2O$ and $FeCl_3.6H_2O$, HCl, NH_4OH, and NaOH were purchased from Sigma-Aldrich (Germany). All reagents are used without purification.

2.2. Preparation of pre-treated graphite waste (PTG)

Graphite electrode waste was dried in oven at 60, 75 and 90°C for 30, 60 and 90 minutes. After that, the dried graphite waste was grinded with a grinder and filtered with a filter size of 75 microns. The prepared graphite waste was immersed and soaked with 0.1 M HCl for 2 h. For convenience, the pretreated graphite waste is abbreviated as PTG.

2.3. Preparation of graphite/Fe_3O_4 composites

Composite was synthesized by a classical co-precipitation method as follows. 5.2 g FeCl2.4H2O and 2 g of FeCl3.6H2O were dissolved in the 10.3 mL of 1N HCl. The mixture was thoroughly stirred, followed by dilution with 15 mL of demineralized water for 15 minutes. The mixture then was poured it into 250 mL of 1.5 M NH_4OH to form a black precipitate which is able to respond to an external magnetic field. The mixture was continuously stirred for 1 hour at room temperature. The Fe_3O_4 magnetite nanoparticles was separated from the solution using a permanent magnet bars. The magnetite nanoparticles black precipitate was washed using demineralized water. After the pH of colloid of the magnetite nanoparticles becomes 7, 1.2 g prepared graphite waste was added. The mixture then was diluted with 100 mL of water and stirred for 4 h. The mixture then was heated in oven at 60°C for 2 hours. Then, precipitate was crushed to obtain the graphite/Fe_3O_4 composite with graphite waste to Fe_3O_4 mass ratio being 1: 1 (w/w). By changing the amount of the starting materials, we prepared the graphite-Fe_3O_4 with various mass ratio to be 1:0.5; 1:1; and 1:2. The graphite/Fe_3O_4 (1:0.5), graphite/Fe_3O_4 (1:1), and graphite/Fe_3O_4 (1:2) composites based PTG is abbreviated GF1, GF2, and GF3, respectively.

2.4. Adsorption performance

Adsorption method for removal of dye follows the method previously reported by Kusrini et al. [7]. The adsorption experimental was performed in batch condition. Mixture of textile dye solution is made with a concentration of 10 mg/L by the way each one mg textile dye was dissolved into 100 mL of distilled water. 20 mL of each selected dye model was mixed with 0.1 g of adsorbent for 5 minutes. The concentration of dyes in their solutions was determined by measuring their UV-Vis spectroscopy at wavelengths of 464 nm for MO, 554 nm for RB, 580 nm for MV, and 664 nm for MB. The mixture then was filtered to separate solid and liquid. The concentration of dyes in filtrate was measured using UV-Vis spectroscopy. Removal of dye was calculated by the following equation.

$$Removal = \frac{(initial\ concentration - final\ concentration)}{initial\ concentration} \times 100\% \quad (1)$$

3. Results and Discussion

3.1 Adsorption studies

We may recall that, from our previous report, the best performance for preparation of pre-treated of graphite waste was obtained at temperature of 60°C and time of 30 minutes. Both variables of temperature and time affected the properties of pre-treated graphite waste for removal of dyes. The SEM-EDX results showed that pre-treated graphite waste is non-porous. The presence of magnetite nanoparticles in surface of pre-treated graphite increased the surface area of graphite waste/Fe_3O_4 composite from 8.44 m^2/g to 64.58 m^2/g.

The heating treatment at temperature higher than 60°C for pre-treated graphite waste can be considered as to clean and remove the organic impurities on surface of graphite waste. As shown in Figure 1, adsorption efficiencies for removal of the textile dyes by the pre-treated graphite waste were obtained to be high. The removal of dyes was found to be 96.32% for MB and 100% for MV at 60°C, whereas it was 89.47% for MO and 76.07% for RB at 75°C. We found that the variation in temperature and contact time were not crucial on the adsorption efficiency. Removal for the MB (95.97%), MO (91.7%) and RB (76.33%) are more effective if the adsorbent of pre-treated graphite waste was prepared at 60°C for 60 minutes, and the adsorption efficiency of MV (100%) was obtained for the PTG at temperature of 60°C and 90 minutes.

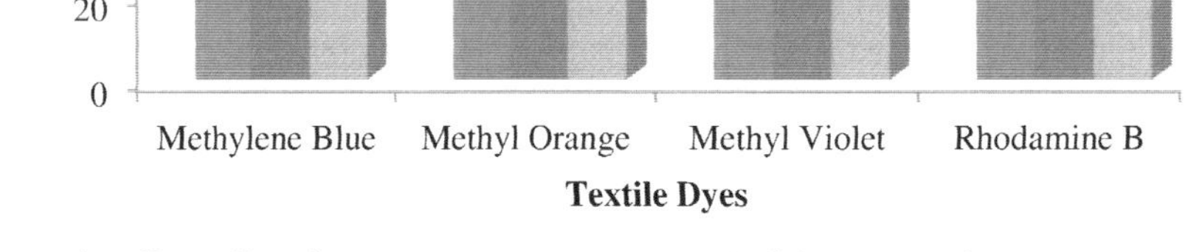

Figure 1. Effect of heating temperatures on removal of dyes by PTG (at temperatures of 60, 75, and 90°C for 30 minutes and textile dye concentrations of 10 mg/L)

3.2. Effect of Acid Activation on PTG

The effect of acid concentration for activation of PTG was evaluated for that heated at 90 °C for 90 minutes. The PTG was activated by acid impregnation using 0.1 and 2 M HCl solution for 2 hours. One could consider that higher acid concentration would remove more impurities and/or undesirable compounds from the graphite, enlarging the pores cavity, so that the surface area becomes larger [8]. However, the PTG which is a carbon material has an acidic or alkaline surface. With the addition of acid on PTG, the adsorbent become more acidic, hydrophilic, and more polar. The effect of acid concentration on adsorbent performance for removal of dyes observed for the GF2 composite (see Figure 2). Removal of dye is slightly increased only for MB and MV, while those for MO and RB are almost unchanged.

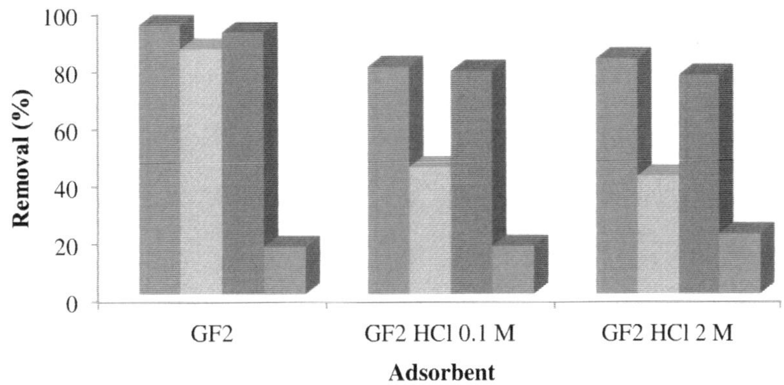

Figure 2. Effect of acid concentrations on removal of dyes using graphite/Fe_3O_4 (1:1 w/w%) composite. MB, MO, MV and RB is denoted by the blue, orange, violet, and pink bars

3.3. Effect of Magnetite Nanoparticles

PTG was modified with magnetite nanoparticles (Fe_3O_4) to increase the surface area and adsorption capacity of the adsorbent for removal of dyes from aqueous solutions. The magnetite nanoparticles is contained in the pores of the adsorbents, thus it is easy to be recovered by an external magnetic field for separation from solutions. Therefore, it facilitates the process of separating the adsorbent from remaining adsorbate in the solution. Due to the nature of the strong magnetic properties of Fe_3O_4, the presence of the magnetite nanoparticles in the adsorbent makes it easier to bind the heavy metal ions. The adsorbent can also be easily regenerated by NaOH and it can be reused repeatedly [6]. The effect of magnetite nanoparticles for removal of dyes is shown in Figure 3. It is clearly seen that the adsorption capacity of the graphite/Fe_3O_4 with ratio being 1:1 w/w% is the better those with different ratios of PTG to Fe_3O_4. However, the removal of dyes in this study using composites is lower that found in PTG.

For PTG showed removal of the textile dyes as high as 95.39%, 87.56%, 83% and 75.13% for MB, MO, MV and RB, respectively. In comparison, the removal of MB, MO, MV and RB on the GF2 composite is 93.36%, 85.01%, 90.85%, and 16.42%, respectively. We may note that increasing mass of Fe_3O_4 magnetite nanoparticles on the composites affected on covering the pores of PTG. At higher ratio of Fe_3O_4, the pore size of the graphites are more covered and blocked by magnetite nanoparticles, leading to lower adsorption efficiency of the dyes.

Adsorption of RB is strongly influenced by pH due to their association with the position of the ionic functiononal groups, which will be optimum at acidic pH and decreased at alkaline pH. At acidic pH, RB dissociates, giving ion Cl^- and positively charged RB. Meanwhile, COOH functional group on the surface of adsorbent dissociates to release H^+ ions, thus the charge becomes more negative. Therefore, it makes stronger electrostatic interaction between RB with the graphite surface. In this sense, graphite/Fe_3O_4 composite with acid activation has a greater adsorption capacity than that without acid activation, and increasing the adsorption capability of the graphite/Fe_3O_4 composite is proportionally increased with acid concentration [9]. Among the textile dyes in this study, however, RB has low adsorption efficiency by both PTG and graphite/Fe_3O_4 composites. We may consider that, since the adsorption mechanism of all the textile dyes, the low adsorption efficiency of RB compared with the other textile dyes is due to the low dissociation and/or the larger size of the RB cation.

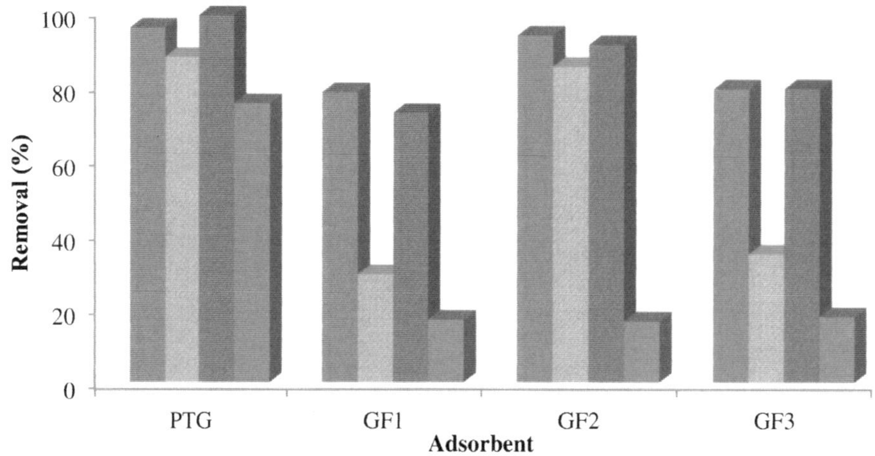

Figure 3. Effect of Fe_3O_4 magnetic nanoparticles for removal of dyes. MB, MO, MV and RB is denoted by the blue, orange, violet, and pink bars

3.4. Selective adsorption of dyes using graphite/magnetite composite

In order to observe the selective adsorption of textile dyes, all the textile dyes were mixed with each concentration of 10 mg/L. Removal of dyes using the graphite/Fe_3O_4 (1:1 w/w%, 2M HCl) composite is shown in Figure 4. As mentioned above, adsorption mechanism and surface structure of the adsorbent strongly influence the selectivity and adsorption efficiency of the dyes. Adsorption mechanism is regulated by various factors, including physical and chemical properties of the adsorbent and textile dyes as well as the process of mass transfer. In particular, MB and MO are planar molecules, therefore they can be adsorbed easily in the adsorbent by electrostatic interaction build up between the dyes and the functional groups on the adsorbent. On the other hand, MV and RB are non-planar molecules, so their adsorption efficiency is lower than those dyes. Thus, we may consider that there is no selectivity towards the textile dyes in this study, however, the molecular sizes of the dyes should be responsible for the different adsorption efficiencies.

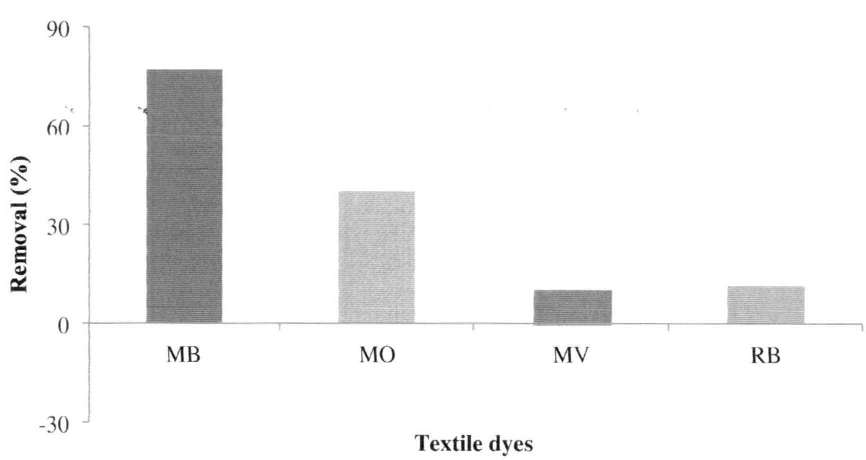

Figure 4. Selective adsorption of MB, MO, MV, and RB textile dyes using GF2 composite

4. Conclusion

In summary, PTG and graphite/Fe_3O_4 composites were successfully utilized as novel adsorbents to remove pollutive textile dyes from aqueous solution. The graphite/Fe_3O_4 composites with different graphite to Fe_3O_4 ratios were prepared by heating, acid activation, and impregnation methods. The adsorption efficiency of the textile dyes is optimized for the adsorbent being the graphite/Fe_3O_4 (1:1 w/w%, 2M HCl) composite. We demonstrated that the adsorption capacity of the dyes is governed by electrostatic interaction between the dyes and the functional groups on the adsorbent. PTG and the graphite/Fe_3O_4 (1:1 w/w%, 2M HCl) composite are promising candidates as low-cost adsorbents for removal of pollutive dyes from waste water.

5. References

[1] Fat'hi M R and Nasab S J H 2018 *International Journal of Biological Macromolecules* **114** 1151– 1160
[2] Kumari P, Shekhar, and Parashara H 2018 *Materials Today: Proceedings* **5** 15473–15480
[3] Long Y, Xiao L, and Cao Q 2017 *Powder Technology* **310** 24–34
[4] Kusrini E, Sasongko A K, Nasruddin, and Usman A 2017 *International Journal of Technology* **8** 1436-1444
[5] Peng L, Qina P, Lei M, Zeng Q, Songa H, Yang J, Shao J, Liao B, and Gu J 2012 *Journal of Hazardous Materials* **209–210** 193–198
[6] Do MH, Phan NH, and Nguyen TD 2011 *Chemosphere* **85** 1269-1276
[7] Kusrini E, Wicaksono B, Yulizar Y, Prasetyanto E A, and Gunawan C 2018 *IOP Conf. Series: Materials Science and Engineering* **316** 012029
[8] Martin-Gullon J M D 2006 *Interface Science and Technology Series* pp 1 - 48
[9] Sumanjit K W 2009 *Journal Surface Science Technology* **24** 3-4

Acknowledgments

The authors would like to acknowledge the financial support provided by the Universitas Indonesia through PITTA Grant No. 2431/UN2.R3.1/HKP.05.00/2018.

Synthesis and Characterization of Natural, Pectin and Activated Carbon as Low Cost Potential Adsorbents from Kepok Banana Peels (*Musa paradisiaca L.*)

E Kusrini*[1], M Aulia[1], AB Widiantoro[1], Y Nurani[1], M Mamat[2]

[1]Department of Chemical Engineering, Faculty of Engineering, Universitas Indonesia, Depok, 16424, Indonesia
[2]Advanced Nano Materials Group, School of Fundamental Science, Universiti Malaysia Terengganu, 21030 Kuala Nerus, Terengganu, Malaysia

Corresponding Author email: ekusrini@che.ui.ac.id

Abstract. The objective of this research was to synthesize of natural, pectin, and activated carbon from Kepok banana peels (*Musa paradisiaca L*). The use of raw banana peels as precursor for synthesis of adsorbent have significant advantages. Kepok banana peels were modified into three types of compounds with different physical and physicochemical treatments. Type 1 is natural, which is made by only thermal and physical treatment. Type 2 is pectin which was made by esterification reaction. Type 3 is activated carbon which was made by chemical activation using NaOH reagent for activation and carbonized at 400 C. The yields of type 1, type 2 and type 3 are 11, 52 and 26%, respectively. FTIR characterization showed that type 1 and type 2 have the hydroxyl and carboxyl groups, while in type 3 has only carboxyl groups. Chemical treatment of dehydrated banana peels with acid (HCl) and alkali (NaOH) agents modified the surface morphological and fundamental physical properties of pectin and activated carbon. Types 1 – 3 are low cost potential adsorbent and these functional groups play major roles in adsorption.

1. Introduction

Adsorbents from biomaterials have recently attracted attention for wastewater treatments. Biosorbents which have been utilized for the separation of lanthanides are mainly marine algae [1], novel hybrid adsorbent [2], novel chelating polysaccharide [3], spirulina biomass [4]. The biosorbents from the peels and shells of the fruits are of interest due to their low cost, abundancy, and utilization of wastes [1-5]. As an important example of fruit peels as an adsorbent is banana peels, which have been utilized to remove Ni, Zn, Cu, Pb, Co metal ions [6-8], Mn^{2+} [9], Sr^{2+} [10], Cr^{6+} [11], cationic dye [12], and palm oil mill effluent [13].

Banana peels are easily and cheaply found from the market. On the other hand, this waste can be generated and can be used for many treatments in wastewater. Considering such the potential of banana peels as adsorbents for metal separation, we focus on Kepok banana (*Musa paradisiaca L.*) as starting material for preparation of low cost potential adsorbents. With the vast production of banana, as it is one of the main fruit commodities all over the world [9-11], the banana peels is abundant. Kepok banana is the main ingredient of banana-based commodities in food industry in Indonesia, and the banana peels can potentially become environmental problem. In Indonesia the banana production in 2016 was about seven million tons.

The aim of this study is to synthesize of natural, pectin, and activated carbon from Kepok banana peels (*Musa paradisiaca L.*) with thermal, physical and physicochemical methods. Banana peels containing carboxylic acid, phospate and hydroxyl groups [9]. These functional groups acted as the active site for the adsorption of metal ions from aqueous medium [9-11]. Thus, utilization of biosorbent from the banana peels could overcome the environmental problem. This biosorbent can be produced in massive industrial scale. It is also noteworthy that banana peels generally consists of organic carbons, such as pectin (10 - 21%), lignin (6–12 %), cellulose (7.6–9.6 %) and hemicellulose (6.4–9.4 %) in w/w dry basis [14].

2. Experimental

2.1. Preparation of natural, pectin and activated carbon

Mature Kepok banana with yellow peels were collected as solid agricultural waste from Pondok Cina, Depok (Indonesia). They were cut into small pieces (1 - 2 cm) and washed once with tap water followed by distilled water to remove dust and particulates. To obtain the dehydrated banana peels, wetted banana peels (642.5 g) were dried in an oven at 110°C for 24 h. After that, the dehydrated banana peels was grinded with pulverize mill – blender and homogenized to 60 mesh with sieve shaker. For convenience, the dehydrated banana peels is abbreviated as natural (Type 1). The average yield of type 1 that made with four times drying was obtained of 11%.

For pectin obtained from the dehydrated banana peels (type 1) was prepared according to the method that reported by Mohammed & Chong [13]. Generally, 9 grams of type 1 was contacted with 633 mL of methanol (99.9%) and 5.4 ml HCl (37%). The solution was stirred for 16 h at 60°C. Then, it was washed using distilled water (20°C) until it reached the pH of 7. The solid material was separated and dried in oven at 110°C for 8 h. The esterified pectin was extracted from dehydrated banana peels in three times. The average yield of the pectin was obtained of 52%. The esterified pectin based type 1 is abbreviated type 2 hereafter.

The dehydrated banana peels (type 1) was carbonized at 400°C for 1 h in furnace. The char was cooled at room temperature. Then, it was activated with the alkaline (NaOH) 45% solution for 16 h at the room temperature. The solution was filtered and dried in oven at 110°C for 24 h. The char was heated in furnace at 500°C for 1 h. The solid materials were then washed and filtered using cold distilled water until pH reach neutral. The wetted materials were dried at 110°C for 8 h. The yield of activated carbon was obtained of 26%, with the initial weight of type 1 is 19.91 g to 5.22 g of activated carbon. The activated carbon based type 1 is abbreviated type 3.

2.2. Characterization

All three types of natural, pectin and activated carbon (Type 1–3) were characterized using FTIR and SEM-EDX. All samples were further identified by Fourier-transform infrared (FTIR; Bruker). FTIR was used to identify the chemical bonding and/or the functional groups of the samples. The surface morphology was determined by scanning electron microscopy and energy dispersive x-ray (SEMEDX) analysis.

3. Results and discussion

3.1. Preparation of natural, pectin and activated carbon

3.1.1. Type 1

Schematic representation for preparing of types 1–3 is shown in Figure 1. Table 1 shows the conversion from wet banana peels to be dehydrated banana peels (type 1) with average yields of 11%. In this study, mature Kepok banana peels were dried in oven at 110°C for 24 h until it reached a constant weight (see

Figure 2). The time is shorter than time that need for drying banana peels (48 h) to produce natural that reported by Mohammed & Chong [13].
Dried banana peels is grounded with pulverize mill and blender, then it was sieved to obtain the mesh size 60 and screened by sieve shaker with size of 250 μm. Blender is used to smooth again pulverize mill results that have not reached the size of 60 mesh. The advantage of using pulverize mill is that it can bear the weight of banana peels and takes a short time (± 1 minute). While the blender only accommodates a little and takes a long time (± 15 minutes). The dehydrated banana peels is crushed to a mesh size 60 aiming to obtain a larger surface area. Natural material included as granular forms (Fig. 3A). For granular materials have a mesh size ≥ 80 μm, it can be used for both liquid and gas applications [12]. The purpose of smoothing is to get a larger surface area.

3.1.2. Type 2
Type 2 would be hydrolysed in aqueous solutions, thus it resulted to carboxyl (COO^-) groups. The esterified pectin was washed using distilled water at 20°C to neutralize the acid. This neutralization is aimed to facilitate for further application, such as as adsorbent for lanthanide adsorption performance by pH initial solution variation. The distilled water (20°C) was used to minimize the loss of carboxylic acid of pectin because the solubility of carboxylic acids in 20°C water is small. In addition, we used distillate water (20°C) for washing is to stop the esterification reaction [10]. The pH of pectin refers to distilled water pH is 5.5. In this study, pectin was dried at 110°C in oven to remove water content. Fig. 3B shows the solid of pectin from Kepok banana peels.

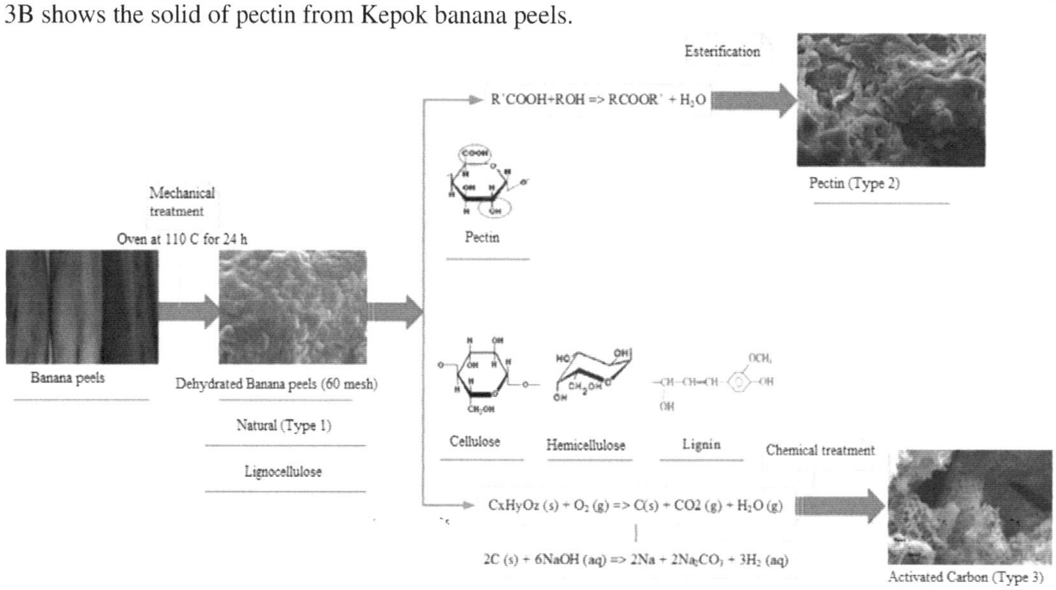

Figure 1. Schematic process for preparation of natural, pectin and activated carbon from Kepok banana peels (*Musa paradisiaca L.*)

Table 1. Conversion from Kepok banana peels as natural material

Number	Initial weight (g)	Dried weight (g)	Yield (%)
0.5	1750	218.0	12
1	3500	380.73	11
2	7000	803.19	11
3	10500	1144.23	11

IOP Conf. Series: Materials Science and Engineering **440** (2018) 012030 doi:10.1088/1757-899X/440/1/012030

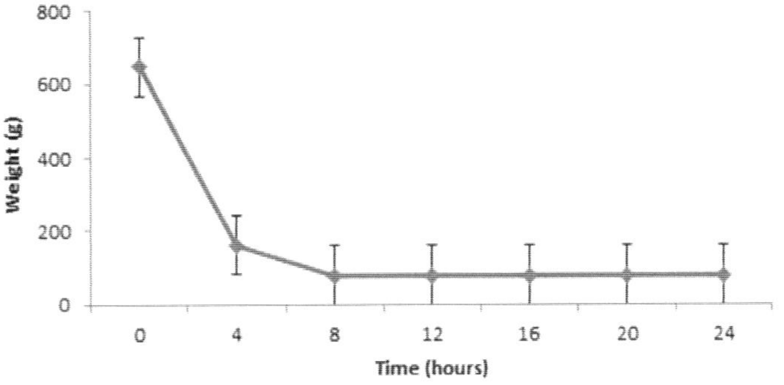

Figure 2. Equilibrium Time on Banana Peels Drying with standard error bar

3.1.3. Type 3

Natural material was activated chemically and thermally to obtain the activated carbon. Natural as starting material was burned in a furnace with minimum oxygen at 400°C for 1 hour, as pyrolytic temperature for precursor decomposition. Burning temperature setting was considered successfully if no ash was formed or minimized. This process is called carbonization [15]. Reaction of carbonization which is given as

$$2C6H10O5\ (s) + O_2(g) \rightarrow 10C\ (s) + 2CO\ (g) + 10H_2O\ (g)$$

Furthermore, the char was chemically activated by NaOH impregnation. This chemical activation is appropriate to degrade materials containing cellulose. Activation reaction using NaOH which is given as

$$6\ NaOH + 2\ C \leftrightarrow 2\ Na + 3\ H_2 + 2\ Na_2\ CO_3$$

After NaOH impregnation, the char was heated at temperature of 500°C, to purify the char from oxygen and hydrogen. This purification includes activation with heat treatment. Activation of carbon thermally is char treatment placed in oxidizing atmosphere in the presence of gases such as water vapour, air, and carbon dioxide or its combination in the temperature range 500–1000°C [15]. The presence of water vapour and carbon monoxide will react further into carbon dioxide. This allows carbon to be further activated by CO or H_2O gases. All the reactions of thermally activation is shown in below.

$$C\ (s) + H_2O\ 9g) \rightarrow CO\ (g) + H_2\ (g)$$
$$C\ (s) + 2H_2O\ (g) \rightarrow CO_2\ (g) + 2\ H_2\ (g)$$
$$C(s) + CO_2\ (g) \rightarrow 2\ CO\ (g)$$

The results from this process produced the activated carbon. Activation of NaOH and heat produce more porous structure for activated carbon [15], which contributes to enlarge the surface area of activated carbon. Activated carbon was washed with 0.1 M HCl to remove the ash. Then, it was neutralized with distilled water pH of 5.5. After being washed, activated carbon was dried in the oven to reduce water content. Fig. 3C shows the activated carbon from Kepok banana peels.

Figure 3. Type 1 natural (A), Type 2 pectin (B) and Type 3 activated carbon (C) from left to right sides

3.2. SEM-EDX Characterization

SEM characterization is used to find out the surface morphology and fundamental physical properties of materials, while EDX characterization is used to determine the element composition on materials. The content of elements obtained from the characterization of EDX is a qualitative result because the data acquisition is based on the shooting of a single data. However, this result can be used as a reference to represent the surface and existing morphology.

Figure 4(A-C) showed adsorbents Types 1–3, namely natural, pectin and activated carbon that origin from banana peels. Natural material has morphology surface of steep roads with small hills – cliff, irregular and surface exhibits a micro rough texture (see Fig. 4A). SEM image of type 1 is similar observed with the banana peel powder that reported by Mahindrakar & Rathod [10]. Pectin compound has a morphological surface of a cloud that congregates or winding road surrounded such circles (see Fig. 4B). The activated carbon has a hollow and pore morphology structure (see Fig. 4C). The activated carbon from Kepok banana peels activated by NaOH seems broken, lost its original structure of natural, and changed to be pore and more sponge. The NaOH is a strong alkaline reagent, thus it be able to interact with carbon atoms, so it support and catalyse the dehydrogenation and oxidation processes, thus increasing the pore size and porosity of activated carbon. Morphology of activated carbon in this study is different with activated carbon banana peels that produced by Mohammed & Chong [13]. This is due to the lower temperatures for carbonized and activated as well as the alkaline reagent used.

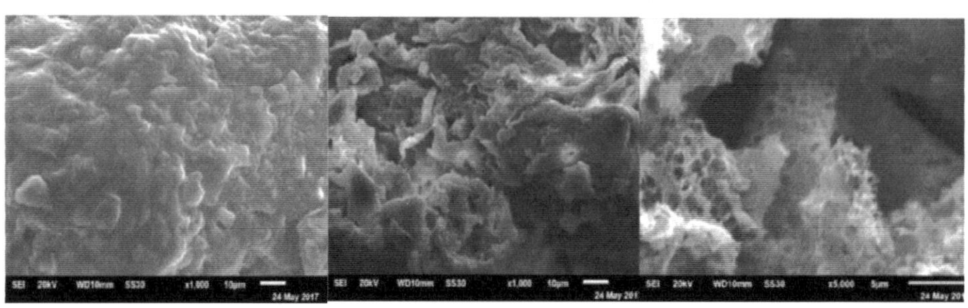

A. Type 1 B. Type 2 C. Type 3

Figure 4. SEM images of natural (A), pectin (B) and activated carbon (C)

The EDX composition is shown in Table 2. In pectin compound, the oxygen percentage of mass increased, this indicated that the formation of COO^- was successfully performed. In the activated carbon, the carbon content increased, indicating the activation process of natural material was successful. Both sodium and chlorine were observed in activated carbon. These elements were detected as residual from NaOH activation and HCl washing.

Table 2. EDX composition of natural, pectin and activated carbon

Element	Type 1 (%w)	Type 2 (%w)	Type 3 (%w)
C	74.02	70.92	90.84
O	20.79	22.92	5.59
Na	-	-	1.05
Cl	-	-	0.82

3.3. FTIR studies

To identify the functional groups on the surface of types 1 – 3, FTIR analysis was carried out. The FTIR spectra depicted in Figure 5. The identified functional groups (CO, COO^-, CH_2 and OH) were observed in types 1 and 2. The presence of OH and CO groups confirmed the existance of acidic oxygen containing functional groups that as active center for heavy metal adsorption. It can be seen that carboxyl (C=O) group at around 1100 cm^{-1} indicated the vibrations in carboxylic acid or ester. The broader and stronger peaks at 3300 cm^{-1} were assigned to hydroxyl groups have been detected for types 1 and 2. In type 3, the band assigned to hydroxyl group does not found because the carbonization process [16]. The CH_2 group at around 3000 cm^{-1} indicate the possibility of the presence of methyl as a branch. In activated carbon (type 3), the CH_2 and OH groups are not observed. This is due to thermally carbon activation released the oxygen and hydrogen in gas form. The reaction can be seen in the subset of preparation of activated carbon. The identified carboxyl group is a source of carboxylic compounds, not carbon monoxide. The identified COO groups showed that banana peels have not been converted to carbon completely. The surface of materials containing functional groups such as hydroxyl and carboxyl play significant roles for adsorption [17].

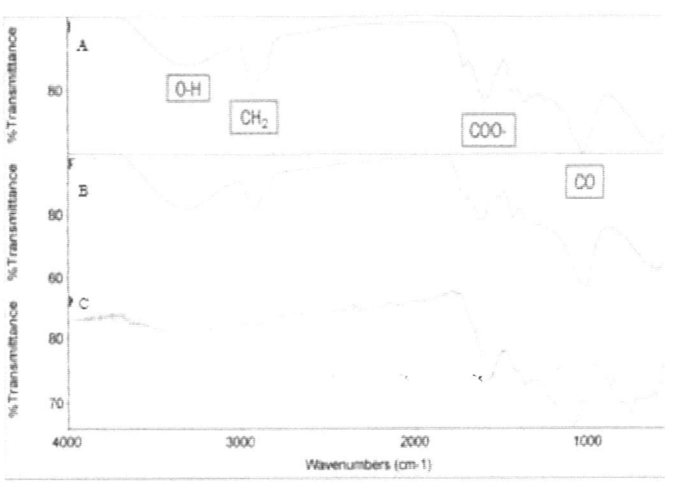

Figure 5. FTIR spectra of natural (A), pectin (B) and activated carbon (C) from Kepok banana peels

4. Conclusion

In this paper, we discussed the synthesis and characterization of natural, pectin and activated carbon from Kepok banana peels (*Musa paradisiaca L.*). The use of banana peels waste as precursor for synthesis of bioadsorbent have significant advantages. Type 1 (natural) was prepared by thermal and physical methods, while for the type 2 (pectin) and type 3 (activated carbon) were synthesised by chemical treatments. Chemical treatment of dehydrated banana peels with acid (HCl) and alkali (NaOH) agents modified the surface morphological and fundamental physical properties of pectin and activated carbon. The alkaline reagent (NaOH) raised its pore size and porosity in type 3. The type 3 is a low cost potential adsorbent for future biosorbent applications in aqueous medium.

5. References

[1] Vijayaraghavan K, Sathishkumar M and Balasubramanian R 2011 *Desalination* **265** 54–59
[2] Awual Md. R, Kobayashi T, Shiwaku H, Miyazaki Y, Motokawa R, Suzuki S, Okamoto Y and Yaita T 2013 Chemical Engineering Journal 225 558–566
[3] Elsalamounya A R, Desouky O A, Mohamed S A, Galhoum A A, and Guibal E 2017 *International Journal of Biological Macromolecules* **104** 963–968
[4] Sadovsky D, Brenner A, Astrachan B, Asaf B, and Gonen R 2016 *Journal of rare earths* **34** 644652
[5] Das N, and Das D 2013 *Journal of Rare Earths* 31, 933-943
[6] Castro R S, Caetano L, Ferreira G, Padilha P M, Saeki M J, Zara L F, Martines M A U and Castro G R 2011 *Ind. Eng. Chem. Res.* **50** 3446–3451
[7] Hossain M A, Ngo H H, Guo W S and Nguyen T V 2012 *Int. J. of GEOMATE* **2** (2) 227-234
[8] Annadurai G, Juang R S and Lee D J 2013 *Water Science and Technology* **4** (1) 185-190
[9] Ali A 2017 *Environmental Nanotechnology, Monitoring & Management* **7** 57–63
[10] Mahindrakar K V and Rathod V K 2018 *Environmental Technology & Innovation* **11** 371–383
[11] Ali A, Saeed K, and Mabood F 2016 *Alexandria Engineering Journal* **55** 2933–2942
[12] Amel K, Hassena M A and Kerroum D 2012 *Energy Procedia* **19** 286 – 295
[13] Mohammed R R, and Chong F M 2014 *The Journal of Environmental Management* **132** 237-249
[14] Mohapatra D, Mishra S and Sutar N 2010 *Journal Of Scientific And Industrial Research* **69** 323 – 329
[15] Cecen F 2014 Activated Carbon [Online] https://www.researchgate.net/publication/263062486
[16] Prachpreeca O, Pipatpanyanugoon K and Sawangwong P 2016 *Journal Environmental Science, Toxicology, and Food Technology* **10** 17-26
[17] Zhou N, Chen H, Xi J, Yao D, Zhou Z, Tian Y, Lu X 2017 *Bioresource Technology* **232** 204–210

Acknowledgments

We acknowledge Universitas Indonesia as financial support through the grant PITTA No. 2430/UN2.R3.1/HKP.05.00/2018.

IOP Conf. Series: Materials Science and Engineering **440** (2018) 012031 doi:10.1088/1757-899X/440/1/012031

Effect of La_2O_3 Dopant on Microstructure and Electrical Nonlinearity of ZnO-$CaMnO_3$ Based Varistor Ceramics

Mohd Sabri Mohd Ghazali[1,2], Norazwa Mohd Shukri[3], Wan Rafizah Wan Abdullah[3], Muhamad Azman Zulkifli[1], Lee Oon Jew[1] and Zahid Rizwan[4]

[1]Advanced Nano Materials (ANoMa) Research Group, Nano Research Team, School of Fundamental Science, Universiti Malaysia Terengganu, 21030 Kuala Nerus, Terengganu, MALAYSIA
[2]Institute of Tropical Biodiversity and Sustainable Development, Universiti Malaysia Terengganu, 21030 Kuala Nerus, Terengganu, MALAYSIA
[3]School of Ocean Engineering, Universiti Malaysia Terengganu, 21030 Kuala Nerus, Terengganu, MALAYSIA
[4]Department of Applied Sciences, Convener Purchase, National Textile University, Faisalabad (37610), PAKISTAN

Corresponding author emails: mohdsabri@umt.edu.my

Abstract. In this research, zinc oxide based varistor are produce by using citrate-gel method compared to conventional solid-state method. The zinc oxide based varistor are fabricate using three different wt.% of ZnO-$CaMnO_3$-La_2O_3 and divided into three system which are 80 mol% ZnO + 20 mol% $CaMnO_3$, 79.5 mol% ZnO + 20 mol% $CaMnO_3$ + 0.5 mol% La_2O_3 and 78.5 mol% ZnO + 20 mol% $CaMnO_3$ + 1.5 mol% La_2O_3. ZnO-$CaMnO_3$-La_2O_3 compound experienced pre-sintering at 500 °C for 2 hours sintering time. In the sintering condition, sintering temperatures are manipulated, which are varied at 1100, 1200 and 1300 °C while, the sintering time is setting at 4.5 hours. X-ray diffraction patterns show this compound is high in crystallinity. The surface morphology with the average grains size between 2.0 ~ 10.0 μm was observed by Scanning Electron Microscopy. Energy Dispersive Spectroscopy confirmed that distribution of element are homogenous with high purity. Density of the sample product is range between 5.0 ~ 5.6 g cm^{-3}. *J-E* characteristic shows the value of non-linearity coefficient in the range of 1.10 to 4.35.

1. Introduction

Zinc oxide (ZnO) is used widely in cosmetics, paint, and textile industries as well as in semiconductor electronic industries [1-4]. In electrical and electronic industries, ZnO based varistor is most-likely used as an electronic component in automobile electronics and also in semiconductor electronic. They are fabricated with different types of dopants such as Bi_2O_3, Co_3O_4, MnO_2, Sb_2O_3, TiO_2 and $CaMnO_3$ [5-12]. Recent studies reported by Vijayanandhini and Kutty [13] $CaMnO_3$ exhibit high nonlinearity coefficient up to 380 and it is remarkably for low-voltage application. Its unique microstructure that can be tailored for application as low, moderate or high-voltage varistors by adding different dopants and liable for nonlinear current-voltage (*I-V*) characteristics of the device [2, 14] and thus, is suitable used

IOP Conf. Series: Materials Science and Engineering **440** (2018) 012031 doi:10.1088/1757-899X/440/1/012031

to protect electrical appliances over-voltage magnitude [15]. *I-V* studies have been colossally investigated by previous researchers in order to improve the sensitivity and nonlinear coefficient of ZnO varistor [7, 16].

In this study, rare earth element-lanthanum (La) used as dopant in ZnO based varistor together with calcium manganate. By doping La, it is expected to increase the nonlinear coefficient (α) of varistor.

2. Materials & Method

Samples were prepared by citrate gel processing based on nominal composition of (80-x) mol% ZnO +20 mol% $CaMnO_3$+x mol% La_2O_3 (x = 0, 0.5 & 1.5). Preparation nano powders of $CaMnO_3$ and La_2O_3 via citrate gel method [17] were start with raw material reagent grade of calcium acetate anhydrous, manganese acetate anhydrous, lanthanum (III) acetate and citric acid. The mixture was place in a beaker and then mixed with ZnO micro-powder and heat stirred for a couple of hours with deionized water to dissolve it. The solution was heated and dried. The complex compound was then granulated and pressed into discs of 10 mm in diameter and 1 mm in thickness at a pressure of 2 tons and sintered at 900, 950 and 1000 °C for 1 and 2 hours sintering time with slow heating and cooling rate of 2.66 °C min^{-1} for sufficient oxygen supply. Silver paste was coated on both faces about 5 mm in diameter of the sample and was fired at 550 °C for 10 minutes soaking time to prepare the Ohmic contact on the both surfaces.

XRD measurement was performed by using Rigaku MiniFlex II in order to analyze and determines the crystalline phase and phase substitution of the sintered compound with different concentration of dopant. The samples scanned directly at 2θ angles between 5o and 80° with X-rays of 1.5406 Å wavelength generated by a CuK_α source. Density of the sintered samples was measure using Electronic Densimeter MDS-300. The surface of samples was ground and polished with SiC paper and 1 micron diamond suspension; respectively, until change into a mirror-like surface. The polished mirror-like samples were thermally etched in order to reveal fine microstructure morphology and observed then by SEM (JEOL JSM-6400).

The *J-E* characteristic of the varistor ceramics were evaluates by using a source measure unit (Keithley 2400). The important parameter like varistor voltage (E_{1mA}) was measured at a current density of 1.0 mA/cm^2 and the leakage current density (J_L) was measured at $0.80E_{1mA}$. Moreover, the nonlinear coefficient, α, was calculated from the following Eq. 1 [18]:

$$\alpha = \frac{\log J_2 - \log J_1}{\log E_2 - \log E} \quad (1)$$

where J_1 = 1 mA/cm^2, J_2 = 10 mA/cm^2, and E_1 and E_2 are the voltages corresponding to J_1 and J_2; respectively.

3. Results and Discussion

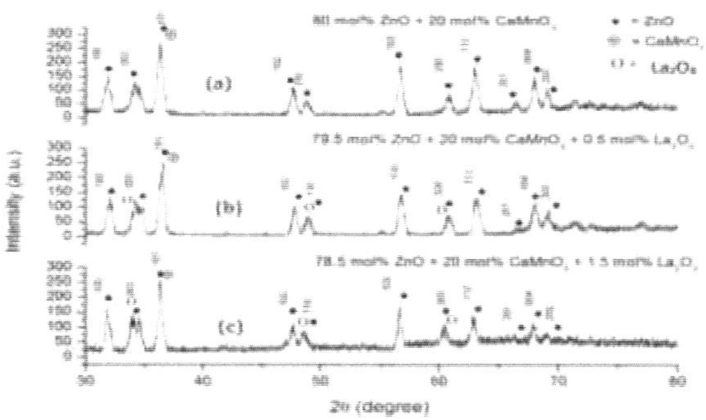

Figure 1. XRD patterns at 1300 °C sintering temperature at 4.5 hours sintering time of (a) without La doping, (b) 0.5 mol% La and (c) 1.5 mol% La

Figure 1 show the XRD patterns after sintering with temperature 1300℃ and soaking time 4.5 hours. Based on the Figure 1, it can be seen that ZnO characteristic which is hexagonal closed pack structure are remain based on ICSD code: 067454 [11] and all peaks was well observed. For Figure 1(a), 80 mol% ZnO + 20 mol% $CaMnO_3$ ceramics was fabricated which is high percentage of ZnO existed and the peaks revealed clearly and Figure 1(b), the sample with 79.5 mol% ZnO + 20 mol% $CaMnO_3$ + 0.5 mol% La_2O_3 was observed and the present of Lathanum was detected. Same trend has been seen in Figure 1(c), 78.5 mol% ZnO + 20 mol% $CaMnO_3$ + 1.5 mol% La_2O_3, the present of lanthanum was observed.

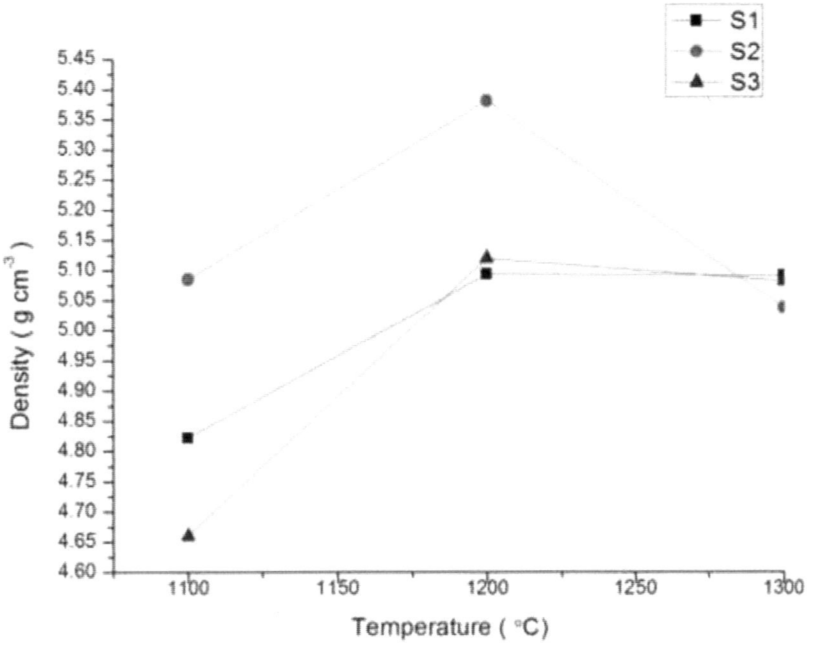

Figure 2. Density of the pellets for S1 (without La doping), S2 (0.5 mol% La) and S3 (1.5 mol% La)

Figure 2 shows the graph density of the sample produced. The higher density were obtain from pellet of S2 with 79.5 mol% ZnO + 20 mol% $CaMnO_3$ + 0.5 mol% La_2O_3, however, at higher temperature, the density dropped dramatically probably due to the trapping of pores and the limited of pores elimination [19]. As the temperature increases up to 1300℃, the density increases as a result in the elimination of pores [20], whereas the further increase of temperature did not affect the density, which saturated at 5.10 gcm^{-3}. At 1200 oC with 4.5 hours soaking time, the density projected the highest value. It can be seen that the density of produced varistor are very similar with the ZnO varistor in commercial which is in range between 5.0 to 5.3 gcm^{-3}[21].

IOP Conf. Series: Materials Science and Engineering **440** (2018) 012031 doi:10.1088/1757-899X/440/1/012031

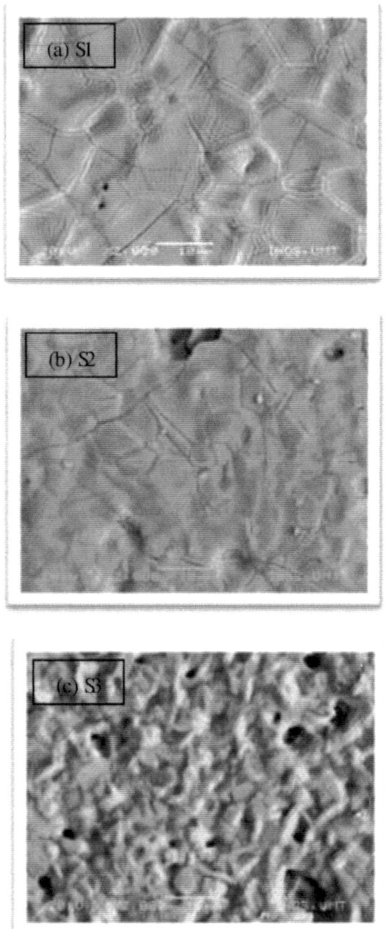

Figure 3.SEM micrographs of sintered varistor ceramics at 1300 °C for 4.5 hours of (a) without La doping, (b) 0.5 mol% La and (c) 1.5 mol% La

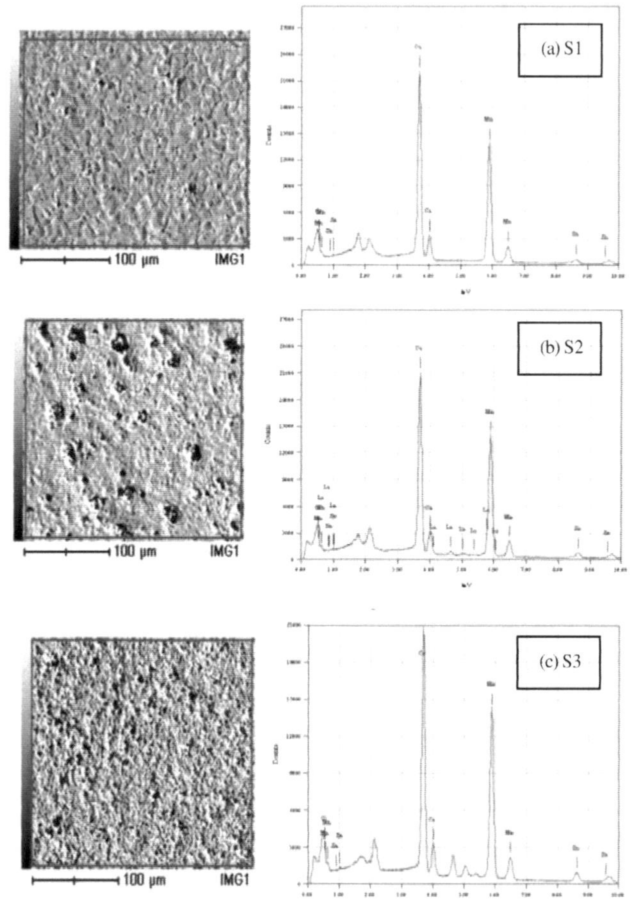

Figure 4. EDX spectra of (a) without La doping, (b) 0.5 mol% La and (c) 1.5 mol% La

Based on SEM micrographs in Figure 3, it is clearly observed that the average grain size decreases from 16, 8 and 3 µm as the concentration of La_2O_3 increases, respectively, which is in agreement with the reported by Nahm (2014) [22]. Other than that, as the La_2O_3 concentration increases, the intergranular phase gradually becomes more concentrated at the nodal points. It is believed that, La ion segregated at grain boundaries due to the difference of ionic radii which is La radii (1.3 Å) is larger as compared to Zn radii (0.74 Å) [23]. It is proved by EDX, there are no evidence of La coexisted in Figure 4(a), meanwhile, La element are traced in the sample S2 and S3 as shows in Figure 4(b) and Figure 4(c).

Table 1. The electrical nonlinearity coefficient (α) of the varistor

System	Sintering Tmax (°C) / Soaking time (hours)	Electrical non-linearity (α)
1	1100/4.5	1.10
2	1100/4.5	-
3	1100/4.5	-
1	1200/4.5	1.44
2	1200/4.5	-
3	1200/4.5	3.68
1	1300/4.5	4.35
2	1300/4.5	1.47
3	1300/4.5	2.19

*The non-linear coefficient (α) was calculated from equation, $\alpha = \log J_2 - \log J_1/\log E_2 - \log E_1$

Table 1 represents the electrical nonlinearity coefficient (α) of ZnO varistor doped with $CaMnO_3La_2O_3$ at temperature 1100, 1200 and 1300℃ for S1 (without La doping), S2 (0.5 mol% La doping) and S3 (1.5 mol% La doping). It can be seen that α initially increases and then decreases with the increases of La_2O_3 concentration. The decrement of α is due to the existence of porosity at higher dopant concentration as observed in Figure 3. The high value of α was obtained from the sample S1 and S3 are 4.35 and 3.68 at 1300 and 1200 °C; respectively. This indicates the possible sintering temperature for doped sample is 1200 °C and it is enough to achieve high α as compared to other sintering temperatures. Other than that, at 1200 °C sintering temperatures gave positive effect for La doping towards density and hence, enhance the α value.

4. Conclusion

The sample with La doping exhibit low porosity (high density) at 1200 °C result in the elimination of pores. The highest α value (3.68) with high density was achieved with La doping at 1200 °C sintering temperature.

5. References

[1] Ammar AH, and Farag AAM 2010. *Phys. B* **405** 1518
[2] Clarke DR 1999. *J. Am. Ceram. Soc.* **82**(3) 485
[3] Feng H, Peng Z, Fu X, Fu Z, Wang C, Qi L, and Miao H 2010. *J. Alloy. Compd.* **497** 304
[4] Look DC 2001. *Mat. Sci. Eng .B-Solid* **80** 383.
[5] Matsuoka M 1971. *Jpn. J. Appl. Phys.* **10**(6) 736
[6] Snow GS, White SS, Cooper RA, and Armijo JR 1980. *Am. Ceram. Soc. Bull.* **59**(6) 617
[7] Eda K 1989. *IEEE Elect. Insul. Mag.* **5** 28
[8] Bai SN, Shieh JS, and Tseng Ty 1995. *Mater. Chem. Phys.* **41** 104
[9] Toplan O, Gunay V, and Ozkan OT 1997. *Ceram. Int.* **23**(5) 251
[10] Fah CP, and Wang J 2000. *Solid State Ionics* **132** 107

[11] Sabri MGM, Azmi BZ, Rafizah WAW, Zaid MHM, and Matori KA 2015. *Advanced Materials Research* **1107** 20
[12] Sabri MGM, Azmi BZ, Rizwan Z, Halimah MK, Hashim M and Sidek HAA 2009. *American Journal of Applied Sciences* **6** (8) 1591
[13] Vijayanandhini, K. and T.R.N. Kutty. 2006. *Applied Physics Letters* **88** 123513
[14] Souza FL, Gomez JW, Bueno PR, Cassia-Santos MR, Araujo AL, Leiti ER, Longo E, and Varela AJ 2003. *Mater. Chem. Phys.* **80** 512
[15] Ma S, Xu Z, Chu R, Hao J, Liu M, Cheng L, and Li G 2014. *Ceram. Int.* **40** (7) 10149
[16] Choon WN, and Byoung-Chil S 2003. *Mater. Lett.* **57** 1322
[17] Philip J and Kutty TRN 2000. *Mater. Chem. Phys* **63** 218
[18] Nahm, CW 2015. *Microelectron. Realib.* **55** (11) 2299
[19] Sabri MGM, Wan Abdullah WR, Azmi BZ, Azman MZ, Zaid MHM and Rizwan Z 2017. *AIP Conference Proceedings* **1885** 020124
[20] Fauzana AN, Azmi BZ, Sabri MGM, Wan Abdullah WR and Hashim M 2013. *Sains Malaysiana* **42** (8) 1139
[21] Lin J, Li S, He J, Zhang L, Liu W and Li J 2017. *J. Eur. Ceram. Soc.* **37** (11) 3535
[22] Nahm CW 2014. *Ceram. Int.* **40** (1) 2477
[23] Li S, Lin J, He J and Liu W 2017. *J. Mater. Sci. Mater. Electron.* **28** (18) 13905

Acknowledgement
The authors gratefully acknowledge the financial support for this work from RAGS (Vot no. 57114) and Talent and Publication Enhancement-Research Grant (TAPE-RG) (Vot no. 55139).

IOP Conf. Series: Materials Science and Engineering **440** (2018) 012032 doi:10.1088/1757-899X/440/1/012032

Development and Testing of Hydraulic Ram Pump (Hydram): Experiments and Simulations

Fatihhi Szali Januddi*[1], Mohd Munsyi Huzni[1], Mohamad Shahrul Effendy[1], Adnan Bakri[1], Zulhaimi Mohammad[1], Zuhaila Ismail[2]

[1]Universiti Kuala Lumpur, Malaysian Institute of Industrial Technology, Persiaran Sinaran Ilmu, 81750 Bandar Seri Alam, Johor Bahru, Johor, Malaysia
[2]Department of Mathematical Sciences, Faculty of Science, Universiti Teknologi Malaysia, 81310 UTM Johor Bahru, Johor, Malaysia

*Corresponding author email: alfatihhi@gmail.com

Abstract. Hydraulic ram pump (hydram) is a rather 'mature' technology as it has been used over the last two centuries with many variations in design and basic configurations have been tried. Pump motor is favoured rather than hydram to transfer the water from one point to another point because of its consistency. However, following the increasing awareness of the adverse impact of global warming and the needs of sustainable technology, interest in hydram for water supply purpose has revived. The purpose of this paper is to study the effectiveness of hydram implementation and the effect of the stroke valve design towards flow rate of water being delivered. There are three methods used in this study; (i) experimental evaluation of hydram; (ii) experimental evaluation of adjustable waste valve; and (iii) computational simulation of adjustable waste valve. The adjustable waste valve was designed, fabricated for experiment and modelled for simulation. The hydram is very good to be installed as a pump at hill because it can supply water more than high of inlet water. The different length of stroke shows different behaviour of water flow and more length of stroke, more flow rate of water at discharge pipe. Hydram has a great potential to be functioning as a simple and reliable pumping device in those hilly remote areas which are of no water supply coverage.

1. Introduction

Water provides the basis for industry, commerce, transportation, agriculture, and energy production. Although water is important in our life, the population growth, urban expansion and growing competition among users worldwide, worsen by periodic drought and possible climate change, are raising concerns over the continued availability of a stable, dependable supply of freshwater for our nation and the world, now and for the foreseeable future. In rural areas where mobilization of water is limited [1, 2]; it gives rise to supress production, poor attendance and school performance of local children, as well as health issues related to poor sanitation and impoverished home garden. Furthermore, some areas have very limited timeframe to access water that coincides with their needs for plantation. Some advancement in technology to accommodate traditional work may help in sustaining life and economy of these communities.

Hydram consists of two moving parts; the waste valve and the delivery valve which are connected to a drive pipe, a delivery pipe and an air chamber (Figure 1) [3]. Hydram works continuously due to the opening and closing cycles of the waste valve and the delivery valve. The closure of the waste valve increases pressure in the drive pipe. The air chamber is needed to ensure the water to flow continuously at adequate pressure. There are six design factors should be considered in designing a hydram [4]. These include the height in between the water source and the hydram, the height in between pump site and storage, the flow quantity, the length of pipe from water source to the pump, the volume of desired water, and the length of drain pipe.

According to Sampth et al. [5] the pressure of water increase when the length of drive pipe increases but the length of pipe can cause pressure loss. Thus, another pressure tank is installed in the middle of drive to ensure the loss of pressure is less. Diameter of the drive pipe play important role to start-up and running the hydram. According to Dumaoal et al. [6] of drive pipe, it can deliver more gallon of water per minute. Furthermore, the small diameter of pipe of pipe limit the flow of water into the pump body, such a result the water deliver from the pump is less [7]. Supply head is a height from the pump to water inlet. The higher the height of water inlet, the higher gravitational force that makes the high velocity water flow in the pipe drive pipe. According to Hamid et al. [8], the insufficient pressure in the main pumps causing a problem to start-up the pump. It is because water hammer effect cannot be happened due to water cannot push the air pressure inside the pressure chamber. High supply head also bring a problem to main pump. There is too much water flow into the pump, so that water flows out from waste valve are much more. Besides that, the turbulence happens at pump body brought a drawback to the pump because it cannot function smoothly [9].

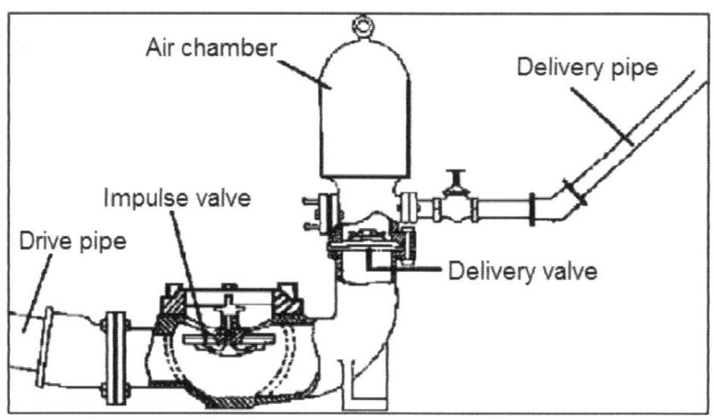

Figure 1. Hydraulic Ram Pump

According to Dumaoal et al. [6], the delivery head depend on the height of supply head. The more height of supply head, the more height of delivery head will be attained. It also supported by Nnene [10], the height of delivery head effect the flow rate enters to the pump. The high flow rate can be increase by increasing the height of supply head. Even, delivery head become higher when increasing the supply head, it needed more maintenance and become more often due to long delivery pipe [8].

Waste valve is a main component to the hydram to allow the water hammer effect to occur [11]. According to Mohamed [12] the waste valve needs to complete its cycle between 1.5 to 2 seconds. The output performance will be worsened if it is either too slow or too fast. The waste valve needed more weight if the cycle is too fast. According to Sheikh et al. [13], there is optimal weight used on the waste valve to control the flow at delivery pipe. If the weight used on waste valve is too high than optimal weight, the water discharged reduce significantly with increasing of weight. The optimal weight is also important to find, to ensure maximum amount of water been delivered. Both Mohamed [12] and Sheikh et al. [13] findings also supported by Diwan et al. [7] and Siswinda [11], which found out the increasing weight on waste valve would decrease the frequency. Thus, it may increase water delivered if optimal frequency is determined. The frequency of waste valve can be controlled by the managing the weight on waste valve. The current design of waste valve by Inthachot et al. [14] is made using galvanized metal which is expensive to be built for small application of hydram. According to Njhia et al. [15] the weight and stroke of the open and close valve can regulate the frequency of waste valve. The high frequency of waste valve can stop the pump function and to low the water flow out in delivery pipe is slow. Further, simulation study conducted by Harith et al. [16, 17] found that improved design of threaded waste valve have reduced water loss.

The aims of the present study are to investigate the effectiveness of hydraulic ram pump towards flow rate and delivery head at discharge pipe. To the best of our knowledge, no study reported on the different behaviour of adjustable valve and its effect towards the output pressure. Thus, this study also aims to develop adjustable waste valve to control the stroke of valve towards the different output pressure.

2. Methodology/Material

2.1 Design and Experimental Setup

The head of water supply was maintained at 1 m height with the size 52 litre of water tank. The process started with the hdyram tests in which the length of drive pipe was evaluated against the flow rate of discharge pipe. Next, the adjustable waste valve was compared to swing waste valve to see it effects towards flow rate of water at discharge pipe. Different length of stroke (5, 7, 10, 13 and 15 mm) was chosen based on the optimum flow rate with an adequate high pressure. It showed that different behaviour of water flow and the more the length of stroke, the higher the flow rate of water at discharge pipe. The hydram was more effective compared to water that was delivered using the siphon, in which it supplied water at higher volume than that of inlet water. Hydram has a great potential to be functioning as a simple and reliable pumping device in those highly remote areas which are of limited water supply.

2.1.1 The hydram test

The parameters used for the experiment were the length of drive pipe and the height of delivery head. The data collected from this experiment were flow rate of water being discharged at delivery pipe and pressure at the inlet of hydram. The data were collected using liquid flow meter and pressure gauge. Figure 3 shows the experimental setup of the hydram.

2.1.2 The valve test

The parameters used for this experiment was the length of stroke waste valve. Data collected from this experiment was the flow rate of water being discharged at discharge pipe.

2.2 Fabrication of Hydram

Figure 2 shows the design of the hydram before fabrication. The hydram used in present study was fabricated with PVC pipe and stainless steel (base to support the hydram). Various pipe sizing were used; 2", ¾", and ½" in diameter. The diameters for inlet and outlet pipe of the hydram were ¾" and ½", respectively.

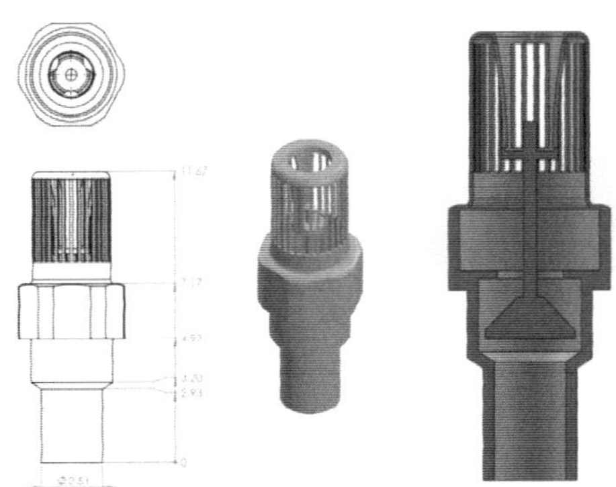

Figure 2. Adjustable waste valve

There were two check valves used in the present study which acted as waste valve and check valve. The size of pressure tank was 2" diameter pipe with 2.5' height in order to contain more air pressure. Figure 3 shows the final product of the hydram with swing waste valve. The figure also shows the sealing process to avoid the hydram from bursting during the operation.

The opening of the valve was 2 cm. The open and closing of the valve was done by a 2.5 cm rubber disc fixed to a 5 cm long threaded rod. The rubber disc has a cone shape to ensure the valve was tightly closed. The rod was used in the valve to support the vertical movement of valve and adjusting the stroke.

The adjustable waste valve as in Figure 4 was fabricated using ¾" PVC pipe to match with the hydram when being installed. The cap of waste valve was simply design with the net cap to ensure the water is not spill out during the opening of valve. So, the weight of water can help open the valve easily.

Figure 3. Hydram setup

Figure 4. Adjustable waste valve installed at hydram

2.3 Simulation of Hydram

Simulation of hydram was to study the behaviour of flow inside the pipe and valve as well as to determine the velocity and pressure profile for different valves design. Figure 5 shows the simulation done using COMSOL 5.3. Then, in the simulation software, the laminar flow was evaluated. The pressure for inlet pipe of model was set to 1.42 Psi. The data collected from the simulation was compared with the data from experiment. The simulation of adjustable waste valve was to study the behaviour of water flow inside the pipe and adjustable of waste valve focusing on the laminar flow inside the pipe in which the velocity and pressure at the adjustable waste valve for each stroke can be obtained. Then, the data was compared with the data from waste valve experiment

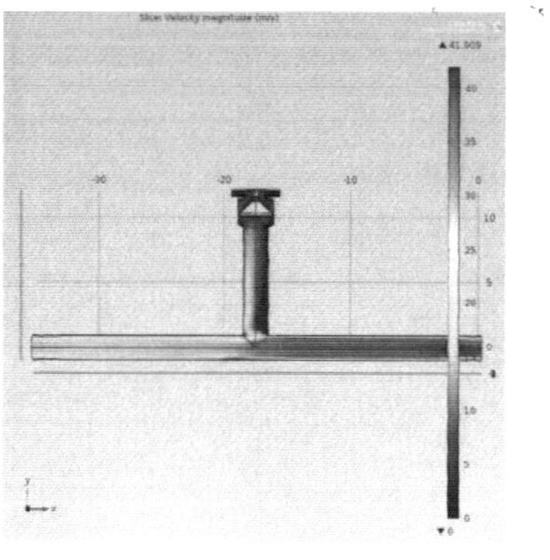

Figure 5. Flow velocity in the model

3. Results and Discussions

The entire flow rate at discharge pipe was found to decline slightly between 1 m and 2 m for each length of the pipe (Figure 6). According to Boatwright et al. [18], it is because the water is starting to climb up more than the height of the supply water. Thus, the siphon effect that gives pressure on the water is loss. The height of supply water is 1 m. The water that flow out at discharge pipe above the 1 m is causing by water hammer effect on the hydram. The water also is affected by the gravity that causing the return flow velocity when the water is at inclined state [19]. As the gravitational effect increases, there is a reduction in the kinetic energy supply to sustain the flow rate of water at discharge pipe.

The flow rate of water at discharge pipes becomes constant from the 2 m and 3 m height of delivery head for all length of drive pipes. According to Camigniani et al. [20], in the hydram there are resonance of wave pumping. There is strong and steady flow for each of the resonance that being created by the hydram. Thus, the flow rate is at 2 m and 3 m is being constant due to the pulsating of air pressure inside the pressure tank as the waste valve opens and closes.

The most optimum length of drive pipe is 2 m. There is similar flow rate of discharge water between 1 m and 2 m drive pipe length at 1.5 m of delivery head height. Based on the observation of experiment the 2 m drive pipe vibrates more than 1 m drive pipe when the hydram operates. There is occurrence of turbulence in pulsating pipe flow [21, 22]. The pressure outlet of drive pipe 1 m and 2 m are 1.10 psi and 1.42 psi respectively. The longer the drive pipe the more pressure at its outlet because the water has built its velocity [23]. Thus, each open and close of waste valve causing more pulsating on the drive pipe 2 m than drive pipe 1 m.

3.1 Effect of drive pipe length towards flow rate of discharge water

From the observation of this experiment (Figure 6), although the 5 mm and 7 mm stroke valve gave low flow rate of discharge water but it maintained the flow for 1 m to 2 m delivery head height. The open and close of the valve is very fast when the length of stroke is short. According to Njihia et al. [15], the pressure inside the pressure tank is being developed by opening and closing of the waste valve. Thus, the water discharged by the pipe can be constant for each delivery of head height.

The 10 mm stroke valve showed a drop of flow rate of water discharge from 1 m to 2 m. Each length of stroke has a different size of gap to water flow out from the valve [20]. The 10 mm stroke valve was observed to give unstable open and close of the valve. The huge changing of pressure inside the pipe makes the water inside pipe turbulence. Thus, it caused low volume of water to pass through the check valve.

The optimum length of stroke valve is 13 mm. The flow rate of discharge water is slightly decreased due to the loss of the siphon effect. The water is also affected by the gravity that causing the return flow velocity when the water is at inclined state [19]. As the gravitational effect is increased, there is a reduction in the kinetic energy supply to sustain the flow rate of water at discharge pipe.

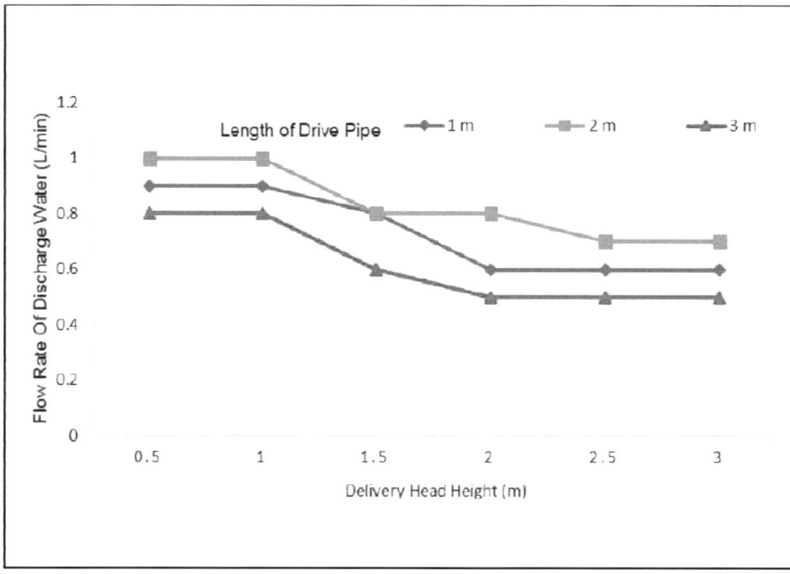

Figure 6. Effect of drive pipe length towards flow rate of discharge water

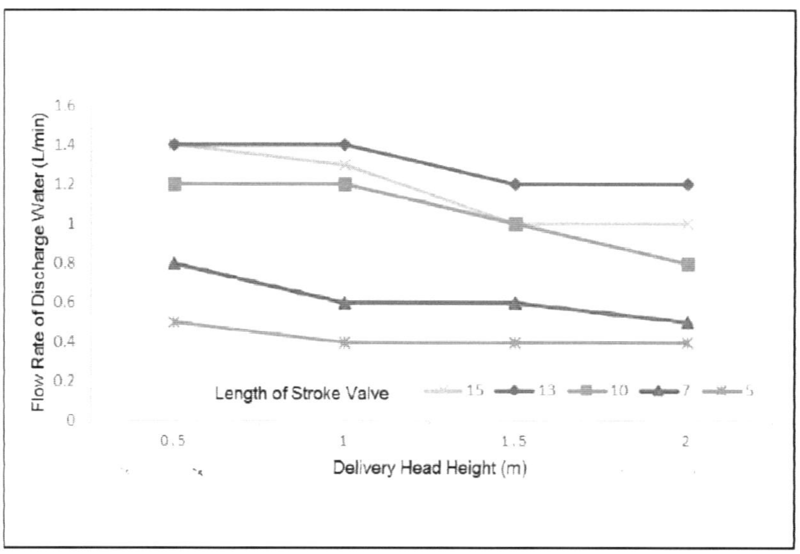

Figure 7. Comparison of swing waste valve and adjustable waste valve.

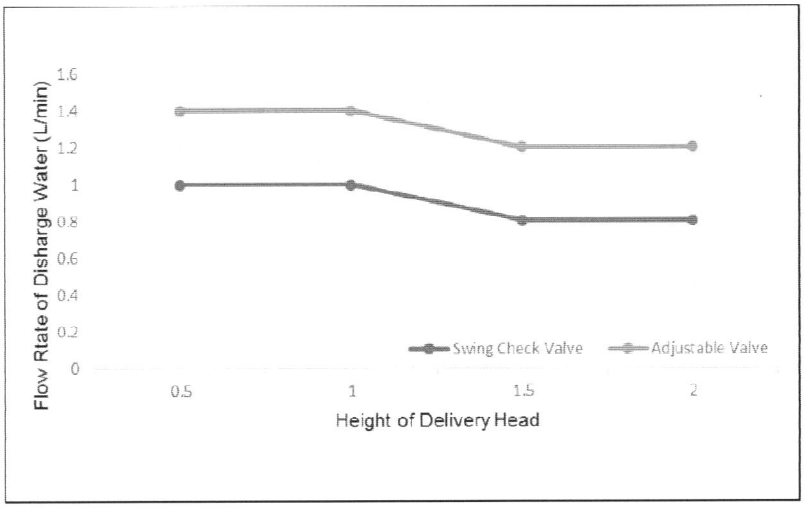

Figure 8. Comparison of swing waste valve and adjustable waste valve

Swing waste valve showed low flow rate of discharge water than adjustable waste valve (Figure 8). Thus may due to the different weight of sealing cap for both valves. The swing waste valve used brass sealing cap while adjustable waste valve used rubber cap that is lighter than that of brass. According to Mohamed [12], Sheikh et al. [13] and Girgio et al. [24], the added mass on the cap of valve can reduce the kinetic energy of water. Thus, the cap of valve is slightly hard to close when it weighs more.

Next, both valves used different type of materials, which have different of surface roughness. According to Shockling et al. [25], the general behaviour of turbulent pipe flow is in the presence of surface roughness. The index of roughness of swing waste valve and adjustable waste valve is 0.002 and 0.0015. As the roughness of valve increase, the flow of water inside the valve becomes transitionally rough. So, the water does not flow out smoothly from the valve.

The entire simulation inlet has been set up to 1.42 psi which equivalent to the pressure at inlet hydram for 2 m length of drive pipe from the experiment. 5 mm adjustable waste valve showed a high velocity of water flow passing through the cap and valve opening (Figure 9). The pressure of water was focussing on the side of cap and at the opening of waste valve (Figure 10). This can be used to explain why the 5 mm length of stroke valve is open and close very fast. According to Polenta et al. [23], the small area of water flow can increase the velocity of water and pressure. High velocity of water on top of the valve allows the high pressure on top of valve which contributes to the valve cap opens and closes very fast. The 13 mm adjustable waste valve is the optimum length of the stroke. Based on the Figure 3, there is low velocity of water at the top of valve cap. The pressure was mostly focus at the bottom of valve cap (Figure 10). According to the Jang et al. [26], the small size gap caused the turbulent of the flow. Thus, the turbulence from the small size gap between cap and valve wall at the bottom make the turbulence under the cap. The smooth flow of water above the cap put additional weight to valve cap to make the valve open. The high pressure on the bottom of cap pushes the valve to close. Both situation, help the valve to open and close at the optimal level.

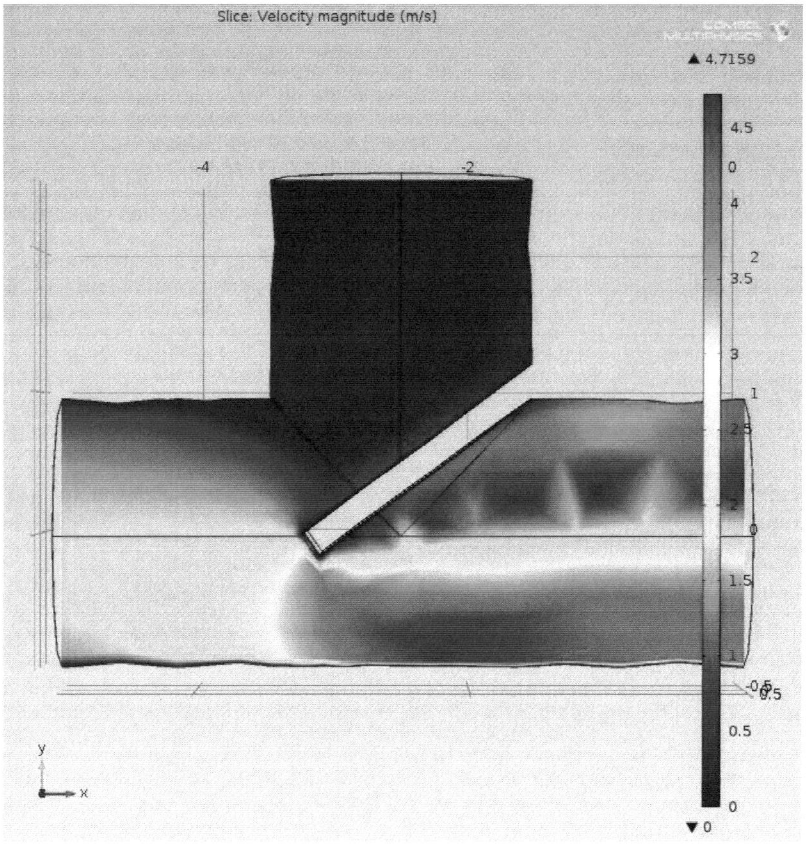

Figure 9. Velocity of water for 5 mm adjustable waste valve

IOP Conf. Series: Materials Science and Engineering **440** (2018) 012032 doi:10.1088/1757-899X/440/1/012032

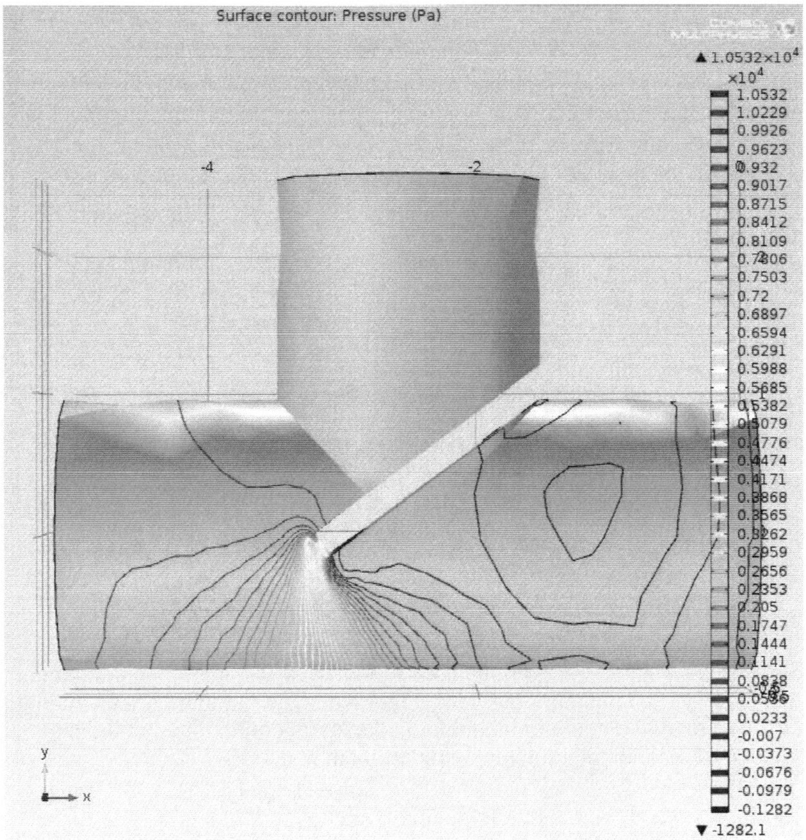

Figure 10. Pressure on the surface for 5 mm adjustable waste valve

4. Conclusion

In conclusion, the implementation of hydram as a pump system for high hill water supply is relevant. It can supply the water above the height of inlet water. It is also environmental friendly and needed minimal cost of fabrication and maintenance. The adjustable waste valves also give huge advantage to the hydram, where it can increase the flow rate of water at discharge pipe. Most of existing hydram are using a swing waste valve which its stroke cannot be controlled. The adjustable waste valve provides variety of stroke length. The most optimum length of stroke valve is 13 mm which it can discharge 1.2 L/min of water at high 2 m. The hydram can be installed in remote area to supply water in between different altitude at constant flow rate.

5. References

[1] Herlambang A and Wahjono H D 2018 *Jurnal Air Indonesia* **2**

[2] Pratowo B 2018 Universitas Bandar Lampung Indonesia

[3] Mondol S S 2017 Department of Mechanical Engineering Heritage Institute of Technology

[4] Pathak A, Pawar M, Deo A, Khune S and Mehroliya S 2016 *International Journal for Innovative Research in Science and Technology* **2** 290

[5] Sampath S, Shetty S, Pendanathu A M and Javaid W 2015 7

[6] Sr A F D, Urbano F A and Pareja B P Ilocos Agriculture Resources Research and Development Consortium Ilocos Norte, Philippines

[7] Diwan P, Patel A and Sahu L 2016 *Int. J. Curr. Eng. Sci. Res.* **3** 5

[8] Hamid Z A, Zain M Z M, Hung F C, Halim M F, Bidin W N W, Chatta I H and Omar N A 2016 *Malaysian Construction Research Journal* **19** 57
[9] Shende P B, Ninawe A and Choudhary S 2015 *International Journal for Innovative Research in Science and Technology* **2** 109
[10] Nnene O, Okoye I and Agunwamba J 2009
[11] Siwinda D E 2011 University of Malawi
[12] Dauda M T 2010 Bachelor of Engineering School of Engineering and Engineering Technology Federal University of Technology Minna, Nigeria
[13] Sheikh S, Handa C and Ninawe A 2013 *International Journal of Mechanical Engineering and Robotics Research* **2** 170
[14] Inthachot M, Saehaeng S, Max J F, Müller J and Spreer W 2015 *Agric. Agric. Sci. Procedia* **5** 107
[15] Njihia R, Maranga S and Mwai M 2015 in *Proceedings of Sustainable Research and Innovation Conference* 6
[16] Harith M, Bakar R, Ramasamy D and Quanjin M 2017 in *IOP Conf. Ser. Mater. Sci.*
[17] Harith M, Bakar R, Ramasamy D, Kardigama K and Quanjin M 2018 in *IOP Conf. Ser. Mater. Sci.*
[18] Boatwright A, Hughes S and Barry J 2015 *Sci. Rep* **5** 16790
[19] Chan C L and Chen C 2010 *J. Fluid Mech.* **647** 91
[20] Carmigniani R A, Benoit M, Violeau D and Gharib M 2017 *J. Fluid Mech.* **811** 1
[21] Xu D, Warnecke S, Song B, Ma X and Hof B 2017 *J. Fluid Mech.* **831** 418.
[22] Tuzi R and Blondeaux P 2008 *J. Fluid Mech.* **599** 51
[23] Polenta V, Garvey S, Chronopoulos D, Long A and Morvan H 2015 *International Journal of Mechanical, Aerospace, Industrial, Mechatronic and Manufacturing Engineering* **9** 278
[24] Giorgio-Serchi F and Weymouth G 2016 *J. Fluid Mech.* **798**
[25] Shockling M, Allen J and Smits A 2006 *J. Fluid Mech.* **564** 267
[26] Jang S J, Sung H J and Krogstad P-Å 2011 *J. Fluid Mech.* **687** 376

Acknowledgement
This research work is supported by the Universiti Kuala Lumpur through Short Term Research Grant (STR15904). The authors would like to thank the research management centre of Universiti Kuala Lumpur, for managing the project. Special thanks go to Dr Zuhaila as given the opportunities to use the facilities in her laboratory.

Characterization on the mechanical and physical properties of chicken skin gelatin films in comparison to mammalian gelatin films

N Suderman[1], M I N Isa[2] and N M Sarbon[1*]

[1] School of Food Science and Technology, Universiti Malaysia Terengganu, 21030 Kuala Terengganu, Terengganu, Malaysia.
[2] School of Fundamental Science, Universiti Malaysia Terengganu, 21030 Kuala Terengganu, Terengganu, Malaysia.

Email address: norizah@umt.edu.my

Abstract. The purpose of this study was to investigate the mechanical and physical properties of chicken skin gelatin film as compared to mammalian gelatin films in terms of tensile strength (TS), elongation at break (EAB), puncture force, water vapour permeability (WVP), Fourier transform infrared spectroscopy (FTIR), X-ray diffraction (XRD), light barrier, thermal properties and microstructure. Three film formulations using 4g gelatin of three different types (chicken skin, bovine, and porcine gelatin) and 1.5g glycerol were prepared under mechanical stirring at a temperature of 45°C. The use of different types of gelatin resulted in different mechanical and physical properties. Results revealed that chicken skin gelatin film was optimal due to its high tensile strength (5.57 MPa) and low WVP (1.29 x 10^{-9} kPa) rate as compared to bovine and porcine gelatin films. The carboxyl group was revealed to be stronger in FTIR assay for chicken skin gelatin film, while XRD revealed amorphous characteristics at a peak $2\Theta=20^{o}$. These results contributed to its superior physical characteristics. These desirable characteristics mean that chicken skin gelatin film has remarkable potential as a biodegradable film material as compared to commercial gelatin. It may become a key preferred alternative for producing gelatin films for edible film purposes.

Keywords: Biodegradable films, gelatin films, chicken skin gelatin, mammalian gelatin, characterization.

1. Introduction

Food packaging is used to preserve product quality while minimizing product–packaging interactions. Recently, a wide variety of packaging materials offering these desired effects have been employed to interact with food [1]. Additionally, studies have found that plastics are widely used as edible film materials due to considerations of low cost and availability [2]. Although conventional packaging materials such as plastics and their derivatives are effective for food preservation, however this materials has led to a serious environmental problems [3]. Therefore, the biodegradable packaging from biopolymers is an effective alternative to synthetic packaging material due to its eco-friendly and non-toxic characteristics. Studies have shown that biodegradable films have more desirable physicochemical characteristics than synthetics [4].

Proteins and polysaccharides are the main biopolymers used in edible packaging films [5]. Among proteins, gelatin has recently been studied extensively in manufacturing packaging films due to its excellent physical and mechanical properties ([4]; [6]). Gelatin is a water-soluble protein derived from animal sources, obtained from the hydrolysis of bone collagen or connective tissues skin of mammalian and fish [7]. The acceptance of gelatin as a ''Generally Recognized as Safe'' (GRAS) substance in the area of food additives by the U.S. Food and Drug Administration (FDA), along with its excellent film forming ability, gel-forming properties around 35°C, excellent versatility due to its alpha-amino acid composition, abundance, low cost, and biodegradability, all make gelatin an attractive protein in the design and development of functional films with potential application in the food sector [8].

In general, films composed of gelatin show good mechanical properties but have also been found to be highly sensitive to moisture and exhibit poor barrier properties against water vapour [9]. Due to the strong cohesive properties of the gelatin polymer, thus the gelatin film produced are brittle and susceptible to crack [4]. Plasticizing additives address this inherent brittleness by reducing intermolecular forces that increase polymeric chain elasticity, thus enhancing film flexibility [10]. Plasticizers also decrease the process temperature, sticking in moulds and enhance wetting. As a final product modifier, plasticizers increase the temperature range of treatment, increase toughness, and lower the glass transition temperature [11]. Good plasticizers used in plasticizing gelatin based film are often cited as polyols. However, the most widely used is glycerol ([12]; [11]).

Gelatin has been extensively studied in terms of tensile strength [10], elongation at break [13], water vapour permeability (WVP) ([14]; [15]), light transmission [16], Fourier transform infrared spectroscopy (FTIR) [17]; [13]), X-ray diffraction [18], and microstructure [19]. [18] reported that film with 60% bovine gelatin exhibited lowest WVP (1.06 x 10-7 kPa) and its X-ray diffraction of the film produced amorphous characteristics, explaining its high biodegradability rate (4.3%). A film derived from pork gelatin was reported to have lower water solubility than those formulated using bovine and fish gelatin [13]. FTIR spectra presented the increased of Amide-A wave number in films as higher gelatin concentrations were used, which resulted from greater interaction occurring between the gelatine's functional groups [13].

Recently, chicken skin has been considered the best alternative in producing gelatin film due to its excellent thermal properties (T_m = 33.57 °C) and high gel strength (355 ± 1.48 g), which is greater than that of bovine gelatin (229 ± 0.71 g) ([20]; [10]). This will improve the strength and permeability of gelatin films. Research on the comparison of potential materials for the manufacture of biodegradable and edible films from different gelatin sources has been carried out ([21]; [13]). However, to date no studies have compared the physical and mechanical properties of films manufactured from alternative gelatin (chicken skin gelatins) and mammalian gelatin (bovine and porcine gelatins). Therefore, the aims of this study was to investigate the properties of gelatin films manufactured from bovine, porcine, and chicken skin gelatin via both mechanical and physical testing.

2. Material and methods

2.1 Materials

Chicken skins were freshly obtained from the local poultry industry (TD Poultry SDN. BHD., Kuala Terengganu) and were kept in ice while transporting to the laboratory. The chicken skins were washed thoroughly in excessive water and stored at -80°C until further use. Commercial type B bovine skin gelatin, type A porcine skin gelatin, and glycerol (MW = 92.09) were obtained from Sigma-Aldrich Company Ltd., Poole, Dorset, UK. All chemicals used in this study were of analytical grade.

2.2 Sample preparation

Visible fat on the skin was mechanically removed. The skin was then thoroughly rinsed in excessive water to remove impurities. Skins were cut into 2-3 cm pieces and freeze-dried. Completely dried skins were grinded before being defatted using the Soxhlet method [22].

2.3 Preparation of chicken skin gelatin

Extraction of chicken skin gelatin was conducted following previous described method [20] by using acid–alkaline pretreatment. The defatted chicken skin was grinded and soaked in sodium hydroxide (0.15%, w/w), sulphuric acid (0.15%, w/v), and citric acid (0.7%, w/w) solutions, consecutively. Gelatin solution was stirred slowly at room temperature (30 min) before centrifugation (Multi-purpose centrifuge, GYROZEN 1580, Korea) at 3500 x *g* (10 min). Each solution was repeated three times and the supernatant was removed followed by thoroughly rinsed with distilled water. The precipitate obtained were then washed with excessive water before extracted in distilled water for overnight at 45 °C. The clear extract was filtered, concentrated by evaporation under vacuum, and lyophilized. Dried 'gelatin powder' obtained was ground, weighed, and stored for further use.

2.4 Development of gelatin films

Gelatin films were prepared using the casting technique as described by [23] with slight modifications. Three filmogenic solutions of gelatin and glycerol were prepared according to the optimized formulation generated by response surface methodology software [23]. For film preparation, 4.0 gelatin powders (chicken skin, bovine and porcine) were mixed with 100 ml of distilled water under mechanical stirring until completely dissolved. The film forming solution was thus obtained. Then, 1.5 g glycerol was added into the film forming solution as a plasticizer. All mixtures were stirred at 45°C for 20 min obtain a homogeneous gelatin film solution. Lastly, approximately 25 g of filmogenic solution was poured into the Petri dish and dried at 45°C in an oven for 2 days. Dried films were then removed from Petri dish and conditioned in a desiccator prior to investigation of tensile strength (TS), elongation at break (EAB), puncture force, water vapour permeability (WVP), differential scanning calorimetry (DSC), light transmission, X-ray diffraction (XRD), Fourier transform infrared spectroscopy (FTIR), and scanning electron microscopy (SEM).

2.5 Determination of tensile strength (TS) and elongation at break (EAB)

Tensile strength (TS) and elongation at break (EAB) of the films were determined using a texture analyser (TA.TX Plus, Stable Micro System, USA) with the ASTM technique 0882-97 [24]. A 20 x 100 mm film strips with uniform thickness of 0.20 mm were prepared using a cutting blade and placed onto grip pairs of AT/G probe which was attached to the texture analyser with a 5-kg load cell with 20 mm of initial gap between the up and down parts. The film strips were stretched by moving the headspace at 50 mm/min until breaking. The TS (MPa) was calculated using the following equation:

$$Tensile\ Strength\ (TS)(MPa) = \frac{F_{max}(N)}{A\ (m^2)} \quad (1)$$

F_{max} is max load (N) needed to pull the sample apart, A is cross sectional area (m^2) of film sample. Meanwhile, the elongation at break (EAB) (%) was calculated as below:

$$Elongation\ at\ break\ (\%) = \frac{l_{max}}{l_o} \times 100 \quad (2)$$

Where l_{max} the film elongation (mm) is at the moment of rupture and l_o is the initial grip length (mm) of sample.

2.6 Determination of puncture force

The breaking force and the breaking deformation of the films were determined via puncture force test of the same texture analyser. Films placed in a 5.6 cm diameter cell were punched to the breaking point with a round-ended stainless-steel plunger (3 mm in diameter) at a cross-head speed of 60 mm/min. Breaking force was expressed in terms of N and breaking deformation in terms of percentage (%).

2.7 Determination of water vapour permeability (WVP)
Water vapour permeability (WVP) was determined by modification of ASTM method as explained by [23]. The films were sealed with silicone vacuum grease onto a cup containing silica gel, and held in place with a rubber band. The cups were individually weighed and placed in desiccators with distilled water at room temperature. The cups were then placed in desiccators containing distilled water at 30 °C. The cups were weighted at 1 hour intervals over an 8 hour period. Three films were used for WVP determination and the measurement was conducted in triplicate. WVP was calculated as follows:

$$WVP\ (Kg.m/m\text{^}2.s.Pa) = w \times x/A\ t\ \times)P_2 - P_1 \tag{3}$$

where w is the weight gained by the cup (g), x is the average film thickness (mm), A is the permeation area (m^2), t is the time gained (h) and $P_2 - P_1$ is the difference of partial pressure (Pa).

2.8 Determination of thermal property
The thermal properties of films were evaluated via differential scanning calorimetry (DSC, TA Q2000 Instrument, USA) according to the method described by [25]. Film samples (3 mg) were weighted in aluminium pans. The pans were sealed with a TA sample encapsulation press. The samples were placed in the sample cells, while an empty pan was placed in the reference cell. Conditioned films were scanned between 0°C to 100°C at a heating rate of 10°C /min. Melting temperature (T_m, °C) and enthalpy (ΔH, J/g) were measured in triplicate. The melting temperature (T_m) was calculated as the temperature at which the endothermic peak occurs. The glass transition temperature corresponded to the temperature where a baseline inflexion occurred, and the melting temperature was determined as the peak temperature of the endothermic event of the DSC curves.

2.9 Determination of light barrier and transparency
The light transmission of films was measured at the ultraviolet (200-800 nm) and visible range (200-800 nm) using a UV-vis spectrophotometer (Pharo 300, USA) according to the method by [15]. The transparency value of film was calculated using the following equation:

$$Transparency\ Value = -logT_{600}\ x^{-1} \tag{4}$$

where T_{600} is the fractional transmission at 600 nm and x is the film thickness (mm). The greater transparency value represents the lower transparency of film.

2.10 Functional group determination by Fourier transform infrared spectroscopy (FTIR)
Following [10], the functional group that arose from the blending of film materials were investigated via infrared spectra recorded by Thermo Nicolet 380 Spectrometer (Fisher Scientific Inc, USA) using deuterated triglycerine sulphate (DTGS) as detector. The sample holder comprised a multi-ounce horizontal attenuated total reflectance (HATR) plate of zinc selenite (ZnSe) crystal. The background spectrum (without sample) was collected after the plate was cleansed with acetone, followed by affixing the film samples to the plate and recording the spectra obtained. A single beam spectrum for each sample was ratioed against an ambient air single beam background spectrum before conversion to absorbance units. The resolutions used varied from 4000 to 650 cm^{-1} over 32 scans. Each sample was conducted in triplicate.

2.11 Determination of X-ray diffraction (XRD)
X-ray diffraction (XRD) analysis was conducted to measure the crystalline and amorphous characteristic of the films. A Rigaku X-ray Diffractometer (Rigaku Corp, USA) with a copper source using 40 kV and 30 mA current at room temperature was used as described by [18]. The measurements were performed by mounting the sample with size of 4 x 4 cm^2 onto a glass slide before placing in the diffractometer chamber. An angle diffraction range used was $2\Theta = 3–80°$, with each sample repeated twice.

2.12 Determination of film microstructure
To study the film microstructure, the cross-sections and surface areas of all gelatin films were viewed at 25,000x magnification employing scanning electron microscopy (SEM) (Hitachi S-4300SE, Hitachi Science System Ltd., Japan) at an accelerating voltage of 5.0 kV. Prior to analysis, films were coated with gold to make the samples conductive. The samples were observed in a superficial position at 100x magnification [13].

2.13 Statistical analysis
All experiments were run in triplicate with different three lots of films. Data were subjected to analysis of variance (ANOVA) and mean comparisons were carried out by the Duncan's multiple range test. Analysis was performed using the SPSS package (SPSS for windows, SPSS Inc., Chicago, IL, USA).

3. Results and discussion

3.1 Tensile strength (TS) and elongation at break (EAB)
The results obtained for tensile strength (TS) of gelatin films are presented in Table 1. The results ranged from 2.97 ± 0.06 MPa to 5.57 ± 0.11 MPa, suggesting significant differences ($p < 0.05$) among the tensile strengths of gelatin film of different gelatin source. The data clearly show that chicken skin gelatin film possesses the highest TS value. This was supported via the compact and organized structure of chicken skin gelatin film, as shown by scanning electron microscopy. Variation in TS value may also be due to the levels water content in the gelatin. In general, different types of gelatin contain different levels of OH group. FTIR results showed that chicken skin gelatin film possessed lower levels of OH group than commercial gelatin. Also, chicken skin gelatin demonstrated higher gel strength (355 ± 1.48 g) than bovine gelatin (229 ± 0.71 g), in which a high gel strength property would also contribute to the tensile strength of the film [20]. These findings are in the agreement with those of [13], who reported that porcine and bovine gelatin films possess lower TS value compared to chicken skin gelatin films, which were 4.04 and 4.46 MPa, respectively.

Elongation at break (EAB), also known as fracture strain, is the ratio between changed length and initial length after breakage of the test specimen. It is an important factor in biodegradable film, as it describes the capability of the film to resist changes of shape without crack formation [26]. The results obtained for EAB of gelatin films are presented in Table 1. The results ranged from 122.78 ± 0.29 % to 135.60 ± 0.44 %, suggesting a significant difference ($p < 0.05$) in the EAB of gelatin films made with different gelatin sources. From the data obtained, it is clearly shown that chicken skin gelatin film possessed the lowest EAB value, while bovine gelatin film possessed the highest EAB value. X-ray diffraction revealed that bovine and porcine gelatin films exhibited broader amorphous peak at $2\Theta = 20^{\circ}$ due to high levels of water content as generated by FTIR spectra. This amorphous character and high water content contributed to the flexible and stretchable film structure. This is in agreement with the results of [15], who found that the amorphous character of gelatin-CMC film incorporated with xanthan gum improved the EAB of the film.

3.2 Puncture test
The results obtained for the puncture test (PT) of gelatin films (Table 1) showed that chicken skin gelatin film possessed the highest PT value. The amount of imino acid proline and hydroxyproline content in the gelatin would also affect the PT of the film. [20] reported that imino acid proline and hydroxyproline content is higher in chicken skin gelatin (proline: 13.42%, hydroxyproline: 12.13%) than in bovine gelatin (proline: 12.66%, hydroxyproline: 10.67%). A high amount of imino acid proline and hydroxyproline content indicated that the film possessed rigid hydrogen bonding and a compact molecular structure [20]. Thus, more force is needed to punch through the film membranes. This finding explains the highest PT values being obtained were in chicken skin gelatin and low PT values obtained in bovine and porcine gelatin.

3.3 Water vapour permeability (WVP)

The results for water vapour permeability (WVP) of chicken skin, porcine, and bovine gelatin films are presented in Table 1. The results ranged from 1.29 x 10^{-9} ± 0.03 kPa to 2.25 x 10^{-9} ± 0.13 kPa, suggesting a significant difference ($p < 0.05$) in the WVP of gelatin films of different gelatin sources. Chicken skin gelatin film possessed the lowest WVP value at 1.29 x 10^{-9} ± 0.03 kPa, while porcine gelatin had the highest WVP value (2.25 x 10^{-9} ± 0.13 kPa). The highest WVP rate indicates a high amount of water vapour passing through the film membrane. High WVP rate is an undesirable characteristic in packaging film. This is because a film with high WVP would hasten the food deterioration process, shortening the shelf life of food products.

Commercial gelatin showed higher WVP than chicken skin gelatin film. High WVP obtained in porcine and bovine gelatin films may be due to its bigger and scattered films molecular structure as compared to chicken skin gelatin film. This is because, bigger film molecular structure would allow more water vapour to pass through the film, thus increasing its WVP. High amount of water content would also increase the WVP of the gelatin films. High water content also would contribute to the hydrophilic character of the gelatin, as it will attract more water, thus increasing the WVP value. As reported by [28], porcine films contained higher amounts of tyrosine (2.6%) and serine (3.5%) than those from chicken skin (tyrosine: 1.22%, serine: 2.20%) and bovine (tyrosine: 1.16%, serine: 2.93%) gelatin [20]. Low amino acid tyrosine and serine content in chicken skin gelatin explains its hydrophobic character, contributing to its low WVP value as compared to commercial gelatin films. The results obtained were in the same agreement with the study by [27] and [29] who found high WVP value of pigskin gelatin film which ranged from 1.38 to 1.95 kPa. In another study, [30] found that film from fish skin gelatin had a low WVP value, ranging from 0.44 to 1.23 kPa. These results may be due to the low water content of fish skin gelatin.

3.4 Thermal property

The results obtained for melting temperature (T_m) and glass transition (T_g) of chicken skin, porcine and bovine gelatin are presented in Table 1. The results for T_m ranged from 60.42 ± 0.34°C to 76.26 ± 0.73°C, while T_g ranged from 47.23 ± 0.69°C to 51.12 ± 0.55°C ($p < 0.05$). Data shows that chicken skin gelatin film had the highest T_m (76.26 ± 0.73°C) and T_g (51.12 ± 0.55°C), while bovine gelatin film showed the lowest T_m (60.42 ± 0.34°C) and T_g (47.23 ± 0.69°C). This demonstrates that chicken skin gelatin film melts at higher temperatures than commercial gelatin films.

These results may be due to the amount of imino acid content in the gelatin. [20] found that the enthalpy change between T_m and T_g depends on the stability of the collagen structure. Imino acids proline and hydroxyproline are believed to influence the rigidity of gelatin through hydrogen bonding that stabilizes the triple-helix structure. Chicken skin gelatin contains high levels of proline (13.42%) and hydroxyproline (12.13%) as compared to bovine (proline: 12.66%, hydroxyproline: 10.67%) and porcine (proline: 13.1%, hydroxyproline: 10.8%) gelatins ([20]; [31]). High imino acids proline and hydroxyproline indicated that the films have strong hydrogen bonding and stable triple helix structure. Thus, more energy is needed to break the hydrogen bonds in the gelatin. The phenomenon explains the T_m and T_g value of chicken skin gelatin film was similar to the commercial gelatin films. These results are supported by the previous findings of [20], who found that the difference in thermostability between chicken and bovine gelatin may be attributed to the higher proline and hydroxyproline content of chicken gelatin. Thus, more energy is needed to break hydrogen bonds and for helix to coil transitions.

Table 1. Tensile strength (TS), elongation at break (EAB), puncture test (PT), water vapor permeability (WVP), glass transition (T_g) and melting point (T_m) of chicken skin, bovine and porcine gelatin films

Gelatin films	TS (MPa)	EAB (%)	PT (N)	WVP (kPa)	T_g (°C)	T_m (°C)
Chicken skin	5.57 ± 0.11[a]	122. 78 ± 0.29[b]	29.06 ± 0.06[a]	1.29×10^{-9} ± 0.03[b]	51.12 ± 0.55[a]	76.26 ± 0.73[a]
Bovine	2.97 ± 0.06[c]	135.60 ± 0.44[a]	18.03 ± 0.09[b]	1.75×10^{-9} ± 0.06[ab]	47.23 ± 0.69[b]	60.42 ± 0.34[b]
Porcine	3.21 ± 0.08[b]	130.06 ± 0.31[ab]	16.55 ± 0.11[b]	2.25×10^{-9} ± 0.13[a]	50.00 ± 0.91[ab]	61.71 ± 0.89[b]

Different letter ([a-c]) indicated significantly different ($p<0.05$) within the column. Data reported are mean value ± standard deviations.

3.5 Light barrier and transparency

Though the films turned to white when plasticizer was added, observation showed that the gelatin films produced were transparent. This may be due to the light-scattering effect of the emulsion formed by the composite gelatin film blends. The result obtained for light barrier and transparency of chicken skin, porcine and bovine gelatin are presented in Table 2. The results showed that there was a significant difference ($p < 0.05$) on the light barrier and transparency of gelatin film of different gelatin source. Results showed that chicken skin gelatin film obtained the lowest transparency (1.31%), whereas the highest transparency value was obtained by porcine gelatin film (1.74%).

Transparency values were directly influenced by the molecular structure of the gelatin used. Scanning electron microscopy revealed that chicken skin gelatin film had a small, compact and well-organized microstructure compared to bovine and porcine gelatin, which possessed bigger and scattered films microstructure. A smaller film molecular structure would increase the opaqueness of the films, thus inhibiting the amount of light passing through the film. This explains the low transparency value obtained for chicken skin gelatin film as compared to the commercial gelatin films. The results obtained was in agreement with a study by [32], who reported that the transparency of gelatin films decreased as the opacity increased.

Table 2. Light barrier and transparency values of chicken skin, bovine and porcine gelatin films

Gelatin films	Wavelength (nm)								
	200	280	350	400	500	600	700	800	T_{600}
Chicken skin	37.02± 1.92[b]	26.37± 3.22[b]	8.60± 4.21[bc]	18.05± 7.05[b]	39.99± 5.59[b]	39.49± 12.86[a]	42.92± 12.92[a]	42.26± 13.34[b]	3.52± 1.16[a]
Bovine	37.31± 6.52[b]	24.79± 6.50[b]	4.49± 0.88[c]	15.50± 2.18[b]	30.83± 2.31[c]	34.69± 2.67[a]	41.03± 2.50[a]	38.36± 5.81[b]	3.78± 0.28[a]
Porcine	31.72± 3.48[b]	20.24± 6.59[b]	3.36± 0.76[c]	11.97± 2.01[b]	30.58± 1.92[c]	34.43± 1.42[a]	37.78± 4.85[a]	38.37± 1.93[b]	2.96± 0.12[a]

Different letter ([a-b]) indicated significantly different ($p<0.05$) within the column. Data reported are mean value ± standard deviations.

3.6 Fourier transform infrared spectroscopy (FTIR)

The FTIR spectra of the chicken skin, bovine, and porcine gelatin films are shown in Table 3. The bands situated around 1624 to 1648 cm^{-1}, 1510 to 1580 cm^{-1} and 3275 to 3320 cm^{-1}, correspond to the amount of water content and also the alcohol, probably from glycerol, which were Amides I, II, and III, respectively. The stretching of COO of proteins amide resulted in Amide I arises, while vibrational of N–H groups and stretching vibrations of C–N groups resulted in Amide II arises. Other than that, the vibrations in plane of C–N and N–H groups of bound amides, or vibrations of C–H_2 groups of glycine corresponds to Amide III arises [33].

A significant difference ($p < 0.05$) were found at the peaks corresponding to Amides I, II and III. The intensity of Amide III bands was the highest in porcine gelatin film (3300 cm^{-1}) as compared to bovine (3292 cm^{-1}) and chicken skin (3288 cm^{-1}) gelatin films. The same trends were found in the Amide I and II bands. Porcine gelatin film generated the highest wave number at Amide I (1646 cm^{-1}) and II (1558 cm^{-1}) bands, while chicken skin gelatin film generated the lowest wave number at Amide I (1626 cm^{-1}) and II (1539 cm^{-1}) bands. Whereas bovine gelatin film generated moderate wave number at Amide I, II and III bands at 1540 cm^{-1}, 1635 cm^{-1} and 3292 cm^{-1}, respectively. The increase of amplitudes of amide I groups could be due to that the intermolecular hydrogen bonds formed between the C-O bands and C-N bands from gelatin.

The aliphatic alcohol group content in the film represent the hydroxyl groups (-OH) and C-O stretching. Therefore, results have shown that all gelatin film possessed an aliphatic alcohol group. The values obtained showed that there were no differences between gelatin films was due to the standardize amount of glycerol was used. These behaviours explain the TS, EAB and WVP values obtained by porcine, bovine and chicken skin gelatin films, as discussed earlier. Higher water content would resulted in more flexible films, thus increasing its EAB and WVP value and lowering in TS value of the films. The results obtained were in agreement with a study by [18], who found that FTIR spectra of Amide I, II and III of bovine gelatin film situated around 1630 cm^{-1}, 1539 cm^{-1} and 3290 cm^{-1}, respectively. While in another study, [33] found that the Amide I, II and III bands of porcine gelatin are situated around 1632 cm^{-1}, 1548 cm^{-1}, and 3288 cm^{-1}, respectively.

Table 3. FTIR spectra of chicken skin, bovine and porcine gelatin films

Gelatin films	Amide III	Amide I	Amide II	Aliphatic Alcohol
	stretching vibration of C-N bands and N–H groups of bound amide, vibrations of C–H_2 groups of glycine	C=O stretching	bending vibration N-H group, stretching vibration of C-N group	hydroxyl groups (-OH), C-O stretching
	3275-3276cm^{-1}	1648-1624 cm^{-1}	1510-1580 cm^{-1}	1250-1020 cm^{-1}
Chicken skin	3288.97±0.20^{b}cm^{-1}	1626.20±0.20^{b} cm^{-1}	1539.13±0.00^{b} cm^{-1}	1033.80±0.20^{a} cm^{-1}
Bovine	3292.49±0.20^{ab} cm^{-1}	1635.64±0.40^{ab} cm^{-1}	1540.77±0.20^{ab} cm^{-1}	1034.63±0.40^{a} cm^{-1}
Porcine	3300.20±0.10^{a} cm^{-1}	1646.79±0.20^{a} cm^{-1}	1558.91±0.30^{a} cm^{-1}	1034.70±0.20^{a} cm^{-1}

Different letter ($^{a-b}$) indicated significantly different ($p<0.05$) within the column. Data reported are mean value ± standard deviations.

3.7 X-ray diffraction (XRD)

The diffractograms of chicken skin, porcine, and bovine gelatin films are shown in Figure 1. The results obtained for all film formulations for various types of gelatin demonstrate similarities. The diffractograms for all samples present an amorphous character, indicating no tendency towards re-crystallization, which probably due to the high stability and high moisture content in gelatin films. The diffractograms of the chicken skin, porcine and bovine gelatin films showed one diffraction peak at approximately $2\theta = 20°$. The more amorphous structure of the composite film at 20° of gelatin film peaks diffraction, attributed to the typical fingerprints for gelatin powder [34]. However, intensity of the peak varied depending to the types of gelatin used. According to the diffractograms, porcine gelatin film possessed the highest intensity at peak $2\theta = 20°$. Chicken skin gelatin film possessed the lowest intensity at peak $2\theta = 20°$.

The results revealed that the amorphousness of the films decreases as the intensity of the peak increased. Therefore, porcine gelatin film has the most amorphous film characteristics as compared to the chicken skin and bovine gelatin film, as it has the highest intensity at peak 2θ =20°. Result obtained was in the same agreement with [10] who found that the diffractograms of chicken skin gelatin film which plasticized with glycerol present an amorphous character.

In addition, the diffractograms of porcine and bovine gelatin films also generated small peaks at 2θ = 8°. The peaks indicate that the films possessed crystalline character. Difractograms of chicken skin gelatin film showed no sign of crystallization, while porcine and bovine gelatin films exhibited small crystalline peak at 8°. This may be due to the molecular structure of the gelatin used. According to [35], the peak at 2θ = 8° indicates the triple helix diameter; thus, the intensity of the gelatin films is associated to the triple-helix content. The intensity of the peak increased as the molecular structure of the gelatin used increased. SEM showed that chicken skin gelatin film possessed the smallest and most organized microstructure than that of commercial gelatin. While porcine and bovine gelatin films possessed bigger and scattered films molecular structure, explaining their crystalline character. A similar effect was also observed in bovine gelatin, with semi-crystalline regions at peaks situated around 2θ = 7 to 8° [36].

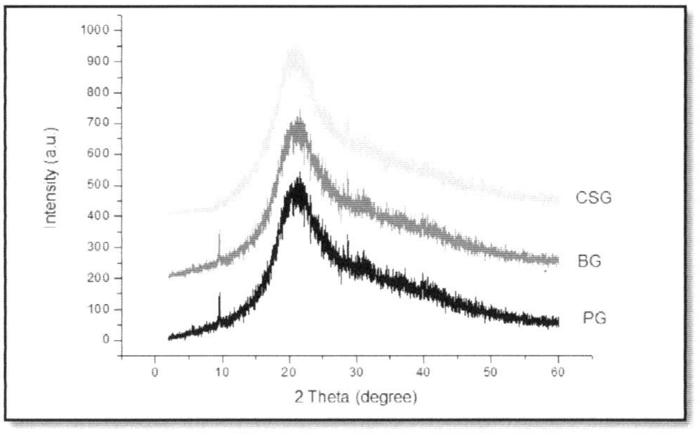

Figure 1. Diffractograms of different gelatin films between 0 – 60°, where CSG -chicken skin gelatin, BG -bovine gelatin and PG -porcine gelatin films.

3.8 Film microstructure

Scanning electron microscopy (SEM) was used to study the microstructural changes in gelatin films and to obtain the surface and cross-section topography of the films. The microstructures of chicken skin, porcine, and bovine gelatin films are presented in Table 4. SEM revealed that the cross section of chicken skin gelatin film possessed small, compact and organized microstructure, while bovine and porcine gelatin films possessed bigger and scattered film microstructures with possible signs of a cracked surface area. This may be due to the high levels of hydroxyl group in bovine and porcine gelatin films, as revealed by FTIR. High levels of water content may disrupt the film network and create films with large and disorganized microstructure. This result was in agreement with studies by [37] and [27], who found that the internal structure of pigskin gelatin films revealed discontinuous zones that were characterized by horizontal cracks randomly distributed along the networks.

Additionally, it was observed that the surface area of the chicken skin gelatin film was homogeneous, with no brittle areas or bubbles. The XRD results also show that X-ray diffraction of chicken skin gelatin film displayed amorphous structure with no sign of crystallization, whereas both bovine and porcine gelatin films displayed crystalline peak at 2Θ=8°. These findings support the homogenous structure obtained in chicken skin gelatin film. The formation of this homogenous and smaller surface area structure is related to the lower water content in chicken skin gelatin than in bovine and porcine gelatins. The structural difference of films may be responsible for the improvement of WVP obtained in the films.

IOP Conf. Series: Materials Science and Engineering **440** (2018) 012033 doi:10.1088/1757-899X/440/1/012033

The results are in agreement with those of [13], who found that bovine and porcine gelatin films displayed the formation of bumps and bubbles, as their bigger microstructures impact the results of oxygen permeability of the films.

Table 4. Cross section and microstructure of chicken skin, bovine and porcine gelatin films

Gelatin films	Surface area	Cross section
Chicken skin	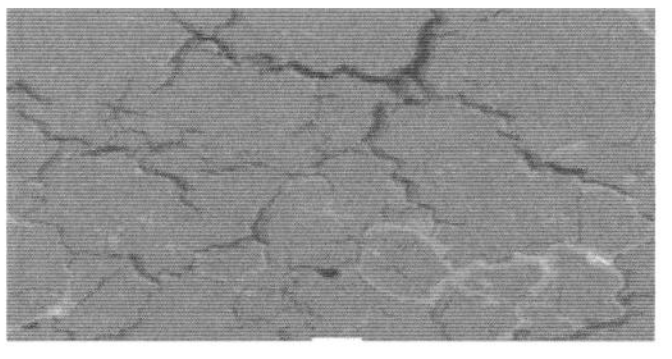	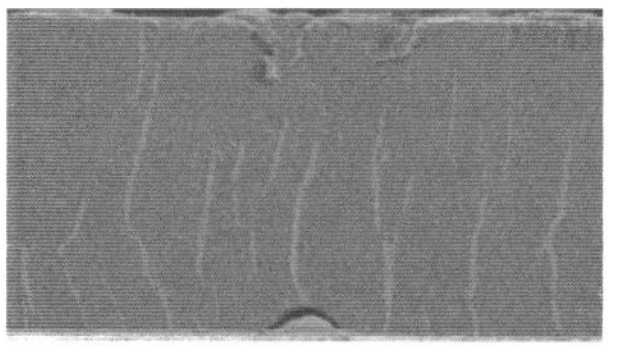
Bovine	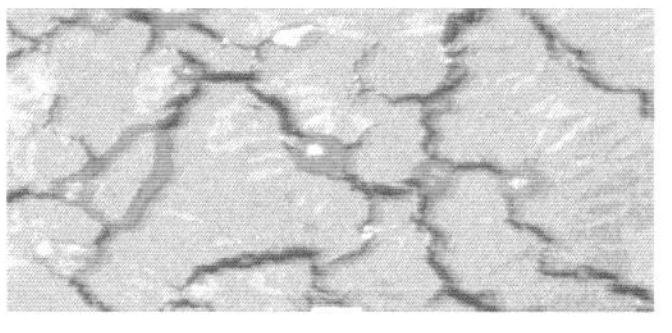	
Porcine		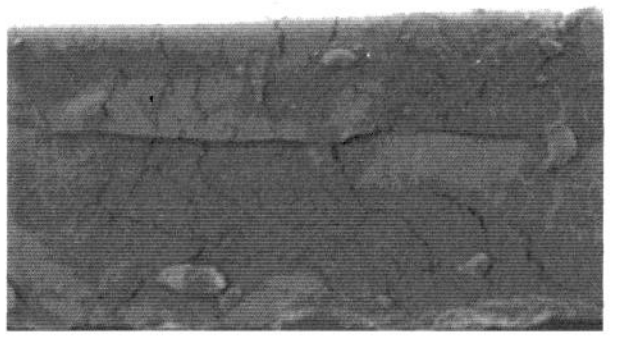

4. Conclusion

In conclusion the mechanical and physical properties of gelatin films from different sources showed that films from chicken skin gelatin offer better properties than commercial (porcine and bovine) gelatin films. This study found that chicken skin gelatin film possessed good properties for tensile strength, elongation at break and water vapour permeability, thermal properties, as well as film's microstructure. Therefore, chicken skin gelatin could be a potential alternative to commercial gelatin, as biodegradable films packaging.

5. References

[1] Cutter C N 2006 Opportunities for bio-based packaging technologies to improve the quality and safety of fresh and further processed muscle foods *Meat science* **74** 1 131-142

[2] Prashanth K H and Tharanathan R N 2007 Chitin/chitosan: modifications and their unlimited application potential—an overview *Trends in food science & technology* **18** 3 117-131

[3] Aider M 2010 Chitosan application for active bio-based films production and potential in the food industry: Review *LWT-Food Science and Technology* **43** 6 837-842

[4] Tongnuanchan P, Benjakul S and Prodpran T 2012 Properties and antioxidant activity of fish skin gelatin film incorporated with citrus essential oils *Food Chemistry* **134** 3 1571-1579

[5] Wu J, Liu H, Ge S, Wang S, Qin Z, Chen L and Zhang Q 2015 The preparation, characterization, antimicrobial stability and in vitro release evaluation of fish gelatin films incorporated with cinnamon essential oil nanoliposomes *Food Hydrocolloids* **43** 427-435

[6] Hanani Z N, Roos Y H and Kerry J P 2014 Use and application of gelatin as potential biodegradable packaging materials for food products *International journal of biological macromolecules* **71** 94-102

[7] Gómez-Guillén M C, Giménez B, López-Caballero M A and Montero M P 2011 Functional and bioactive properties of collagen and gelatin from alternative sources: A review *Food Hydrocolloids* **25** 8 1813-1827

[8] Martucci J F and Ruseckaite R A 2010 Biodegradable three-layer film derived from bovine gelatin *Journal of Food Engineering* **99** 3 377-383

[9] Gómez-Guillén M C, Pérez-Mateos M, Gómez-Estaca J, López-Caballero E, Giménez B and Montero P 2009 Fish gelatin: a renewable material for developing active biodegradable films *Trends in Food Science & Technology* **20** 1 3-16

[10] Nor M M, Nazmi N N M and Sarbon N M 2017 Effect of different hydrophilic plasticizer concentrations on functional properties of chicken skin gelatin films *International Food Research Journal* **24** 5 1910-1918

[11] Bakry N.F, Isa M I N and Sarbon N M 2016 Effect of sorbitol at different concentrations on the functional properties of gelatin/carboxymethyl cellulose (CMC)/chitosan composite films *International Food Research Journal* **24** 4 1753-1762

[12] Al-Hassan A A and Norziah M H 2012 Starch–gelatin edible films: water vapour permeability and mechanical properties as affected by plasticizers *Food Hydrocolloids* **26** 1 108-117

[13] Hanani Z N, Roos Y H and Kerry J P 2012 Use of beef, pork and fish gelatin sources in the manufacture of films and assessment of their composition and mechanical properties *Food Hydrocolloids* **29** 1 144-151

[14] Hosseini S F, Rezaei M, Zandi M and Ghavi F F 2013 Preparation and functional properties of fish gelatin–chitosan blend edible films *Food chemistry* **136** 3 1490-1495

[15] Nur Hazirah M A S P, Isa M I N and Sarbon N M 2016 Effect of xanthan gum on the physical and mechanical properties of gelatin-carboxymethyl cellulose film blends *Food Packaging and Shelf Life* **9** 55-63

[16] Nunez-Flores R, Gimenez B, Fernandez-Martin F, Lopez-Caballero M E, Montero M P and Gomez-Guillen M C 2013 Physical and functional characterization of active fish gelatin films incorporated with lignin *Food Hydrocolloids* **30** 163-172

[17] Ahmad M, Benjakul S, Prodpran T and Agustini T W 2012 Physico-mechanical and antimicrobial properties of gelatin film from the skin of unicorn leatherjacket incorporated with essential oils *Food Hydrocolloids* **28** 1 189-199

[18] Jahit I S, Nazmi N N M, Isa M I N and Sarbon N M 2016 Effect of drying temperature on physical properties of gelatin-based film blended with CMC and chitosan as potential edible film *International Food Research Journal* **23** 3 1068-1074

[19] Pereda M, Ponce A G, Marcovich N E, Ruseckaite R A and Martucci J F 2011 Chitosan-gelatin composites and bi-layer films with potential antimicrobial activity *Food Hydrocolloids* **25** 5 1372-1381

[20] Sarbon N M, Badii F and Howell N K 2013 Preparation and characterisation of chicken skin gelatin as an alternative to mammalian gelatin *Food Hydrocolloids* **30** 1 143-151

[21] Gómez-Estaca J, Gómez-Guillén M C, Fernández-Martín F and Montero P 2011 Effects of gelatin origin, bovine-hide and tuna-skin, on the properties of compound gelatin–chitosan films *Food Hydrocolloids* **25** 6 1461-1469

[22] AOAC 2006 *Official methods of analysis of AOAC international* 18th ed. Virginia, USA: Association of Official and Analytical Chemists International

[23] Suderman N, Isa M I N and Sarbon N M 2016 Effect of drying temperature on the functional properties of biodegradable CMC-based film for potential edible film *International Food Research Journal* **23** 3 1075-1084

[24] American Society for Testing and Materials ASTM 1997 Designation D 882-97: standard test method for tensile properties of thin plastic sheeting *Annual book of ASTM standards*. Philadelphia: American Society for Testing and Materials p 159
[25] Gómez-Estaca J, Bravo L, Gómez-Guillén M C, Alemán A and Montero P 2009 Antioxidant properties of tuna-skin and bovine-hide gelatin films induced by the addition of oregano and rosemary extracts *Food Chemistry* **112** 1 18-25
[26] Cheng M, Deng J, Yang F, Gong Y, Zhao N and Zhang X 2003 Study on physical properties and nerve cell affinity of composite films from chitosan and gelatin solutions *Biomaterials* **24** 17 2871-2880
[27] Thomazine M, Carvalho R A and Sobral P J 2005 Physical properties of gelatin films plasticized by blends of glycerol and sorbitol *Journal of Food Science* **70** 3 E172-E176
[28] Nhari R, Hafidz R M, Che Man Y, Ismail A and Anuar N 2011 Chemical and functional properties of bovine and porcine skin gelatin *International Food Research Journal* **18** 2 813-817
[29] Sobral P D A, Menegalli F C, Hubinger M D and Roques M A 2001 Mechanical, water vapour barrier and thermal properties of gelatin based edible films *Food Hydrocolloids* **15** 4 423-432
[30] Jongjareonrak A, Benjakul S, Visessanguan W, Prodpran T and Tanaka M 2006 Characterization of edible films from skin gelatin of brownstripe red snapper and bigeye snapper *Food Hydrocolloids* **20** 4 492-501
[31] Hermanto S and Fatimah W 2013 Differentiation of bovine and porcine gelatin based on spectroscopic and electrophoretic analysis *Journal of Food and Pharmaceutical Sciences* **1** 3
[32] Bertan L C, Tanada-Palmu P S, Siani A C and Grosso C R F 2005 Effect of fatty acids and 'Brazilian elemi'on composite films based on gelatin *Food Hydrocolloids* **19** 1 73-82
[33] Bergo P and Sobral P J A 2007 Effects of plasticizer on physical properties of pigskin gelatin films *Food Hydrocolloids* **21** 1285–1289
[34] Lavorgna M, Piscitelli F, Mangiacapra P and Buonocore G G 2010 Study of the combined effect of both clay and glycerol plasticizer on the properties of chitosan films *Carbohydrate Polymers* **82** 2 291-298
[35] Bigi A, Panzavolta S and Rubini K 2004 Relationship between triple-helix content and mechanical properties of gelatin films *Biomaterials* **25** 25 5675-5680
[36] Andreuccetti, C, Carvalho R A, Galicia-García T, Martínez-Bustos F and Grosso C R 2011 Effect of surfactants on the functional properties of gelatin-based edible films *Journal of Food Engineering* **103** 2 129-136
[37] Vanin F M, Sobral P J A, Menegalli F C, Carvalho R A and Habitante A M Q B 2005 Effects of plasticizers and their concentrations on thermal and functional properties of gelatin-based films *Food Hydrocolloids* **19** 5 899-907

IOP Conf. Series: Materials Science and Engineering **440** (2018) 012034 doi:10.1088/1757-899X/440/1/012034

Optical, Structural and Electrical Studies of Biopolymer Electrolytes Based on Methylcellulose Doped with $Ca(NO_3)_2$

A M S Nurhaziqah[1], I Q Afiqah[2], Mohd Fazrul Hisam Abd. Aziz[2], N A Nik Aziz[2,a] and S Hasiah[3]

[1]School of Ocean Engineering, Universiti malaysia Terengganu, 21030 Kuala Nerus, Terengganu, Malaysia
[2]School of Fisheries and Aquaculture Sciences, Universiti Malaysia Terengganu, 21030 Kuala Nerus, Terengganu, Malaysia
[3]Center for Foundation and Liberal Education, Universiti MalaysiaTerengganu, 21030 Kuala Nerus, Terengganu, Malaysia

Corresponding author: [a]nikaziz@umt.edu.my

Abstract. The intention of this research is to study effect of the different weight percent (wt%) of calcium nitrate, $Ca(NO_3)_2$ (CN) salt into biopolymer of Methylcellulose (MC). MC-CN solid biopolymer electrolyte was prepared by using casting technique. The concentrations of CN salt were prepared between 5 to 25 wt%. The optical and electrical characteristics of the samples were study by using Fourier Transform-Infrared (FTIR) spectroscopy, X-ray Diffraction (XRD) and Electrical Impedance Spectroscopy (EIS). The highest conductivity of the samples was containing 25 wt% of CN is $2.21X10^{-7}$ Scm^{-1}. The optical study by using FTIR shows that the CN salt act as the conducting ions devoid of effect the molecular structure of the polymer. Additionally, the data of XRD revealed that MC-CN an amorphous phase.

1. Introduction

The dependency of science and technology according by study of materials and application stand of ionically conducting solid known as "Solid-State Ionics". Besides, solid states ionic are dividing into liquid electrolyte and solid electrolyte. The liquid electrolyte gives a major weakness such as poor electrochemical stability, leakage and reaction with electrode indirectly not suitable to apply in electrochemical devices [1]. To overcome this weakness, the potential application in various technological areas focuses on solid polymer electrolytes which act as an authority over liquid electrolytes because of being strong [2]. Polymer electrolyte are one of the significant part in solid state ionic based on several properties which are good electrode-electrolyte contact [3], high ionic conductivity with wide surface area to operate high energy density while ease fabrication into thin film and capable to accommodate a large range of doping salt compositions [4].

Furthermore, technology nowadays emphasize of "green materials" aspect to avoid the all profanation environment and energy disaster. Besides, the new biopolymer electrolytes (BPE) product be found by many researchers help to lower the cost and being environmentally green compare synthetic polymer electrolytes [5]. The polymer electrolyte could unfold as solid electrolytes comparable perfectly intercalation electrodes. The polymer are also the most crucial categorization of conjugated polymers appropriate in a various type of applications such as light emitting diodes, conducting polymers and field effect transistors because of their great optical and electrical properties [6]. However, common polymer without blend with other polymer not guaranteed good chemical or physical characteristic to meet extensive variety of electrochemical devices [7]. The blend polymers have better conductivity compare with common polymer. While the conductivity for common polymer can increase by mix with salt materials involving dissimilar types of transporting ions [8].

Meanwhile, biopolymer are make more intention among researcher due to their properties which are biodegradable polymer and known as environmentally friendly products [9]. Biopolymer is also inexpensive of cost materials and abundance in nature which act as the main reason to choose Methylcellulose (MC) as the biopolymer electrolyte in this research. MC is one of the simplest cellulose derivatives [10]. Besides, MC is also good solubility in water with low temperature while act as biodegradable cellulose ether [11]. MC present large-strength films that are transparent, water soluble, oil and grease resistant and low oxygen with transmission rates of moisture vapour [12]. In addition, with this research, the biopolymer of MC as the host polymer are mix with CN salt to increase the conductivity and achieve the standard characterize of electrolyte which help in the system of application such as solar cell, battery and others.

2. Methodology/Materials

2.1Preparation of Biopolymer Electrolyte Films

Solid biopolymer electrolyte thin films were prepared by using cast technique. 2 g of MC from Sigma Aldrich were dissolved in distilled water then different amount of wt% of CN salt from Merck KGaA was added and stirred continuously until the salt was dissolved completely. Then, make sure the salt and polymer complex were homogenously mixed and were placed into different polystyrene petri dishes and bare to evaporate the mixtures under room atmosphere.

2.2Fourier transform-infrared (FTIR) spectroscopy

Fourier transform-infrared (FTIR) spectroscopy, model of Thermo Nicolet Avatar 380 FT-IR spectrometer are functionate to study the synergy of the biopolymer provide and the doping salts. While, it is also to investigate the complex nature and functional groups present in complex electrolyte. The sample was analyzed by used spectrophotometer helmeted with an Attenuated Total Reflection accessory with a germanium crystal. On the surface of germanium, crystal placed of the sample and infrared light passed through the sample, which used range of wavenumber at 4000 to 400 cm^{-1} and a resolution at 1 cm^{-1}.

2.3X-ray Diffraction (XRD) Study

The sample was analysed by using X-ray Diffraction (XRD). It was also investigated the XRD patterns at room temperature and the amorphousity structure of the sample by used a MiniFlex II diffractometer equipped with an X'celerator, using Cu Kα radiation based on scale of 2θ from 5° to 80°.

2.4Electrical Impedance Spectroscopy (EIS)

The biopolymer electrolytes were sandwiched in the middle blocking stainless steel electrode that was connected to a computer in the form of small circle with 2 cm diameter of electrolyte. The sample were measured the conductivities of MC-$Ca(NO_3)_2$ by using a Hioki 3532-50 LCR Hi-Tester interfaced to a computer with used a frequency range between 50Hz to 1MHz. The Nyquist/Cole-Cole plots obtain negative imaginary impedance ($-Z_i$) versus real impedance (Z_r) values. While can measure the bulk resistance, R_b from the plot and conductivity was calculated using equation follows:

$$\sigma = \frac{t}{R_b A} \qquad (1)$$

Where, t is the thickness of sample (cm) and A is the contact area (cm^2) between the samples and the electrode.

3. Results and Discussion

3.1 FTIR Analysis

FTIR analysis has been operated to investigate the compositional, structural and interaction bonding between MC and CN in BPE system [5]. Besides, the FTIR spectra of MC-CN biopolymer electrolytes shown in Fig.1 with wavenumbers between 4000 to 500 cm^{-1}. From the figure, the peak at 1055 cm^{-1} is described as the antisymmetric stretching of asymmetric oxygen bridge in the cyclohexane ring with range between 1150 to 1000 cm^{-1} which is C-O-C bond [13]. The increment with the 25 wt% of CN, the peak is observed shifted to lower wavenumber show the complexation between MC and CN in the system. Consequent to addition of CN salt, Ca^{2+} will provide as cation which will attract at the oxygen atom at C-O-C bond of the ether group in the identical polymer to form polymer salt complexes [14]. However, from the Fig.1 show 1055 cm^{-1} and 1049 cm^{-1} peaks are raise in the percent of transmittance as the weight percent of salt is increased. The changes in this system augmented due to the coordination interaction of C-O-C in MC with Ca^{2+} of CN salt [15].

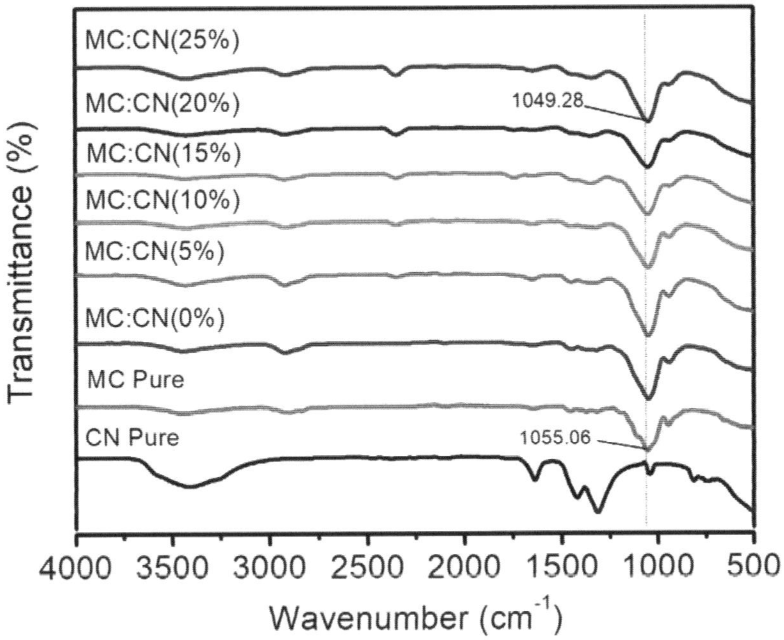

Figure 1. Fourier transform-infrared (FTIR) spectroscopy analysis.

3.2 XRD Analysis

The amorphous on crystallinity of the polymer film with respect to pure MC and CN complex have been examine by x-ray diffraction. XRD is a multi-purpose tool to monitor the transition in structural properties to both crystalline and amorphous region of BPEs system. The pattern of six different samples which are 0, 5, 10, 15, 20, and 25 wt% with pure MC based on an x-ray diffraction at room atmosphere from $2\theta = 5^{\circ}$ to 80° are shown in Fig.2. Based that figure, the broad diffused band at $2\theta = 21^{\circ}$ while the starter peak at $2\theta = 8^{\circ}$. The two peaks, at $2\theta = 8^{\circ}$ with 21° was appeared and showed as pure MC pattern [16]. The existence based on peak at 8° is declaration with cellulose adaptation [9]. The increment of the width peaks of MC at $2\theta = 8^{\circ}, 21^{\circ}$ due to addition of CN salt and proportionate intensity decrease. While the strongest cellulose peak at 21° with MC:CN(25%). However, the Fig.2 below can be observed that at 25 wt% CN salt to pure MC relative intensity of broad peak at $2\theta = 21^{\circ}$ increase prove the increment of the conductivity.

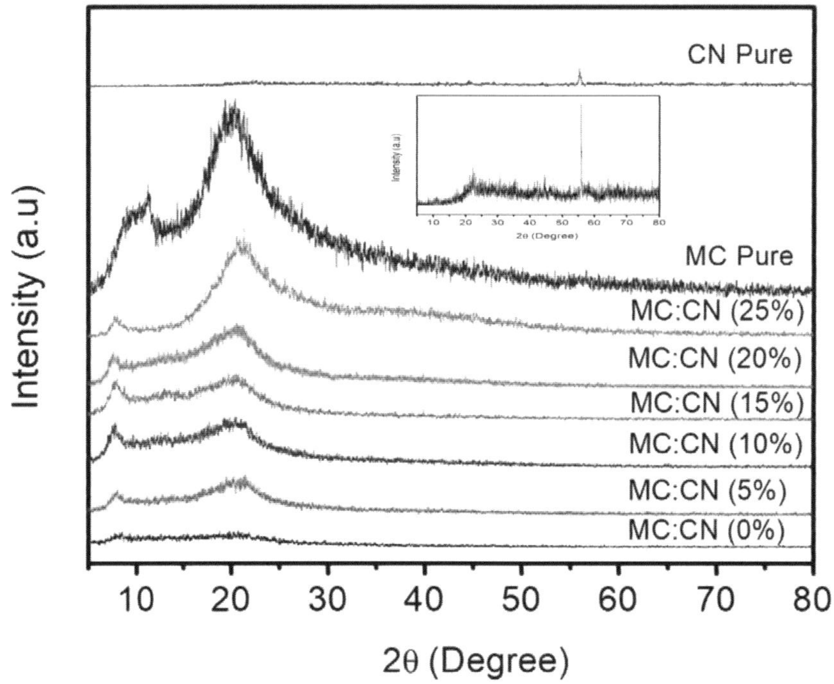

Figure 2. X-ray Diffraction (XRD) analysis.

3.3 EIS Analysis

The conductivity of MC without salt is $\approx 10^{-10}$ Scm^{-1} and it increase distinctly to $\approx 10^{-7}$ Scm^{-1} on complexing the MC with CN. However, the conductivity of BPEs relies on diverse determinant, which are ionic conducting types, the temperature and variety of charge carriers (cationic or anionic). In prevalent, the higher frequency semicircle depict of complex impedance plot as bulk resistance of polymer electrolytes, while the low frequency is because of the capacitance of the double-layer formed at the electrode/electrolyte interface [17].

Besides, based on Fig.3 show that the decrease in the high frequency semi-circular part consequent to doping CN from 0 to 25 wt%. In this case can relate that the current carriers are ions and the total conductivity mostly the result of ion conduction [18]. Furthermore, the value of conductivity occur escalation correlative with inclusion of the CN and achieve maximum value of 2.21×10^{-7} Scm^{-1} for sample with 25 wt%.

Moreover, increment of the conductivity continuously with increasing of wt% CN configuration while it claimed to the FTIR analysis supported by the ion association. In MC-CN, the Ca^{2+} interacted with the O at C-O-C bond of the ether group of the MC host. The increment of salt may have increase the number of ions or developed new conducting pathway which increase the movement of the protons [19]. While, the more salt added to the host polymer endorse a decline in conductivity because of the complexation between MC and CN reduces the movement of the charge carriers [20].

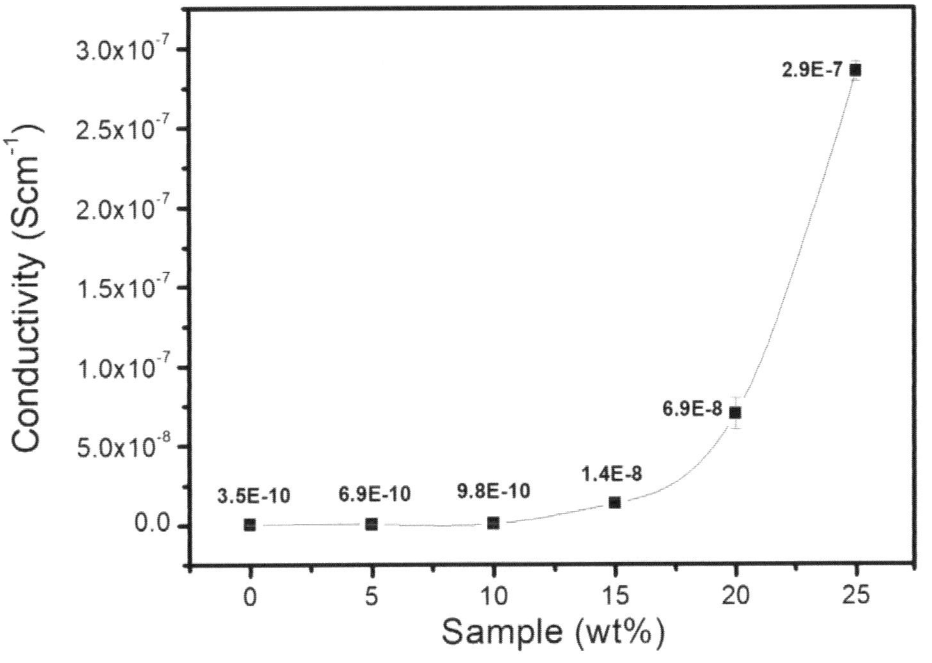

Figure 3. Electrical Impedance Spectroscopy (EIS) analysis of conductivity (Scm^{-1}).

4. Conclusion

The discovery of new BPE has been consummate in this work by incorporating different concentrations (0-25 wt%) of CN with biopolymer materials MC via solution casting technique. The conductivity biopolymer electrolyte of MC-CN increases due to function of 0-25 wt% of CN. The highest conductivity obtained of this polymer-salt complex was 2.9×10^{-7} Scm^{-1}, for sample 25 wt% CN. The increment in movement of proton ions and malleable structure are accountable for the augmentation of conductivity. Its evidence that effect of complexation of CN in MC polymer common, influenced by the charge carrier concentration due to the shifted of the peak at 1055 cm^{-1} of wavenumber and confirms composite nature of electrolyte with intensity of broad peak at $2\theta = 21^{\circ}$ increase show the increment of the conductivity from FTIR analysis and as well as XRD analysis.

5. References

[1] Idris N et al. 2009 Ionic conductivity studies of chitosan-based polymer electrolytes doped with adipic acid *Ionics* **15** (5) 643-646

[2] Kumar M, Tiwari T and Srivastava N 2012 Electrical transport behaviour of bio-polymer electrolyte system: Potato starch+ ammonium iodide *Carbohydrate Polymers* **88** (1) 54-60

[3] Aziz N N 2010 Proton conducting polymer electrolytes of methycellulose doped ammonium fluoride: Conductivity and ionic transport studies *Thesis dissertation* Universiti Malaysia Terengganu

[4] Rani M et al, 2015 Conductivity and dielectric behavior studies of carboxymethyl cellulose from kenaf bast fiber incorporated with ammonium acetate-BMATFSI biopolymer electrolytes. *International Journal of Polymer Analysis and Characterization* **20** (3) 250-260

[5] Monisha, S, et al. 2017 Investigation of bio polymer electrolyte based on cellulose acetateammonium nitrate for potential use in electrochemical devices *Carbohydrate Polymers* **157** 38-47

[6] Salleh H et al. 2014 Study the Electrical Properties and the Efficiency of Polythiophene with Dye and Chlorophyll as Bulk Hetero-Junction Organic Solar Cell in *Advanced Materials Research* Trans Tech Publ.

[7] Rudhziah S et al. 2015 Biopolymer electrolytes based on blend of kappa-carrageenan and cellulose derivatives for potential application in dye sensitized solar cell *Electrochimica Acta* **175** 162-168

[8] Agrawal R. and Pandey G 2008 Solid polymer electrolytes: materials designing and all-solidstate battery applications: an overview *Journal of Physics D: Applied Physics* **41** (22) 223001

[9] Nik Aziz N, Idris N and Isa M I N 2010 Solid polymer electrolytes based on methylcellulose: FT-IR and ionic conductivity studies *International Journal of Polymer Analysis and Characterization* **15** (5) 319-327

[10] Lin S-Y et al. 2007 Temperature effect on water desorption from methylcellulose films studied by thermal FT-IR microspectroscopy *Surface science* **601** (3) 781-785

[11] Kanimozhi K, Basha S K and Kumari V S 2016 Processing and characterization of chitosan/PVA and methylcellulose porous scaffolds for tissue engineering *Materials Science and Engineering: C* **61** 484-491

[12] Jingjid S, Damrongsakkul S and Rimdusit S 2006 Characterization of Methyl Cellulose/Montmorillonite Nanocomposites in *The Fourth Thailand Materials Science and Technology Conference* Thailand

[13] Abiddin, J F B Z and Ahmad A H 2015 Fourier transform infrared spectroscopy and electrical characterization of methylcellulose based solid polymer electrolyte doped with sodium iodide *Jurnal Teknologi* **76** (3) 41-45

[14] Sahli N B 2012 Effect of lithium triflate salt concentration in methyl cellulose-based solid polymer electrolytes in *Humanities, Science and Engineering (CHUSER,* IEEE 2012 Colloquium)

[15] Samsudin A, Lai H and Isa M I N 2014 Biopolymer materials based carboxymethyl cellulose as a proton conducting biopolymer electrolyte for application in rechargeable proton battery *Electrochimica Acta* **129** 1-13

[16] Tang Y et al. 2010 Production and characterisation of novel injectable chitosan/methylcellulose/salt blend hydrogels with potential application as tissue engineering scaffolds *Carbohydrate Polymers* **82** (3) 833-841

[17] Baskaran R et al. 2006 ac impedance, DSC and FT-IR investigations on (x) PVAc–(1− x) PVdF blends with $LiClO_4$ *Materials Chemistry and Physics* **98** (1) 55-61

[18] Jacob M, Prabaharan S and Radhakrishna S 1997 Effect of PEO addition on the electrolytic and thermal properties of PVDF-$LiClO_4$ polymer electrolytes *Solid State Ionics* **104** (3) 267-276

[19] Majid S and Arof A 2007 Electrical behavior of proton-conducting chitosan-phosphoric acidbased electrolytes *Physica B: Condensed Matter* **390** (1) 209-215

[20] Alves R et al. 2013 Study and characterization of a novel polymer electrolyte based on agar doped with magnesium triflate *Molecular Crystals and Liquid Crystals* **570** (1) 1-11

6. Acknowledgement

I would like to thank those who have helped me in writing my journal. My first sincere gratitude goes to my main supervisor, Dr. Nik Aziz Bin Nik Ali who has given valuable knowledge and help during my research in Universiti Malaysia Terengganu. Their generosity and support have encouraged me to spend most of my time in this research. They have provided all the benefits not only in the scope of the research but also in life including inspiring me to do the original work.

IOP Conf. Series: Materials Science and Engineering **440** (2018) 012035 doi:10.1088/1757-899X/440/1/012035

Identification of functional group present in *Andrographis paniculata* (kalmegh) leaves by FTIR analysis

M F F Maria[1], W M K W M Ikhmal[1], M G M Sabri[1,2], M S C Ibrahim[2] and A Adnan[1]

[1]*Advanced Nano Materials (ANoMa) Research Group, Nano Research Team, School of Fundamental Science, Universiti Malaysia Terengganu, 21030 Kuala Nerus, Terengganu, MALAYSIA*
[2]*Institute of Tropical Biodiversity and Sustainable Development, Universiti Malaysia Terengganu, 21030 Kuala Nerus, Terengganu, MALAYSIA*

Email: mohdsabri@umt.edu.my

Abstract. This present study aims to analyze the functional group present in the *Andrographis paniculata* leaves or also known as kalmegh by using Shimadzhu Fourier Transform Infrared (FTIR) with the absorbance range of 500 cm^{-1} - 4000 cm^{-1}. *Andrographis paniculata* (Nees) is a valuable traditional medicinal plant and it has many important bioactive compounds. In this study, kalmegh leaves powder was soaked in two different solvent which were 80% aqueous methanol and distilled water. After soaked for 24 hours, the soaked kalmegh leaves was extracted using rotary evaporator to get the crude extract before being subjected to Fourier Transform Infrared (FTIR) spectroscopy for characterization of functional group contains in the leaves. Results obtained from this present research show six major compound group which are alkene, amine, aromatic, nitro, and alcohol. This six functional group were identified from various wavenumber of bending and stretching type of alkene group (894.97 cm^{-1} and 1645.10 cm^{-1}), stretching type of amine (1080.11 cm^{-1}), stretching type of aromatic group (1409.96 cm^{-1}), stretching type of nitro group (1543.05 cm^{-1}), stretching type of alcohol group (3323.35 cm^{-1}) and phenol group (3749.62 cm^{-1}). This study attempts to display the use of FTIR while simultaneously creating interest among related researchers in natural plant analysis.

1. Introduction

Natural plant resources are widely distributed around the world for several thousands of years as traditional medicine [1]. The important value of these medicinal plant lies in its great importance for the individual and community's health care. Also known as herb, these materials have been a crucial catalyst for the development of disease resistant drug to counter against countless type of negative metabolic activities in human body [2]. Based on a report released by the world health organization (WHO), up to 80% of the human population in the third world countries rely on medicine fabricated from plant for their health needs [3]. The active bio-constituents that contribute to the beneficial properties of these plants are many i.e. flavonoids, tannins, steroids and polyphenols [4]. Furthermore, some of this group such as flavonoids have been extensively studied and confirmed to possess the capability to act as antioxidant, antibacterial, antiviral and anti-carcinogen agent on multiple condition [5]. Due to this reason, it is estimated that approximately 25% of commercialized medicine incorporated plant extract as the ingredient.

IOP Conf. Series: Materials Science and Engineering **440** (2018) 012035 doi:10.1088/1757-899X/440/1/012035

Generally, there are two type of metabolite present in plant which are primary and secondary metabolites. Both are classified based upon their function in plant metabolism [6]. Primary metabolites directly affect the physiological function in plant and also commonly called as central metabolites. Conversely, secondary metabolite does not directly entail in the growth of plant but it still plays a significant role in combating diseases [7]. Therefore, a comprehensive analysis on the constituent is of utmost importance since through the understanding of compound present will the researchers be able to fully utilize the potential of each plant.

An example of such plant is kalmegh (*Andrographis paniculata*) where it is categorized as an herbaceous plant originated from the family of Acanthaceae. It is also commonly known as the 'king of bitter' due to the extremely bitter taste. Furthermore, the plant can be easily found growing in southeastern Asia like Indonesia, India, Sri Lanka and Pakistan. In this region, kalmegh can been found in moist shady place such as at the hills, village, forest and also along the roadside. It is an annual herbs plant that has feature characteristics which is branched, growing 1-2 meters straight upward, have sharply quadrangular stems and also have dark green color leaves. The seeds are small in size with an ovoid shape and yellow, green and brown in color [8]. Kalmegh is one of a wonderful herbal plant that have many benefit for medicinal purposes since the plant possesses the properties of analgesic, anti-inflammatory, antidiabetic, antifertility, antibacterial, antipyretic and antihepatotoxic [9,10]. Kalmegh leaves produce a variety of compounds such as flavonoid and andrographolide that show lower side effect when compares to other commercialize medicine for the same purpose since it contains phytochemical constituent possessing the capability to do so [11].

Up till now, several techniques have been introduced for analytical process of plant extract. They are UV spectrometry [12], gas chromatography [13], thin layer chromatography (TLC) [14], and nuclear magnetic resonance (NMR) and high performance liquid chromatography (HPLC) [15]. Although all of the method mention are accurate and highly valued for their precise performance, but they consume too much time and require several reagents to operate. Thus, this study tries to implement a simple and less time consuming technique to accomplish the same task.

Fourier Transform Infrared Spectroscopy is one of the popular techniques used for physiochemical analysis to determine unknown composition of compound by identifying their functional groups [16]. The technique makes used of the intensity of absorption spectra and frequency by measuring the effectiveness of photon energy transferred and vibrational frequency of molecules when exposed to infrared light source [17]. The advantage of this technique is that it is cheaper since it does not require reagents to work, less time consuming, highly accurate in its determination and eco-friendly. At present, the application of FTIR have been successfully applied in various field peculiarly in medicine for the identification of Pharmacopoeia [18].

2. Materials and methods

Material: Distilled water was taken from Physical Laboratory University Malaysia Terengganu, Methanol (Sigma Aldrich), Kalmegh leaves powder was plucked from kalmegh tree that buy from botanist at Selangor.

2.1. Preparation of Plant Extract

The leaves material for this study were collected from plantation in Selangor, Malaysia. The kalmegh leaves were plucked and collected early in the morning to ensure the freshness of the leaves. Then, it was washed properly for 3-4 times under running tap water to remove the residue before dried under room temperature and stored in air tight polyethylene bags. Later, the leaves were crushed into finely powder form by using grinder. For the extraction, 20g of the leaves powder was weighed before soaked in 400 ml of distilled water (aqueous extract) and 80% of aqueous methanol with the ratio of powder and solvent is 1:20. The extracts were soaked and shook by using shaker for 24 hours which is the optimal time for the extract and solvent to reach the equilibrium state. After 24 hours, the extract was filtered using vacuum pump using Whatman No.1 filter paper. The filtrate extract then was evaporated using rotary evaporator at 40℃ water bath and cooling compressor at 10℃ in order to get the dried crude. The dried crude then stored into the freeze dryer to ensure all the water is removed and stored properly before use it for FTIR analysis.

2.2. Identification of Functional Group
Fourier transform infrared (FTIR) spectrometry was used to analyze and characterize the functional group present in the leaves [19]. A small amount of crude extract was placed at the germanium piece of infrared spectrophotometer with the absorbance collected over the wave number ranged from 500 cm^{-1} to 4000 cm^{-1}. The spectra were compared with previous journal as reference. The data collected from the FTIR analysis were plotted by using OriginPro 2016 32 bit software. The peak present on the graph represent the functional group in both methanol and aqueous kalmegh leaves extract base on the wavelength value. This method are taken from [20] with a slight modification.

3. Results and Discussion
The infrared spectroscopic analysis gives the information about the possible functional groups of active principles. The functional group analysis of the dried powder, methanol and aqueous extract of kalmegh leaf was performed by using solid FT-IR spectroscopic analysis using KBr disc method. Figure 1 shows the graph of FTIR spectrum of kalmegh leaves extract. Results obtained for both solid and liquid form was compared. Graph plotted in Figure 1 have been used for comparison of different extracts using methanolic and aqueous. The analysis of the functional group was made by compared the result of absorption peak produced with the previous research. All the functional group represented by the peak absorption was recorded in Table 1.

Table 1. Wavenumber and functional group of kalmegh leaves for aqueous, methanolic and dried powder extract

Wavenumber (cm^{-1})	Functional Group
894.97	Alkene =C-H bending
1080.11	Amine C-N stretching
1409.96	Aromatics C=C stretching
1543.05	Nitro N-O, stretching
1645.10	Alkenes C=C stretching
3323.35	Alcohol O-H stretching
3749.62	Phenol O-H stretching

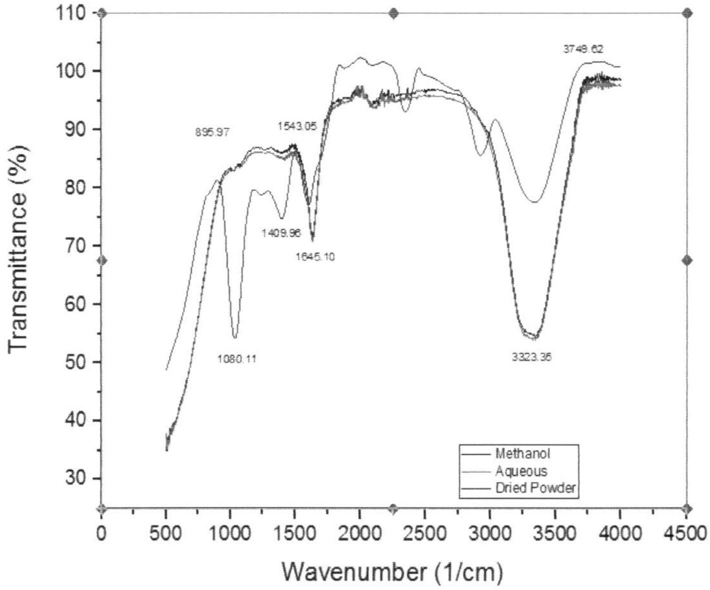

Figure 1. FTIR spectrum from different solvent of kalmegh leaves extract

Based on Figure 1 and Table 1, major compound that have been found contained in the kalmegh leaves extract are some specific functional group which are alkene, amine, aromatic, nitro, alcohol and phenol. At wavenumber 3323 cm^{-1} which is the broadest peak produced represent a strong absorption of alcohol group with stretching vibration. The appearance peak at 1645.10 cm^{-1} represent the alkenes stretching group while strong bond at 1543.05 cm^{-1} wavenumber was reveals a nitro asymmetric stretch. According to peak value 1409.96 cm^{-1}, this represent C=C stretching which is called as an aromatics group. The spectrum obtained from wavenumber 1080.11 cm^{-1} represent a stretching vibration of amines group while at frequency 894.97 cm^{-1} shows another alkene group but with bending vibration. At peak 3749.62 cm^{-1} stretching mode display the presence of phenol compound [12]. From Figure 1, the wavelength intensity of the methanol and aqueous extract is almost similar with slight difference, but the dried powder frequency have a huge difference of frequency.

4. Conclusion

The preliminary result of FTIR give an idea to extend this research for their further research. In this present study, preliminary screening confirmed the presence of six functional group present in kalmegh leaves extract. Hence the kalmegh leaves extract need to be further manipulated to investigate bioactive compound content in kalmegh leaves that maybe it can be used for various field applications that give many benefit in future. Purification and chromatography techniques ae required to isolate bioactive compound in future work

5. References

[1] Abu-Rabia A 2005 *J. Ethnobiol. Ethnomed.* **1** 1
[2] Srujana M, Hariprasad P, Hindu M J, Sravani P and Raju V 2012 *Int. J. Pharm. Sci. Rev. Res.* **14** 21
[3] Igoli J O, Ogaji O G, Tor-Ayiin T A and Igoli N P 2006 *Afr. J. Tradit. Complement. Altern.Med.* **2** 134
[4] Cook N C and Samman S 1996 *J. Nutr. Biochem.* **7** 66
[5] Wulandari L, Retnaningtyas Y and Lukman H 2016 *J. Anal. Methods Chem.* **2016** 1

[6] Visveshwari M, Subbaiyan B and Thangapandian V 2017 *Int. J. Pharmacogn. Phytochem. Res.* **9** 914
[7] Banso A and Adeyemo S O 2007 *Nig. J. Biotech.* **18** 27
[8] Niranjan A, Tewari S K, and Lehri A 2010 *Ind. J. of Nat. Prod. Res.* **1** 125 [9] Rafat A, Philip K & Muni S 2010 *J. of Med. Plants Res.* **4** 197.
[10] Mishra S K, Sangwan N S, and Sangwan R S 2007 *Pharmacog. Rev.* **1** 283
[11] Sithara N V, Komathi S, and Rajalakshmi G 2017 *J. of Med. Plants* **5** 192
[12] Jain P K, Soni A, Jain P and Bhawsar J 2016 *J. Chem. Pharm. Res.* **8** 1
[13] Kanthal L K, Dey A, Satyavathi K and Bhojaraju P 2014 *Pharmacogn. Res.* **6** 58 [14] Gujjeti R P and Mamidala E 2013 *Int. J. Innov. Res. Sci. Eng. Technol.* **2** 5275
[15] Sasidharan S, Chen Y, Saravanan D, Sundram K M and Latha L Y 2011 *Afr. J. Tradit. Complement. Alter. Med.* **8** 1
[16] Bobby M N, Wesely E and Johnson M 2012 *Int. J. Pharm. Pharm. Sci.* **4** 293
[17] Surewicz W K, Mantsch H H and Chapman D 1993 *A Crit. Assess. Biochem.* **32** 389
[18] Liu H X, Sun S Q, Lv G H and Chan K K 2006 *Spectrochim. Acta A Mol. Biomol. Spectrosc.* **64** 321
[19] Ashok Kumar R, Ramaswamy 2014 M *Int. J. Curr. Microbiol. App. Sci.* **3** 395
[20] Sule A, Ahmed Q U, Samah O A, Omar M N, Hassan N M, Kamal L Z M, Yarmo M A 2011 *Am J Appl. Sci.* **8** 525

Acknowledgement

The project was supported by the Ministry of Higher Education Malaysia through Niche Research Grant Scheme (NRGS/2015/53131/14). An acknowledgement for Fundamental Physics Laboratory and Central Laboratory staff who are willing to guide and help us during the preparation of sample.

Calcination effect to the physical and optical properties of Zn_2SiO_4 composite prepared by impregnation of ZnO on SiO_2 amorphous nanoparticles

E A G Engku Ali , K A Matori , E Saion , S Hj Abdul Aziz , M H M Zaid and I M Alibe

School of Fundamental Sciences, Universiti Malaysia Terengganu Kuala Nerus, 20130 ,Terengganu Malaysia.

Corresponding author email: engku_ghapur@umt.edu.my

Abstract. In this study, Zn_2SiO_4 composite-based ceramic was synthesised using amorphous SiO_2 nanoparticles as a silicon source. The amorphous SiO_2 nanoparticles which obtained from a simple precipitation process were mixed with aqueous zinc nitrate. Amorphous SiO_2 nanoparticles were encapsulated by the zinc source in aqueous solution, dried, and subjected to calcination. The underwent calcination showed the changing of phases, morphology, and size with increased temperatures. During calcination, ZnO phase appeared at the beginning of heating temperature and Zn_2SiO_4 phase started to emerge at 800 °C onwards, as shown by XRD patterns. The optical band gap analysis of Zn_2SiO_4 composite was determined to be within the range of 3.12 to 3.19 eV. The diffusion of zinc ions into SiO_2 nanoparticles with high surface area also reduced the phase formation temperature for Zn_2SiO_4, compared to a conventional solid state method. This optical characteristic is expected to be a potential candidate for phosphor materials in opto-electronic devices application.

1. Introduction

The synthesis of zinc silicate (Zn_2SiO_4) composite has gained more interest due to the combination of the optical properties of both zinc oxide (ZnO) and silica (SiO_2) components. This composite system merges the low band gap ZnO with the high band gap SiO_2. The formation of ZnO phase during the synthesis depends on the preparation method, sintering temperature, and the ratio of ZnO and SiO_2. As for silicate-based phosphor, the selection of silica precursors can certainly influence the final properties of the targeted materials.

Conventionally, ZnO and SiO_2 powders are used with appropriate stoichiometry and heated at a high temperature to form Zn_2SiO_4. Several precursor materials of SiO_2 have been used whether from natural source or laboratory grade SiO_2. Soda lime silica produced from waste materials [1, 2], silicon alkoxide, or tetraethyl orthosilicate (TEOS) [3], and commercialized mesoporous SiO_2 [4] are among common SiO_2 sources. The conventional melt quenching of soda lime silica, mixed with ZnO, produces a glass precursor, which is then subjected to heat treatment to form Zn_2SiO_4-based glass-ceramic. The drawback of using soda lime silica is the presence of other elements in the glass, which can affect the properties of the final materials [5]. TEOS is a liquid SiO_2 precursor that needs to be added into a mixture of water and alcohol solvent for the Si ions to react with zinc ions. A suitable ratio of water to alcohol solvent is required for the hydrolysis of TEOS to occur and to form Zn_2SiO_4. Consequently, the hydrophobic properties [6] and toxicity of TEOS [7] become drawbacks to those that have specific requirements during production. With a simple route of reaction, porous silica powder can be synthesised from the reaction of water glass with weak acids, such as alcohol (ethanol), as demonstrated by [8] and [9]. The resulting porous silica particles can offer an advantage when used as a precursor during the formation of the Zn_2SiO_4 phase. Porous silica made of water glass has different porosity characteristics due to the reaction with a weak acid.

Several methods for synthesising Zn_2SiO_4 composite have been reported by previous researchers. Sohn, Ramakrishna and Diao [10-12] performed solid state reaction that involving high temperature heat treatment with certain soaking time in order to form whether single phase or composite of Zn_2SiO_4. Sol- gel method that used zinc salts, such as nitrate and acetate, and alkoxy derivatives of silicon, such as tetraethyl orthosilicate are often reported in preparation of Zn_2SiO_4 composite [13]. There are also variations of synthesis arrangements for the sol-gel method. Some researchers may produce solid SiO_2 first, then mix it with zinc precursor, or vice versa, or combine a mixture of ZnO and SiO_2 sources together. Silica in the form of tetraethoxysilane (TEOS) is usually mixed with the aqueous solution of zinc, particularly zinc nitrate ($Zn(NO_3)_2$) and zinc acetate ($Zn(CH_3COO)_2$). The prepared nanoparticles would appear amorphous prior to calcination. The simultaneous mixing of precursors will cause the ZnO nanoparticles to be homogeneously dispersed in the silica matrix and retained at a certain temperature. Calcination of the amorphous phase of ZnO-SiO_2 composite start to form Zn_2SiO_4 composite phase at 800 °C that as in the form of thin films [14] and at 900 °C for powder form [15,16]. Hydrothermal and vapour method are both often been used for Zn_2SiO_4 composite preparation. In hydrothermal method, water, or an aqueous mixture is used as the reaction medium. The experiments is regularly conducted in a Teflon-lined autoclave to which raw materials or precursors are loaded and heated at a temperature that ranges between 100 to 370 °C. The high pressure in the Teflon-lined container can reduce the formation temperature of Zn_2SiO_4 phase [17]. A gas phase is used as the reaction medium in a vapour method. The raw materials, or precursors are sprayed, or vaporized into a gas at temperatures that range between 300 to 1300 °C to produce solid materials. The vapour method contributes to low crystalline, or amorphous-like α-phase Zn_2SiO_4 and calcination should be done to turn it into a crystalline phase [18].

The application of impregnation techniques using silica begin with the preparation of metal-supported catalysts. Using this method, the filling of the surface pores can initiate the adsorption of metal ions. The impregnation or absorption of metals on silica as a substrate have been reported by Clause, Chaneac and Bonneviot [19-21]. They emphasized that the interactions between metal ions in a solution, or mobile medium with the solid surface of support material can strongly influence the size distribution and stability of the oxide particles during calcination process. The interaction of metal ions in a solution at the solid-liquid interface lead to deposition precipitation with addition of electrostatic adsorption of cations and counteranion without modification of support surface.

This study was focused on synthesising Zn_2SiO_4 composite using a wet impregnation method. This method applied amorphous silica nanoparticles as a morphology-controlling template to perform chemical reactions or for new phase formation when combined with other ions; in this case, zinc ions.

2. Materials and method

Zn2SiO4 composite powders were produced by mixing amorphous silica nanoparticles that were produced based on the method established elsewhere [22], with zinc nitrate hexahydrate. First, 1.20 g of amorphous silica nanoparticles was dispersed in 200 mL of deionized (D.I.) water using an ultrasonic agitator for 30 minutes. Next, the amorphous silica nanoparticle dispersion was mixed with a solution of 11.89 g of zinc nitrate hexahydrate (Sigma Aldrich), which was already dissolved in 200 mL of deionized water. This mixture was stirred for 2 hours using a magnetic stirrer. The ratio of the precursor concentration of Zn:Si is 2:1. Then, the mixed solution was poured into a petri dish and dried in an oven for 24 hours at 120 °C. The dried powder was ground using a mortar grinder and finally, heat treated at 600, 700, 800, 900, and 1000 °C at 10 °C/min heating rate for 3 hours.

To study the structural characteristic of synthesized Zn2SiO4 composite powders, the sample powder were characterized by X-ray diffraction (XRD) (PANalytical X'pert PRO PW 3040 MPD), thermal analysis (Mettler Toledo TGA/DSC 1HT), field emission scanning electron microscopy (FESEM) (FEI NOVA NanoSEM 230) and transmission electron microscopy (TEM) (Hitachi H-7100). In addition, the optical property was characterized using UV-Visible (Shimadzu UV-3600) spectrometer for optical absorption and band gap determination.

3. Results and discussion

3.1. XRD analysis

The effect of calcination on the structure of ZnO-Zn_2SiO_4 composite was investigated using XRD spectrum of each sample at temperatures ranging from 600 to 1000 °C. Figures 1 shows the XRD spectra of the ZnO-Zn_2SiO_4 composite for Zn:Si ratios of 2:1. The wurtzite structure of the ZnO nanocrystals was observed at 600 °C. Nine major diffraction peaks were positioned at 2θ values of 31.79°, 34.44°, 36.27°, 47.56°, 56.61°, 62.88°, 66.39°, 67.96°, and 69.10°, registered as JCPDS No. 89-0510. The lattice constant was obtained using the standard formula for tetragonal crystal system $1/d^2 = ((h^2+k^2)/a^2)+(l^2/c^2)$, where h,k,l are Miller index, d is inter planar spacing obtained from Bragg's law, and a,b,c are the lattice constant. The lattice constant calculated from the XRD spectrum gave the value of a = 3.249 Å and c =5.206 Å, which were similar to ZnO hexagonal wurtzite structure [23]. With the progression of sintering temperature from 600 to 700 °C, the diffraction peaks had broadened. Usually, the temperature gain will increase the crystallite size due to the agglomeration. The changes in FWHM can be influenced by the changes of crystallite size and strain due to the temperature increase. This was shown by the reduction of FWHM peak intensity. From the data, the progressive sintering temperature, from 600 to 700 °C, had expanded the FWHM of the diffraction peaks. Therefore, the diffraction peak broadening of ZnO nanocrystals was not affected by the increasing particle size. Strain could also promote the broadening of XRD peaks when the temperature was increased. As a result, the diffraction peaks of ZnO were shifted towards the higher angle side, with increases in temperature from 600 to 700 °C. Thus, the broadening and shifting of the diffraction spectrum peaks of ZnO nanocrystals could be linked to the strain induced in the ZnO nanocrystals by the surrounding silica network [24].

The formation of Zn_2SiO_4 phase was initiated at 700 °C due to the presence of peak shoulder and peak broadening, but appeared clearly at 800 °C. The major diffraction peaks were positioned at 21.81°, 22.19°, 25.67°, 31.64°, 34.13°, 38.94°, 42.26°, 43.73°, 45.16°, 46.30°, 47.05°, 54.43°, 56.67°, 57.72°, 59.64°, 61.00°, 62.94°, 66.42°, and 68.03°, registered as α-Zn_2SiO_4 crystal phase of JCPDF No. 37- 1485. The formation of Zn_2SiO_4 peak from 700 °C onwards was influenced by the ZnO phase that became a dominant diffusing phase during calcination [25]. This formation temperature is lower than conventional solid state reaction temperature that required the heating temperature between 900 to 1500 °C for several hours for the mixture of precursor to melt, followed by heat treatment in order to get the crystalline phase [17].

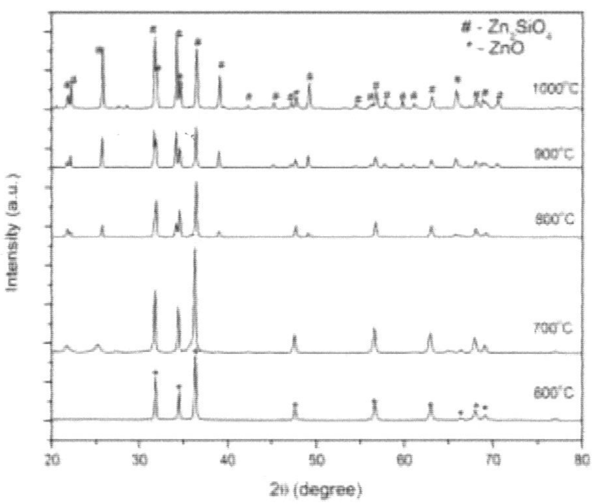

Figure 1. XRD pattern of calcined ZnO-Zn_2SiO_4 composite between 600 to 1000 °C.

3.2. Thermal analysis

Figure 2 show the TG and DTG curves for Zn_2SiO_4 composite. Weight loss became an approximate constant value beginning at 300 °C. Additionally, similar four sharp peaks were observed in the DTG curves for all samples with different ratios of zinc. The first and second peak at approximately 78 to 88 °C and 123 to 136 °C can be attributed to dehydration water and dehydroxylation of zinc hydroxide, which would normally evaporate at this temperature. The third and fourth peak, which appeared in the range of 219 to 220 °C and 278 to 291 °C, were caused by the decomposition of nitrate to nitrogen oxide, and the release of oxygen gas, as identified by Biswick [26]. The decomposition of these substances was accompanied by the abrupt weight loss at approximately 300 °C for all samples.

The ZnO phase was already formed at this temperature after the decomposition of the organic substances, up to 600 °C, as shown by the XRD patterns. The particle weight remained constant from 600 °C up to 1200 °C. The subsequent minor weight losses before constant weight began were observed between 300 °C to 600 °C. These weight losses could be related to the decomposition of oxygen due to the initial formation of Zn_2SiO_4 phase.

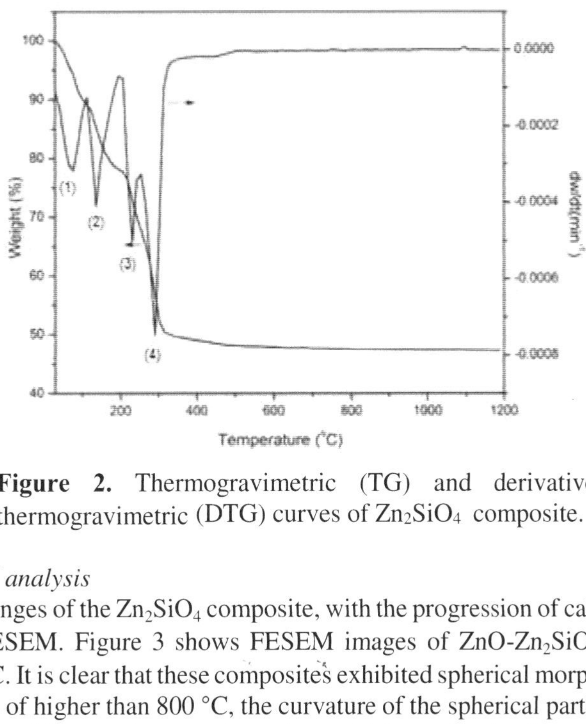

Figure 2. Thermogravimetric (TG) and derivative thermogravimetric (DTG) curves of Zn_2SiO_4 composite.

3.3. FESEM and TEM analysis

The morphological changes of the Zn_2SiO_4 composite, with the progression of calcination temperature, were studied using FESEM. Figure 3 shows FESEM images of ZnO-Zn_2SiO_4 composite calcined between 600 to 1000 °C. It is clear that these composites exhibited spherical morphology at 600 and 700 °C, but at temperatures of higher than 800 °C, the curvature of the spherical particles that are touching each other will be diffused first, and start to form the neck-like or dumbbell structures [27]. The small spherical particles have a vast surface area compared to a similar volume of bulk solid. The surface properties of bulk and nanosize particles are different because a nanosize particle has a large surface to volume ratio and small radius of curvature. The FESEM images were analysed using image processing and analysis software, ImageJ in order determine the particles size distribution. The overall morphology of Zn_2SiO_4 composite, upon annealing at 600 °C, showed the spherical form with an average diameter of 69.68 ± 3.20 nm. Some parts of the composite displayed individual particles, while other parts showed agglomeration. Burned off residual organic groups, such as nitrates, could lead to the encapsulation of ZnO on the surface of SiO_2 nanoparticles [28]. Further increased of temperature lead to diffusion and particle size cannot be precisely estimated from annealing at 700 °C (Figure 3 (b)). Eventually, the pores would also be closed by the diffusion of zinc ions at higher than 700 °C.

Figure 4 shows the TEM images of Zn_2SiO_4 composite at different calcination temperatures. These images show a nearly dense core that became darker than the layer near the surface of the particles. The zinc oxide encapsulated the silica core, which can be seen as a semi-transparent layer at the edge of the particle boundary, at sintering temperature of 600 °C. With further increase of sintering temperature, the cores of the particles became darker and the semi-transparent layer near the surface was reduced [29]. Concurrently, interparticle diffusion was observed, with the near-to-polyhedral shape morphology forming at 8000 °C (Figure 3 (c)). Subsequently, the continuous diffusion between particles led to the formation of larger particles at 1000 °C (Figure 4 (b)).

Figure 3. FESEM images of $ZnO-Zn_2SiO_4$ composite sintered at temperature of (a) 600 °C, (b) 700°C and (c) 800 °C.

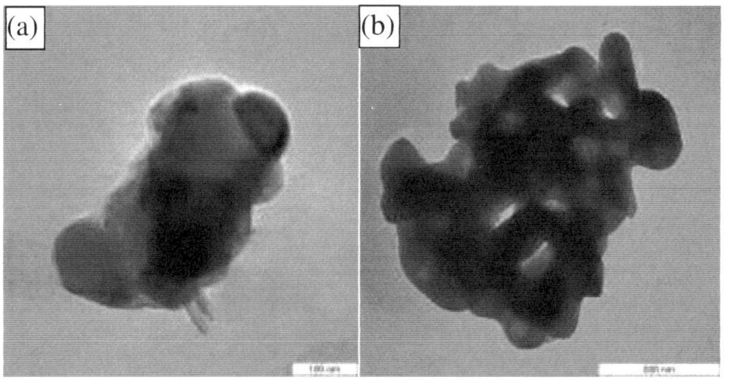

Figure 4. TEM images of $ZnO-Zn_2SiO_4$ composite sintered at temperature of (a) 600 °C and (b) 1000 °C.

3.4. UV-vis analysis

The UV-visible spectroscopy measurement of the Zn_2SiO_4 composite samples heat treated at various temperatures were recorded in the spectral region of 200 to 800 nm. Figure 5 show the absorption spectra of the $ZnO-Zn_2SiO_4$ composite sintered at various temperatures. The maximum absorption edge of the heat treated samples was at approximately 380 nm. This sharp absorption edge corresponds to the characteristics of the crystalline phase of ZnO. For sample calcined at 600 °C, lower absorption intensity was observed. When the calcination temperature was increased higher than 800 °C, a decrease was observed in the absorption peak intensity, along with the shift of the peak towards lower wavelength. Another absorption edge was observed between 260 and 280 nm for the sample sintered at higher than 800 °C. Uegaito [31] linked the emerging shoulder peak at 255 nm as absorption edge from Zn_2SiO_4 phase after annealing at higher than 800 °C. Omri and Sharma [32,33] also made the same reference for this particular absorption edge. However, in this study, the minimum observed wavelength in the spectrum was at 220 nm, which made the existence of the small absorption edge possible. Based on the previously discussed XRD, FESEM, and TEM results, the decrease in the UV- vis absorption peak intensity can be attributed to the reducing of ZnO phase and the formation of Zn_2SiO_4 crystal [34].

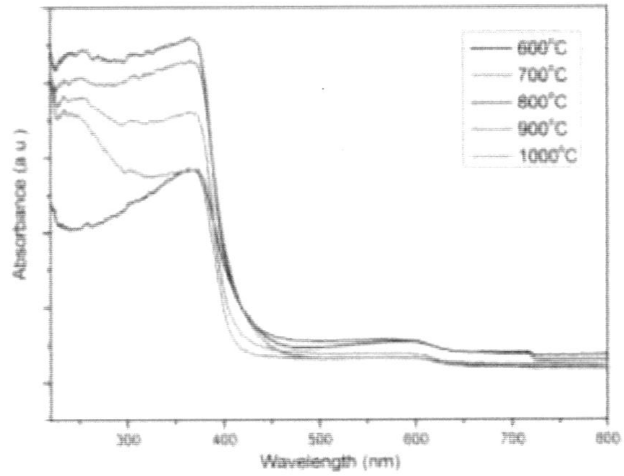

Figure 5. UV-Vis absorption spectra for Zn_2SiO_4 composite sample, with Zn:Si ratio of 2:1 using SiO90m, sintered at 600 to 1000 °C.

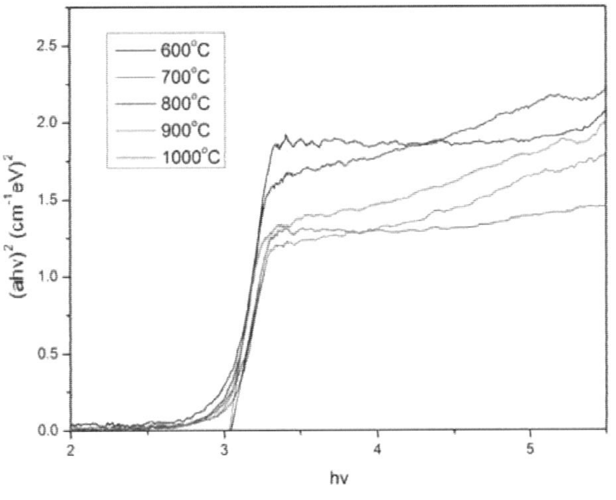

Figure 6. Tauc plot of $(\alpha h\upsilon)^2$ vs. energy for Zn_2SiO_4 composite, with Zn:Si ratio of 2:1, sintered at 600 to 1000 °C using SiO90m.

The optical band gap was determined from the relationship between the absorption coefficient and photon energy, which is associated with an electronic transition. The relationship is given by Equation $(\alpha h\nu)^{1/n} = A(h\nu - Eg)$, where A is a constant, the photon energy is denoted by $h\nu$, and Eg is the optical energy band. The n = ½ which is direct allowed transition is used to determined optical band gap value from the best linear fit by extrapolating the linear parts of $(\alpha h\nu)^{1/n}$ vs. $h\nu$ curves [35,36]. The extrapolation of linear region from individual plot lines to x-axis intercept will give optical band gap value as shows in Figure 6. The optical band gaps of Zn2SiO4 composites had ranged between 3.12 and 3.19 eV, which was lower compared to the bulk band gap for ZnO crystal phase at 3.37 eV. The lower optical band gap value obtained in this study was due to the optical confinement effect of the formation of ZnO layers on the silica nanoparticles [37,38]. The increasing heat treatment temperature would definitely affect the ZnO layer, which would start to diffuse within the SiO2 nanoparticles. In this study, higher calcination temperature had increased the particle size and the porous particles became denser, as shown in FESEM and TEM micrographs. This condition would lead to higher absorbance of photons [39].

4. Conclusions

In this study, ZnO-Zn_2SiO_4 composite were successfully prepared by the wet chemical impregnation method followed by calcination. Amorphous SiO_2 nanoparticles were chosen as the precursors and mixed with aqueous zinc nitrate. Results from the XRD analyses showed the formation of ZnO-Zn_2SiO_4 composite when the temperature was increased to higher than 800 °C, which was lower than conventional solid state sintering temperature. The impregnation of zinc ions with amorphous SiO_2 nanoparticles had significantly reduced the formation temperature of Zn_2SiO_4 phase. Based on the morphological analyses from FESEM and TEM, as the temperature increases, the spherical particles exhibited interconnections between the particles. The UV-Vis measurements showed the values of the calculated optical band gap of ZnO to be in the range of 3.12 to 3.19 eV. Conclusively, the synthesised Zn_2SiO_4 composite had shown a promising properties for phosphor materials. Its optical properties appeared suitable to be further developed for applications in photonic and opto-electronic materials.

5. References

[1] Samsudin N F, Matori K A, Abdul Wahab Z, Josephine Y C L, Fen Y W, Hj Ab Aziz S and Zahid M H M 2016 *Optik* **127** 8076
[2] Zaid M H M, Matori K A, Abdul Aziz S H, Kamari H M, Wahab Z A, Fen Y W and Alibe I M 2016 *J. Mater. Sci.: Mater. Electro.* **27** 11158
[3] Lu Q, Wang P and Li J 2011 *Mater. Res. Bull.* **46** 791
[4] Li X and Chen F 2013 *Mater. Res. Bull.* **48** 2304
[5] Zaid M H M, Matori K A, Abdul Aziz S H, Zakaria A and Ghazali M S M 2012 *Inter. J. Mol.Sci.* **13** 7550
[6] Zeng J H, Fu H L, Lou T J, Yu Y, Sun Y H and Li D Y 2009 *Mater. Res. Bull.* **44** 1106
[7] Nakashima H 1994 *Arch. Toxic.* **69** 59
[8] Buining P, Liz-Marzan L M and Philipse A P 1996 *J. Coll. Inter. Sci.* **179** 318
[9] Godoi R H M, Fernandes L, Jafelicci Jr M, Marques R C, Varanda L C and Davolos M R 1999 *J.Non-Cryst. Sol.* **247** 141
[10] Sohn K, Cho B, Dong H, Gyu Y and Hon K 2000 *J. Euro. Ceram. Soc.* **20** 1043
[11] Ramakrishna P V, Murthy D B R K, Sastry D L and Samatha K 2014 *Spec. Acta A: Mol. Bio.Spec.* **129** 274
[12] Diao C C and Yang C F 2010 *Ceram. Inter.* **36** 1653
[13] Petrovykh K A, Rempel A A, Kortov V S and Buntov E A 2015 *Inorganic Mater.* **51** 152
[14] Kandula K R, Sarkar A and Bhaktha B N S 2013 *Optic. Mater. Exp.* **3** 2078
[15] Cannas C, Casu M, Lai A, Musinu A and Piccaluga G 1999 *J. Mater. Chem.* **9** 1765
[16] He H, Wang Y and Zou Y 2003 *J. Phys. D: App. Phy.* **36** 2972
[17] Takesue M, Hayashi H and Smith R L 2009 *Pro. Cry. Growth Char. Mater.* **55** 98
[18] Wang W C, Tian Y T, Li K, Lu E Y, Gong D S and Li X J 2013 *App. Surf. Sci.* **273** 372
[19] Clause O, Kermarec M, Bonnevio L, Villain F and Che M 1992 *J. Am. Chem. Soc.* **114** 4709
[20] Chaneac C, Tronc E and Jolivet J P 1996 *J. Mater. Chem.* **6** 1905
[21] Bonneviot L, Clause O, Che M, Manceau A and Dexpert H 1989 *Catalysis Today* **6** 39
[22] Engku Ali E A G, Matori K A, Saion E, Sidek H A A, Zaid M H M, Alibe I M 2017 *Dig. J.Nanomater. Bios.* **12** 441
[23] Omri K, Ghoul J E, Alyamani A, Barthou C, Mir L E 2013 *Phys. E Low-Dimensional Syst.Nanostructures.* **53** 48
[24] Raevskaya A E, Panasiuk Y V, Stroyuk O L, Kuchmiy S Y, Dzhagan V M, Milekhin A G, Yeryukov N A, Sveshnikova L A, Rodyakina E E, Plyusnin V F and Zahn D R T 2014 *RSC Adv.* **4** 63393
[25] Yang Y, Kim D S, Knez M, Scholz R, Berger A, Pippel E, Hesse D, Gosele U and Zacharias M 2008 *J. Phys. Chem. C* **112** 4068
[26] Biswick T, Jones W, Pacuła A, Serwicka E and Podobinski J 2007 *J. Solid State Chem.* **180** 1171
[27] Carter B and Norton G 2007 *Ceram. Mater. Sci. Eng.*. New York: Springer Science+Business Media

[28] Xu J, O'Keefe E S and Perry C C 2004 *J. Mater. Chem.* **14** 1744
[29] El Mir L, Amlouk A, Barthou C and Alaya S 2007 *Phy. B: Cond. Matt.* **388** 412
[30] Gun'Ko V M, Bogatyrov V M, Oranska O I, Borysenko L I, Skubiszewska-Zieba J, Ksiazek A and Leboda R 2013 *App. Surf. Sci.* **276** 802
[31] Uegaito K, Hosokawa S and Inoue M 2012 *J. Luminescence* **132** 64
[32] Omri K, Ghoul J El, Alyamani A, Barthou C and Mir L El 2013 *Phys. E: Low-Dim. Syst.Nanostructures* **53** 48
[33] Sharma P and Bhatti H S 2009 *J. Alloys Comp.* **472** 483
[34] Ali A M, Ismail A A, Najmy R and Al-Hajry A 2014 *J. Photochem. Photobio. A: Chem.* **275** 37
[35] Zaid M H M, Matori K A, Aziz S H A, Kamari H M M, Wahab Z A, Effendy N and Alibe I M 2016 *J. Non-Cryst. Solids* **449** 107
[36] Tan G L, Lemon M F, Jones D J, French R H 2005 *Phys. Rev. B – Cond. Matt. Mater. Phys.***72** 1
[37] Viezbicke B D, Patel S, Davis B E and Birnie D P 2015 *Phys. Stat. Sol. (B)* **11** 1700
[38] Foo K L, Hashim U, Muhammad K and Voon C H 2014 *Nanoscale Res. Lett.* **9** 429
[39] Kumar S and Sahare P D 2012 *Optics Comm.* **285** 5210
[40] Carter B and Norton G 2007 *Ceram. Mater. Sci. Eng.* New York: Springer Science+Business Media
[41] Xu J, O'Keefe E S and Perry C C 2004 *J. Mater. Chem.* **14** 1744
[42] El Mir L, Amlouk A, Barthou C and Alaya S 2007 *Phy. B: Cond. Matt.* **388** 412
[43] Gun'Ko V M, Bogatyrov V M, Oranska O I, Borysenko L I, Skubiszewska-Zieba J, Ksiazek A and Leboda R 2013 *App. Surf. Sci.* **276** 802
[44] Uegaito K, Hosokawa S and Inoue M 2012 *J. Luminescence* **132** 64
[45] Omri K, Ghoul J El, Alyamani A, Barthou C and Mir L El 2013 *Phys. E: Low-Dim. Syst.Nanostructures* **53** 48
[46] Sharma P and Bhatti H S 2009 *J. Alloys Comp.* **472** 483
[47] Ali A M, Ismail A A, Najmy R and Al-Hajry A 2014 *J. Photochem. Photobio. A: Chem.* **275** 37
[48] Zaid M H M, Matori K A, Aziz S H A, Kamari H M M, Wahab Z A, Effendy N and Alibe I M 2016 *J. Non-Cryst. Solids* **449** 107
[49] Tan G L, Lemon M F, Jones D J, French R H 2005 *Phys. Rev. B – Cond. Matt. Mater. Phys.***72** 1
[50] Viezbicke B D, Patel S, Davis B E and Birnie D P 2015 *Phys. Stat. Sol. (B)* **11** 1700
[51] Foo K L, Hashim U, Muhammad K and Voon C H 2014 *Nanoscale Res. Lett.* **9** 429
[52] Kumar S and Sahare P D 2012 *Optics Comm.* **285** 5210

Acknowledgments

The authors gratefully acknowledged the financial support for this study from the Ministry of Education Malaysia, Universiti Malaysia Terengganu and Universiti Putra Malaysia.

Effect of Sintering Temperature on Zinc Oxide Varistor Ceramics

S Muhamad Syaizwadi[1], S Siti Noradilah[2], M G Mohd Sabri[1,3], W A Wan Rafizah[2], K Syara[1] and O J Lee[1]

[1] *Advanced Nano Materials (ANoMa) Research Group, Nano Research Team, School of Fundamental Science, Universiti Malaysia Terengganu, 21030 Kuala Nerus, Terengganu, MALAYSIA*

[2] *School of Ocean Engineering, Universiti Malaysia Terengganu, 21030 Kuala Nerus, Terengganu, MALAYSIA*

[3] *Institute of Biodiversity and Sustainable Development, Universiti Malaysia Terengganu, 21030 Kuala Nerus, Terengganu, MALAYSIA*

Email: mohdsabri@umt.edu.my

Abstract: Zinc oxide (ZnO) varistors is a solid state electronic ceramic component whose function to protect electronic devices against over voltage surges due to their highly nonlinear electrical properties and ability in controlling the energy. In this study, ZnO varistor is prepared via solid state reaction method. The ZnO based varistor ceramics undergoes sintering temperature at 900 °C, 1100 °C and 1300 °C for 45 minutes. The used of sintering temperature from 900 to 1300 °C, has increased the grain size of ZnO and improved their crystallinity as shown from the peaks intensity of X-ray diffraction (XRD). Current-Voltage properties exhibit the value of nonlinear coefficient in the range of 1.44 to 1.76. Different sintering temperature is used to adjust the microstructure and improve the varistor ceramic characteristics where it shows a potential difference in various application of electronic devices. Increasing in sintering temperature would lead to improved grain growth for better electrical properties.

Keywords: Varistor, ZnO ceramics, sintering

1 Introduction

ZnO varistors are electronic component with highly nonlinear properties and commonly used as overvoltage protector in electronic circuits. ZnO varistor are used to protect a variety of electronic devices against overvoltage surges due to nonlinear current-voltage (I-V) properties [1]. ZnO ceramics as a based varistor usually study due to its capability and good performance [2]. The improvement of ZnO varistor capabilities is accomplish by testing their effect alone first and second was added with some metal oxide additives to improve their electrical properties. ZnO varistors are sintered with various metal oxide additives in order to enhance their nonlinear electrical properties [3]. However, complex formulation where more than five doping elements are needed to improve nonlinear properties of varistor ceramics. The microstructure and electrical characteristics of ZnO is not only influenced by the type of additives, but also by the different stages of sintering. The microstructure are related to electrical characteristics, where the varistor voltages can be significantly controlled by means of the sintering temperature [4]. The additives and the processing conditions are the most important factor of success in the development of the surge protection device [5]. The use of different sintering temperature will show the potential difference in the development of microstructure and electrical properties in doped zinc oxide. Sintering temperature gives rise to a particular microstructure of ZnO [6]. Electrical characteristics of ZnO varistor ceramics are studied intensively due to its capability and it is important to investigate the electronic states of ZnO ceramic in different sintering temperatures. Sintering has a prominent effect on the microstructure of ZnO varistors [7]. The charge transfer basically controls the electrical conduction within ZnO varistors during sintering [8].

ZnO varistor, due to their high electrical nonlinearity and energy absorption abilities have been used in electronic industries. It is important to obtain electronic samples with high energy handling capabilities for efficiency protection in electronic industries from overvoltage that can damage the electrical systems [9]. Varistor components are important in protecting various electrical equipment from overvoltage and it can be operate constantly without damage [10]. In varistor component, the most important parameter is the nonlinear coefficient (α) where it defines how fast the change is from the high resistive to the low resistive condition, by absorb the dangerous surge to flow through the ground [11]. Significant differences in parameter used in the processing of varistor ceramics will give a better insight to the limitation of the material, therefore in this study the effect of different thermal treatment on the performance of ZnO varistor is investigate. This varistor ceramis are fabricated via solid state reaction method and analyze by XRD and I-V source measure unit.

2 Materials and Methods

ZnO micro powders with 99.0% purity (Merck) was used as the base material. Mass of the powders that used for ceramic preparation was weighed according to their weight ratio using digital analytical balance. The powder was mixed with the 1.75 wt.% polyvinyl alcohol (PVA) powder until it is fully wet and is drying by using oven. PVA powder acts as a binder to give strength to the pressed compact pellets and avoid micro-cracks. After drying, the powder was grounded using agate mortar and then sieved using a 75 micron sieve. The powder for each sample were measure again by using digital analytical balance. The powders were pressed to form a pellet with 13.0 mm diameter and 1.3 mm thickness using Hydraulic Press machine at a pressure 2.6 ton/5min. Finally, the pellet were exposed to temperature at 900 °C, 1100 °C and 1300 °C for 45 minutes in a box furnace with the heating and cooling rate at 2.66 °C/min. The crystalline phase of ZnO powders before and after sintering were measured using X-ray diffraction (XRD). A small amount sample was carried out at 2θ scan over the diffraction angles from 5° to 80° at the speed of 2.00°/min. The sample pellets at different sintering temperature and duration are selected and silver conductive paint is applied on both sides of samples. Electrical characteristic of sample pellets were measured using Current-Voltage source measure unit. The nonlinearity of the samples were determined from the Equation (1),

$$\alpha = \frac{\log_{10} J_2 - \log_{10} J_1}{log_{10} E_2 - log_{10} E_1} \tag{1}$$

where α defines the degree of nonlinear property of varistor materials. E_1 and E_2 are the electric fields corresponding to the current densities J_1 and J_2.

3 Results and Discussion

3.1 XRD

The XRD patterns of ZnO samples before and after sintering are shown in Figure 1. From the XRD analysis of the samples, the pattern reveals can be identified as the hexagonal wurtzite structure of ZnO ceramic based on ICSD code: 067454 [12]. The plotted graph of figure 1 shows the change in crystallinity or grain size in ZnO ceramics after undergoes different sintering temperatures. Microstructure properties are important factors that can be modified by the means of sintering parameter which directly affects the microstructure of ZnO varistor ceramics [13]. The peaks intensity is sharp and narrow, it shows that the sample are in good crystallinity and fine grain size. The effects of different sintering temperature on ZnO microstructure was compared based on the change in their crystallinity or grain size. The XRD technique reveals the primary phase of ZnO ceramic materials and no other phases are detected confirming that the pure ZnO are crystalline in nature.

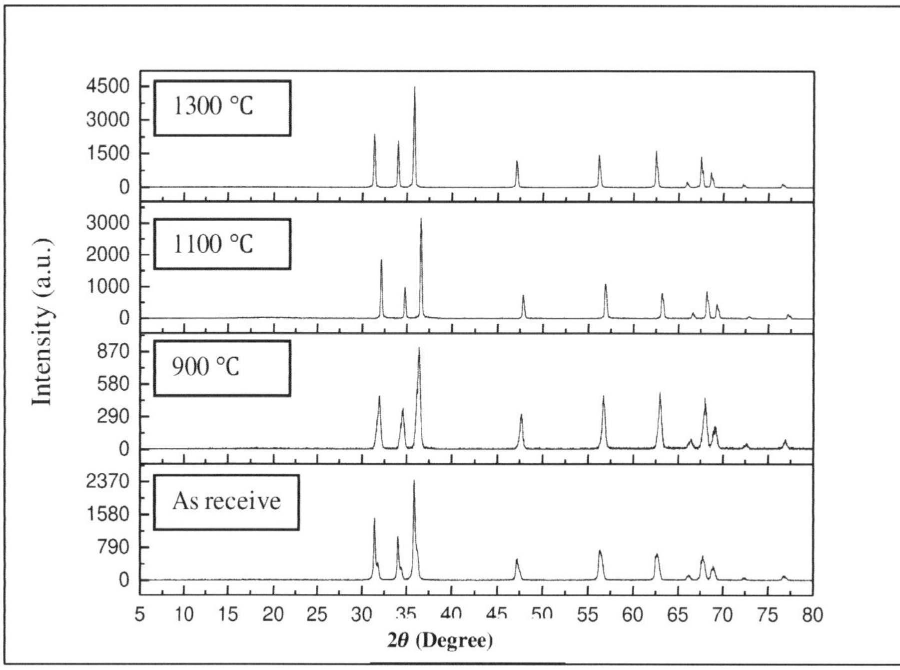

Figure 1: X-ray diffraction (XRD) pattern of pure ZnO.

Based on the XRD spectra it was shown that as the sintering temperature increased the diffraction peaks became sharper and stronger, indicating the crystallinity improvement of the samples. As the sintering temperature increases, the average grain size abruptly increased, indicating that the microstructure became more compact and less grain boundaries [14]. The average grain size of ZnO increases due to the grain growth and pore elimination. The rapid grain growth resulted from capillary driving forces at high temperatures the grain boundaries of ZnO can drastically change the physical properties [15].

3.2 I-V

The electrical characteristics of varistor are result from the fabrication process, particularly sintering temperature. Figure 2 shows the electrical characteristics of ZnO based varistor ceramics sintered at different temperatures. The influence of sintering temperature on the nonlinear electrical properties of ZnO varistor ceramics were investigate at 900 °C, 1100 °C and 1300 °C. The change from the electrical behaviour with increased sintering temperature was assigned to the improved microstructure.

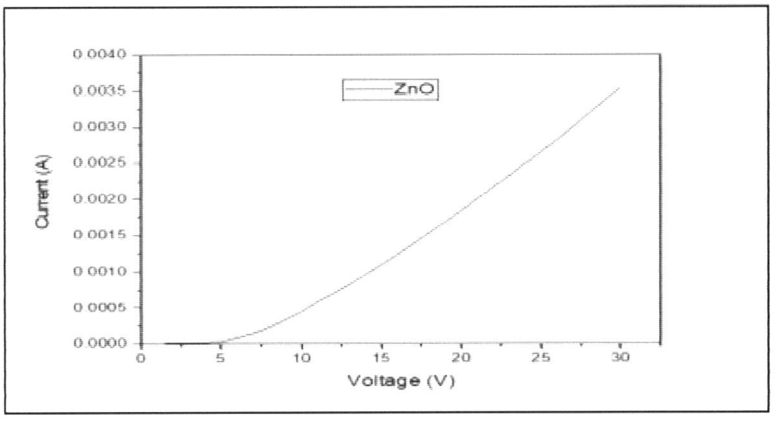

(a) 900 °C

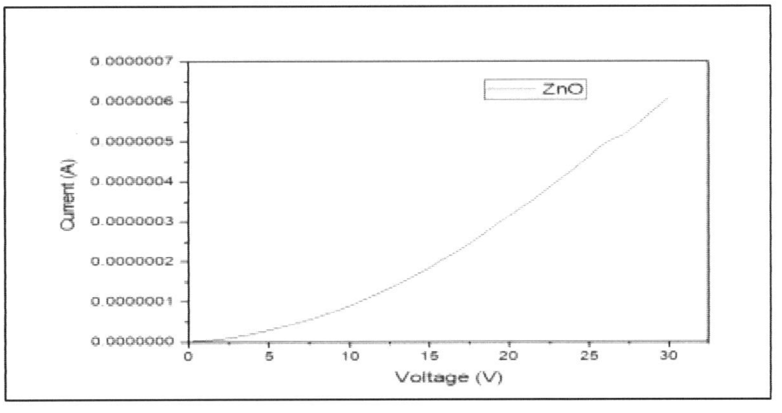

(b) 1100 °C

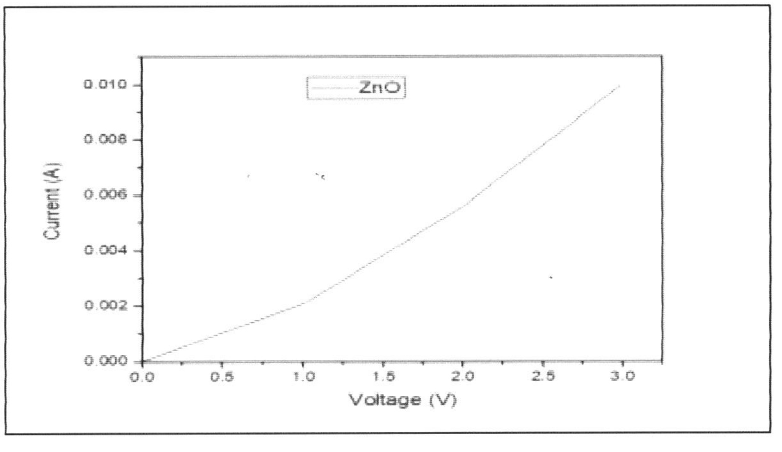

(c) 1300 °C

Figure 2: Current-Voltage (I-V) characteristics of sintered samples at 900 °C, 1100 °C and 1300 °C.

The good characteristics found suitable for varistor behaviour was presented by the sample sintered at 1100 °C. The variation of α as a function of grain size is exhibited in Table 1.

Table 1: Comparison of nonlinear coefficient, α

Compound	Sintering Temperature	Nonlinear coefficient (α)
ZnO	900 °C	1.71
	1100 °C	1.76
	1300 °C	1.44

It was observed that α value in Table 1 increases when the temperature increased and then decreases with further increase in sintering temperature and induced the grain size of varistor ceramics. The increment of α value with the sintering temperature was due to the increase of grain size and improve their microstructure. The increase in nonlinear coefficient of ZnO varistor correspond to an increase in the grain size which suggests the existence of larger grain size in the grain boundaries at higher sintering temperature [16].

4 Conclusion

Preliminary studies on ZnO based varistor ceramics give a broad idea to researcher for their further research. In this present study, the XRD pattern shows the change in crystallinity through increasing the sintering temperature. The nonlinearity of ZnO varistor ceramics as a function of sintering temperature was monitored by using I-V source measurement. The α value increased proportionally with the increased in sintering temperature. However, a sudden drop was observed in α value after sintering above 1100 °C, which suggests the existence of larger grain size in the grain boundaries at higher sintering temperature. Finally, the sintering process gives rise to a microstructure and improves the electrical properties of varistor ceramics.

Acknowledgement

The project was supported by the Ministry of Higher Education Malaysia through Research Acculturation Grant Scheme (RAGS) (Vot. No. 57114). An acknowledgement to Fundamental Physics Laboratory and Central Laboratory for facilitating the equipment's for the research.

References

[1] Higashi Y and Koga E 2017 *J. Ceram. Soc. Jpn.* **125** 468

[2] Zulkifli M A, Ghazali M S M, Abdullah W R W, Zakaria A, and Ahmad Z 2016 *J. Teknol* **78** 327

[3] Rečnik A, Bernik S, and Daneu N 2012 *J. Mater. Sci.* **47** 1655

[4] Akinnifesi J O and Akinwunmi O O 2015 *J. Mater. Sci. Res.* **4** 40

[5] Kharchouche F, Belkhiat S, and Belkhiat D E C 2013 *IET Sci. Meas. Technol.* **7** 326

[6] Lakin I I, Zakaria A, Abdollah Y, and Umaru D 2015 *Dig. J. Nanomater. Biostruct* **10** 189

[7] Desouky O A and Rady K E 2016 *Chin. Phys. B* **25** 2

[8] Sendi R K 2018 *Solid State Sci.* **77** 54

[9] Khalaf A A 2012 *Energy Procedia* **19** 116

[10] Ghazali M S M, Abdullah W R W, Zakaria A, Kamari H M, and Rizwan Z 2016 *J. Phys.* **776** 012057

[11] Masteghin M G and Orlandi M O 2016 *Mater. Res.* **19** 1286

[12] Ghazali M S M, Zakaria A, Abdullah W R W, Hafiz M Z M, and Matori K A 2015 *Adv. Mater. Res.* **1107** 20

[13] Lee W S, Chen W T, Lee Y C, Yang T, Su C Y, and Hu C L 2007 *Ceram. Int.* **33** 1001

[14] Chen G H, Chen X, Kang X L, and Yuan C L 2015 *J. Mater Sci.: Mater in Electron.* **26** 2389

[15] Gao L, Li Q, Luan W, Kawaoka H, Sekino T, and Niihara K 2002 *J. Am. Ceram. Soc.* **85** 1016

[16] Norailiana A R, Azmi B Z, Ismayadi I, Hashim M, Ibrahim I R, Nazlan R, Mohd Idris F 2017 *J. Mater. Sci. Surf. Eng.* **5** 528

IOP Conf. Series: Materials Science and Engineering **440** (2018) 012038 doi:10.1088/1757-899X/440/1/012038

Ionic Conductivity and Structural Analysis of 2-hyroxyethyl Cellulose Doped with Glycolic Acid Solid Biopolymer Electrolytes for Solid Proton Battery

M A Ramlli, N A A Bashirah and M I N Isa*

School of Fundamental Science, Universiti Malaysia Terengganu, 21030 Kuala Nerus, Terengganu, Malaysia.

Corresponding author email: ikmar_isa@umt.edu.my

Abstract. Cellulose derivatives have high potential for use as solid biopolymer electrolytes in proton batteries because they are biodegradable, affordable, have good mechanical properties and their ionic conductivity can be enhanced with addition of ionic dopant. In this work, we developed a new type of solid biopolymer electrolytes (SBEs) based on 2-hydroxyethyl cellulose (2HEC) doped with glycolic acid (GA) as an ionic dopant. A solution casting technique was used to prepare the SBEs. The ionic conductivity and structural properties of the SBEs were analysed using Electrical Impedance Spectroscopy (EIS), Fourier Transform Infrared spectroscopy (FTIR) and an X-ray Diffractometer (XRD). The highest ionic conductivity achieved was 3.80 $\times 10^{-4}$ S cm^{-1} for the sample with 40 wt.% GA concentration at room temperature. FT-IR analysis showed that complexation occurred in the polymer system from the shifting of υC-O and υ_sCOO$^-$ band of 2HEC and GA. FTIR deconvolution revealed the increasing pattern of percentage free mobile ions with the addition of GA concentration until 40 wt.% GA , which can be related to the high ionic conductivity of the sample. In XRD analysis, all SBEs shows amorphous nature. Based on the results obtained, GA is a good ionic dopant because it succeed in improve the ionic conductivity of the 2HEC film.

1. Introduction

Solid biopolymer electrolytes (SBEs) is one of the most active areas of study in material research for several years. They caught the attentions of many researchers because of their excellent properties to be used in small electrical devices such as batteries. They are mechanically stable, lightweight, environment friendly and has no leakage problem when used as electrolyte. SBEs were prepared by doping the host polymer with various type of ionic dopant which, can hugely increase the electrical properties of the host polymer [1, 2]. Some application that are suitable for use of SBEs are in solid-state batteries, supercapacitors, power windows and many other solid-state electrochemical devices [3-5]. SBEs usually consist of a crystalline phase-that possess a regular structure with immobile ions, and an amorphous phase having no regular structure with mobile ions [6-10]. A few research groups have demonstrated this phenomenon and revealed the biphasic structure of SBEs [11, 12].

The arrangement of the polymer molecules greatly affects the physical properties, chemical behaviour and reactivity of the SBEs. In solid state, the movement of ions mostly occurred in the amorphous phase and moving from one site to another by hopping mechanism [13-15]. Proton conducting polymer is a special class of solid electrolytes with hydrogen ions acting as charged carriers, which was first suggested by Rogers and Ubbelohde, in 1950 [16] and many more researchers subsequently [17-19]. The hopping of ions lead to the increase in ionic conductivity of the material. This suggests that there is a close interplay of the structural nature and conductivity in ionic conducting polymer [20-22].

In this work, a new type of SBEs based on 2-hydroxyethyl cellulose (2HEC) doped with glycolic acid were prepared. 2HEC is one of the cellulose derivatives created by means of etherification. This polymer is widely used in cosmetic, latex paints, building materials, cleaning product and even used as an oilfield chemical as a lubricant, along with many other specialty applications [23-25].
Pharmaceuticals also used 2HEC combined with hydrophobic drugs in capsule formation to improve dissolution [26]. 2HEC is a water-soluble polymer that acts as thickener, binder, emulsifier, stabilizer, film former, and exhibits pseudo-plastic solution behaviour, tolerates salts and retains water. As a good sorbent, 2HEC has a good affinity of ions due to the presence of many polar hydroxyl functional groups that can serve as coordination sites for permanent or temporary ion attachment [27-32]. The ionic conductivity of a polymer can be enhanced by introducing a good ionic dopant into the polymer system, which promotes the dissociation of proton (H^+).

GA is a weak acid base that is a naturally occurring material and it is environmentally friendly. Although many reported that ammonium salt is a good ionic dopant, it is interesting to investigate the GA behaviour as an ionic dopant. This is because GA can undergo partial dissociation, which produces proton (H^+) as a dissociated charge together with glycolate anion ($RCOO^-$), and the charged monomer attached to the backbone of GA [33-35] that eventually can enhance the ionic conductivity of the polymer host. In addition, it is also environmentally friendly. Hence, GA was chosen as an ionic dopant to 2HEC for this work. The 2HEC, GA and 2HEC-GA SBEs were characterised using EIS, FT-IR and XRD to investigate its ionic conductivity and structural properties. To the best of our knowledge, there are no published works on the 2HEC-GA SBEs; therefore, it is a great opportunity to investigate the effect of GA on the properties of 2HEC biopolymer.

2. Materials and Methods

Materials. The materials used to prepare the SBEs are 2-hydroxyethyl cellulose (2-HEC) from Sigma Aldrich, Glycolic acid (GA) from Merck and distilled water as solvent. Firstly, dissolved 2HEC in distilled water, then doped with a GA from 10 wt.% to 50 wt.% at 10 wt.% increments. Stir the solution until homogenous and by using solution casting technique, the solution were dried in oven at 60°C. Completely dried samples were stored in desiccators to prevent moisture contamination. Table 1 lists the sample designation in this work.

Table 1. Sample designation

Sample	Carboxymethyl cellulose (CMC) (g)	Glycolic acid (GA) (wt.%)
GA0		0
GA10		10
GA20		20
GA30	2	30
GA40		40
GA50		50

Experimental techniques. The SBEs were characterised using Electrochemical Impedance Spectroscopy (EIS), Fourier transform infrared (FT-IR) spectroscopy and X-ray diffractometer (XRD). The ionic conductivity of the SBEs were measured using HIOKI 3532-50 LCR Hi-Tester interfaced with a computer in the frequency range from 50 Hz to 1 MHz. The measurement setup was as depicted in Figure 1. The ionic conductivity of the SBEs was calculated using Equation (1).

$$\sigma = \frac{t}{R_b A} \tag{1}$$

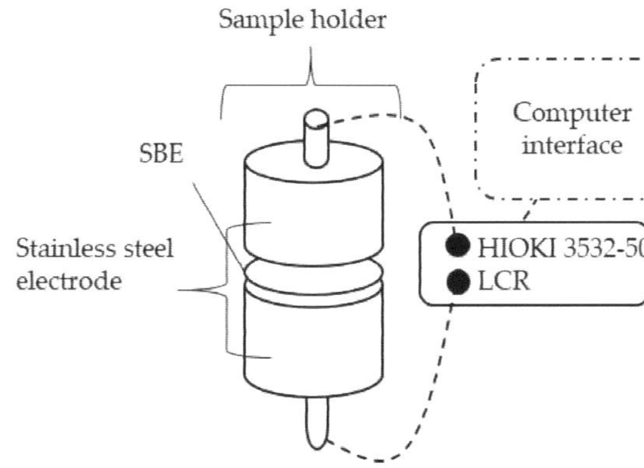

Figure 1. EIS measurement setup.

In the equation, σ is the ionic conductivity, t is the thickness of SBE (cm), R_b is a bulk resistance (Ω), and A is the contact area of the SBE and electrode (cm^2) [2, 11]. A Thermal Nicolet 380 FTIR spectrometer equipped with an Attenuated Total Reflection (ATR) accessory was used to analyse the structure of the SBEs. They were scanned at frequency that ranged from 4000 cm^{-1} to 700 cm^{-1} with a resolution of 4 cm^{-1} at ambient temperature. The percentage of free ions were determined from FT-IR deconvolution that were utilised using Origin 8 fitting software based on the Gaussian-Lorentz function [21]. The free ions (%) can be calculated by using Equation (2)

$$percentage\ of\ free\ ions\ (\%) = \frac{A_f}{A_f + A_c} \times 100\% \tag{2}$$

XRD analysis was done by using a Ringaku MiniFlex II Diffractometer. Before the procedure, 2HEC-GA SBEs were cut into suitable size, then adhered onto a glass slide before being placed in the sample holder of the diffractometer. The SBEs were directly scanned at Bragg's angles of 2θ, between 5° and 70° with CuKα source radiation (wavelength =1.5406 Å).

3. Results and Discussion

3.1 Ionic Conductivity Analysis

Figure 2a shows the Cole-Cole plots for selected samples GA0 and GA40 (inset). Cole-Cole plot for GA0 has semicircle and spike region where they can be related to the SBE being partially resistive and capacitive. Cole-Cole plot for GA40 shows no semicircle and it can be a sign of increased ionic mobility (Ahmad & Isa, 2015). Bulk resistance R_b, value can be determined from the interception of the semicircle and inclined spike with the Z_r-Axis. Figure 2a shows the examples on the determination of R_b value from Cole-Cole plot for sample GA0 and GA40 respectively.

The ionic conductivity of the SBEs was plotted in a graph of ionic conductivity against the GA wt. % and depicted in Figure 2b. From the graph, the ionic conductivity of 2HEC SBE increased with the addition of GA concentration. Pure 2HEC SBE has ionic conductivity of 3.43 x 10^{-6} S cm^{-1} and with the addition of 40 % GA, the ionic conductivity increased to 4.01 x 10^{-4} S cm^{-1}, which is almost two degrees more. Early observation shows that GA is a good ionic dopant to 2HEC. The increased ionic conductivity can be related to the increased number of free mobile ions in the polymer system, thus increasing the number of ionic conduction, which is a significant factor in enhancing the ionic conductivity of the SBEs [44, 46].

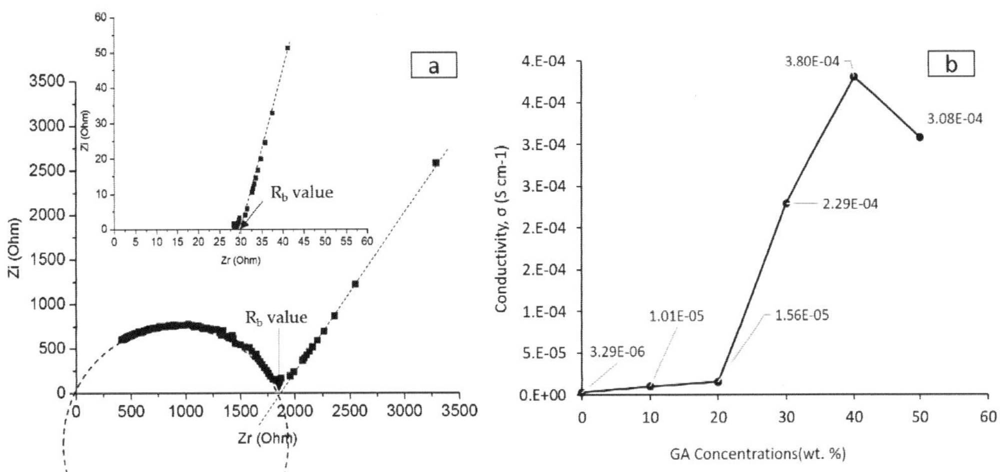

Figure 2. Cole-Cole plot; (a) GA0 and GA40 (inset), and (b) Ionic conductivity versus GA concentrations plot.

3.2 FTIR Analysis of Simple Organic Acid

The FTIR spectra of GA, 2HEC and 2HEC-GA in a range of 1200-1800 cm^{-1} were shown in Figure 3. From the figure, it can be seen that the addition of GA concentrations yielded two main absorption features in the 2HEC that are associated with carboxyl group. According to Hay *et al.* (2007) [36], protonated carboxylic acid (RCOOH) resulted in absorption band that correspond to the carbonyl stretch of υC=O (1690 cm^{-1} - 1750 cm^{-1}) and hydroxyl vibration, υC-OH (1200 cm^{-1} - 1300 cm^{-1}) which comprised of C-O and C-OH group. This vibration often resulted in a single and broad absorption band. According to Cabaniss and McVey, (1995) [37], these changes can be dependent on a few factors, two of which include; i) Electron density of carboxyl that was affected by the presence of electron donating or withdrawing functional group; and ii) Inter/intra-molecular hydrogen bonding involving a carboxylic oxygen or a proton. These two factors are typically affected more consistently and predictably in the $υ_{as}$ mode relative to the others, making it a useful indicator of structural complexation. However, in this work, the appearance of a $υ_{as}$ peak cannot be observed due to the overlapping of naturally absorbed water band of 2HEC (~1647 cm^{-1}) and the changes are most consistently observed at the $υ_s$ mode of carboxyl group [46].

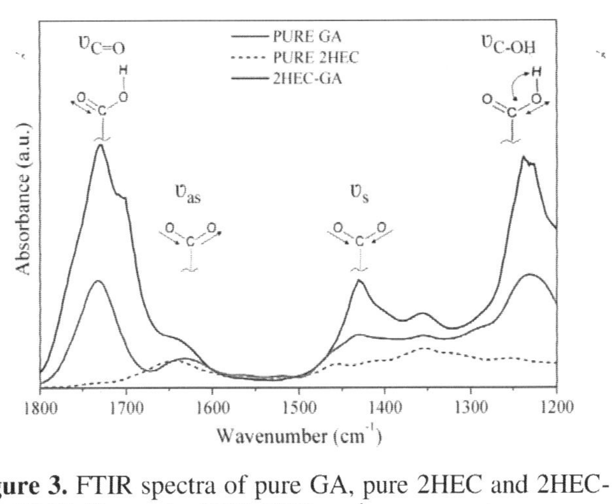

Figure 3. FTIR spectra of pure GA, pure 2HEC and 2HEC-GA (40%) film at the 1200-1800 cm^{-1} spectra region.

3.3. FTIR Analysis of 2HEC-GA SBEs

The complexation of GA and 2HEC was observed in FTIR analysis. Figure 4 represents the FTIR spectrum of pure 2HEC and 2HEC-GA system separated by four region (a) 1150-1350 cm^{-1}, (b) 15001800 cm^{-1}, (c) 900-1200 cm^{-1} and (d) 1400-1500 cm^{-1}, respectively. From the figure, the peak intensity for both υC-OH and υC=O of GA in the spectrum of 2HEC-GA system experienced a reduction in intensity, which due to the deprotonation state of GA (dissociation of H^+) in the SBE system. The absorption bands localized at ~1237 cm^{-1} and ~1730 cm^{-1} were due to the stretching vibrations of single and double CO bonds of the carboxylate group of GA, respectively. They are the characteristic features of the IR spectrum of GA and the increased intensity with higher GA concentration is expected. In Region c, the peak at ~1055 cm^{-1} is attributed to the υC-O of 2HEC. With the addition of GA, the band shifted to a higher wavenumber to ~ 1086 cm^{-1}. This confirms the complexation of both the host polymer and ionic dopant.

From Region (d), it can be observed that a new peak emerged at ~1440 cm^{-1} for sample GA10, which is believed due to the interactions between 2HEC and GA. The band corresponds to symmetric $υ_s$COO, which was affected by the increment of GA concentration. With the addition of GA, the band shifted to a lower wavenumber to ~1436 cm^{-1} and the intensity of the peak increased. These significant changes can be associated with the formation of anionic and cationic species through higher dissociation of ionic dopant, thus allowing protonation on 2HEC chain structures. The 2HEC and GA peaks observed in this work are comparable with other works where all the peaks were found at their own specific range [2434].

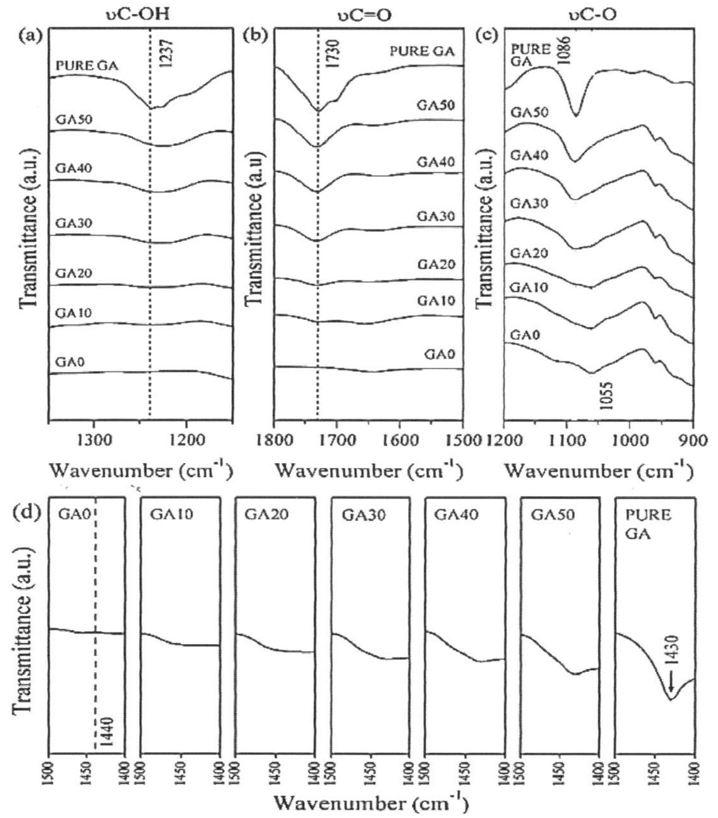

Figure 4. FTIR spectrum of 2HEC-GA system in the (a) 1150-1350 cm^{-1}, (b) 1500-1800 cm^{-1}, (c) 9001200 cm^{-1} and (d) 1400-1500 cm^{-1} region.

IOP Conf. Series: Materials Science and Engineering **440** (2018) 012038 doi:10.1088/1757-899X/440/1/012038

3.4 FTIR Deconvolution

FTIR deconvolution was performed to isolate the possible peaks presence, intensity change and shifting of the bands, hence further determine the contribution of H^+ ions as a charge carrier in 2HEC-GA SBE system [21, 22, 38]. The deconvolution was based on the Gaussion-Lorentz function and was done using Origin 8 fitting software. The sum of the intensity of all the deconvoluted peaks was ensured to fit the original spectrum where regression value of each peak is approximately unity (R^2 = 0.9998). The deconvolution peaks of $\upsilon_s COO^-$ were depicted in Figure 5.

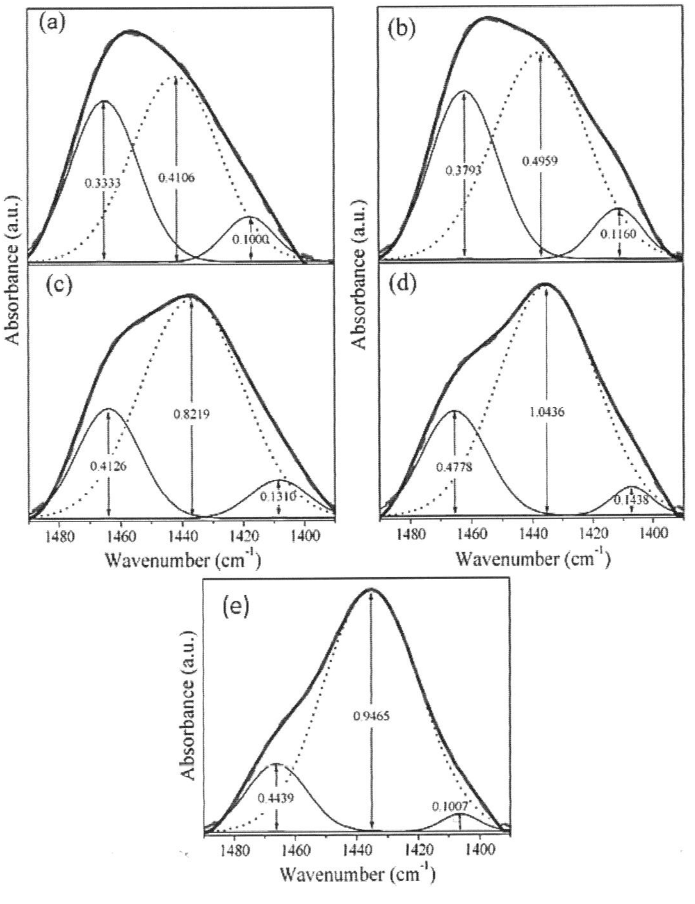

Figure 5. FTIR deconvolution of 2HEC: (a) 10%, (b) 20%, (c) 30%, (d) 40%, (e) 50% of GA in 13901490 cm^{-1} spectra region.

The band representing the free ions can be found in region between ~1390 cm^{-1} and ~1490 cm^{-1} where the symmetric vibration of carboxylate anion ($\upsilon_s COO^-$) is at, this is comparable to work done by Hay & Myneni (2007) [36]. From Figure 4, the peak located at ~1407 cm^{-1} can be assigned to contact the ions peak, meanwhile the peak at ~1465 cm^{-1} is the C-H characteristic band of 2HEC. It can be observed that peak intensity of glycolate anion increased with increasing GA concentration with maximum concentration at 40 wt. % and then slightly decreased at 50 wt. %. This increment could indicates the increase of number of mobile ions and this was shown in Figure 6 where the percentage of free ions alongside with ionic conductivity in a function of GA concentrations were plotted. Here, it is observed that the percentage of free ions increased until GA40 and then decreased at GA50, which in agreement with the hypothesis mentioned earlier.

With the addition of GA concentrations, more free ions dissociated into the polymer system, promotes ionic conduction and then enhanced the ionic conductivity of the SBE. Whereas at GA50, the percentage decreased and it can be attributed to the free ions start to associates back creating clusters of ions. This led to decreased in ionic conductivity of the SBE. From the observation, it can be simplified that the high ionic conductivity of GA40 was due to the high number of free ions. High contact ion percentage can be related to the ions that did not dissociate, hence explaining the low ionic conductivity of the SBE [45].

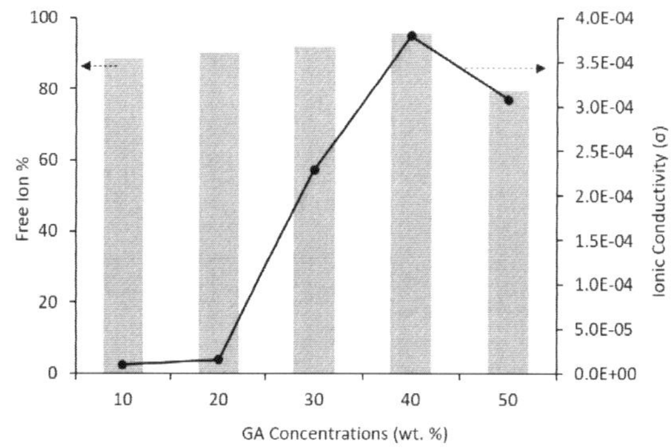

Figure 6. Percentage of free ions and ionic conductivity in a function of GA content.

3.5 XRD Analysis

2HEC is a cellulosic material that consists of crystalline and amorphous domains in various proportions [39, 43]. The reactivity and chemical behaviour are strongly influenced by its structural nature where most of the reactant penetrates only the amorphous regions. It has been reported that cellulose has the most intense reflection peak centred approximately at 2θ~22.00°. Figure 7 shows the X-ray pattern for all SBEs and pure GA. Here, it is observed that the pure GA has multiple crystalline peaks that reveal the crystalline nature of the material. Meanwhile, there are no distinctive crystalline peaks observed in the SBEs, which indicated that GA has completely solvate in the polymer system. In a work done by Attia and Elkader (2013), they reported that the peak attributed to crystalline domain of 2HEC was situated at 2θ = ~22.00°. In this work, pure 2HEC film shows intense reflection peak at ~21.28°, which represent the strongest cellulose peak of the crystalline domain in the polymer host [24]. The broadened cellulose peak with the incorporation of GA concentrations indicates the SBEs are in amorphous state.

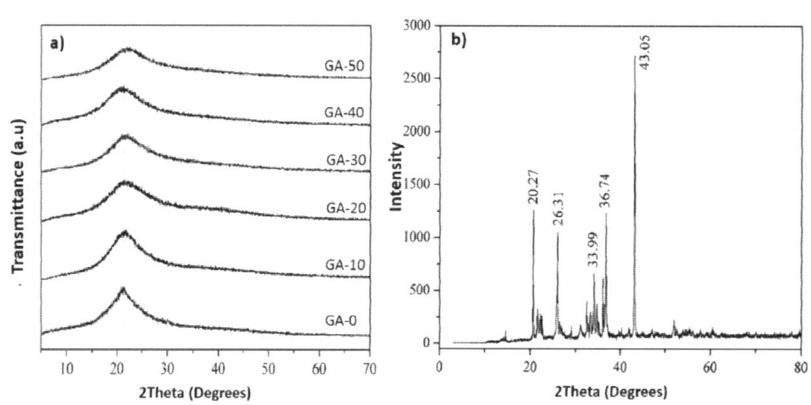

Figure 7. X-ray diffractograms (a) all SBEs, (b) pure GA.

4. Conclusions

Solid biopolymer electrolytes based on 2HEC doped with various GA concentrations was successfully made via solution casting technique. The highest ionic conductivity achieved was 3.80 x 10^{-4} Scm^{-1} for a sample with 40 wt.% GA concentration at room temperature. FTIR revealed the interactions between the host polymer and ionic dopant was from the shifted peaks of υC-O and υC-OH vibration mode. The highest conductive sample possess the highest number percentage of free ions hence confirms the relations that a high number of free ions helps in increasing the ionic conductivity. X-ray diffraction analysis shows all the SBEs are in an amorphous state with no crystalline peak detected. It can be concluded that GA is an effective ionic dopant because it successfully increased the ionic conductivity of the 2HEC.

5. References

[1] Aziz S B, Abidin Z H Z **2013** *Journal of Soft matter* 1-8
[2] Chai M N, Isa M I N **2013** *J. Cryst. Process Technol* **3** 1-4
[3] Padma Kumar P, Yashonath S, **2006** *J. Chem. Sci.* 118 135-154
[4] Rudhziah S, Rani M S A, Ahmad A, Mohamed N S **2014** *Polymers* **6** 2371-2385
[5] Samsudin A S, Isa M I N **2012** *Journal of Applied Sciences* **12** 174-179
[6] Hull S **2004** *Rep. Prog. Phys.* **67** 1233-1314
[7] Ciolacu D, Pitol-Filho L, Ciolacu F **2012** *Cellulose* **19** 55-68
[8] Ciolacu D, Popa, V I, Ritter H **2006** *J. Appl. Polym. Sci.* **100** 105-112
[9] Ciolacu D, Gorgieva S, Tampu D, and Kokol V **2011** *Cellulose* **18** 1527-1541
[10] Agrawal RC, Pandey G P 2008 *J. Phys. D: Appl. Phys.* **41** 1-18
[11] Kamarudin K H and Isa M I N 2013 *Int. J. Phys. Sci. 8* 1581-1587
[12] Samsudin A S. Lai H M and Isa M I N 2014 *Electrochim. Acta* **129** 1-13
[13] Kadir M F Z, Aspanut Z, Majid S R and Arof AK 2011 *Spectrochim Acta A Mol Biomol Spectrosc* **78** 1068-1074
[14] Buraidah M H and Arof A K 2011 *J. Non-Cryst. Solids* **357** 3261-3266
[15] Ramesh S and Wong K C 2006 *Ionics* **15** 249-254
[16] Rogers S E and Ubbelohde A R 1950 *Transactions of the Faraday Society* **46** 1051-1061
[17] Weiber E A, Takamuku S and Jannasch P 2013 *Macromolecules* **46** 3476–3485
[18] Rikukawa M and Sanui K 2000 *Prog. Polym. Sci.* **25** 1463-1502
[19] Simon P and Gogotsi Y 2008 *Nat. Mater.* **7** 845-854
[20] Golodnitskya D, Straussc E, Peleda E and Greenbaumd S 2015 *J. Electrochem. Soc.* **162** A2551A2566
[21] Rahman M H A, Khanaker M U, Khan Z R, Kufian M Z, Noor I S M and Arof A K 2014 *Phys. Chem. Chem. Phys.* **16** 11527-11537
[22] Arof A K, Amirudin S, Yusof S Z and Noor I M 2014 *Phys. Chem. Chem. Phys.* **16** 1856-1867
[23] Sudhakar Y N, Selvakumar M and Krishna Bhat D 2015 *Mater. Renew. Sustain. Energy* **4** 1-9
[24] Attia G and Abd Elkader M F H 2013 *Int J Electrochem Sci.* **8** 5672-5687
[25] Rosi M, Iskandar F, and Abdullah Khairurrijal M 2014 *Int J Electrochem Sci.* **9** 4251-4256
[26] Wang J and Somasundaran P 2006 *J. Colloid Interface Sci.* **293** 322–332
[27] El-Sheikh M A, El-Rafie S M, Abdel-Halim E S and El-Rafie M H 2013 *Journal of Polymers* 1-11
[28] Elidrissi A, El Barkany S, Amhamdi H and Maaroufi A 2010 *J. Mater. Environ. Sci.* **3** 197-204
[29] Kua S K, Chan C Y, Lim Y Y and Leong Y K 2000 *Colloid Polym. Sci.* **278** 485-489
[30] Seki Y, Altinisik A, Demircioğlu B and Tetik C 2014 *Cellulose* **21** 1689-1698
[31] Himmelein S, Lewe V, Stuart M C A and Ravoo B J 2014 *Chem. Sci.* **5** 1054-1058
[32] Lo J T, Yen H T, Tsai C C, Chen B and Hou S S 2014 *J. Phys. Chem. B* **118** 6922-6930

[33] Kister G, Cassanas G and Vert M 1997 *Spectrochim. Acta Mol. Biomol. Spectrosc.* **53** 1399-1403
[34] Sheena M Y, Suprabhan S, Varghese H M and Panicker C Y 2011 *Orient. j chem.* **27** 215-220
[35] Borukhov I, Andelman D, Borrega R, Cloitre M, Leibler L and Orland H 2000 *J. Phys. Chem. B* **104** 11027-11034
[36] Hay M B and Myneni S C B 2007 *Geochim. Cosmochim.* **71** 3518-3532
[37] Cabaniss S E and McVey I F 1995 *Spectrochim. Acta Mol. Biomol. Spectrosc.* **51** 2385–2395
[38] Cabaniss S E, Leenheer J A and McVey I F 1998 *Spectrochim. Acta Mol. Biomol. Spectrosc.* **54** 449-458
[39] Kauppinen J K, Moffatt D J, Mantsch H H and Cameron D G 1982 *Appl. Spectrosc.* **35** 271-276
[40] Rambo M K D and Ferreira M M C 2015 *J. Braz. Chem. Soc.* **26**
[41] S, Hirankumar G, Sakunthala A, Arunkumara D and Nithyaa H 2010 *Spectrochim. Acta Mol. Biomol. Spectrosc.* **75** 474–478
[42] Sohaimy M I H and Isa M I N 2015 *AMM* **719-720** 67-72
[43] Ciolacu D, Ciolacu F and Popa V I 2011 *Cellulose Chemistry and Technology* **45** 13-21
[44] Ramlli M A and Isa M I N 2015 *Polymers from Renewable Resources* **6** 55-63
[45] Ramlli M A and Isa M I N 2016 *J. Phys. Chem. B* **120** 11567−11573
[46] Hafiza M N and Isa M I N 2017 *Carbohydr Polym.* **165** 123-131
[47] Ahmad N H and Isa M I N 2015 *International Journal of Plastics Technology* **18** (2) 1-11

Acknowledgments: The authors would like to thank Malaysian Ministry of Higher Education for the funding through FRGS and ERGS Grant (Vot 59271 & Vot 55101). A big appreciation also to School of Fundamental Science, Universiti Malaysia Terengganu for the facilities provided.

Conduction mechanism via correlated barrier hopping in EC-plasticized 2-hydroxyehyl cellulose-ammonium nitrate solid polymer electrolyte

M N Hafiza and M I N Isa*

Advanced Nano Materials (ANOMA) Research Group, Advanced Materials Team, Ionic State Analysis (ISA) Laboratory, School of Fundamental Science, Universiti Malaysia Terengganu, 21030 Kuala Nerus, Terengganu Darul Iman, Malaysia.

*corresponding author

Abstract. This present work discussed the ionic conductivity and ionic conduction mechanism of solid polymer electrolyte (SPE). The SPE was prepared from 2-hydroxyethyl cellulose incorporated with ammonium nitrate and plasticized with ethylene carbonate (2-HEC-AN-EC) via well-known solution casting method. All SPEs were characterized using electrical impedance spectroscopy (EIS). It was found that the ionic conductivity at elevated temperature obeys Arrhenius rule. The plot of frequency exponent s against temperature shows that the correlated barrier hopping (CBH) was the best model to explain the ion conduction through 2-HEC-AN-EC SPE.

Keywords: 2-hydroxyethyl cellulose, ethylene carbonate, solid polymer electrolyte, correlated barrier hopping model.

1. Introduction

Solid polymer electrolytes (SPE) are of growing important in the field of materials science. An ideal SPE must have high ionic conductivity so that it can be utilized as a conducting medium in solid-state proton battery. A good approach for ionic conductivity improvement is by understanding the ionic conduction mechanism in SPE. The variation of ionic conductivity can be divided into two parts, the frequency dependent and frequency independent component.

As proposed by Jonscher in 1977, variation of ionic conductivity in frequency dependent region can be analyzed based on Universal Power Law (UPL) or so-called Universal Dielectric Responses [1]. There are four well-known theoretical hopping models have been proposed; i) quantum mechanical tunneling (QMT) [1-4], ii) small polaron hopping (SPH) [5,6], iii) correlated barrier hopping (CBH) [7-10] and overlapping large polaron tunneling (OLPT) model [10].

Previous paper has reported the ionic conductivity of 2-HEC–AN–EC SPE at room temperature, where the highest ionic conductivity of (1.17±0.01) × 10-3 Scm-1 was obtained after plasticized with 16 wt.% EC [11]. Hence, the aim of this study is to determine the best conduction model for that system. Electrical impedance spectroscopy (EIS) was used to study the ionic conductivity on the effect of frequency. The result obtained from this study was then used to correlate the conduction mechanism and ionic conductivity enhancement of the SPE.

2. Experimental method

2.1. Preparation of solid polymer electrolyte

2-HEC–AN solution was prepared follows the work done by Hafiza and Isa, 2017b. Various amount of ethylene carbonate (EC) were then added into the solution and continuously stirred until complete dissolution. The homogenous solution was poured into glass petri dishes and left in oven at 50 °C for drying process. The dried SPEs were then kept in desiccator to remove any residual moisture. The designation and composition of 2-HEC–AN–EC SPE was shown in **Table 1**.

Table 1. Designation and composition of 2-HEC–AN–EC SPE.

Designation of SPE	**Composition**				
	2-HEC (g)	AN (wt.%)	AN (g)	EC (wt.%)	EC (g)
P1	2.0	12	0.273	0	0.000
P2				4	0.095
P3				8	0.198
P4				12	0.310
P5				16	0.433
P6				20	0.568
P7				24	0.718
P8				28	0.884

2.2. Characterization of solid polymer electrolyte

2-HEC–AN–EC SPE was characterized using electrical impedance spectroscopy (EIS) model HIOKI 3532-50 LCR Hi-Tester. Prior to characterize, the SPE film was cut into 2 cm diameter and placed between two stainless steel blocking electrodes of the sample holder. The SPEs were tested in various frequency range (50 Hz to 1 MHz) at elevated temperature (303 K to 373 K).

3. Results and discussion

Figure 1 shows the plot of ionic conductivity at elevated temperature (303 K-373 K). As expected, the ionic conductivity measurement increases linearly with increasing temperature, confirming the Arrhenius type SPE where the regression value obtained is close to unity (refers **Table 2**). The increased of ionic conductivity at higher temperature may be attributed by the increase in free volume and polymer segmental motion [12,13]. When the SPE is exposed to higher temperature, the polymer chain segment receives sufficient energy to vibrate and create a bigger free volume around the polymer [14]. This, in turn increases migration of mobile proton through the polymer matrix. In Arrhenius type film, the nature of proton (H^+) transport at elevated temperature is almost similar to that occurring in ionic crystal, where ion is able to hop into neighbouring vacant site, hence increased the ionic conductivity [15]. The Arrhenius equation is expressed as Equation 1 [1, 16].

$$\sigma = \sigma_o \, exp\, (^{-E_a}/_{kT}) \quad (1)$$

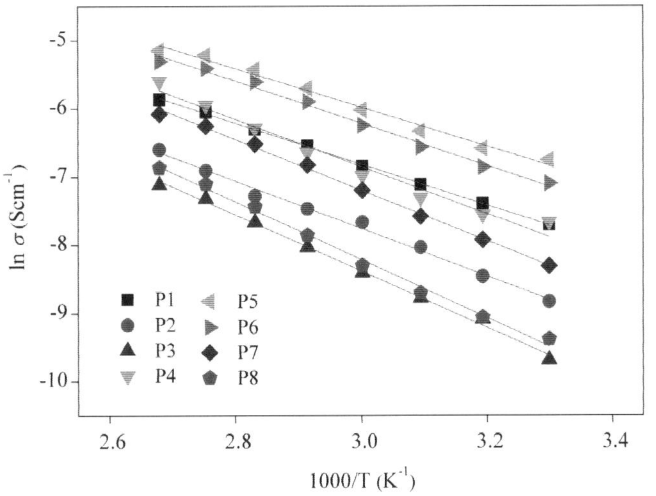

Figure 1. Ionic conductivity of 2-HEC–AN–EC SPE at elevated temperature.

Table 2 lists the regression value and activation energy for all SPEs. The activation energy, E_a can be determined from the Arrhenius plot of ln σ against 1000/T. According to Sohaimy and Isa (2016), activation energy relates to the energy requires by the ion to release from it bond and migrate to another coordinating site through conduction process [12]. Based on this result, the SPE that exhibits the highest ionic conductivity value has the lowest E_a value. It is believed that, at lower E_a value, the H^+ ion need less energy to detach from it bond and coordinate onto the end chain of 2-HEC. The releasing and coordinating process of ion will enhance the conduction activity inside the SPE, which later increased the ionic conductivity in plasticized system.

Table 2. The regression value, R^2 and activation energy, E_a of 2-HEC–AN–EC SPE.

Sample	R^2	E_a (eV)
System 2		
P1	0.99	0.26
P2	0.99	0.35
P3	0.99	0.36
P4	0.97	0.30
P5	0.98	0.25
P6	0.99	0.27
P7	0.99	0.32
P8	0.99	0.37

Figure 2 depicts the plot of ln ε_i against ln ω for the highest conductivity SPE at elevated temperature. The study of conduction mechanism is important in SPE to describe the ion hopping between the coordinating sites. The ionic conduction mechanism was analyzed based on the Jonscher's Universal power law relationship at the higher frequency region where the effect of space charge polarization is assumed less or can be negligible [7, 9]. In this study, the acceptable frequency range is between $13 \leq \ln \omega \leq 16$. The power law relation is described as Equation 2 [1], as follow:

$$\sigma(\omega) = \sigma_{dc} + \sigma_{ac} \tag{2}$$

where,

$$\sigma_{ac} = A\omega^s = \varepsilon_o \varepsilon_i \omega \tag{3}$$

Consolidating and applying logarithm rule to Equation (3) gives,

$$ln\, \varepsilon_i = ln \frac{A}{\varepsilon_o} + (s-1)\ln\omega \tag{4}$$

From Equation 2 to 4, $\sigma(\omega)$ is the total of dc conductivity, σ_{dc} and ac conductivity, σ_{ac}, A is a parameter dependent on temperature, ω is equal to $2\pi f$ and s is the power law exponent in the range of $0 < s < 1$. The value of s can be calculated from the slope, m of ln ε_i against ln ω, where m = s-1.

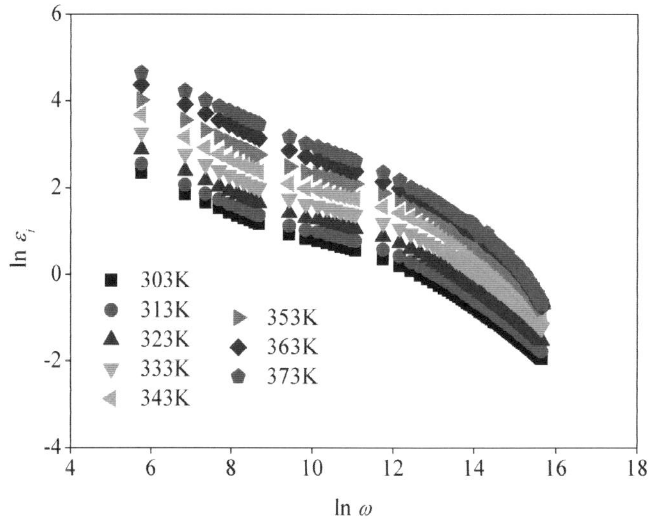

Figure 2. Plot of ln ε_i against ln ω for 2-HEC–AN–EC SPE with the highest ionic conductivity.

Figure 3 shows the exponent s against temperature. The value of exponent s was obtained from the slope of ln ε_i against ln ω discussed before, also was used to determine the possible conduction model for the SPE system. From **Figure 3**, it is observed that the value of exponent s decreases with increasing temperature. The appropriate conduction model can be interpreted using correlated barrier hopping (CBH) model at which the fitted equation of s = -0.00302T + 1.28294 gave a negative gradient in the plot. The temperature dependent of the frequency exponent s also found to obey the relation stated in Equation 5, where s is unity when T is approaching 0K [17].

$$s = 1 - \frac{6k_bT}{[W_m + (k_bT\ln(\omega\tau_o)]} \tag{5}$$

The conduction model can be further verified by comparing the value of E_a and maximum barrier height, W_m. This method was done following exactly with the work reported by Aziz et al., (2017). The value of W_m can be retrieved from the plot of 6kbT against (1-s) (inset plot of **Figure 3**). The W_m value obtained is 0.13 eV which is lower than E_a value (0.25 eV). This result indicates that the ions travel by way of CBH model due to the smaller barrier that exist between the two complexation sites. In this model, the H^+ ions are believed to have sufficient energy to hop across the barrier and coordinate onto another site. The presence of EC plasticizer provides a new shorter pathway between the 2-HEC and

AN which would make the ion easier to hop and require a lower E_a value. Therefore, it enhances the coordination of ion at the end chain of 2-HEC and subsequently increased the ionic conductivity.

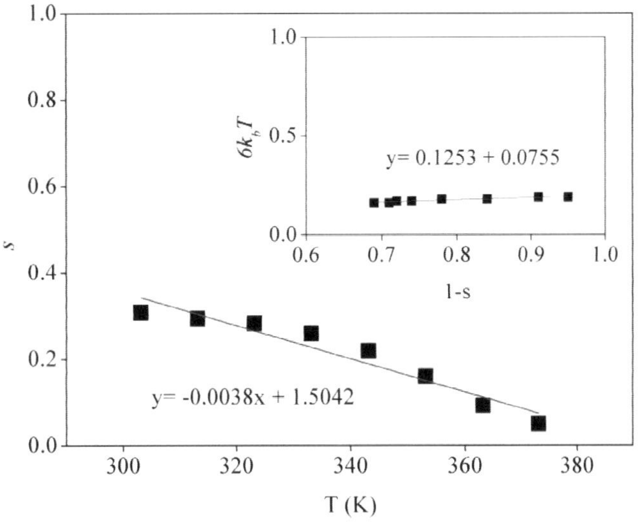

Figure 3. Plot of power law exponent s against *T* for 2-HEC–AN–EC SPE with the highest ionic conductivity. The inset represent the plot of 6kbT against (1-s).

4. Conclusion

Solid polymer electrolytes from 2-hydroxyethyl cellulose doped with ammonium nitrate and plasticized with different ethylene carbonate compositions were successfully prepared using solution casting method. The temperature dependence of ionic conductivity plot proved that all SPEs are Arrhenius type film. The SPE with the highest ionic conductivity (P5) exhibits the lowest activation energy value of 0.25 eV. From dielectric loss-frequency plot, the exponent power law, *s* was obtained. The plot of frequency exponent *s* against temperature indicated that the ionic conduction mechanism of the highest conducting SPE follows correlated barrier hopping (CBH) model.

References

[1] Isa M I N and Noor N A M 2015 *Proceeding of SPIE* **9668** 96685U-1.
[2] Shukur M F, Ithnin R, Illias H A and Kadir M F Z 2013 Optical Materials 35 1834–1841.
[3] Shukur M F, Ithnin R and Kadir M F Z 2014 *Electrochimica Acta* **136** 204–216.
[4] Hafiza M N and Isa M I N 2017b *Journal of Sustainability Science and Management* **2017** 65-70.
[5] Buraidah M H and Arof A K 2011 *Journal of Non-Crystalline Solids* **357** 3261–3266.
[6] Chai M N and Isa M I N 2012 *International Journal of Advanced Technology & Engineering Research* **2** 36-39.
[7] Buraidah M H, Teo L P, Majid S R and Arof A K 2009 *Physica B* **404** 1373-1379.
[8] Rosli N H A, Chan C H, Subban R H Y and Winie T 2012 *Physics Procedia* **25** 215 – 220
[9] Khiar A S A and Arof A K 2011 International Journal of Physical and Mathematical Sciences **5** 1663-1666.
[10] Rani M S A, Mohamed N S and Isa M I N 2015 *International Journal of Polymer Analysis and Characterization* **20** 491-503
[11] Hafiza M N and Isa M I N 2017a AIP Conference Proceeding 1885 020098.
[12] Sohaimy M I H and Isa M I N 2016 *Polymer Bulletin* **74** 1371-1386.

[13] Moniha V, Alagar M, Salvasekarapandian S, Sundaresan B and Boopathi G 2018 *Journal of Non-Crystalline Solids* **481** 424-434.
[14] Hafiza M N and Isa M I N 2017c Journal of Sustainability Science and Management 2017 58-64.
[15] Kumar M, Tiwari T and Srivastanava N 2012 *Carbohydrate polymers* **88** 54-60.
[16] Alves R, Sentanin F, Sabadini R C, Pawlicka A and Silva M M 2018 *Journal of Non-Crystalline Solids* **482** 183-191.
[17] Aziz S B, Abdullah R M, Rasheed M A and Ahmed H M 2017 *Polymers* **9** 338.

Acknowledgments
The authors would like to acknowledge Ministry of Higher Education Malaysia (MOHE) and Universiti Malaysia Terengganu (UMT) for ERGS (vot. 55101) and FRGS (vot. 59452) grant, MyBrain15 for MyPhD awarded. Special thanks Advanced Materials Team for the support given.

IOP Conf. Series: Materials Science and Engineering **440** (2018) 012040 doi:10.1088/1757-899X/440/1/012040

Relationship between Non-Ionizing Radiation (NIR) Exposure Level and the Base Station Tower Installation around Kuala Nerus

A R S N Dianah[1], S N Hazmin[1,2]*, H N Syafiqah[3], M K A Kamarudin[3], R Umar[3]

[1]School of Fundamental Science, Universiti Malaysia Terengganu, 21030 Kuala Nerus, Terengganu, Malaysia
[2]Pusat Islam Sultan Mahmud, Universiti Malaysia Terengganu, 21030 Kuala Nerus, Terengganu, Malaysia
[3]East Coast Environmental Research Institute (ESERI), Universiti Sultan Zanal Abidin, Kampus Gong Badak, 21300 Kuala Nerus, Terengganu, Malaysia

Corresponding author email: norhazmin@umt.edu.my

Abstract. The rapid development of wireless technology nowadays has led to the increases of erection of base station telecommunication (BST). However, the exposure transmitted from BST generated the public concern on health problem because it is said can causes high risk to the human and surroundings. Therefore, this study will focused on the influence of selected BST installed around Kuala Nerus towards the ambient NIR exposure value at surroundings. By using spectrum analyzer connected to receiving omni-directional antenna, the NIR exposure level in terms of electric field (EF) strength is measured horizontally at 50 m until 500 m around the BSTs. The comparison between the measured values and the international reference levels of ICNIRP were done. Besides, the exposure values were compared among the BSTs area by considering the other influenced factors at surroundings. The result shows that the highest exposure level was recorded at BST located in sub-urban area and has the higher number of antennas installed on the top. It can be concluded that the number of BST nearby, the number of antennas, the direction of the antenna and the presence of other sources influenced the ambient NIR exposure at that area.

1. Introduction

The used of electromagnetic (EM) wave spread very fast nowadays, radio frequency especially used in electronic equipment and it almost used in human gadgets [1][2]. Nowadays radio frequency interference (RFI) is everywhere, to understand this RFI, the studies must be done to proof that there are relation between biological effect to human those are highly exposed to radio wave. However there is no proven that the effect of this EM wave to the human. Many researches have been done in worldwide and some finding shows that the most of people who highly exposed to radio frequency has higher chance to diagnose to have brain tumour and affect the reproductive system. As we know in lowfrequency window is usually used for industrial or commercial activities, and also in military systems [2][3]. So that peoples always exposed to this frequency in daily activity. The industry sectors which are active users, they can also transmit signals as well as receive them. The difference between the two users (active and passive) are active users will radiate a strong radio wave signal while passive users only detect weak radio signals from space[4][5].

In Universiti Kebangsaan Malaysia (UKM), the policy is developed based on the Occupational Safety and Health Act 1994 (Act 514) (2006), safety guidelines, supplemented with exemplary nonionising radiation policies of other well-known international institutions of higher education. As for the monitoring activities, results obtained from the six identified hotspots suggest that both RF and ELF exposure levels are far lower than the suggested ICNIRP (International Commission on Non-Ionizing Radiation Protection) exposure limits, at a ratio percentage of RF and ELF of 0.2% and 2.8%, respectively. With the development of the UKM Non-Ionising Radiation policy and the progress of the monitoring activities being carried out periodically, the final element calls for the reintroduction of an academic subject on non-ionising radiation, as was offered before the revision of the Bachelor of Science (Nuclear Science) programme. Health concern if overdose NIR BST distribution and distance form residential area.

In this study, the focus will be on how the base stations influence the ambient NIR exposure value at surroundings. This research findings will provide the ambient NIR exposure value at various distances from the BST. Nearby sources will be determined to identify other influence factors contribute to the exposure level. This is important to determine the influence of the base stations on the production of NIR. This will help Malaysia Malaysian Communications and Multimedia Commission (MCMC) to organize and plan the suitable location for base stations and transmitter approvals for the spectrum provider. This is essential to make sure the recommended public exposure value by International Radiation Protection Association (IRPA) is complied. Moreover, the public health monitoring could be conducted and future clinical studies will be assisted by data provided such in this study.

2. Methodology

The measurement of NIR exposure level for frequencies between 65.2 MHz and 4 GHz has been conducted at distances ranging from 50m to 500m from every 14 BSTs. The flow of methodology is as shown in figure 1, begin with site selection for data collection. The selection of locations was chosen based on different locations such where located close to the residential, campuses and public areas.

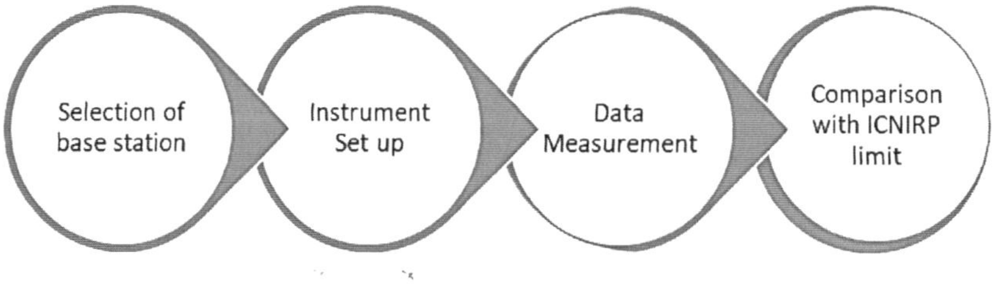

Figure 1. Flow of methodology

Before the observation was performed, the instrument was set up and installed at distances of 50m, 100m, 150m, 200m, 250m, 300m, 350m, 400, 450m, 500m from each BST. The coordinate of latitude and longitude for each distance were recorded using Global Positioning System (GPS) model GPSMAP® 62sc for mapping process as shown in Table 1. The measurement was recorded 10 minutes for each distance. After that, the measurement obtained was extracted from spectrum analyser to computer for the analysis process. The data obtained were compared with the permitted international standard public exposure limit recommended by ICNIRP limit. Then, the relationship between the highest NIR exposure values and the BST existed around the study area were discussed. In overall, this study is conducted to prove whether BST installation affects the increase of radiation in an area.

Table 1. Description of locations.

BST Sites	Latitude (N)	Longitude (E)	Category of Area
Kubang Badak	5°24'17.94"	103°5'40.40"	Sub-urban
UMT	5°24'17.66"	103°5'18.33"	
Kampung Jati	5°23'56.39"	103°5'9.50"	
Kampung Pak Tuyu	5°24'49.13"	103°4'59.90"	
Masjid UniSZA	5°23'52.69"	103°4'58.81"	
UniSZA	5°24'20.86"	103°4'55.13"	
Gong Badak	5°23'57.60"	103°4'18.60"	
Mengabang Telipot 1	5°25'23.26"	103°3'20.02"	
Mengabang Telipot 2	5°25'14.49"	103°3'15.89"	
Batu Rakit 1	5°26'48.52"	103°3'2.73"	
Batu Rakit 2	5°26'55.67"	103°2'57.51"	
Batu Rakit 3	5°26'59.19"	103°2'51.50"	Rural
Batu Rakit 4	5°26'57.36"	103°2'47.39"	
Batu Rakit 5	5°26'57.97"	103°2'44.73"	

3. Results and Discussion

In this measurement, the NIR exposure data were collected roughly from 10.00 am until 4.00 pm with the averaging time is 10 minutes for each survey point. Based on the ICNIRP postulate, the temporal average value of exposure over any 6 minutes interval must be used for the comparison with the reference levels suggested [6].

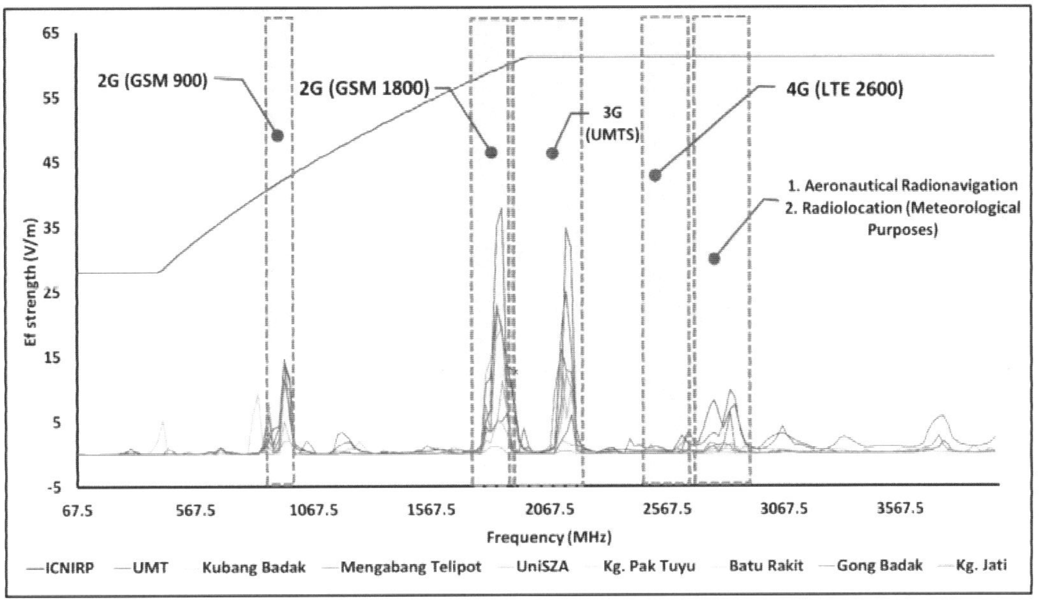

Figure 2. The comparison of NIR exposure trend with ICNIRP reference levels [6]-[8]

The maximum NIR exposure values (electric field (EF) strength) obtained in each study areas were then compared with the international reference levels for general public exposure provided by ICNIRP as shown in Figure 2. Figure 2 reveals that all the maximum values recorded are below the reference levels.

Even though residents in this study areas do not exposed to the higher radiation level, but a repetitive exposure may cause the negative health effect such headaches, insomnia, dermatitis and other effects. Some studies also proved that the long-term adverse effects from long duration exposure to NIR can cause the thermal effect thus the changing of Deoxyribonucleic Acid (DNA) structure may be occurred [8][9].

Since this study areas were classified as rural and suburban area, thus, it can be seen that the telecommunication system, especially for the second-generation (2G) and third-generation (3G) be the main sources to the highest EF strength level and the high quality of signal coverage could be provided. Some of previous studies also mentioned that GSM 900 (2G) is more often installed in rural areas meanwhile GSM 1800 (2G) and UMTS (3G) technologies are found more in the denser area [10][11]. However, the higher EF strength level means that the higher the exposure level was around the area. Therefore, it can be concluded that the installation of BST and the existence of users nearby might be one of the factors that believed causing strong exposure. Then, the maximum exposure value was compared among the 14 BSTs as shown in Figure 3.

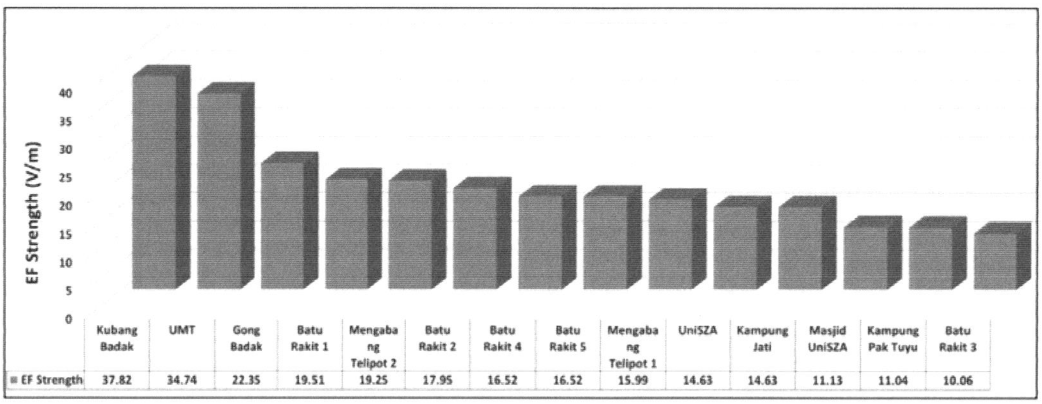

Figure 3. The highest NIR exposure values among 14 BSTs

Based on Figure 3, it is clearly shown that Kubang Badak performs the highest maximum exposure value which is about 64% approaching the reference levels for public exposure of ICNIRP, then followed by UMT. Meanwhile, the lowest maximum exposure value was shown by Batu Rakit 3 where the tower is surrounded by trees with dense and depth foliage at lower altitude than the BST antennas. Since VHF and UHF is a kind of LOS propagation characteristic, thus these radio signals had a strong attenuation rate due to the leaves as the main contributor to the received signal for short distances [12]. Meaning that, the foliage canopy is said to be obstructed either in line of sight or pointto-point communication [13]. It is also observed that the slow rate of attenuation occurred even at rural areas due to the less obstruction at surrounding.

According to the previous studies that have been reviewed, the radiation exposure level around BST is not only influenced by the power transmitted from the antenna of BST but is also influenced by the antenna position on top of BST. This is because the tilt and direction of antennas on the BST are related to the direction of the main beam radiation and the distances of highest level of power transmission distributed on the surface of the earth [14]. Since the power radiated from the main beam of antenna is higher than the side beam, thus, resulted the radiation exposure level varies according to the direction and distance of around BST that influenced by the beam. Table 2 shows the description of each BST selected in this study.

According to the table 2, it can be proved that each BST that has been studied has maximum exposure value at different distance varied from 150 m to 400 m. As expected, it is was found that this case shared the same finding with previous NIR level assessment conducted in Sri Lanka where the main beam to reach ground level typically between 50 m and 300 m from the foot of BST by considering the heights and the tilt of antennas. Awad, (2014) also found from analysing the radiation pattern of a mobile phone tower that the high exposure level from main beam propagation reached the ground level in about 50 m to 300 m. In addition, the exposure level becomes more increasingly high when the locations are facing toward the BST antennas.

Meanwhile, the low exposure area is found when out from the main beam radiation either further from the main beam or directly under the BST. This low exposure is known as side beam radiation [15]. Radiation power level are higher in several kilowatts within the main beam transmission area and will be much lower once distant from the main beam [9][16].

Table 2. The description of the surveyed BST and the highest NIR exposure values.

BST Sites	Height of Tower (m)	Number of Antennas	Highest Exposure Value (V/m)	At Distances (m)	ICNIRP Comparison (%)
Kubang Badak	61.18	23	37.82	150	64
UMT	36.73	6	34.74	200	60
Gong Badak	72.00	13	22.35	250	38
Batu Rakit 1	60.00	5	19.51	250	33
Mengabang Telipot 2	61.19	8	19.25	150	32
Batu Rakit 2	40.99	12	17.95	400	30
Batu Rakit 4	50.53	18	16.52	300	28
Batu Rakit 5	14.61	7	16.52	300	28
Mengabang Telipot 1	75.52	17	15.99	250	26
UniSZA	34.00	5	14.63	150	25
Kampung Jati	46.60	18	14.63	150	25
Masjid UniSZA	40.00	10	11.13	350	19
Kampung Pak Tuyu	44.67	14	11.04	250	19
Batu Rakit 3	58.40	16	10.06	200	17

Besides, in the presence of multiple antennas on single BST, the total exposure level radiated is referred to the total power produced by all the functioning antennas installed on BST [17]. The more transmitting antennas installed, the possibility of increased exposure level is higher. Referring back to the Table 2, the highest exposure value belongs to Kubang Badak BST which it has the largest number of antennas installed compared to other BST. However, the number of antennas is not a major factor affecting the value of radiation because there are many other factors as discussed previously. Table 3 shows the number of BST detected at each of the highest exposure location.

4. Conclusions

The exposure levels in terms of EF strength were measured at 50 m until 500 m surroundings the selected BSTs. The exposure levels were below the international limit of reference levels set by ICNIRP. The highest exposure level was found at 150 m from Kubang Badak's BST area which is has the largest number of transmitting antennas on top of it compared to other BSTs. Besides that, this BST located at sub-urban area where the highest exposure level probably due to the other radiation sources nearby such the moving vehicles from heavy road network traffic or the higher number of mobile users. Meanwhile, the lowest exposure level was found at Batu Rakit 3 at distance of 250 m from BST. This area was categorized as rural area where less human activities and mobile users. However, the exposure level at Batu Rakit area is higher compared to Kampung Pak Tuyu which even also categorized as rural area. This is because there are five BSTs erected at Batu Rakit site. Therefore, it is well known that the BST erected may influenced the ambient NIR exposure level surroundings. It is important to take note that cancer caused by NIR is not instantaneous but would develop over years. Since the radiation from BST can't turn off, thus, it is crucial to reduce the NIR exposure from other sources.

5. References

[1] Umar R, Abidin Z Z, Ibrahim Z A, Hassan M S R, Rosli Z and Hamidi Z S 2012 *Population density effect on radio frequencies interference (RFI) in radio astronomy. Int. Conf. On Physi and Its Applications* :(ICPAP 2011) vol 1454 (1) 39-42
[2] Umar R, Abidin Z Z, Ibrahim Z A, Rosli Z and Noorazlan N 2014 *Research in Astronomy and Astrophysics* **14** (2) 241
[3] Abidin Z Z, Umar R, Ibrahim Z A, Rosli Z, Asanok K and Gasiprong N 2013 *Publications of the Astronomical Society of Australia* **30** 47
[4] Umar R, Hazmin N S, Abidin Z Z, Ibrahim Z A, Azid A, Juahir H, Toriman M E and Kamarudin M K A 2015 *J. Teknologi* **75** No 1
[5] Hazmin N S, Umar R, Mokhtar W Z A W, Abidin Z Z, Ibrahim Z A, Azid A, Juahir H, Toriman M E and Kamarudin M K A *J. Teknologi* vol **75** No 1
[6] ICNIRP, International Commission on Non-Ionizing Radiation Protection 1998 *Guidelines for limiting exposure to time varying electric, magnetic, and electromagnetic fields (up to 300 GHz), Health Physics* **74** (4) 494–522
[7] Malaysian Communications and Multimedia Commission (MCMC) 2014 *Spectrum plan* (Selangor: MCMC)
[8] Dianah A R, Hazmin S N, Umar R, Kamarudin M K and Dagang A N 2017 *J. Fundam. Appl. Sci.* **9** (5S) 367–380
[9] Dianah A R, Hazmin S N, Umar R, Kamarudin M K and Dagang A N 2017 *J. Fundam. Appl. Sci.* **9** (2S) 182–98
[10] Dianah A R S N, Hazmin S N, Umar R, Jaafar H, Kamarudin M K A, Dagang A N, and Syafiqah H N 2018 *J. Fundam. Appl. Sci.* **10** (1S) 52 –540
[11] Dianah A R S N, Hazmin S N, Umar R, Kamarudin M K A and Dagang A N 2017 *J. Fundam. Appl. Sci.* **9** (5S) 367–380
[12] Schwering F, Violette E and Espeland R 1988 *IEEE Transactions on Geoscience and Remote Sensing* **26** 355–367
[13] Amajama J 2016 *Int. J. Eng. Research & General Sci.* **4** (3) 481-494
[14] Awad A H A and Habeeballah M T 2014 *Met., Env. & Arid Land Agric. Sci.* **25** No. 2 51-65
[15] Al-Ruwais A S 2001 *Sci. & Techn.* **6** 45-51
[16] Al-Akhras M A, Albiss B A, Alqudah M S and Odeh T S 2015 *Jordan J. Earth and Env. Sci.* **7** (2) 77-85
[17] Yong C Y, Rahman T A and Chew K M 2015 *ARPN J. Eng. Appl. Sci.* **10** (20) pp 9800–9807

Acknowledgments

This study is made possible by the usage of grant FRGS/1/2015/SG02/UNISZA/02/1, TPM/68006/2016/79 and UMT 68006/INSENTIF/60. Our sincere gratitude for all assistance and help from Assoc. Prof Dr. Ahmad Nazri b Dagang and Dr. Hajar Ja'far. The authors also thank to Universiti Malaysia Terengganu, Universiti Sultan Zainal Abidin and Universiti Teknologi MARA for the facilities provided.

IOP Conf. Series: Materials Science and Engineering **440** (2018) 012041 doi:10.1088/1757-899X/440/1/012041

The Modelling of Tropical Weather Effects on Ultra-High Frequency (UHF) Radio Signals Using SmartPLS

R Mat[1,4], S N Hazmin[1,5]* , R Umar[2], S Ahmad[3], S N A S Zafar[2] and **M S Marhamah[2]**

[1]Advanced Nano Materials (ANoMa) Research Group, School of Fundamental Science, Universiti Malaysia Terengganu, 21030 Kuala Nerus, Terengganu, Malaysia
[2] East Coast Environmental Research Institute (ESERI), Universiti Sultan Zanal Abidin, Kampus Gong Badak, 21300 Kuala Nerus, Terengganu, Malaysia
[3]School of Informatics and Applied Mathematics, Universiti Malaysia Terengganu, 21030 Kuala Nerus, Terengganu, Malaysia
[4]Institut Pendidikan Guru Kampus Sultan Mizan, 22200 Besut, Terengganu, Malaysia
[5]Pusat Islam Sultan Mahmud, Universiti Malaysia Terengganu, 21030 Kuala Nerus, Terengganu, Malaysia

Corresponding author email: norhazmin@umt.edu.my

Abstract. The interference occurs in radio signals transmission in many communication technologies are attributed to weather variations as well as other environmental factors. This work evaluates some atmospheric variables that have a dominating impact on temporal signal strength fluctuations. The Received Power Level (RPL) of Ultra High Frequency (UHF) communication signals (1800 MHz and 2160 MHz) and three tropical weather predictors (humidity, temperature and wind) were collected using spectrum analyser (KEYSIGHT N9915A) and weather station (Vantage Pro 2) respectively for 24 hours in rainy days. SmartPLS 3.2.6 was used to determine the strongest predictors influence the RPL for UHF frequency. It has been found that changes in weather conditions have affected the signal in which all weather predictors provide a significant relationship to the signal where R^2 (coefficient of determination) value is 0.314 for frequency 1800 MHz and 0.254 for frequency 2160 MHz. The findings also show that humidity, temperature and wind are anticorrelations to RPL. However, humidity is found to be the strongest predictor influences the RPL of communication signals for frequency 1800 MHz (β= -0.449, p=0.000). This model may benefit many sectors such as telecommunication service provider, radio and TV transmission, radio astronomy study, Electromagnetic (EM) researcher and satellite broadcasting.

1. Introduction

Nowadays, the use of communication technologies such as smartphone, laptop and computer are increasing day by day in our lives. Most of us depend on these various technologies to achieve certain tasks in our daily activities especially in getting news, making online or banking business and also connecting with social media. According to the Malaysian Communications and Multimedia Commission (MCMC) Internet Users Survey 2014, the smartphone is the most widely-used device to access the Internet at 74.3% [1]. As the numbers indicate, users fully embrace the concept of getting connected everywhere at any time. However, a poor mobile phone signal reception is a dilemma experienced by almost every one of us at some point in time. A recent survey by MCMC has shown that 48.1% of Malaysian owners experience some form issue due to poor cellular network strength [1]. These can result due to some failures in cell tower and overcrowding etc. However, the weather condition also plays an important role in affecting the cellular signal. Understanding the effects of the weather on radio wave propagation will help to improve the quality of radio signals in communication. Radio waves in Ultra High Frequency (UHF) range, is very widely used in modern technology, particularly in telecommunications. UHF is the part of the electromagnetic spectrum, ranging from 300 to 3000 MHz [2]. When the radio wave is interrupted by the bad weather, the quality of the signal will be reduced. For a tropical weather like Malaysia, environmental factors such as temperature, rainfall, wind and humidity will change, depending on the seasons. The weather of Malaysia has a nearly uniform temperature, high humidity, abundant rainfall and the wind is weak because its location is close to the equator [3].

Several factors that can affect radio wave propagation such as terrain, building, vegetations, CO_2 concentration and weather parameters (such as rain, solar radiation and dust) have been studied by many authors [4-7]. Most of the research and observations were made in another part of the tropical countries had shown some discrepancy which did not apply to Malaysia weather [6-8]. Therefore, more data are required to confirm the effect of weather on radio waves in tropics. Research in radio wave propagation is a new field in Malaysia and only a very few have examined the impact of tropical weather on UHF radio signals There is only a few direct or indirect measurements of attenuation of weather have been made in Malaysia and they have not looked into the correlation between radio signals and weather parameters so far [9, 10]. In this study, we used statistical analysis to identify the correlation between the variables and finally we used regression to determine the strongest predictor influences the signal. The usage of SmartPLS is a novel idea, and it is a first-hand study of its kind. Findings could be enhanced with the study of using this method. Therefore, this study is hoped to address the above problems.

1.1 Literature review

Weather factors such as rain, wind, temperature and humidity are one of the Radio Frequency Interference (RFI) sources that disturb the propagating signal and cause loss of data which have been studied by many authors. [11] investigated the influence of air temperature, relative humidity and atmospheric moisture on the UHF radio propagation in South Western Nigeria. The results showed that air temperature, relative humidity and liquid water density of the cloud relative (LWDC) have a significant influence on UHF signal propagation within the troposphere region of southwest Nigeria. They observed that as the air temperature increases, relative humidity decreases, hence a proportional decrease in UHF path loss while the received signal strength (RSS) shows a proportional increase. [12] found that an inverse relationship between the path loss and the relative humidity was obtained from their results. That is, a decrease in relative humidity brings about a slight increase in path loss. Another finding research from Nigeria, [13] also indicated that radio signal strength is inversely proportional to atmospheric temperature, pressure and humidity; provided that for any of the giving components, others were observed constant, including the wind speed and direction, since it has been erected that wind has a marked effect on a radio signal. The correlation of the signal strength and atmospheric temperature, pressure and humidity were respectively r = -0.94, -0.99 and -0.93 and the equation $S = K(\alpha t)/TPH$ constant wind speed and direction was postulated, where S, T, P, H and K are Signal strength, Atmospheric temperature, Atmospheric pressure, Relative humidity and constant respectively. [14] studied the effects of temperature and humidity on radio signal strength in outdoor wireless sensor and also showed that changes in weather conditions affect received signal strength in western Finland. However, they found that temperature seems to have a significant negative influence on signal strength in general, while high relative humidity may have some effect on it, particularly below $0°C$. Same as [8] also have found that a linear relationship between passive UHF Radio Frequency IDentification (RFID) tags' resonant frequency and temperature at relative humidity (RH) of 50% and 80% in their study. They also found that there is no significant change in the resonant frequency was observed as a result of humidity variation at a temperature ranging from 20^oC to 80^oC. Analysis using regression was also used in previous studies. [15] has conducted a study on the impact of the weather conditions on the wireless link quality. Mathematical models extracted fit different transmission problems over a link IEEE 802.11b/g as a function of different meteorological variables. This research has been carried out by symbolic regressions since it is a powerful and reliable method to analyse difficult hidden relations among variables. Results show that grouping weather conditions entail a significant increment in the correlation coefficients in comparison with previous studies.

1.2 Objectives and hypothesis
The study aims to examine and evaluate the impact of three tropical weather factors (temperature, humidity and wind) in predicting the Received Power Level (RPL) of telecommunication signals transmission in the UHF band (Peak3=1800 MHz and Peak4=2160 MHz). The main objective of this research is to identify the strongest weather predictor influences the signal in the UHF band for Peak3 and Peak4.

Past researches [9-15] has shown that weather influences signal directly. Thus, we hypothesize that:

H1: Humidity will have an effect on RPL signals of Peak3
H2: Temperature will have an effect on RPL signals of Peak3
H3: Wind will have an effect on RPL signals of Peak3
H4: Humidity will have an effect on RPL signals of Peak4
H5: Temperature will have an effect on RPL signals of Peak4
H6: Wind will have an effect on RPL signals of Peak4

2. Methodology

2.1 Site description
The data measurement was conducted at Balai Cerap KUSZA (BCK), Universiti Sultan Zainal Abidin, Terengganu. BCK is located in Merang (5 Â° 32 '10 "N and longitude 102 Â° 56' 55" E), a coastal area of East Coast Peninsular Malaysia which is 28 km away from the main city of Kuala Terengganu. BCK is near the Merang beach located on top of Merang Hill, 160 m above sea level (Figure 1). This observatory has a tropical climate with uniform temperature, high humidity and constant moist with no dry and cold [16]. It is very suitable for the study because the Radio Frequency Interference (RFI) due to other factors are low because it is located in rural areas, far from the commercial and residential area, top of the hill so it receives a direct signal from base station transmission (reduced blockage) [17].

Figure 1. Location of Balai Cerap KUSZA

2.2 *Data collection and data analysis*
The procedure framework for statistical modelling of meteorological factors on UHF communication signals transmission is as illustrated in Figure 2. To determine the impact of the weathers on a radio signal, the study begins with instrument testing and calibration using the measurement setup as shown in Figure 3. The observation started with a preliminary study to get the overall view of the study. Then the real data measurement takes place for a selected date in the presence of rain. Finally, the data

obtained was analysed and the statistical model was developed. The instrument set up was illustrated in Figure 3. This method has been adapted from [5, 18-20] that this study has been using a discone antenna. The RPL at frequencies (namely Peak3-1800 MHz and Peak4-2160 MHz) and tropical weather variation (humidity, temperature and wind) were collected using spectrum analyser (KEYSIGHT N9915A) and weather station (Vantage Pro 2) respectively for 24 hours in rainy days during January - December 2017 (selected date). The description of Peak3 and Peak4 and their allocations were illustrated in Table 1 which are owned by cellular phone.

Figure 2. The procedure framework of Modelling of Tropical Weather effects on UHF Signals.

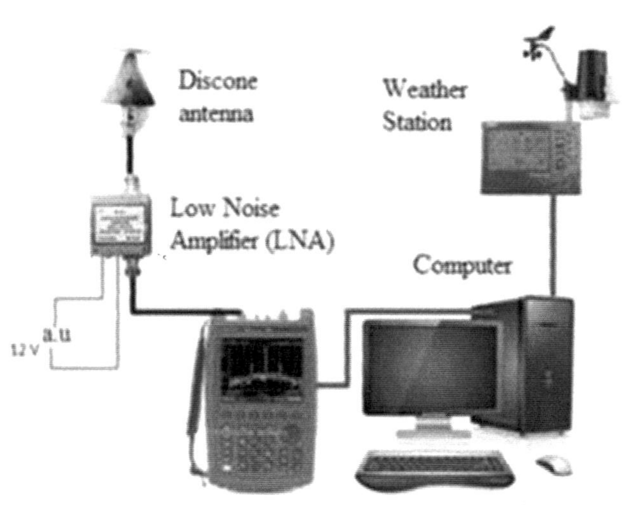

Figure 3. The instruments set up for observation work [5, 18-20]

Table 1. The description of Peaks and its allocation

Frequency MHz	Peaks	Malaysia Allocation	Sources
1800	Peak3	FIXED MOBILE 5.384A 5.388A MLA53 MLA89 MLA91 MLA92 MLA99 5.149 5.341 5.385 5.386 MLA3 MLA81 MLA90	Cellular mobile services [EGSM/GSM]/ [IMT]
2160	Peak4	FIXED MOBILE 5.388A MLA53 MLA92 5.388	Cellular mobile services IMT

Based on the data obtained from both a spectrum analyzer and weather station, data is then analyzed to investigate the relationship between signal and weather factors. We applied linear regression to explore the magnitude of these effects using PLS-SEM. The partial least squares path modelling (PLS-SEM) method to structural equation modelling allows estimating complex cause-effect relationship models with latent variables. It is a component-based estimation approach that differs from the covariance-based structural equation modelling. We used the Partial Least Squares (PLS) technique using the SmartPLS 3.2.6 software to analyze the research model and determine the strongest weather predictor influences the signal [21]. Following the recommended two-stage analytical procedures by [22], we tested the measurement model (validity and reliability of the measures) and followed by an examination of the structural model in order to test the hypothesized relationship [23-26]. Besides, to test the significance of the path coefficients and the loadings a bootstrapping method (5000 resamples) was used [23].

3. Results and Discussion

Figure 4 shows the validity of the Measurement Model Results. To assess the measurement model two types of validity were being examined, first the convergent validity and the discriminant validity. All the values passed as suggested in the literature (see Figure 4) indicating that the converging validity and discriminant validity of this model have been ascertained. The SRMR value was 0.000 (< 0.08) and NFI was 1.000 (>0.90) indicating the data fit the model well. The SRMR is defined as the difference between the observed correlation and the model implied correlation matrix whereby values less than 0.08 [27] are considered a good fit. [28] introduced the SRMR as a goodness of fit measure for PLS-SEM that can be used to avoid model misspecification. The second fit index is a normed fit index (NFI) an incremental fit measure which computes the Chi-square value of the proposed model and compares it against a meaningful benchmark [29]. NFI values above 0.9 usually represent acceptable fit.

Figure 5 illustrates the bootstrapping results of the model and Table 2 shows the results of the hypothesis testing, to assess the structural model, [21] suggested looking at the R^2, beta (β) and the corresponding t-values via a bootstrapping procedure with a resample of 5,000. R^2 indicates how well all weather factors explain RPL of Peak3 and Peak4. For example, a value of 0.314 indicates that 31.4% of the variance in the RPL of Peak3 is explained by the weather factors (humidity, temperature, wind). Accordingly, our primary way to compare the strength of relationships is looking at the path coefficients, β. Furthermore, we could calculate the effect strength f^2 for each predictor, which indicates to which extent this predictor contributes to the explanation of a RPL. Based on Table 2, we found that for Peak3 (1800 MHz) predictors, humidity (β = -0.449, t = 40.337, p< 0.01, f^2 = 0.2818), temperature (β = -0.375, t = 34.018, p< 0.01, f^2 = 0.1920) and wind (β = -0.096, t = 8.125 p< 0.01, f^2 = 0.0123) negatively influenced by 31.4 % of the radio signal strength of Peak3. This supports H1. H2 and H3. Next we looked at the weather predictors on the Peak4 (2160 MHz), humidity (β = -0.370, t = 30.586, p < 0.01, $f2$ = 0.1757), temperature (β = -0.370, t = 33.908, p < 0.01, f^2 = 0.1718) and wind (β = -0.0.089, t =

7.402 $p<0.05$, $f^2 = 0.096$) were significant and giving 25.4 % contribution to this signal. The findings also support H4, H5 and H6.

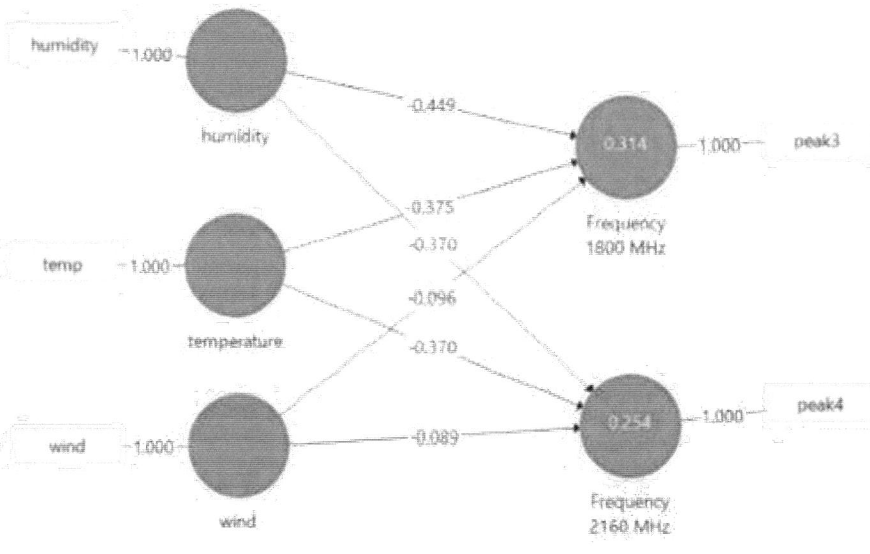

Figure 4. The Measurement Model Results

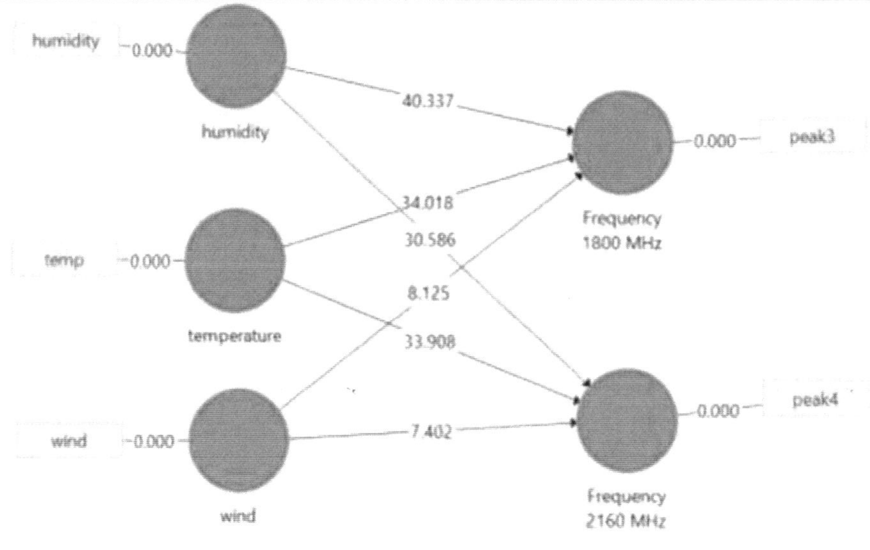

Figure 5. Bootstrapping Result

Table 2. The results of the hypothesis testing

Hypothesis	Relationship	Std beta, β	p-value	T value	Decision	f^2	R^2
H_1	Humidity -> Peak3	-0.449	0.000	40.337	supported	0.2818	
H_2	Temp-> Peak3	-0.375	0.000	34.018	supported	0.1920	0.314
H_3	Wind -> Peak3	-0.096	0.000	8.125	supported	0.0123	
H_4	Humidity -> Peak4	-0.370	0.000	30.586	supported	0.1757	
H_5	Temp> Peak4	-0.370	0.000	33.908	supported	0.1718	0.254
H_6	Wind -> Peak4	-0.089	0.000	7.402	supported	0.0096	

4. Conclusions

From the results of the hypothesis testing and boostrapping, we can conclude that weather factors (temperature, humidity and wind) play a pivotal role in UHF radio wave propagation. It is found that the change in weather conditions affects the RPL where R^2 was 0.314 for Peak3 and 0.254 for Peak4 (Table 2). Peak3 at frequency 1800 MHz was affected the most due to tropical weather variation with contribution 31.4% whereas weather factors contribute 25.4% to Peak4 (2160 MHz). This confirms our prediction that weather factors (humidity, temperature and wind) are important in UHF radio signal. Humidity is found to be the strongest predictors influence the RPL of communication signals especially for frequency 1800 MHz (β=-0.449, p=0.000). The model of the potential impact of weather factors on radio frequency interference may benefit telecommunication service provider and makes it possible to propose alternative solutions to assure the better service suitable for all weather condition.

5. References

[1] Malaysian Communications and Multimedia Commission 2014 Spectrum management. [Online]. Available: http://www.mcmc.gov.my/ Spectrum/ Spectrum-Management.aspx

[2] Cohen J, Titus S, Ambrosini R and Driel W V 2005 *CRAF Handbook for Radio Astronomy* 3rd ed Cohen J, Spoelstra T, Ambrosini R and Driel W V Eds, Strasbourg, (France: European Science Foundation)

[3] Climate of Malaysia (2011) [Online] Available: http://www.eoearth.org/view/article/151260

[4] Umar R, Sulan S S, Azlan A W, Ibrahim Z A, Mokhtar W Z A W and Sabri N H 2015 Radio frequency interference: The study of rain effect on radio signal attenuation *Malaysian J. of Analytical Sci.* **19** 1093-1098

[5] Sabri N H, Zafar S N A S, Umar R and Mokhtar W Z A W 2015 Radio frequency interference: The effect of ambient carbon dioxide (CO2), concentration on radio signal for radio astronomy purposes *Malaysian J. of Analytical Sci.* **19** 1065-1071

[6] Meng Y S, Lee Y H and Ng B C 2009 The effects of tropical weather on radio-wave propagation over foliage channel *IEEE Transactions on Vehicular Technol* **58** 4023-4030

[7] Alade M O 2013 Investigation of the effect of ground and air temperature on very high frequency radio signals *J. of Information Engineering and Applications* **3** 16-21

[8] Zhenzhong L, Nezih M, George X, Yuu O, Guocheng L and Dayan B 2011 Effects of temperature and humidity on UHF RFID performance *Proc. CANSMART CINDE IZFP'11 1*

[9] Sabri N H, Azlan A W, Umar R, Sulan S S, Ibrahim Z A and Mokhtar W Z A W 2015 The effect of solar radiation on radio signal for radio astronomy purposes *Malaysian J.of Analytical Sci.* **19** 1374-1381

[10] Umar R, Sulan S S, Azlan A W, Ibrahim Z A, Mokhtar W Z A W and Sabri N H 2015 Radio frequency interference: The study of rain effect on radio signal attenuation *Malaysian J. of Analytical Sci.* **19** 1093-1098

[11] Segun A A, Olusope A M and Kofoworola A H 2015 Influence of air temperature, relative humidity and atmospheric moisture on UHF radio propagation in South Western Nigeria *Int. J. of Sci. and Research* **4** 588-592

[12] Suleman K O, Bello I T, Tijani L O, Ogunbode A O and Olayiwola W A 2017 Effect of Temperature and Ground Water on VHF Radio Wave Propagation in Tropical Climate *Int.J. of Sci. and Engineering Research* **8** (1) 1391-1396

[13] Amajama J 2016 Impact of Weather Components on (UHF) Radio Signal *Int. J. of Engineering Research and General Sci.* **4** (3) 481-494

[14] Luomala J and Hakala I 2015 Effects of temperature and humidity on radio signal strength in outdoor wireless sensor networks *Proc. IEEE FedCSIS'15* 1247-1255

[15] Bri D, Ramos F, Lloret J and Garcia M 2016 Model Fitting to Account for the Weather's Impact on Wireless Propagation at 2.4 GHz *Natl. Acad. Sci. Lett* [online] DOI 10.1007/s40009-016-0520-7

[16] Climatemps.com 2015 Kuala Terengganu climate and temperature [Online]. Available: http://www.terengganu.climatemps.com/.

[17] Umar R, Abidin Z Z, Rosli Z A and Noorazlan N 2013 Selection of radio astronomical observation sites and its dependence on human generated RFI *Research in Astronomy and Astrophysics* **14** 241-248

[18] Umar R, Sabri N H, Abidin Z Z, Ibrahim Z A, Azid A, Juahir H, Toriman M E and Kamarudin M K A 2015 Rain preliminary study of radio astronomical lines effect of rain below 2.9 GHz *Jurnal Teknologi* vol **75** 7-11

[19] Mat R, Marhamah M S, Ahmad S, Umar R, Yew Been Seok, Sabri N H 2016 Temperature Effect on The Tropospheric Radio Signal Strength for UHF Band at Terengganu, Malaysia *Int. J. On Advance Sci. Engineering Information Technol.* **6** 770-774

[20] Sabri N H, Umar R, Marhamah M S, Zafar S N A S, Mat R, Ahmad S, and Ibrahim Z A 2017 Correlation Analysis of Tropical Rainforest Climate Effect on Radio Signal Strength at KUSZA Observatory, Terengganu*" Advanced Sci.Letters* **23** 1268-1271

[21] Ringle C M, Wende S & Becker J-M 2015 "SmartPLS 3" www.smartpls.com

[22] Anderson J C & Gerbing D W 1988 Structural Equation Modeling in Practice: A Review and Recommended Two-Step Approach *Psychological Bulletin* **103** 411-423

[23] Hair J, Hult F, Ringle, G T M, C M, & Sarstedt M 2017 *A Primer on Partial Least*

[24] Ramayah T, Lee J W C, Boey, J C I 2011 Network collaboration and performance in the tourism sector *Service Business* **5** (4) 411 -428

[25] Ramayah, T, Yeap J A L & Ignatius J 2013 An empirical inquiry on knowledge sharing among academicians in higher learning institutions. *Minerva: A Review of Science, Learning and Policy* **51** (2) 131 -154

[26] Rahman S A, Amran A, Ahmad N H & Taghizadeh S K 2016 Enhancing the wellbeing of base of the pyramid entrepreneurs through business success: the role of private organizations. *Social Indicators Research* **127** (1) 195-216

[27] Hu L T and Bentler P M 1998 Fit Indices in Covariance Structure Modeling: Sensitivity to Underparameterized Model Misspecification *Psychological Methods* **3** (4) 424-453

[28] Henseler J, Hubona G & Ray P A 2016 Using PLS path modeling in new technology research: updated guidelines *Industrial Management & Data Systems* **116** (1) 2-20

[29] Bentler P M and Bonett D G 1980 Significance Tests and Goodness-of-Fit in the Analysis of Covariance Structures *Psychological Bulletin* 588-600

Acknowledgments

This study is made possible by the usage of the grants RACE/F1/ST1/UNISZA/15-RR118, FRGS/1/2015/SG02/UNISZA/02/1, UM-0000033/HRU.PT.CG(CR008-2015) and 8006/INSENTIF/60. The authors also thank to Universiti Malaysia Terengganu and Universiti Sultan Zainal Abidin for the facilities provided.

IOP Conf. Series: Materials Science and Engineering **440** (2018) 012042 doi:10.1088/1757-899X/440/1/012042

Effects of extrusion pressure and printing speed of 3D bioprinted construct on the fibroblast cells viability

Khalida Fakhruddin[1], Mohd Syahir Anwar Hamzah[2] & Saiful Izwan Abd Razak[1,3,*]

[1]Faculty of Biosciences and Medical Engineering, Medical Devices & Technology Group (MEDITEG), Universiti Teknologi Malaysia, 81310, Skudai, Johor, Malaysia

[2]Department of Chemical Engineering Technology, Faculty of Engineering Technology, Universiti Tun Hussein Onn, 86400 Batu Pahat, Johor, Malaysia.

[3]IJN-UTM Cardiovascular Engineering Centre, Institute of Human Centered Engineering,Universiti Teknologi Malaysia,81310 Skudai, Johor, Malaysia

Corresponding author email: saifulizwan@utm.my

Abstract. In 3D bioprinting system, there are varies set of printing parameter involved that will affect the structure of bioprinted construct and cell viability. This study reports the effect of extrusion pressure and printing speed on the structure of construct and cell viability. This study contains two parts which first, to see the effect of the morphological structure of the construct by printing using Pluronic F127 and second, too evaluate the viability of the cell on the bioprinted construct by printing using fibroblast cell-laden Lithium phenyl-2,4,6-trimethylbenzoylphosphinate (LAP) and Gelatin Methacrylate (GelMa). The construct was printed by using extrusion based 3D bioprinter with various ranges of extrusion pressure and printing speed. Based on the experiment, the result shows that printing at lower extrusion pressure requires a lower printing speed whereas printing at higher extrusion pressure requires a higher printing speed. For the cell viability of the bioprinted construct, printing at higher extrusion pressure will reduce the percentage of cell viability.

1. Introduction

Bioprinting is one facet of tissue engineering which has been emerging in recent years to build a three dimensional constructs containing biological cells [1]. 3D printing technique also has been primarily used to create acellular 3-Dimensional (3D) scaffolds and molds which then be seeded with cells post-fabrication [2]. Sacrificial ink likes pluronic F127 can be used for temporary support or to create channels, vessel or vasculature and can be washed away from the target structure which then the cell will take over the structure [3]. In general, bioprinting is the process that builds a 3D construct using computer-controlled 3D printing devices by depositing the cells and biomaterials precisely into geometries that create anatomically correct biological structures. It prints by outputting layer-upon-layer of living cells in hydrogels or viscous fluid, or cell-seeded microcarriers which is term as bioink [1]. 3D construct generated by 3D bioprinting technology can be used in regenerative medicine applications, and also as tissue models for drug screening or as a disease models [4].

In 3D bioprinting system, there are varies set of printing parameter involved, such as pressure and printing speed, that will affect the bioprinted construct. It is essential to understand how the system parameters affect the structure of the printed construct and the viability of the cell on the bioprinted construct as cell are very sensitive to environmental changes. There are two parts of experimental in this study. The first part is to evaluate the effect of the extrusion pressure and printing speed on the morphology structure of the printed construct. The second part is to evaluate the effect on extrusion pressure on viability of the fibroblast cells on the bioprinted construct after printing process.

2. Experimental

2.1. Effect of extrusion pressure and printing speed on structure

Polyoxyethylene–polyoxypropylene–polyoxyethylene (Pluronic F127, BioBots) was allow to cool at 4°C before used in printing to make sure the material in liquid form and was transferred into a metal extruder and capped with 30 gauge needle to be used for printing. For printing process, commercial 3D bioprinter (BioBots) was utilized. The desired printing model was prior designed by using CAD software (SolidWorks). The .stl file was converted to .gcode file by loaded into repetier host (Slic3r) to adjust the preferred printing parameters. The 3D construct was printed at pressure of 60, 80 and 100 psi, and the printing speed from 2, 4, 8 and 12 mm/s. Morphology characterization of the extruded line was made by observing the 3D printed construct under the stereo microscope (Leica Zoom 2000) at 15x and 30x magnification.

2.2. Effect of Extrusion Pressure on Cell Viability

Figure 1 shows an illustration on how the process of printing with cell. First, human skin fibroblast cells were cultured in Dulbecco's Modified Eagle Medium (DMEM) supplemented with 10% Fetal Bovine Serum (FBS), 100 mg/ml of penicillin and 10 mg/ml streptomycin and were maintained in the incubator at 5% CO2 and 37°C. For bioink preparation, 0.5% w/v Lithium phenyl-2,4,6-trimethylbenzoylphosphinate (LAP, photoiniator, BioBots) was mixed in supplemented DMEM until fully dissolved. Then 10% w/v Gelatin Methacrylate (GelMa, BioBots) was added to the LAP DMEM mixture solution and mixed at 60°C. The mixture solution of DMEM, LAP and GelMa were then be heated to 90°C and filtered with 0.2 μm syringe. The solutions then were allowed to cool down to 37°C prior to addition of fibroblast cells of concentration $2x10^6$ cells/ml. The mixture then were transferred to sterile extruder and capped with sterile 27 gauge tapered needle. Printing process was the same as printing with pluronic F127. The construct was print by extruding the line with 3, 4, 5, 6, 7, 8, 9, 10, 15, 30, and 45 psi with 4 mm/s printing speed under the sterile condition. The construct then will be evaluated for cell viability by staining the construct with Live/Dead assay kit (Ethidium homodimer-1 and calcein-AM, Thermo Fisher Scientific) and incubated it at room temperature for 30 minutes, then examined by using inverted fluorescence microscope. The number of live and dead cell then was calculated by using ImageJ software. The percentages of live cells were calculate as the number of green-staining cells divided by the total number of cells and the percentages of dead cells were calculated as the number of red-staining cells divided by the total number of cells.

3. Results and discussion

3.1. Effect of extrusion pressure and printing speed on structure

From observation, it can be clearly seen that the extrusion pressure and the printing speed affect the structure of the printed construct. At lower pressure (60 psi), the flow of the pluronic F127 was not continuous resulting beading of the material. At 80 psi, the flow of the pluronic F127 was continuous which create a more constant thickness of line. However, printings with 80 psi extrusion pressure at higher printing speed (8 and 12 mm/s) create a beading construct. Whereas excessive outpour was observed at constructs that printed at highest pressure (100 psi) with lower printing speed (2 and 4 mm/s). From this observation, we can state that printing at lower extrusion pressure require a lower printing speed whereas printing at higher extrusion pressure require a higher printing speed. Figure 2 shows the printed construct print at vary extrusion pressure and printing speed under a stereo microscope at 15x magnification.

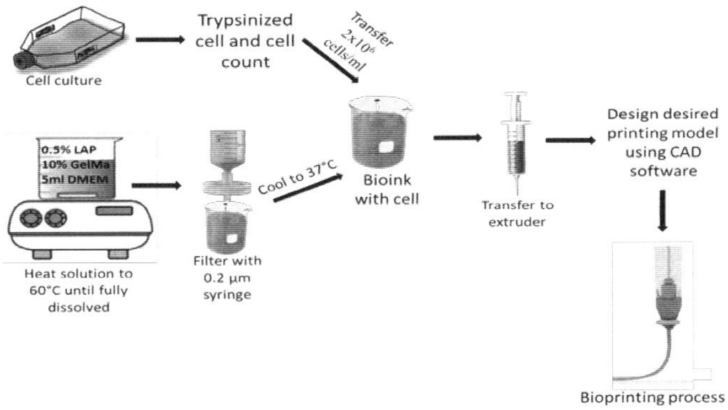

Figure 1. Illustration of bioprinting process

3.2. Effect of extrusion pressure on cell viability

Previous study shows that when printing with cell, it is essential to print at a lower extrusion pressure as print at high extrusion pressure will stress the cell [5, 6]. Because of that, we had changed the material use for printing from Pluronic F127 to GelMa as printing using Pluronic F127 requires high extrusion pressure. The constructs were print with various extrusion pressure ranging from 3 psi to 45 psi and printing speed of 4 mm/s as print at this speed will not produce excessive outpour and build-up of beading of the material.

Cell viability is an important experimental endpoint in this study. The cell viability on the bioprinted construct was determined by staining the construct with Live/Dead assay kit (Ethidium homodimer-1 calcein-AM, Thermo Fisher Scientific) where green staining cell indicates live cell and red staining cells indicate dead cells (Figure 3). A major hallmark of viable cells is an intact plasma membrane and intracellular enzymatic activity [7]. Generally, a green fluorescence will be generated if the live cells are identified on the basis of intracellular esterase activity, while the dead cells are identified by the lack of esterase activity and non-intact plasma membrane which allows red dye staining [8]. The results shows decreasing of the percentage of live cells when print with increasing extrusion pressure. Whereas printing at higher extrusion pressures resulting in increase in number of dead cells (Figure 4). According to Kong et al., 2003, higher pressure will increase the shear stress in the nozzle, which subsequently damage the cell membrane and cause lower cell viability after extrusion. Normally, cells will respond to stress by enduring adaptation. However, when the cells are unable to adapt, it will cause injury to the cell and followed by cell death [6].

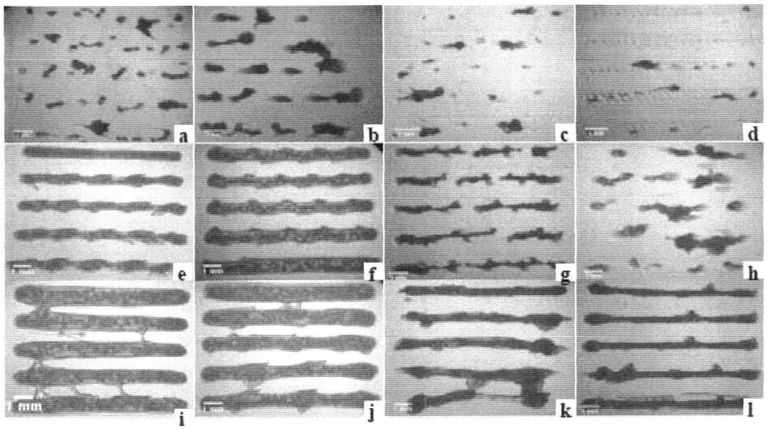

Figure 2. Image of the printed construct at 15x magnification, printing with 60 psi at 2mm/s (a), 4mm/s (b), 8mm/s (c), 12mm/s (d), 80 psi at 2mm/s (e), 4mm/s (f), 8mm/s (g), 12mm/s (h), and 100 psi at 2mm/s (i), 4mm/s (j), 8mm/s (k), 12mm/s (l).

Figure 3. The image of bioprinted construct after staining with Live/Dead assay kit printing with extrusion pressure of (a) 3 psi, (b) 4 psi, (c) 5 psi, (d) 6 psi, (e) 7 psi, (f) 8 psi, (g) 9 psi, (h) 10 psi, (i) 15 psi, (j) 30 psi, (k) 45 psi.

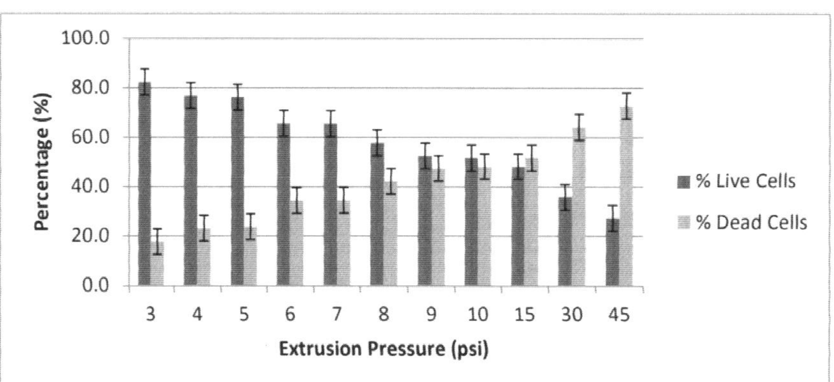

Figure 4. The percentage of cell viability when print at different extrusion pressure

4. Conclusion

Herein, we have presented and described the effect of extrusion pressure and printing speed on the structure of the printed construct and the effect of extrusion pressure on fibroblast cell viability of the printed construct. Our initial finding show that printing at lower extrusion pressure require a lower printing speed whereas printing at higher extrusion pressure require a higher printing speed and suggest 80 psi of extrusion pressure and 4 mm/s of printing speed will produce the best printed construct structure when printed with pluronic F127 using extrusion-based bioprinter. As for cell viability, percentages of live cells are decreasing when printing at higher extrusion pressure because higher extrusion pressure will produce higher shear stress at the nozzle hence damaging the integrity of the membrane cell thus promoting toward cell death. Therefore, to print viable cell, it is essential to print at lower extrusion pressure.

5. References

[1] Atala A and Yoo J J 2015 *Essentials of 3D biofabrication and translation* Academic Press 1
[2] Kolesky D B, Truby R L, Gladman A, Busbee T A, Homan K A and Lewis J A 2014 3D bioprinting of vascularized, heterogeneous cell- laden tissue constructs. *Advanced materials* **26** (19) 3124-3130
[3] Malda J, Visser J, Melchels F P, Jüngst T, Hennink W E, Dhert W J, Groll J and Hutmacher D W 2013 25th anniversary article: engineering hydrogels for biofabrication *Advanced materials* **25** (36) 5011-5028
[4] Zhang L G, Fisher J P and Leong K 2015 *3D Bioprinting and nanotechnology in tissue engineering and regenerative medicine* Academic Press 61
[5] Kong H J, Smith M K & Mooney D J 2003 Designing alginate hydrogels to maintain viability of immobilized cells *Biomaterials* 24(22) 4023-4029
[6] Nair K, Gandhi M, Khalil S, Yan K C, Marcolongo M, Barbee K and Sun W 2009 Characterization of cell viability during bioprinting processes *Biotechnology journal* 4(8) 1168-1177
[7] Kepp O, Galluzzi L, Lipinski M, Yuan J, & Kroemer G 2011 Cell death assays for drug discovery *Nature reviews Drug discovery* **10** (3) 221-237
[8] Haugland R P 2005 *The handbook: a guide to fluorescent probes and labeling technologies* Molecular probes (11th ed.) Carlsbad, California: Invitrogen

Acknowledgment
The authors would like to acknowledge Ministry of Higher Education Malaysia for funding through Fundamental Research Grant Scheme.

Loading and release of drugs inside the cavity of halloysite nanotubes

I F Wahab[1], N Zailani[1], S I A Razak[1,2*] and N H M Nayan[3]

[1]Faculty of Biosciences and Medical Engineering, Universiti Teknologi Malaysia, 81310 Skudai, Johor, Malaysia.
[2]IJN-UTM Cardiovascular Engineering Centre, Universiti Teknologi Malaysia, 81310 Skudai, Johor.
[3]Faculty of Engineering Technology, Universiti Tun Hussein Onn Malaysia, Pagoh Education Hub, KM 1, Jln Panchor, 84600 Panchor, Johor.

Corresponding author email: saifulizwan@utm.my

Abstract. Recent years have witnessed a huge growth of research in pharmaceutical sciences using nanoparticles as drug transport to reduce toxicity and for more specific drug targeting. Conventional application of drugs has major limitation and drawback since they have limited effectiveness, unsustainable release, and poor biodistribution. Thus this paper aims to produce a suitable material loaded with drugs that can provide an efficient effect of the drug at the site of action and maintain them throughout the treatment. Halloysite nanotubes were used as the material since it fulfills all the criteria needed for sustained drug release. Here, benzalkonium chloride was used as an antiseptic along with methylene blue and methyl orange as the cationic and anionic drug models. All three drugs were loaded into the halloysite lumen by vacuum suction. Result shows the most successful loading efficiency is halloysite loaded with methyl orange where the loading is almost 82% and also shows an excellent sustained release for 19 hours when released in 0.3M hydrochloric acid.

1. Introduction

Drugs are the main existing treatment to stop or reverse the effects of diseases. In the recent years, a number of materials have been extensively studied in order to deliver drugs into the body efficiently. A huge interest has been developed in nanoparticle as they are capable of targeting drugs to the site of action while minimizing the toxicity and the side effects. There is an enormous type of nanoparticles available varying from biological substances like albumin, gelatine and liposomes, and substances of chemical nature like polymer and also clay [1]. Halloysite nanotube (HNT) is an ultramicroscopic multi-layered hollow cylinder which its walls are formed from alternating layers of alumina and silica. Halloysite is capable of being loaded with chemically active agents such as drugs, antibiotic, flameretardant, anticorrosion, and protein. HNT becomes one of the favorable excipients because of its biocompatibility and easily available. R. R. Price et al. (2001) reported that halloysite clay could be used as a low-cost alternative to more traditional microencapsulation system [2].

Silica, SiO_2 made up the outermost layer of halloysite nanotubes while alumina, $Al(OH)_3$ construct as the inner lumen [3]. Due to this composition difference of the outer and inner layer, the surface of halloysite has different electrical ξ -potential. The outermost surface is negatively charged while inner lumen is positively charged. This made halloysites unique since most of the typical non-clay nanotubes, such as carbon nanotubes, have the same inner and outer chemistry. Besides, halloysite nanotubes have porosity of 50 to 60 $cm^2\ g^{-1}$ and cationic exchange capacity of 30 to 50 × 10^{-2} mol kg^{-1} [3]. Studies showed this porosity signifies the availability of the inner and outer tube surfaces for adsorption.

The release of bioactive agents from halloysite tubules usually takes place in two stages: the fast release of drugs from the external surface and the tube ends and slow release from the inner lumen [4]. Work by Dzamukova, M. R., and her colleague (2015) have shown a comparison of the release profile of antiseptic, brilliant green [5]. It dissolves in water within 5 minutes and has limited in time action, while it is released from halloysite during several hours. The initial stage of the release lasts about two hours and constitutes about 60% of the loaded active agents. Complete release of brilliant green is extended over 30 hours, which allows for drastic enhancement in its antiseptic functions adsorption.

The ideal mechanism of a targeted drug delivery is to have the drug releases in a determined manner and degrade within a period of time. However, traditional drugs have limited effectiveness, unsustainable release, and poor biodistribution. Halloysite is believed to be the most promising material for drug loading and sustained release. This project aims to prepare a drug loaded halloysite for drug delivery use. Three model drugs were used; benzalkonium chloride (BAC) as an antiseptic, methylene blue (MB) as cationic drug and methyl orange (MO) as the anionic drug model. Furthermore, this project also aims to have a profound and remarkable knowledge of the loading and release profile of drugs inside halloysite.

2. Materials and Method

2.1. Preparation of HNT-drug

Drug-loaded HNT were prepared according to the following procedures. 0.3 g of HNT powder was suspended in 150 mL of 0.3mg/mL drug in distilled water followed by stirring for 1 hour. The mixture was then transferred to vacuum desiccator for 15 minutes and cycled back to atmospheric pressure for 5 minutes in order to replace the entrapped gas with the saturated solution. This process was repeated two additional times to ensure that the halloysite is filled with the maximum amount of drugs possible. Next, the solution was centrifuged at 3000 rpm for 1 hour to remove the excess drug. The supernatant was collected and analyzed via a UV spectrophotometer. The halloysite-drug powder was washed three times with distilled water and then air dried.

2.2. Characterization

2.2.1. Morphology.

Transmission electron microscopy (TEM) image of the loaded halloysite is taken to see the presence of successful drug loaded inside the tubes.

2.2.2. Drug Loading Efficiency.

Drug entrapment efficiency (EE) was determined with the following equation:

$$EE(\%)=C_{HNT}=(V_iC_i-V_sC_s)/V_{HNT}\times 100 \tag{1}$$

where C_{HNT} = concentration of drug associated to halloysite
C_i = initial drug concentration
C_s = drug concentration in the supernatant after incubation
V_i = volume of initial drug solutions
V_s = volume of the supernatant
V_{HNT} = volume of halloysite respectively

2.2.3. In-Vitro Drug Release.

1 mg of dried drug-loaded halloysite was mixed with 15 ml of 0.3M HCl. The mixture is left to settle and kept in 37°C. At one hour interval, 3mL release media was collected and replaced with the same volume of fresh media. The same steps applied for drug release in Phosphate Buffer Solution (PBS) pH

7.4. Concentration is determined by measuring the adsorption of the solubilized active agents in a UV/VIS spectrophotometer (GENESYS 10S UV-Vis Spectrophotometer) at 662 nm (MB), 463 nm (MO), and 263 nm (BAC).

3. Results

TEM image of HNT-drug is expected to show the presence of drug inside the lumen. Methylene blue is used because it has color compared to BAC which is colorless. However, result from the TEM image does not show any sign of drug presence (Figure 1). Nevertheless, there is evidence that able to validate the entrapment of drug inside the HNT. Slight fizzing and bubbling of the suspension during vacuum suction can be seen which signifies air is being removed from the lumen. When the fizzing is seen to stop, the suspension is cycled back to atmospheric pressure to substitute the entrapped gas with the saturated drug solution.

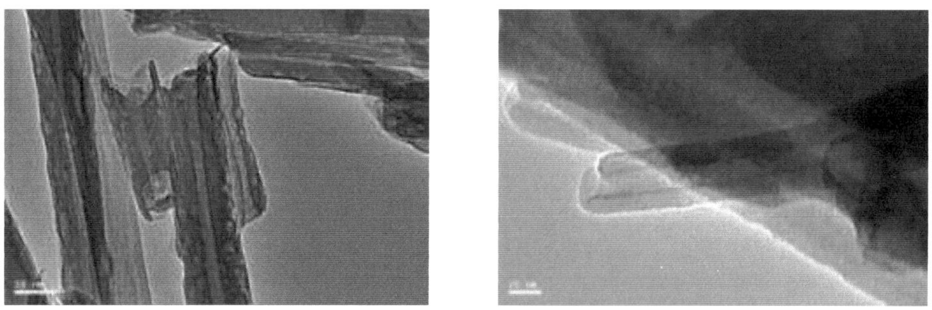

(a) **(b)**

Figure 1. (a), (b) TEM image of MB loaded halloysite

Figure 2 shows the entrapment efficiency of three type of drugs. MO shows the highest efficiency entrapment of about 82.29% compared to MB and BAC which are about 70.45% and 66.58% respectively. MO has higher loading efficiency due to the electrostatic attraction between the negative charge of MO and positive charge of HNT inner lumen. Meanwhile, for positively charge MB and BAC, the drug are more attracted to be deposited at the negatively charged HNT surface. According to Vergaro et al., higher loading efficiency also can be contributed by higher solubility of the drug [6].

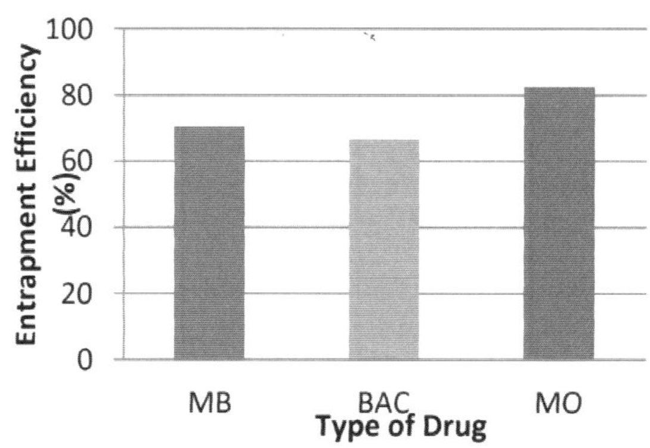

Figure 2. Entrapment efficiency of three type of drugs

IOP Conf. Series: Materials Science and Engineering **440** (2018) 012043 doi:10.1088/1757-899X/440/1/012043

Besides, with the aid from vacuum suction technique, it provides a great force to push the liquid drug inside the halloysite cavity. R. R. Price et al., (2001) in his work suggest that vacuum cycling is the recipe for successful entrapment into the very small lumen of the halloysite cylinder [2]. Cycling the suspension from vacuum to atmospheric pressure for multiple times also enhanced the replacement of air in the internal cavities with the drug.

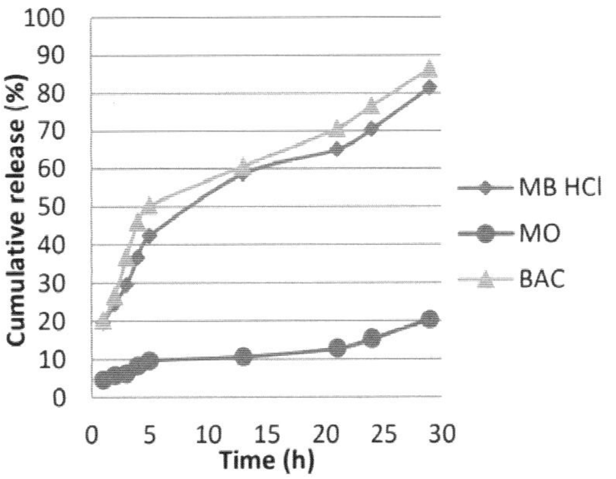

Figure 3. Cumulative release of drugs in 0.3M HCL

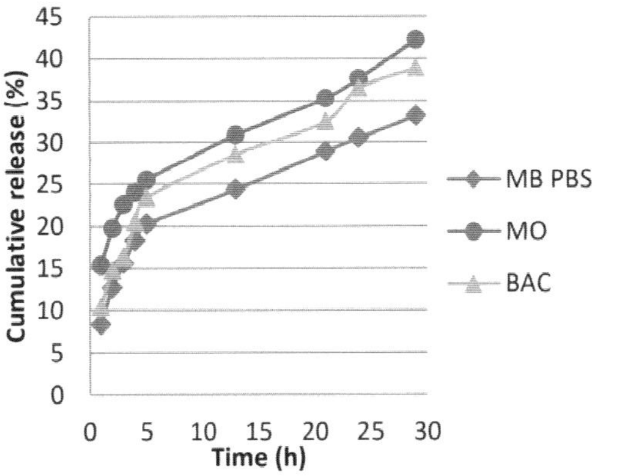

Figure 4. Cumulative release of drugs in PBS

In Figure 3, the extended release profile of three drugs from halloysite nanotubes in 0.3M HCl is illustrated. In 0.3M HCl, MO shows the slowest release with 15.30% release in 24 hours. A number of factors can affect the drug release rate, primarily involving the structure of the container and the chemical properties related to both the clay and the drug.

Chemically, halloysite nanotubes composed of different composition of the outer and inner layer which results in different electrical zeta-potential. In low pH, both HNT lumen and MB/BAC are positively charged. Therefore, a strong electrostatic repulsion accelerates the release of drug from HNT. For anionic MO case, the retention or the slow release could be explained on the basis of strong interaction of opposite charge molecules.

In comparison, Figure 4 shows an extended release of three drugs in PBS pH 7.4. MO gives the fastest release (37.58%) than MB (30.44%) and BAC (36.43%) in 24 hours. These rates contradict with the result in 0.3M HCl due to the chemical properties of the drug molecule in pH neutral. In neutral pH, both MB and MO are in their neutral form. Thus, there is no repulsion or attraction force between the molecules and the HNT charge. Hence, the release kinetics was evaluated using the Korsmeyer-Peppas model:

$$M_t/M_\infty = kt^{1-n} \qquad (2)$$

where M_t / M_∞ is fraction of drug released at time t, k is the release constant, and n is the release exponent. The value of n indicates the type of diffusion dominating the process.

Table 1 summarized the release kinetics of three drugs. The release exponent (n) for all three drugs is up to 0.5 which conclude the drug transport mechanism under Fickian diffusion. Fickian diffusion is associated with concentration gradient, diffusion distance, and the degree of swelling. In fact, MO has the highest solubility of all three drugs and can travel through the pores of halloysite through diffusion and degradation.

Table 1. Drug release kinetics.

Type of drug	Release exponent (n)
Methylene blue	0.3691
Methyl orange	0.2716
Benzalkonium chloride	0.3679

4. Conclusion
This project has successfully synthesized halloysite nanotubes as the drug nano-container. Benzalkonium chloride, methylene blue, and methyl orange are used as the drug to be loaded inside the halloysite lumen using the vacuum suction technique. Numerous findings have demonstrated that halloysite nanotubes are proven suitable as drug delivery system. In this study, measurement of drugloading efficiency has highlighted that methyl orange (anionic) interacts preferentially with the HNT lumen, compared to methylene blue and BAC (cationic). It is shown that the loading of drug is affected by chemical properties associated with both HNT and the drug. As for the release of drug, the process is greatly controlled by diffusion and degradation rate.

5. References
[1] De Jong W H and Born P J 2008 *Int. J. Nanomedicine* **3** 133–49
[2] Price R R, Gaber B P and Lvov Y 2001 *J. Microencapsul* **18** 713–22
[3] Lvov Y M, Shchukin D G, Mohwald H and Price R R 2008 *ACS Nano* **2** 814–20
[4] Abdullayev E and Lvov Y 2013 *J. Mater. Chem. B* **1** 2894-903
[5] Dzamukova M R, Naumenko E A, Lvov Y M and Fakhrullin R F 2015 *Sci. Rep.* **5** 10560
[6] Vergaro V, Lvov Y M and Leporatti 2012 *Macromol. Biosci.* **12** 1265–71

Acknowledgement
The authors would like to acknowledge Ministry of Higher Education Malaysia for funding through Fundamental Research Grant Scheme.

IOP Conf. Series: Materials Science and Engineering **440** (2018) 012044 doi:10.1088/1757-899X/440/1/012044

Effect of different nitrogen source combinations on microbial cellulose production by *Pseudomonas aeruginosa* in batch fermentation

J M Chai[1] and A Adnan[2]

[1,2]School of Fundamental Science, Universiti Malaysia Terengganu, 21030, Kuala Nerus, Terengganu, MALAYSIA

Email: azila.adnan@umt.edu.my

Abstract. Microbial cellulose (MC) is a type of polysaccharide that has high purity properties, which can be produced by different types of microorganism such as *Pseudomonas, Komagataeibacter, Sarcina* and *Azotobacter*, and widely used in industrial applications. However, the main drawback in microbial cellulose production is low yield, which unable to meet the requirements of the industry. This study aims to determine the effects of different nitrogen sources combinations on microbial cellulose production by *Pseudomonas aeruginosa* and to maximize microbial cellulose production using the best nitrogen sources combination. To improve cellulose production, five different nitrogen sources combinations, which were yeast extract (YE) + $(NH_4)_2SO_4$, YE + NH_4NO_3, YE + urea, YE + tryptic soy broth and YE + beef extract were added into the media with the composition of 50 g/L glucose, 3/L g KH_2PO_4 and 0.05 g/L $MgSO_4$. The batch fermentation was done in incubator shaker with 150 rpm at 30°C for five days. The samples were harvested every 24 hours, centrifuged, purified and dried for microbial cellulose concentration determination. The results obtained indicated that YE and beef extract combination produced the highest microbial cellulose concentration of 1.7 g/L. To attain the second objective, optimization study was carried out by using different concentrations, which were 5 g/L, 10 g/L, 15 g/L and 20 g/L of YE and beef extract combination. An investigation using 10 g/L YE and beef extract has produced 1.7 g/L of microbial cellulose, which proved to be the potential enhancer for microbial cellulose synthesis. It is recommended that additional morphology analyses should be done to investigate the influence of fermentation conditions and low-cost nitrogen sources media components on microbial cellulose morphology and its mechanical properties. This will help to synthesize desired and reliable microbial cellulose pellicles.

1. Introduction

Cellulose is a renewable and biodegradable polysaccharide found richly in the earth. This means it is a kind of energy source, which is not depleted when used and can be decomposed by other living organisms. This undoubtedly can overcome the problems such as energy crisis, population growth and environmental pollutions [1]. These renewable and biodegradable features of cellulose make it being vital topic in the study by many researchers and applied in various aspects. Cellulose has been given an enormous attraction on the common topic of widespread investigations in the field of macromolecular chemistry [1]. It is the major constituent of both about over 50% of the wood and approximately over 94% of the cotton [2]. A French chemist, Anselme Payen, discovered a fibrous substance, which is called cellulose in early 19th century during the experiment of the plant tissues treated with acids and ammonia [3].

Cellulose is not only can be produced by plants, but also can be synthesized by various bacteria. It can be synthesized by acetic acid bacteria such as *Pseudomonas aeruginosa*. These bacteria are used in this study because it can synthesize microbial cellulose (MC) production. Despite the plant cellulose has the same molecular formula with microbial cellulose, their physicochemical features are significantly different with each other [4]. In fact, the cellulose that is synthesized from the cell wall of the plants is usually not pure because it is commonly contaminated with lignin and hemicelluloses. It is also difficult to purify the plant cellulose from the lignin and hemicelluloses due to the complexity of the cell wall structures [5].

On the contrary, microbial cellulose has a high chemical purify as compared to plant cellulose because it was not associated with the impurities [6,7]. Meanwhile, MC also has high tensile strength, high biocompatibility, high water holding capacity, high crystallinity and high degree of polymerization [8, 9,10,11,12]. Due to these desirable properties of MC, it has been used as an alternative resource instead of plant cellulose in order to synthesize high quality cellulose. Hence, the cellulose can be derived from fermentation process via a microbial system [12].

Although microbial cellulose can be used to replace the plant cellulose, the high processing cost of microbial cellulose had become a great attention in the issues of microbial cellulose synthesis. The expensive nitrogen sources used even increase the cost of microbial cellulose production. Hence, this study is elucidating the effects of different nitrogen sources combinations on microbial cellulose synthesis. The nitrogen sources combination was used as parameter in this study and evaluated with intention to find the optimum concentration to generate the maximum yield of microbial cellulose synthesis [8].

2. Methodology

2.1 Culture media

Five different nitrogen sources combinations, which were 5 g/L yeast extract (YE) and 5 g/L ammonium sulphate $(NH4)_2SO_4$, 5 g/L YE and 5 g/L ammonium nitrate (NH_4NO_3), 5 g/L YE and 5 g/L urea, 5 g/L YE and 5 g/L tryptic soy broth and 5 g/L YE and 5 g/L beef extract were added into the culture media with the composition of 50 g/L glucose, 3 g/L potassium hydrogen orthophopsphate (KH_2PO_4) and 0.05 g magnesium sulphate ($MgSO_4$). Each 80 mL of culture media were added to 250-mL shake flasks. The pH of the media was adjusted to 6.8 and sterilized at 121 °C for 15 minutes.

2.2 Microorganism

Pseudomonas aeruginosa provided by Microbiology Laboratory, School of Fundamental Science was used throughout the study for microbial cellulose production.

2.3 Fermentation conditions and sampling

Five different nitrogen sources combinations, which were YE and $(NH_4)_2SO_4$, YE and NH_4NO_3, YE and urea, YE and tryptic soy broth and YE and beef extract used in this study. Eighty mililitre of culture media was added to 250-mL shake flasks. All trials were done at 30°C at pH 6.8 for five days. On each day of fermentation, the broth from shake flasks was harvested and then homogenized at 100 rpm for 5 minutes. The samples were centrifuged at 4000 rpm for 20 minutes. Then, they were washed gently with distilled water and centrifuged again to remove culture broth. The washing procedure was repeated three times. After centrifugation process, the washed pellets were used for microbial cellulose purification and determination. The data analysis was done by plotting the graph.

2.4 *Microbial cellulose determination and purification*
The washed pellets from one sample were treated with NaOH at 90°C for 30 minutes in order to dissolve cells. The cellulose obtained was centrifuged at 4000 rpm for 20 minutes. Next, they were washed with distilled water. After that, cellulose obtained was dried at 80°C for 24 hours and then weighed. Total weight of cellulose on each day of fermentation was measured.

2.5 Optimization of cellulose
The optimization study was carried out after identifying the best nitrogen sources combination. Different concentrations which were 5 g/L, 10 g/L, 15 g/L and 20 g/L of the best nitrogen sources combination were used in this study. Trials were done in 250-mL shake flasks. Fermentation was done at 30°C with 150 rpm for five days at 6.8. The samples, which taken every 24 hours were centrifuged at 4000 rpm for 20 minutes. Then, they were washed gently with distilled water and centrifuged again to remove culture broth. The washing procedure was repeated three times. After centrifugation, the washed pellets were used for cellulose purification and determination. The data was done by plotting the graph.

3. Results
This section explains the results on the effects of different nitrogen sources combinations and also the optimization of MC by varying the best combination of nitrogen source concentrations.

3.1 Effect of Different Nitrogen Sources Combinations on Microbial Celullose Production by Pseudomonas aeruginosa After Five Days of Fermentation
Based on figure 1, the bar graph depicted the average cellulose concentration produced for the combination of YE and $(NH_4)_2SO_4$, YE and NH_4NO_3, YE and urea, YE and tryptic soy broth and YE and beef extract from day 0 until day 5. YE and $(NH_4)_2SO_4$ combination acts as control in this study. For this combination, the amount of microbial cellulose produced was increased gradually from day 0 to day 5 as shown in figure 1. There was the least amount of cellulose of 0.1 g/L on the first day whereas celluloses production reached the highest concentration of 0.7 g/L on the fifth day.

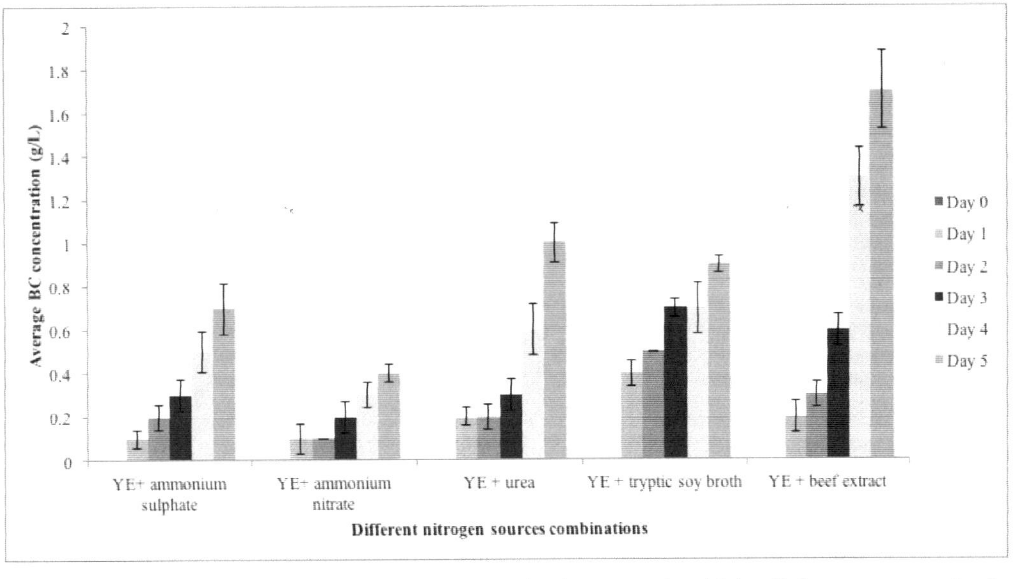

Figure 1. Effect of various nitrogen sources combinations on microbial cellulose concentration by *Pseudomonas aeruginosa*

For YE and NH_4NO_3 combination, the graph indicated that the microbial cellulose production was increased initially from day 0 to day 1 and then increased steadily from second day to fifth day. There was no increment in the amount of cellulose on second day of fermentation. The combination of YE and ammonium nitrate produced the least cellulose concentration of 0.1 g/L on first day and second day while it synthesized the highest amount of microbial cellulose of 0.4 g/L on fifth day. Although the cellulose concentration was increased constantly until the fifth day, this combination gave negative effect on cellulose synthesis since it produced lower cellulose concentration than the control medium.

According to the graph of YE and urea combination, there was an initial increase in the amount of cellulose from day 0 to day 1 and then the increment of cellulose concentration was continued significantly from second day to fifth day. The amount of cellulose produced on second day was same with the first day with 0.2 g/L of cellulose synthesized. This graph indicated that 0.2 g/L of cellulose produced on first day and second day were the least cellulose concentration whereas 1.0 g/L of cellulose produced on fifth day was the highest cellulose concentration. This combination gave positive effect on the cellulose production since it produced higher cellulose concentration than the control medium. Based on YE and tryptic soy broth combination, MC production was increased greatly from 0 g/L of MC on day 0 to 0.4 g/L of MC on day 1 and then it was rose gradually from first day to fifth day. There was no increment of MC occurred on fourth day in which same MC concentration of 0.7 g/L synthesized on both third day and fourth day. The least amount of MC produced was 0.4 g/L on first day whereas the highest amount of MC produced was 0.9 g/L on fifth day. This combination also gave positive effect on the MC synthesis since it produced more amount of MC than the control medium.

From the graph of YE and beef extract combination, the average MC production was increased significantly from day 0 to day 5. This combination produced the least MC concentration of 0.2 g/L on first day while it reached the maximum MC concentration of 1.7 g/L on fifth day. It could not be denied that this combination also gave positive effect on MC production as it synthesized the maximum MC concentration as compared to the control medium.

Based on Figure 1, the combinations of YE and urea, YE and tryptic soy broth and YE and beef extract produced 1.0 g/L, 0.9 g/L and 1.7 g/L of MC respectively, which were higher than the MC concentration of control medium (0.7 g/L) at the end of fermentation. In contrast, YE and NH_4NO_3 combination synthesized only 0.4 g/L of MC on the fifth day of fermentation, which lower than the MC concentration of control medium. Furthermore, the combination of YE and urea produced higher MC concentration than the combination of YE and NH_4NO_3. However, it was lower than the combination of YE and urea and YE and beef extract. From this graph, YE and beef extract combination produced the highest MC concentration of 1.7 g/L of MC and then followed by the combination of YE and urea, YE and tryptic soy broth and YE and NH_4NO_3. Therefore, YE and beef extract combination was considered as the best nitrogen sources combination in this study.

3.2 Volumetric Productivity of MC on Various Nitrogen Sources Combinations by Pseudomonas aeruginosa

From table 1, it showed the MC volumetric productivity of various nitrogen sources combinations by *Pseudomonas aeruginosa*. The combination of YE and NH_4NO_3 yielded the lowest MC productivity with 0.027 g/L/day. Meanwhile, YE and beef extract combination gave the highest MC productivity with 0.113 g/L/day and followed by the combination of YE and urea (0.067 g/L/day), YE and tryptic soy broth (0.060 g/L/day) and YE and $(NH_4)_2SO_4$ (0.047 g/L/day). Hence, the combination of YE and beef extract was considered as the best nitrogen sources combination as it obtained the highest MC productivity as compared to other nitrogen sources combinations.

Table 1. Effects of various nitrogen sources combinations on microbial cellulose volumetric productivity by *Pseudomonas aeruginosa*

Various Nitrogen Sources Combinations	MC Volumetric Productivity (g/L/day)
YE + $(NH_4)_2SO_4$ (control)	0.047
YE + NH_4NO_3	0.027
YE + urea	0.067
YE + tryptic soy broth	0.060
YE + beef extract	0.113

3.3 Effect of Different Concentrations of YE and Beef Extract Combination on MC Production by Pseudomonas aeruginosa After Five Days of Fermentation

According to figure 2, the graph showed the MC concentration produced by different concentrations of YE + beef extract combination, which were 5 g/L, 10 g/L, 15 g/L and 20 g/L of YE and beef extract from day 0 until day 5. For 5 g/L YE and beef extract combination, the average MC concentration was increased initially from day 0 to day 1 and then it was rose significantly from 0.1 g/L on first day to 0.4 g/L on second day. After that, the MC production was continued to increase gradually from third day to fifth day. Although there was an increment on MC concentration until fifth day, the amount of MC was remained constant on third day with 0.4 g/L of MC produced. This combination produced the least MC concentration of 0.1 g/L on first day while it synthesized the highest MC concentration of 0.7 g/L on fifth day.

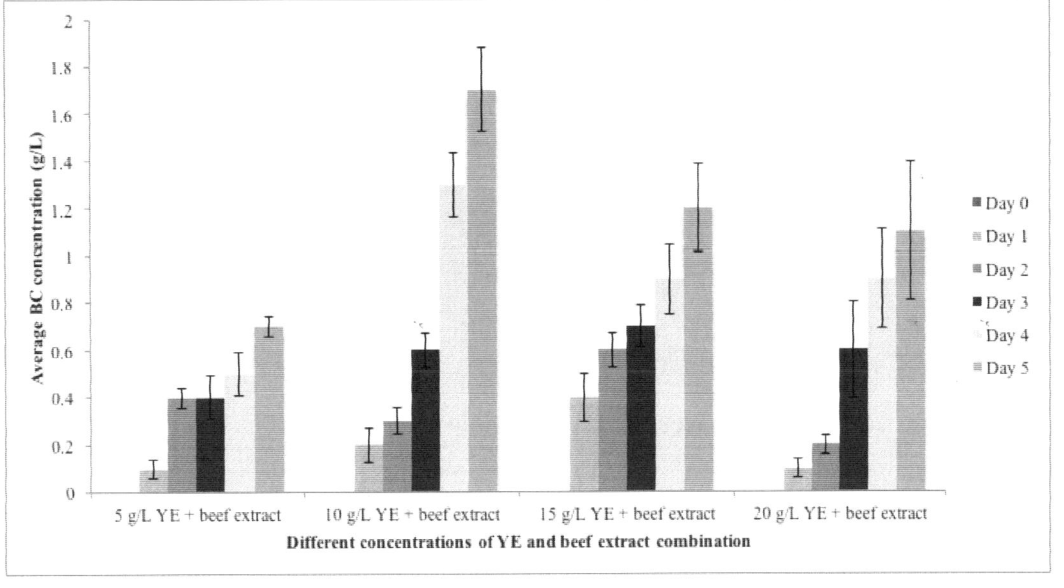

Figure 2. Effects of different concentration of YE and beef extract combination on MC production by *Pseudomonas aeruginosa*

Based on the graph of 10 g/L YE and beef extract combination, the MC production was increased significantly from the initial day to fifth day. This combination synthesized the least amount of MC with 0.2 g/L on first day whereas it produced the highest amount of MC with 1.7 g/L on fifth day.

According to 15 g/L YE and beef extract combination, the MC production was increased greatly from day 0 to day 1 and then it was continued to increase gradually until fifth day. For this combination, the lowest concentration of MC was 0.4 g/L on first day while the highest MC concentration was 1.2 g/L on fifth day.

For 20 g/L YE and beef extract combination, the graph showed that the average MC concentration was rose slowly from day 0 to day 2 and then it increased significantly until fifth day of fermentation. This combination synthesized 0.1 g/L of MC on first day which was the lowest MC production, whereas it produced 1.1 g/L of MC on fifth day, which was the highest MC production.

Based on figure 2, the combinations of 10 g/L, 15 g/L and 20 g/L YE and beef extract produced 1.7 g/L, 1.2 g/L and 1.1 g/L of MC respectively, which synthesized above 1.0 g/L of MC after five days of fermentation. However, the combination of 5 g/L YE and beef extract synthesized only 0.7 g/L of MC, which produced below 1.0 g/L of MC concentration on the fifth day of fermentation. The combination of 15 g/L YE and beef extract synthesized higher MC concentration than the combination of 5 g/L and 20 g/L YE and beef extract but it produced lower MC concentration than the combination of 10 g/L YE and beef extract. According to Figure 2, the combination of 10 g/L YE and beef extract produced maximum MC production of 1.7 g/L, followed by the combination of 15 g/L YE and beef extract, 20 g/L YE and beef extract and 5 g/L YE and beef extract. Hence, 10 g/L YE and beef extract combination was preferred as the best concentration of YE and beef extract combination to maximize the MC production.

3.4 Volumetric Productivity of MC on Different Concentrations of YE and Beef Extract Combination by Pseudomonas aeruginosa

According to table 2, it showed the MC volumetric productivity of different concentrations of YE and beef extract combination by *Pseudomonas aeruginosa*. The combination of 5 g/L YE and beef extract gave the lowest MC productivity with 0.047 g/L/day. Meanwhile, the combination of 10 g/L YE and beef extract reached the highest MC concentration with 0.113 g/L/day and followed by the combination of 15 g/L YE and beef extract (0.080 g/L/day) and 20 g/L YE and beef extract (0.073 g/L/day). Therefore, the combination of 10 g/L YE and beef extract was considered as the best concentration of YE and beef extract combination as it obtained the highest MC productivity as compared to other concentrations of YE and beef extract combination.

Table 2. Effects of different concentrations of YE and beef extract combination on MC volumetric productivity by *Pseudomonas aeruginosa*

Different Concentrations of YE and Beef Extract Combination	MC Volumetric Productivity (g/L/day)
5 g/L YE + beef extract	0.047
10 g/L YE + beef extract	0.113
15 g/L YE + beef extract	0.080
20 g/L YE + beef extract	0.073

4. Discussion

The effect of various nitrogen sources on the productivity of MC have been studied by previous researchers. For example, Hungund & Gupta [6] investigated that the effect of different nitrogen sources such as peptone, casein hydrolysate, beef extract, malt extract, sodium nitrate, ammonium chloride, ammonium sulphate, potassium nitrate, ammonium nitrate and urea on the MC production. They concluded that peptone, casein hydrolyzate, beef extract and malt extract gave the MC yield of above

2.0 g/L. A previous study also indicated that the use of different nitrogen sources such as yeast extract, peptone, malt extract, tryptone and sodium nitrate gave different concentration of the MC [13]. Nevertheless, these studies only used a single nitrogen source in the medium to investigate the effectiveness of MC productivity by bacteria. The single nitrogen source exhibited a significant reduction in the MC production [14].

However, there are limited studies on the production of MC by using nitrogen sources combination. In this study, the combination of YE and urea, YE and tryptic soy broth and YE and beef extract were the new nitrogen sources combination being investigated. YE used in all combinations because it was considered as highly potential nitrogen source to enhance the MC synthesis. It increased and enhanced the growth of bacteria [15, 12]. It also provided many nitrogen compounds, carbon and growth factors such as amino acids and vitamins, especially vitamin B complex, which required by the bacteria for growth and also serves to stimulate the productivity of MC [12]. In addition, YE plays an important role in the biosynthesis of essential molecules such as protein and nucleic acids [11]. It was also stated that the combination of both YE and other nitrogen sources improved the MC synthesis effectively [14]. According to figure 1, the average MC concentration was generally increased from day 1 to day 5. This is because the MC concentration was relative to the amount of nitrogen and glucose consumed. When the time taken for fermentation was longer until optimum periods, the amount of nitrogen and glucose consumed was higher and thus increased the MC production by *Pseudomonas aeruginosa* [15]. The bacteria inoculated into the medium containing nitrogen sources, needed certain time to incorporate and respond it for MC production. Thus, there was no MC produced in day 0 because the inoculated bacteria were not responded to the medium within a few minutes to produce MC. Apart from that, there was low MC yield in day 1 in which the combination of YE and $(NH4)_2SO_4$ and YE and NH_4NO_3 were produced 0.1 g/L of MC only while the combination of YE and urea and YE and beef extract were produced 0.2 g/L of MC. This is because the bacteria were adapted at a slow rate to the medium and so they synthesized low amount of MC on initial day. On the contrary, the combination of YE and tryptic soy broth was produced a higher MC concentration of 0.4 g/L on the first day because it contained digests of casein and soybean meal, which supply excessive amino acids and other nitrogenous substances to stimulate the bacteria in order to enhance the MC production at the initial stage.

In figure 1, there was the same average MC production on day 1 and day 2 for the combination of YE and NH_4NO_3. The similar situation also occurred in day 1 and day 2 for the combination of YE and urea and day 3 and day 4 for the combination of YE and tryptic soy broth. This indicated that there was no increase in the amount of MC on the latter day. This is because the media placed in the incubator shaker, were affected by the oxygen transfer rate. The limited amount of incubator shakers in the laboratory were used by many students and the repeated step of opening and closing the incubator shakers influenced the availability of oxygen inside it.

$(NH_4)_2SO_4$ and NH_4NO_3 used in this study were inorganic nitrogen sources whereas urea, tryptic soy broth and beef extract were organic nitrogen sources. From figure 1, it showed that the combination of both organic nitrogen sources such as YE and urea, YE and tryptic soy broth and YE and beef extract gave higher MC concentration than the combination of both organic and inorganic nitrogen sources such as YE and $(NH_4)_2SO_4$ and YE and NH_4NO_3. Embuscado and colleagues [16] stated that organic nitrogen sources enhanced higher MC concentration than inorganic nitrogen sources. It was similar with the results of the research carried out by Abdelhady and coworkers [17]. Nitrogen sources support rapid growth and high cell yields of bacteria than inorganic nitrogen sources [18]. In addition, it is also indicated that the combination of both inorganic nitrogen sources gave little or no MC produced and vice versa [16]. Therefore, from the result in figure 1, it can be concluded that the combination of both organic nitrogen sources gave the highest MC concentration and then followed by the combination of both organic and inorganic nitrogen sources.

The combination of YE and NH_4NO_3 gave negative effect to the MC synthesis since it synthesized lower MC concentration than the control medium because NH_4NO_3 provided less nitrogen than

$(NH_4)_2SO_4$. $(NH_4)_2SO_4$ is composed of nitrogen and sulphur whereas NH_4NO_3 is the chemical compound of nitric acid and ammonia. Nitric acid is not required by *Pseudomonas aeruginosa* and its acidic property gave effect of significant reduction on MC production. Yeast extract incorporated into the medium containing NH_4NO_3 also could not improve much the amount of MC to be produced. In previous study, it has been found that NH_4NO_3 produced the least amount of MC as compared to other inorganic nitrogen sources [6]. Apart from that, the combination of YE and tryptic soy broth synthesized lower MC concentration than the combination of YE and urea because many nutrients such as amino acids and other nitrogenous substances in the tryptic soy broth has been utilized at the beginning and there was no much nutrients left to stimulate the bacteria to produce much MC when the fermentation process was continued.

The combination of YE and beef extract produced the maximum MC concentration as compared to other nitrogen sources combinations and it was considered as the best nitrogen sources combination to enhance the MC production. Beef extract provided various sources of nutrients and it was a mixture of minerals, organic acids, nucleotide fractions, vitamins, amino acids and peptides [19]. In addition, its function could be described as complementing the nutritive properties of peptone by providing adequate phosphates, minerals and energy sources. Various nutrients contributed by beef extract became the requirements of the bacteria for cell growth and MC yields. At the same time, these nutrients also enhanced the biosynthesis pathway of bacteria and thus increased the average MC production. According to table 1, the combination of YE and beef extract also gave the highest MC productivity of 0.113 g/L/day as it produced the highest amount of MC during the fermentation.

In further optimization study, based on figure 2, the amount of MC produced also increased gradually from day 0 to day 5 since the MC concentration was directly proportional to the time taken for fermentation. Lina et al. [1] indicated that the fermentation time gave a significant effect on the MC production. There was also low MC synthesized in the first day of fermentation for all the combination of different concentration of yeast extract and beef extract except 15 g/L, because the bacteria needed time to respond with media for MC production. However, the MC concentration was higher in day 1 for the combination of 15 g/L YE and beef extract because the bacteria were adapted well initially in this concentration to produce much MC at the beginning. For 5 g/L YE and beef extract combination, there was same MC concentration in day 2 and day 3 because it was affected by the availability of oxygen in the incubator shaker which had mentioned on above.

Based on figure 2, 5 g/L YE and beef extract combination produced the least MC concentration because the little concentration of beef extract used gave insufficient nutrients and nitrogen for *Pseudomonas aeruginosa* to enhance the production of MC. Meanwhile, the combination of higher concentrations of YE and beef extract such as 15 g/L and 20 g/L also not synthesized the higher production of MC. This is because the excess nitrogen provided by the higher concentration of beef extract could not increase the MC synthesis. The excess nitrogen would disrupt the cell metabolism of bacteria and caused them could not absorbed properly the other vital nutrients such as carbon and salts. Although the higher concentration of beef extract provided excess nitrogen, the insufficient of other nutrients could not enhance the productivity of MC. In addition, Ashjaran & Sharabiyani [20] stated that the extra nitrogen favours the biomass production but reduced the MC synthesis. Furthermore, 10 g/L YE and beef extract combination produced the highest MC concentration and it was the potential concentration of beef extract to produce the maximum MC concentration. This is due to the adequate nitrogen provided by optimum concentration of beef extract would not affect the absorption of other nutrients by bacteria. *Pseudomonas aeruginosa* obtained sufficient nitrogen and other crucial nutrients simultaneously and thus eventually improved the MC production. Moreover, Hungund & Gupta [21] investigated that 6 g/L of beef extract produced the highest MC concentration of 5.89 g/L. According to their research and this study, it could be deduced that the range of 6 g/L to 14 g/L of YE and beef extract would be the optimum concentration of beef extract to maximize the MC production. Beyond the optimum concentration, the biosynthesis pathway of bacteria was influenced and thus reduced the average MC production. Based on table 2, 10 g/L YE and beef extract combination achieved the highest MC productivity of 0.113 g/L/day as this combination synthesized the highest amount of MC during fermentation process.

5. Conclusion

In conclusion, YE and beef extract combination gave the highest MC concentration and volumetric productivity. Varying this combination at different concentrations influenced MC production. In addition, 50 g/L of YE + beef extract combination yielded highest MC production. However, increasing YE + beef extract concentration beyond 50 g/L inhibits MC synthesis and its volumetric productivity.

6. References

[1] Lina F, Yue Z, Jin Z and Guang Y 2011 *J. Biomed. Eng.* **5** 250–27
[2] Keshk S M 2014 *J. Bioprocess Biotech.* **4** 1–10
[3] Chen X I 2014 '*Degradation Studies on Plant Cellulose and Bacterial Cellulose by FT-IR and ESEM' Master Thesis*, University of Birmingham, United Kingdom
[4] Tsouko E, Kourmentza C, Ladakis D, Kopsahelis N, Mandala I, Papanikolaou S, Paloukis F, Alves V and Koutinas A 2015 *Int. J. Mol. Sci.* **16** 14832–14849
[5] Bian J, Peng F, Peng X P, Peng P, Xu F and Sun R C 2012 *Biores.* **7** 4626–4639
[6] Hungund B S and Gupta S G 2010a *J. Microbiol. Biochem. Technol.* **2** 127-133
[7] Anna Z, Dorota P, Rafał R, Maciej K, Marian K and Karol F 2015 *Acta Sci. Pol. Zootechnica.* **14** 137-150
[8] El-Saied H, El-Diwany A I, Basta A H, Atwa N A and El-Ghwas D E 2008 *Biores.* **4** 11961217
[9] Torres F G, Commeaux S and Troncoso O P 2012 *J. Funct Biomater.* **5** 64-78
[10] Nainggolan H, Gea S, Bilotti E, Peijs T and Hutagalung S D 2013 *Beilstein J Nanotechnol.* **4** 325-329
[11] Al-Kalifawi E J and Hassan I A 2014 *Baghdad Sc. J.* **11** 1420-1428
[12] Kazim A R 2015 *World J. Exp Biosci.* **3** 89–93
[13] Cheng K C, Catchmark J M and Demirci A 2009 *J. Biol. Eng.* **3** 1-10
[14] Yodsuwan N, Owatworakit A, Ngaokla A, Tawichai N and Soykeabkaew N 2012 *First Mae Fah Luang Univ. Int. Conf.* **61** 1-6
[15] Coban E P and Biyik H 2011 *African J. Biotech.* **10** 5346–5354
[16] Embuscado M E, Marks J S and BeMiller J N 1994 *Food Hydrocolloids* **8** 407–418
[17] Abdelhady S S, Hassan E A, Hemmat M, El-Salam A and Abdullah S M 2015 *British Microbiol. Research J.* **9** 1-13
[18] Costa E, Teixido N, Usall J, Atare´s E and Vin˜as I 2002 *Letters in Appl. Microbiol.* **35** 117120
[19] Mahendra K T, Gopal N, Shrikant P, Rama M T, Snehasis J and Rakesh M 2015 *J. Nutr. Food Sci.* **5** 1-8
[20] Ashjaran A and Sharabiyani N M 2014 *Res. J. Pharm. Biol. Chem. Sci.* **5** 1324-1334
[21] Hungund B S and Gupta S G 2010b *World J. Microbiol. Biotechnol.* **26** 1823–1828

Acknowledgement

This research is supported by TPM (Tabung Penyelidik Muda) grant. The authors fully acknowledged Universiti Malaysia Terengganu for the approved fund, which makes this important research viable and effective.

Dream over life: Psychedelic terphenyl derivative induce hallucination via cannabinoid receptor 1

F A Fauzi[1], M S Goh[1], S A T T Johari[1,2], F Hashim[1], M F N Hassim[1,*]

[1]School of Fundamental Science, Universiti Malaysia Terengganu, 21030 Kuala Nerus, Terengganu, Malaysia
[2]Faculty of Bioresources and Food Industry, Universiti Sultan Zainal Abidin Besut Campus,
22200, Besut, Terengganu, Malaysia

[*]Author to whom any correspondence should be addressed.
E-mail:muhamad.fairus@umt.edu.my

Abstract. For ages, natural psychedelic resources have been used by ancient tribes for religious inspiration. In modern medicine, these compounds were prescribed to relieve severe distress and depression on cancer patients. Despite medical benefit, abuse of these compounds have become prevalent in our modern society. These compounds usually interacted with cannabinoid receptor 1 (CNR1) on neuron cell causing hallucination, and on other cell-types. In this study, chemically synthesized terphenyl derivative, 1,4-di(phenyl)benzene (13-BPB) interaction with human and its animal model were assessed. This derivative is an analogue found in fungi although their functional molecular mechanism is unknown. Terphenyl derivative known to have pharmacological activities - antifungal, anti-cancer, anti-coagulant. Our study designed includes in-vitro assessment and in-silico model of 13-BPB interaction to the molecular mechanism in human and its animal model, mice. Cytotoxicity assessment using MTT has shown that treatment of 13BPB on NIH-3T3 and RAW 264.7 have significant reduction in cell viability at 0.016mM and 0.08mM, respectively. Virtual database screening based on homologous compounds identified possible interaction with 15 different proteins from receptors, enzymes and transcription factors, in human and mice. Further docking analysis shows terphenyl derivatives binding affinities (pKd/pKi) are the highest with CNR1 and oestrogen receptors (ESRs).

1. Introduction

For ages, psychedelic plants and fungi were used by shamans of ancient tribe during religious ritual. They believe that these plants assist their ascension to the spiritual dimension [1]. Psychedelic plants have psilocybin that can alter the states of consciousness or hallucination. In modern medicine, psilocybin was prescribed by the cancer patients as it can relieve severe distress and depression [2].

Depression is synonym with cancer patients due to prolonged hospital admission and having a limited day of being alive. Psilocybin which can be found in *Psilocybe cubensis,* magic mushroom can alter the emotions and perceptions of the patients by causing hyperactivity of frontal regions of brain followed by other hallucinogenic effects such as euphoria. Other effect includes increase body temperature and blood pressure as well as having difficulty in thinking. This due to hypermetabolism on right hemisphere of brain caused by the consumption of psilocybin [3,4]. At the molecular level, the hallucination is caused by psilocybin interaction with cannabinoid receptors (CNRs) of neuronal cells [5]. Hallucination also can be caused by terphenyl derivatives which mainly found in fungi.

In this study, pharmaceutical potential of a terphenyl analogue found in fungi, 1,4-di(phenyl)benzene (13-BPB) was assessed. Terphenyl derivative, 13-BPB is chemically synthesised which is more eco- friendly that mimic the characteristics of natural terphenyl. Our study designed included in-vitro assessment and in-silico model of 13-BPB interaction to the molecular mechanism in human.

2. Materials and method

2.1. Cell propagation assay

The 500 mL culture media Dulbecco's Modified Eagle's Medium (DMEM) from Nacalai Tescue and Roswell Park Memorial Institute (RPMI, Gibco) were used as the main culture media for the NIH 3T3 fibroblast cells and RAW 264.7 macrophage cells, respectively. Fetal bovine serum and penicillin/streptomycin (Gibco) were added into the media with the volume of 50 mL and 5 mL respectively. Cells were incubated at 36.5°C, 5%CO_2.

2.2. 1,4-di(phenyl)benzene

The 1,4-di(phenyl)benzene (13-BPB) is chemically synthesised terphenyl derivative based on derivative found in mushroom. The crystalized compound was dissolved with DMSO (as a vehicle) at 10mM stock concentration and stored in -20°C.

2.3. Cytotoxicity assessment

Tetrazolium salt, 3-4,5 dimethylthiazol-2,5 diphenyl tetrazolium bromide (MTT) assay was used to determine the cytotoxicity of 13-BPB. NIH 3T3 fibroblast cells and RAW 264.7 macrophage cells were treated for 24 hours' prior addition of 20 µL of 5 mg mL-1 MTT. The incubated for three hours for formazon crystal formation. After that, MTT dye was removed from the wells carefully to avoid disturbance on the cell monolayer. The formed formazon that remained in the were dissolved with DMSO. Absorbance were measured using microplate reader at 570 nm wavelength.

2.4. Virtual screening of potential protein target

Identification of possible proteins targeted by 13-BPB was screened using online database, SwissTargetPrediction (http://www.swisstargetprediction.ch). The prediction is based on similarity in 3-dimensional structure of ligands that are known to be interacted with the protein [6]. From the list, top three human proteins from the prediction were selected for docking analysis.

2.5. Molecular docking

The crystal structure of selected proteins was obtained from RSCB protein database. Type of interaction and binding affinities of 13-BPB with the selected proteins were assessed using online SystemsDock application (http://systemsdock.unit.oist.jp).

3. Result and discussion

3.1 Cytotoxicity assessment of 13-BPB to NIH 3T3 fibroblasts and RAW264.7 macrophages

The cytotoxicity of 13-BPB was determined by treating NIH 3T3 murine embryonic fibroblast (MEF) and RAW264.7 macrophage cell line for 24 hours at various concentration. Cytotoxic level of vehicle (DMSO) has to be determined to avoid false positive. Both cell-type were treated with vehicle at equivalent concentration to the 13-BPB concentration ratio. The data revealed that the vehicle caused the reduction in cell viability measurement at 0.4mM (4%v/v) and 1mM (1%v/v) for NIH 3T3 MEF and RAW264.7 macrophage cell lines, respectively (figure 1a). 24 hours' treatment with 13-BPB on the NIH 3T3 and RAW 264.7 at gradient concentration, shows significant reduction (two-tailed T-test, $p_{value} < 0.01$) in cell viability at 3.2μM for both cell-type, lower than the vehicle. This indicates that the 13-BPB can induce cell death at very low concentration, less than an order of magnitude micromolar (figure 1b). Previous study revealed that other terphenyl derivatives treatment caused significant reduction in cell viability at concentration from 0.87μM up to 0.91μM when tested on human cancerous cell lines, lung and hepatocellular carcinoma cells however cytotoxicity was not observed when tested on normal human hepatocytes cell lines [7,8].

3.2 Database screening of protein targeted by 13-BPB

Screening of proteins being targeted by 13-BPB using online database shows 15 potential proteins in human (figure 2a) and mouse (figure 2b). Ranging from transcription factors, enzymes and membrane receptors. The potential interaction was predicted based on similarity in the three-dimensional structure of compound that known to interact with the listed proteins [6]. Almost the same proteins in human and in mice are being targeted by 13-BPB especially in the top five. The top four proteins are membrane receptors that are presence in majority of cell-type including neuronal cells. These receptors not just responsible for neuronal signalling but also will influence activity of cyclin dependent kinases, hence affecting the fate of cell in cell cycle phases and cell death [9,10].

3.3 Molecular docking analysis

A total of five proteins were selected for docking analysis to determine binding affinities and to characterise their interaction structure. The selection is based on the highest probability scores and availability of crystallised structure in the RSCB protein database. Results from 13-BPB docking were compared with the native ligand of selected proteins (figure 2c, d). The model predicted that 13-BPB highest bind affinity is with the CNR1 at 5.622 pKd/pKi, even stronger than the native ligands (Taranabant) which is at 3.4 pKd/pKi. Taranabant is a cannabinoid-1 receptor reverse agonist that used as anti-obesity where it can induce the feeling of satiety [11]. Structural inspection on the docking site shows that 13-BPB docks on the same site as Taranabant (figure 3). A total of nine amino acid residues hydrophobically interacted with 13-BPB compared to 12 residues with Taranabant (table 1). Both share eight amino acid residues thus indicate potential competitor or could cause very similar physiological outcome. Higher binding affinity (pKd/pKi) of 13-BPB is owing to its structure is more stable and has higher surface area than Taranabant, which has open structure thus unstable to allow optimal hydrophobic interaction.

13-BPB also interacted with receptors for steroid hormones such as androgen receptor (AR), estrogen receptor 1 (ESR1) and estrogen receptor 2 (ESR2) which affecting the cell cycle. This interaction would affect MAPK signalling pathway that involved in cell proliferation and cell survival [9]. For example, 13-BPB also interacted with ESR2, the second highest protein after CNR1. However, the binding affinity of 13-BPB with ESR2 at 5.037 pKd/pKi is lower compared to its native ligand, Genistein at 5.6 pKd/pKi. Genistein is an isoflavone that had been used in cancer therapy together with protein tyrosine kinase where it activates MAPK pathway thus inhibits the inflammatory action of cancer cells [12]. Structural inspection on the docking site shows that 13-BPB docks on the same site as Genistein (figure 4). A total of eleven amino acid residues form hydrophobic interaction with 13-BPB compared to eight residues with Genistein (table 2) and both share five amino acid residues. Higher binding affinity of Genistein than 13-BPB is due to the formation of hydrogen bond of oxygen atom of Genistein with the arginine-346 and Histidine-475. While 13-BPB only interacted hydrophobically with the residues which is weaker compared to the hydrogen bond.

Interaction of 13-BPB with CNR1 and receptors for steroids that directly connected to the MAPK signalling pathway would have significantly affect homeostasis of this pathway and the downstream physiological components. Notably affecting the homoestasis of cyclin-dependent kinases of cell cycle hence may causes the cell-death as observed in cytotoxicity assessment (figure 1).

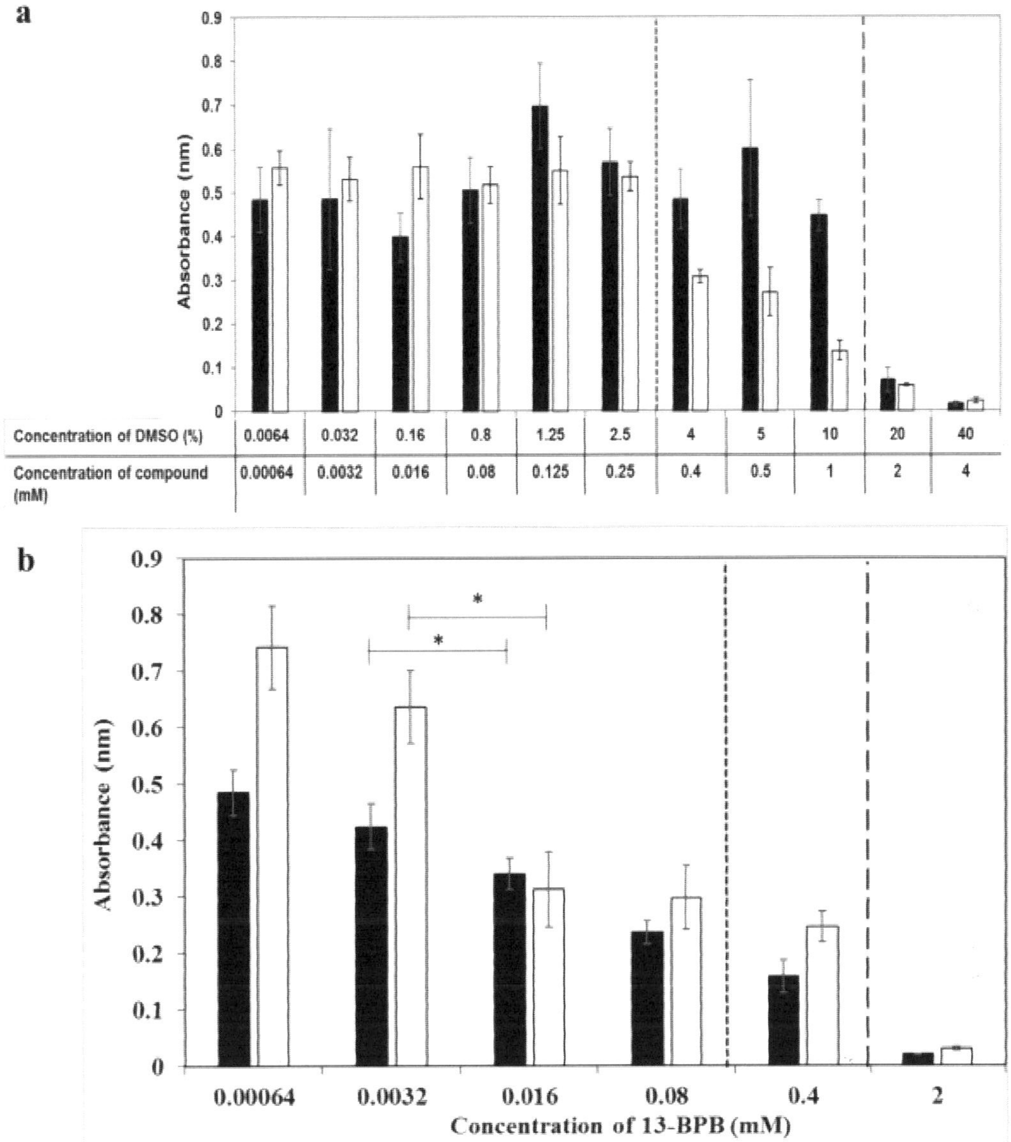

Figure 1. MTT assay for assessment of 13-BPB cytotoxicity. Treatment of NIH-3T3 murine embryonic fibroblast cell line (□) and RAW264.7 macrophage cell line (■) with DMSO (a) and 1,4-di(phenyl)benzene (13-BPB) (b). Dotted line and dashed line indicate assay limit threshold for NIH3T3 fibroblast and RAW264.7 macrophage, respectively. Both cell density; 1x10^4 cell/well in 96 wellplate. Error bars mean ± SE; n=8, * indicate $p_{value} < 0.01$ for two-tailed Student's T-test.

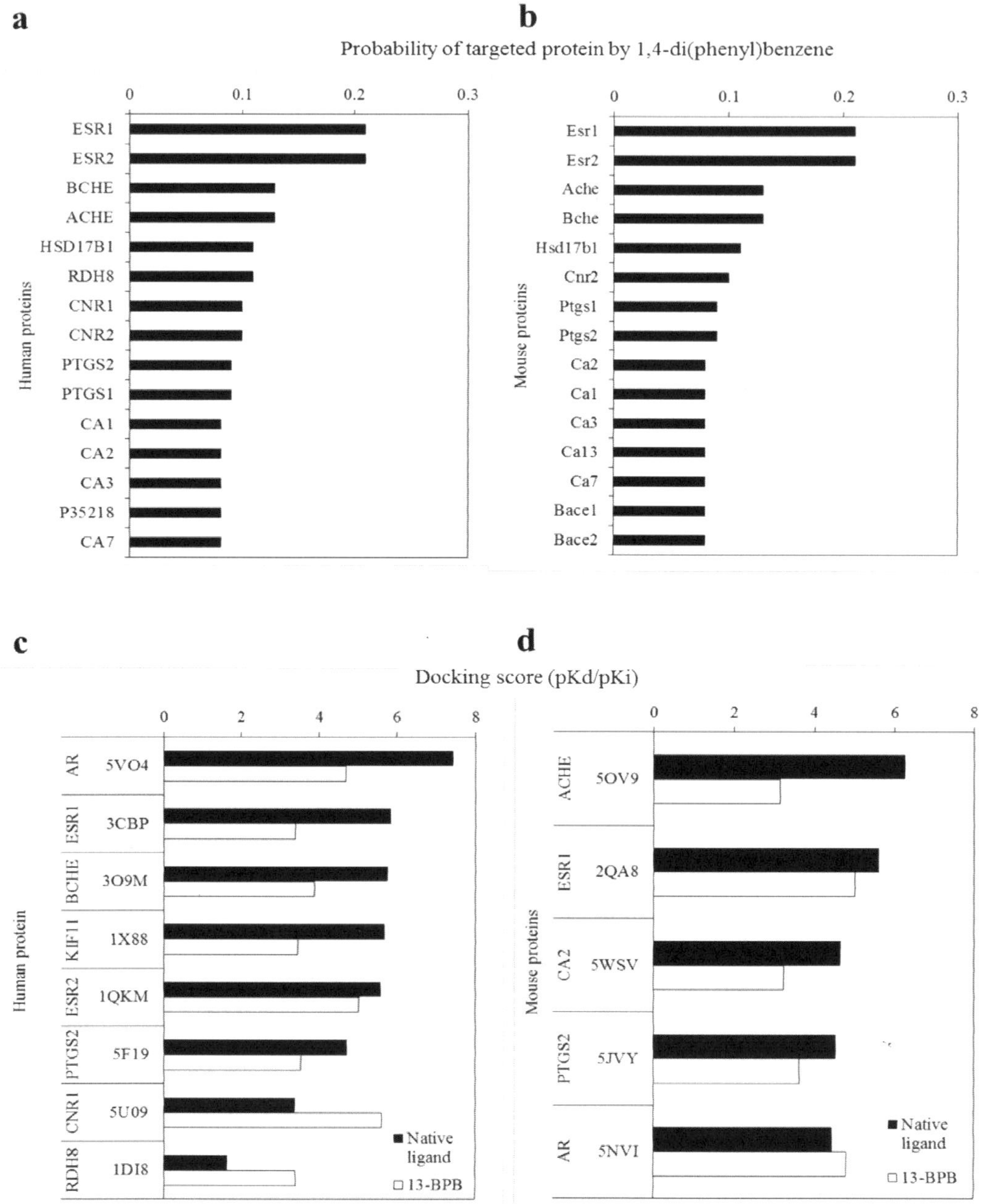

Figure 2. List of potential proteins to be targeted by synthesized 1,4-di(phenyl)benzene in human (**a**) and mouse (**b**). The prediction is based on similarity in 3-dimensional structure of known ligands available in www.swisstargetprediction.ch. The protein abbreviation is based on GeneCard ID or UniProt ID. Binding strength of 1,4-di(phenyl)benzene interaction with human (**c**) and mouse (**d**) proteins in comparison with their native ligands. The binding strength was determined based on docking score. Unlisted proteins indicate the database could not determine type of interaction with the predicted proteins in fig. 2. ACHE is acethylcholinesterase; AR is androgen receptor; BCHE is cholinesterase; CA2 is carbonic anhydrase; CNR1 is cannabinoid receptor 1; ESR1 is estrogen receptor; ESR2 is estrogen receptor beta; KIF11 is kinesin-like protein KIF11; PTGS2 is prostaglandin G/H synthase 2; RDH8 is retinol dehydrogenase 8. The protein ID is based on GeneCard ID and PDB ID.

IOP Conf. Series: Materials Science and Engineering **440** (2018) 012045 doi:10.1088/1757-899X/440/1/012045

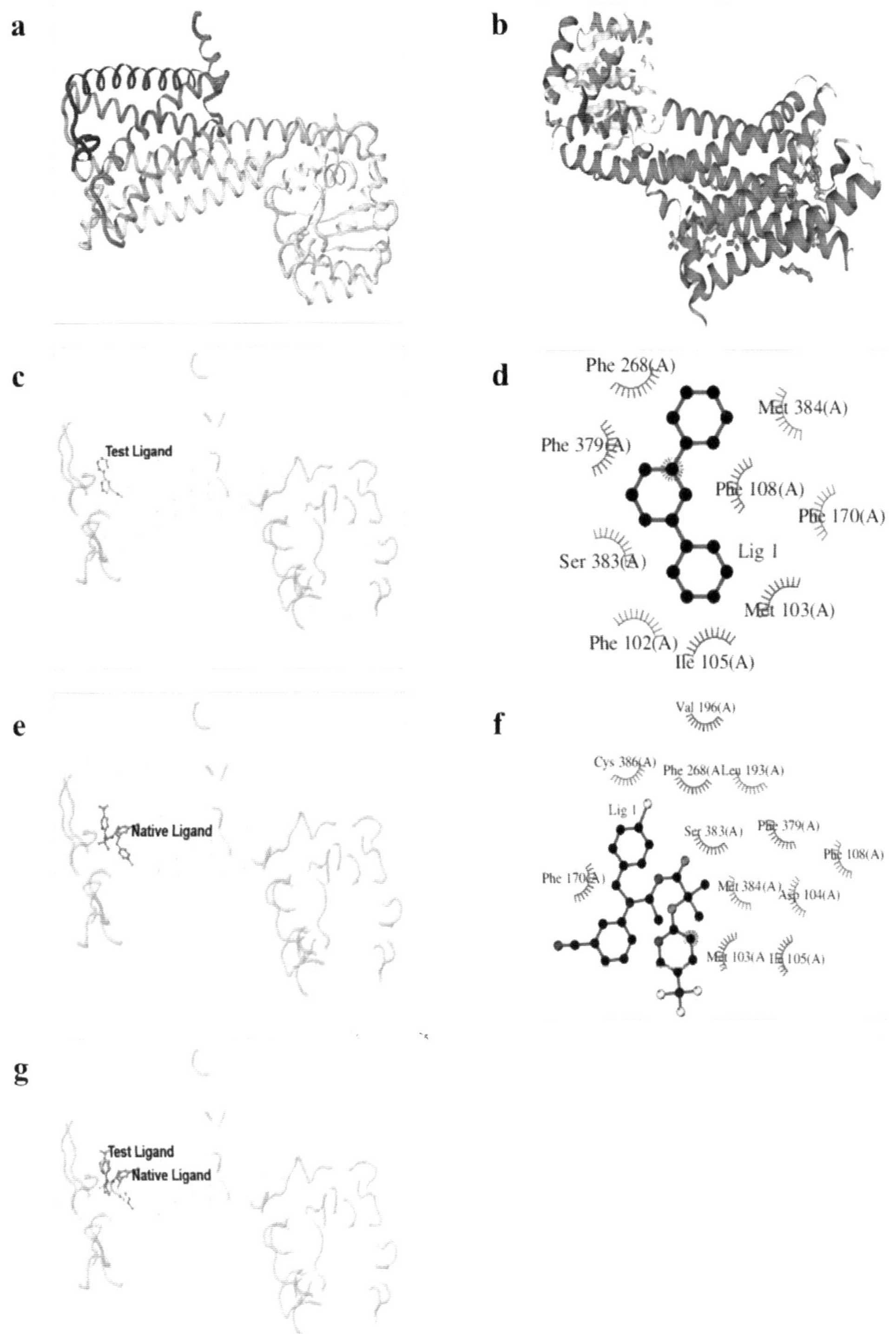

Figure 3. Docking simulation between cannabinoid receptor 1 (CNR1; PDB ID: 5U09) and 1,4-di(phenyl)benzene. (**a**) 1,4-di(phenyl)benzene (**c**) shows the synthesized compound dock at the same binding Taranabant (**b** and **e**), the native compound (**e**). 2D views of CNR1 docking result for native ligand (**F**) and 1,4-di(phenyl)benzene (**d**). (**g**) Merge 3D view of synthesized compound and Taranabant at CNR1 docking site.

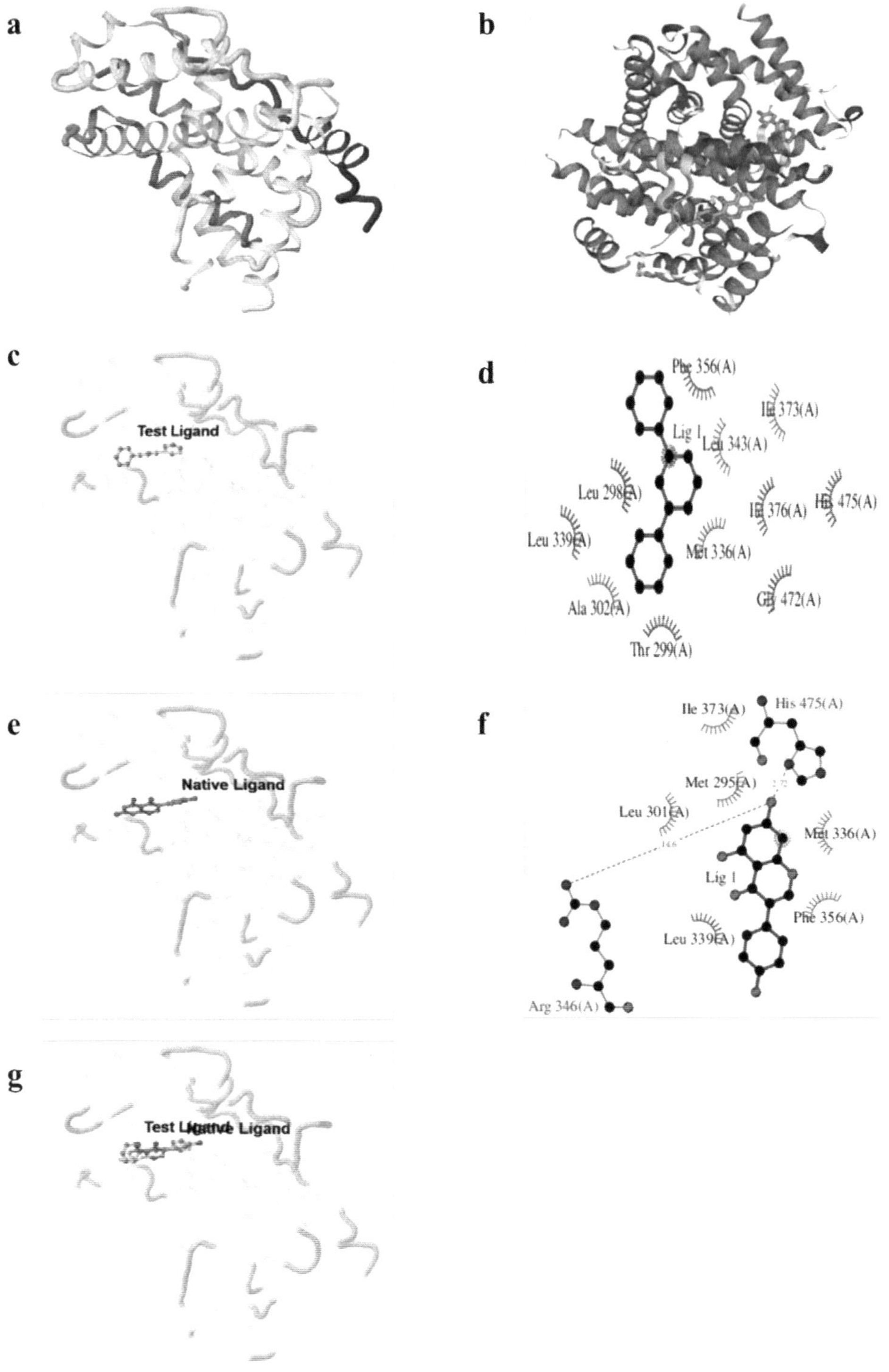

Figure 4: Docking simulation between estrogen receptor beta (ESR2), PDB ID: 1QKM (**a**) 1,4-di(phenyl)benzene (**c**) shows the synthesized compound dock at the same binding site as Genistein (**b** and **e**), the native compound (**e**). 2D views of ESR2 docking result for native ligand (**f**) and 1,4-di(phenyl)benzene (**d**). (**g**) Merge 3D view of synthesized compound and Genistein at ESR2 docking site.

IOP Conf. Series: Materials Science and Engineering **440** (2018) 012045 doi:10.1088/1757-899X/440/1/012045

Table 1: A total of 9 amino acid residues form an interaction with 1,4- di(phenyl)benzene. In comparison, taranabant (native ligand) interacted with 12 residues at the CNR1 docking site, although only 8 out of 12 amino acid residues are listed in this table.

1,4-di(phenyl)benzene	Taranabant
MET_{103}	MET_{103}
ILE_{105}	ILE_{105}
PHE_{170}	PHE_{170}
PHE_{268}	PHE_{268}
PHE_{379}	PHE_{379}
SER_{383}	SER_{383}
PHE_{108}	PHE_{108}
MET_{384}	MET_{384}
PHE_{102}	-

Table 2: A total of 11 amino acid residues form an interaction with 1,4-di(phenyl)benzene. In comparison, genistein (native ligand) interacted with 8 residues at the ESR2 docking site, although only 5 out of 8 amino acid residues are listed in this table.

1,4-di(phenyl)benzene	Genistein
HIS_{475}	HIS_{475}
LEU_{339}	LEU_{339}
MET_{336}	MET_{336}
PHE_{356}	PHE_{356}
ILE_{373}	ILE_{373}
ILE_{298}	-
THR_{299}	-
ILE_{376}	-
GLY_{472}	-
ALA_{302}	-
LEU_{343}	-

4. Conclusion

Our study shows potential hallucinogen based on terphenyl derivatives found in mushrooms due to interaction with the receptor associated with hallucination. The compound, 13-BPB form a strong hydrophobic interaction with cannabinoid receptor 1 (CNR1) at the same docking site of known agonist, Taranabant with higher binding affinity. The interaction almost similar to Taranabant but with higher affinities indicate more potent inverse agonist for CNR1, potentially could cause similar side effect but worse. The compound also interacted strongly with other receptors such as ESRs family. Both CNR1 and ESRs are connected with cyclin-dependent kinases through MAPK signalling pathway thus may affecting the cell cycle and cell death. Cytotoxicity assessment reveals high toxicity of 13BPB at micromolar range.

5. Reference

[1] Sayin H U 2016 Psychoactive Plants Used during Religious Rituals Neuropathology *of Drug Addictions and Substance Misuse* (Elsevier) pp 17–28

[2] Carhart-Harris R L, Bolstridge M, Rucker J, Day C M J, Erritzoe D, Kaelen M, Bloomfield M, Rickard J A, Forbes B, Feilding A, Taylor D, Pilling S, Curran V H and Nutt D J 2016 Psilocybin with psychological support for treatment-resistant depression: an open-label feasibility study *The Lancet Psychiatry* **3** 619–27

[3] Padavala S and Devaraj E 2016 Natural Psychodysleptic compounds: Sources and Pharmacology *Asian Journal of Pharmaceutical and Clinical Research* **9** 40

[4] Halberstadt A L 2015 Recent advances in the neuropsychopharmacology of serotonergic hallucinogens *Behav. Brain Res.* **277** 99–120

[5] Carhart-Harris R L, Muthukumaraswamy S, Roseman L, Kaelen M, Droog W, Murphy K, Tagliazucchi E, Schenberg E E, Nest T, Orban C, Leech R, Williams L T, Williams T M, Bolstridge M, Sessa B, McGonigle J, Sereno M I, Nichols D, Hellyer P J, Hobden P, Evans J, Singh K D, Wise R G, Curran H V, Feilding A and Nutt D J 2016 Neural correlates of the LSD experience revealed by multimodal neuroimaging *Proceedings of the National Academy of Sciences* **113** 4853–8

[6] Gfeller D, Grosdidier A, Wirth M, Daina A, Michielin O and Zoete V 2014 SwissTargetPrediction: a web server for target prediction of bioactive small molecules *Nucleic Acids Res* **42** W32–8

[7] Kamal A, Tamboli J R, Nayak V L, Adil S F, Vishnuvardhan M V P S and Ramakrishna S 2014 Synthesis of a terphenyl substituted 4-aza-2,3-didehydropodophyllotoxin analogues as inhibitors of tubulin polymerization and apoptosis inducers *Bioorganic & Medicinal Chemistry* **22** 2714–23

[8] Norikura T, Fujiwara K, Narita T, Yamaguchi S, Morinaga Y, Iwai K and Matsue H 2011 Anticancer Activities of Thelephantin O and Vialinin A Isolated from Thelephora aurantiotincta *J. Agric. Food Chem.* **59** 6974–9

[9] Finn R S, Aleshin A and Slamon D J 2016 Targeting the cyclin-dependent kinases (CDK) 4/6 in estrogen receptor-positive breast cancers *Breast Cancer Research* **18** 17

[10] Ofek Orr, Attar-Namdar Malka, Kram Vardit, Dvir-Ginzberg Mona, Mechoulam Raphael, Zimmer Andreas, Frenkel Baruch, Shohami Esther and Bab Itai 2011 CB2 cannabinoid receptor targets mitogenic Gi protein–cyclin D1 axis in osteoblasts *Journal of Bone and Mineral Research* **26** 308–16

[11] Addy Carol, Rothenberg Paul, Li Susie, Majumdar Anup, Agrawal Nancy, Li Hankun, Zhong Ling, Yuan Jinyu, Maes Andrea, Dunbar Stephanie, Cote Josee, Rosko Kim, Dyck Kristien, Lepeleire Inge, Hoon Jan, Hecken Anne, Depré Marleen, Knops Annemie, Gottesdiener Keith, Stoch Aubrey and Wagner John 2013 Multiple-Dose Pharmacokinetics, Pharmacodynamics, and Safety of Taranabant, a Novel Selective Cannabinoid-1 Receptor Inverse Agonist, in Healthy Male Volunteers *The Journal of Clinical Pharmacology* **48** 734–44

[12] Harris J 2017 The anti-inflammatory and anti-angiogenic effect of Genistein on MCF-7 breast cancer cell lines

Acknowledgments

This project is funded by Research Endowment Scheme from School of Fundamental of Science, Universiti Malaysia Terengganu.

IOP Conf. Series: Materials Science and Engineering **440** (2018) 012046 doi:10.1088/1757-899X/440/1/012046

Thidiazuron amends the organ development of endangered aquatic plant *Cryptocoryne elliptica* Hook. F

S Norhanizan[1,2] and A Aziz[2*]

[1]Fisheries Research Institute, Freshwater Fisheries Research Division, Glami Lemi, 71650 Titi, Jelebu, Negeri Sembilan, Malaysia.
[2]School of Fundamental Science, Universiti Malaysia Terengganu, 21030 Kuala Nerus, Terengganu, Malaysia.

Corresponding author email: aaziz@umt.edu.my

Abstract. *Cryptocoryne elliptica* is a Malaysia endogenous aquatic plants that has been categorized as endangered species. The plant received a highly demand for aquarium trade and being collected from wild. Slow growth and propagates only through runners has obstructs the plantlets production. Micro propagation is the best option. However, study on the effects of cytokinin particularly thidiazuron (TDZ) on the proliferation rate of this plant *in vitro* is still limited. Thus, the objective of this study was to determine the effects of thidiazuron on multiple shoots proliferation under *in vitro* conditions. A complete randomized design was used. TDZ was tested at the concentrations of 0.2 – 1.0 mg/l in two media: Murashige and Skoog (MS) and Linsmaier and Skoog (LS), respectively. Single shoot explant was used for each culture vessel with thirty cultures per-treatment. Results showed that the highest shoots number generated per explant was obtained in MS + 0.8 mg/l TDZ, which was 7.33 ± 3.08 after 8 weeks of cultures. The protocorm-like-bodies (PLBs) were produced in MS + 0.4 – 0.8 mg/l TDZ and LS + 0.2 – 1.0 mg/l TDZ, after 28 weeks of culture, respectively. The inflorescences were also formed on the explants cultured on TDZ containing medium. This finding suggested that TDZ had alters the proliferation and differentiation on *C. elliptica* cells.

1. Introduction

Cryptocoryne belongs to the family of Araceae. The plant is commercially important for aquarium trade [1]. Nonetheless, nursery production of many *Cryptocoryne* species is restricted due to low seed production and slow rhizome propagation [2]. Thus, most of the *Cryptocoryne* available in market or for aquarium trade are collected from natural environment. *Cryptocoryne elliptica* is an aquatic or amphibious plant that is endemic to Malaysia. This plant can be found only at three localities in Peninsular Malaysia, which are the Gunung Pongsu Forest Reserve, Pondok Tanjung Forest Reserve, and Bukit Panchor, Penang [3, 4]. This *Cryptocoryne* species is also facing the same problem of being collected from the wild for aquarium trade and also by other treat such as deforestation. The plants already fall under endangered categories and the need to conserve and to increase production is seriously needed [4]. Micro propagation technique is the best alternative technique to increase the plant production and high quality plantlets. Cytokinin is commonly used to induce the shoot proliferation. Thidiazuron (TDZ) a synthetic cytokinin, which is a urea-derived cytokinin has been widely used in tissue culture for proliferation of woody plant [5]. Unfortunately, report regarding the efficiency of TDZ on aquatic plants, especially the *Cryptocoryne* is lacking. The current study objective was to determine the effects of TDZ on *in vitro* culture of *C. elliptica*.

2. Materials and Methods

2.1. Materials

The established tissue culture plantlets of *C. elliptica was* obtained from the Biotech Laboratory, School of Fundamental Science, Universiti Malaysia Terengganu (UMT), Malaysia for used in this study.

2.2. Culture medium and treatment

Two culture media, the Murashige and Skoog, (MS) [6] and Linsmaier and Skoog, (LS) [7] basic salt were used this study. Both medium were added with 20g/l sucrose, 0.1 g/l myo-inositol and solidified with 2g/l Gelrite®. The pH of the medium was adjusted to 5.6 prior to autoclaving at 121°C (1.4 kgcm^{-2}) for 15 min. TDZ was added into medium before being autoclaved at concentration of 0, 0.2, 0.4, 0.6, 0.8 or 1.0 mg/l, respectively. Shoots tips, approximately 1.0 - 1.5 cm long were excised from plantlets and inoculated onto treatment medium. A complete randomized design was used in this experiment. Single shoot explant was used for each culture vessel with thirty cultures per-treatment. TDZ-free medium was used as control. Subsequently, all cultures were incubated in growth room at temperature of 25 ± 2 °C, 9-h photoperiod with an irradiance of 50 μmol m^{-2}s^{-1} provided by cool white fluorescent lamps. Numbers of shoots per explant were recorded after 8 weeks of culture. Culture were kept for 28 weeks to observe the effect of TDZ on shoots after being exposed for a longer period of time. Data were statistically analysed using SPSS Ver. 21 (SPSS Inc. Chicago, USA).

3. Results

Results showed that the highest shoots number regenerated per explant (7.33 ± 3.08) was obtained in MS + 0.8 mg/l TDZ after 8 weeks of cultures (Table 1, Figure 1A and 1B). The inflorescences were also formed on explants cultured on TDZ containing medium after 8 weeks of cultures (Figure 1C and 1D). Interestingly, the protocorm-like-bodies (PLBs) were developed on explants cultured in MS + 0.4 – 0.8 mg/l TDZ and LS + 0.2 – 1.0 mg/l TDZ, after 28 weeks of culture, respectively (Table 2, Figure 1B).

Table 1. Effect of TDZ on multiple shoot regeneration from shoot tips of *Cryptocoryne elliptica* in MS and LS medium after 8 weeks of culture.

TDZ concentration (mg/l)	MS medium		LS medium	
	shoots regeneration (Mean ± SE)	Flower occurrences	Shoots regeneration (Mean ± SE)	Flower occurrences
0	1.71 ± 0.91^{b}	nil	3.44 ± 1.74^{b}	nil
0.2	3.50 ± 2.43^{a}	yes	6.10 ± 2.38^{a}	yes
0.4	6.25 ± 1.28^{a}	yes	6.90 ± 3.28^{a}	yes
0.6	5.11 ± 2.42^{a}	yes	6.75± 1.98^{a}	yes
0.8	7.33 ± 3.08^{a}	yes	5.88 ± 2.10^{a}	yes
1.0	3.33 ± 0.58^{a}	yes	5.30 ± 1.95^{a}	yes

Note: Values represent means ± standard errors of two independent experiments of ten replicates. Means followed by the same superscript letter within columns did not significantly different by Tukey's test ($p = 0.05$).

Table 2. Occurrence of protocome-like-bodies (PLBs) of *Cryptocoryne elliptica* in MS and LS medium after 28 weeks of culture.

TDZ Concentration (g/ml)	TDZ Concentration (g/ml)	LS medium
0	nil	nil
0.2	nil	PLB
0.4	PLB	PLB
0.6	PLB	PLB
0.8	PLB	PLB
1.0	nil	PLB

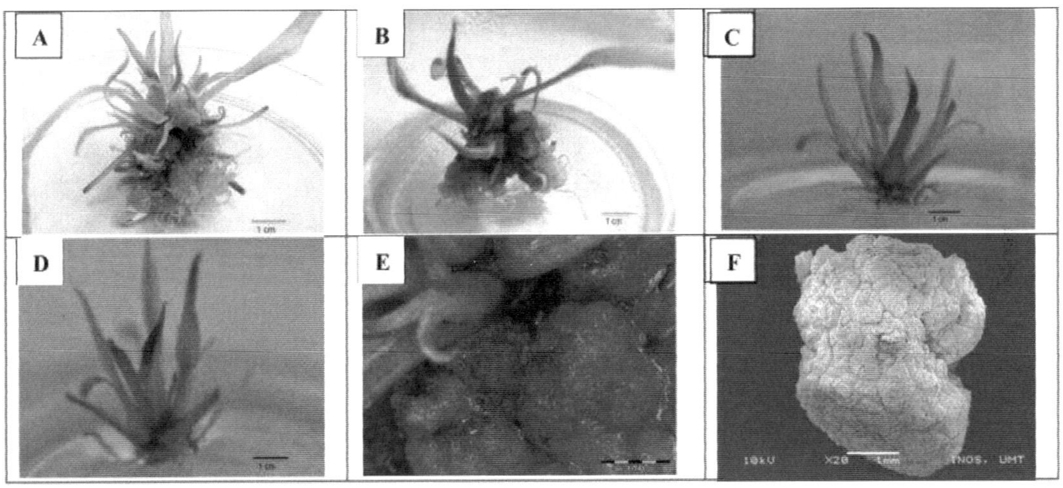

Figure 1. Effects of TDZ on organ development of *Cryptocoryne elliptica* cultures; (A) multiple shoots formation on cultures in MS medium + 0.8 TDZ after 8 weeks, (B) multiple shoots and PLBs on cultures LS medium + 0.6 TDZ after 28 weeks, (C) multiple shoots and flowers buds on cultures in MS medium + 0.2 TDZ after 8 weeks, (D) multiple shoots and flowers buds on cultures in LS medium + 0.8 TDZ after 8 weeks, (E) protocorm-like bodies (PLBs) and shoots under microscope, x4 magnification, (F) protocorm-like bodies PLBs under Scanning Electron Microscope (SEM), x20 magnification.

4. Discussion

To our knowledge this is the first results on the effects of TDZ on *Cryptocoryne* plant. Application of TDZ in the culture medium was found to increase the length of shoots, leaves width and green leaves colours. As shown in Figures 1A to 1D above, TDZ has managed to induce an explant to increase the number of shoot tips, formation of flower buds and the protocorm like bodies. Effects of TDZ on clustered short shoots with very small leaves was also reported in *Ludwigia repens* [8], *Cercis canadensis* var *alba* [9] and *Hibiscus rosa-sinensis* [10]. Cytokinins commonly stimulate shoot proliferation and inhibit their elongation, therefore; inhibition of shoot bud elongation by TDZ was consistent with its high cytokinin activity. The formation of flower in TDZ treatments were observed after 2 months of culture in growth regulator-free medium. The same observation was reported in *Dendrocalamus strictus* Nees [11]. The formation of PLBs on the TDZ treated explant was recorded after 28 weeks of culture in both media types. The capability of TDZ to induce the PLBs was also reported on Doritaenopsis orchid after 5 weeks culture [12] and *Dendrobium nobile* orchid after 8 weeks [13].

5. Conclusion

TDZ induced multiple shoots and inflorescence of explant shoot tips of *C. elliptica* when cultured for 8 weeks. TDZ also induced PLBs of *C. elliptica* when cultured for longer period of time (28 weeks). This finding suggested that TDZ had alters the proliferation and differentiation on *C. elliptica* cells.

6. References

[1] Rataj R and Horeman T J 1977 *Aquarium Plants*. T.F.H. Publications, Inc., Neptune City, NJ

[2] Windelov H 1987 *Aquarium Plants*. T. F. H. Publications, Inc., Neptune City, NJ

[3] Othman A S, Jacobsen N and Mansor M 2009 *Cryptocoryne of Peninsular Malaysia: A Portrait of an Enchanting Plant Genus*. Penerbit Universiti Sains Malaysia

[4] Mansor M & Masnadi M 1994 *Cryptocoryne elliptica*, an endangered amphibious plant in Pondok Tanjung forest reserve, Peninsular Malaysia *Aquatic Botany* **47** **(1)** 91-96

[5] Huetteman C A and Preece J E 1993 Thidiazuron: a potent cytokinin for woody plant tissue culture *Plant Cell Tissue Organ Culture* **33** 105-119

[6] Murashige T and Skoog F 1962 A revised medium for rapid growth and bioassays with tobacco tissue culture *Physiol Plant* **15** 473-479

[7] Linsmaier E M and Skoog F 1965 Organic growth factor requirements of tobacco tissue cultures *Physiol Plant* **18** 100-127

[8] Ozturk M, Khawar K M, Atar H H, Sancak C and Ozcan S 2004 In-vitro propagation of the aquarium plant Ludwigia repens *Asia-Pacific J Mol Biol Biotecnol* **12** 21-25

[9] Yusnita S, Geneve R L and Kester ST 1990 Micropropagation of white flowering eastern redbud (*Cercis canadensis* var. alba L) *J. Environ. Hort.* **8** 177-179

[10] Preece J E, Huetteman C A, Puello C H and Neuman M C 1987 The influence of thidiazuron on *in-vitro* culture of woody plants *Hort. Science.* **22** 1071

[11] Singh M, Jaiswal U and Jaiswal V S 2000 Thidiazuron – induced in-vitro flowering in *Dendrocalamus strictus* Nees *Current Science* **79** 11 p 1529-1530

[12] Park S Y, Murthy H N and Paek K Y 2003 Protocorm-like body induction and subsequent plant regeneration from root tip cultures of *Doritaenopsis Plant Science* **164** 919-923

[13] Bhattacharyya P, Kumaria S, Diengdoh R and Tandon P 2014 Genetic stability and phytochemical analysis of the in vitro regenerated plants of *Dendrobium nobile* Lindl., an endangered medicinal orchid *Meta Gene* **2** 489-504

Acknowledgement

The authors wish to express their gratitude to the Director of the Fisheries Research Institute Glami Lemi, and Dean of School of Fundamental Science, Universiti Malaysia Terengganu for the support and encouragement during the study.

IOP Conf. Series: Materials Science and Engineering **440** (2018) 012047 doi:10.1088/1757-899X/440/1/012047

Effect of Six Month Storage on Physicochemical Analysis and Antioxidant Activity of Several Types of Honey

M B Monggudal[1], M N F M Radzi[2], M M Ismail[3] and W I W Ismail[1,2,*]

[1]Faculty of Pharmacy, Universiti Teknologi MARA (UiTM) Selangor, Puncak Alam Campus,42300 Bandar Puncak Alam, Selangor, Malaysia
[2]School of Fundamental Science, Universiti Malaysia Terengganu (UMT), 21030 Kuala Nerus, Terengganu, Malaysia
[3]Institute of Agricultural and Food Policy Studies, Universiti Putra Malaysia, 43400 Serdang, Selangor, Malaysia

Corresponding author emails: basirmonggudal@yahoo.com ; naim.fadhli@gmail.com ; mkk@upm.edu.my ; waniryani@umt.edu.my

Abstract. Honey composition could change during storage, which might affect the quality of honey. Subsequently, it could reduce the honey commercial value. However, data on the effect of the storage on several types of local honey is still lacking. Thus, in this study, moisture content, pH and antioxidant level of honey stored at 4°C were analysed after six-month of the storage. Initially, physicochemical properties (moisture, electrical conductivity, ash content, pH, free acidity and hydroxymethylfurfural (HMF) level) and antioxidant levels of five local honey samples (Tualang, Gelam, Pineapple, Acacia and Kelulut honey) and Manuka honey as a control were determined. Afterwards, honey samples were stored at 4°C for six months and their moisture content, pH and antioxidant levels were monitored monthly. Physicochemical properties were measured based on the standard method set by the International Honey Commission (IHC). Antioxidant activity was determined based on previous studies. Most of the honey samples exhibited physicochemical properties within the range set by the Codex Standard for Honey but displayed lower antioxidant level compared to other studies. On average, honey moisture content increased by 0.1 to 0.4% while their antioxidant level increased by 2.6 mg GAE/kg to 6.3 mg GAE/kg monthly. In contrast, pH level of honey decreased by 0.03 to 0.09 monthly. Storage of honey at low temperature increases its moisture and antioxidant level while decreases its pH level. However, further study needs to be carried out to observe whether the pattern will continue even after a lengthy storage period exceeding six months.

1. Inroduction

Honey consists of a mixture of carbohydrates, mainly simple sugars such as glucose and fructose and other important constituents such as amino acids, enzymes, minerals, vitamins, organic acids, phenolic acids and volatile chemicals [1], [2]. It is unique as its characteristic differs based on its geographical origin, floral sources and bee species. Additionally, the quality of honey is also affected by the weather conditions, packaging, processing and storage time [3].

Honey can be categorized as food and is susceptible to spoilage or degradation. It can deteriorate rapidly without proper care and under average condition it can be considered as a readily perishable product [4]. Proper storage is crucial to preserve the quality of honey since honey composition could change during storage through chemical reactions such as oxidation and fermentation [5].

Changes in honey composition always reflect to low quality of honey and will affect its commercial value. Temperature is one of the factors that plays an important role during prolong storage of foods. Storage at low temperature could slow down food degradation by lowering the activity of enzymes and chemical reactions and preventing the growth of microorganism.

Several studies have been conducted to investigate the effect of low temperature on honey [6], [7]. However, most of them are interested on the crystallization of honey and very few studies analyse the changes in the physicochemical properties and stability of compounds in the form of antioxidant level throughout storage at low temperature. Thus, in this study two physicochemical properties of honey; moisture content and pH as well as antioxidant level of several types honey stored at low temperature of 4°C were observed monthly for 6 months period.

2. Materials and Methods

2.1. Honey collection

Six types of honey samples, namely Tualang, Gelam, Pineapple, Acacia, Kelulut and Manuka honey were used in this study. Tualang honey was harvested from Kedah, Gelam honey was collected from Terengganu, Pineapple, Acacia and Kelulut were obtained from Johor. Two honey samples (from two different hives) were collected from each of the location and were harvested as mentioned by Ismail [8]. Manuka honey was purchased from the local Guardian pharmacy. Tualang and Manuka honey samples were used as controls in the study. All honey samples except of Kelulut honey were produced by *Apis* species of bee. Meanwhile, Kelulut honey was produced by Meliponini bee, i.e. *Heterotrigona itama*. All samples were kept in glass jars, wrapped with aluminium foil and stored in a refrigerator at 4°C throughout the experiment for six months.

2.2. Physicochemical analyses

Right after sample collection and prior to store as mentioned above, physicochemical analyses were carried out according to the methods outlined in the International Honey Commission (IHC) [10] and Samat et al. [9]. These include moisture, electrical conductivity, ash content, pH, free acidity and hydroxymethylfurfural (HMF) level.

2.2.1 Moisture content

The moisture content of honey was measured using MISCO BKPR-2 digital refractometer (MISCO refractometer, United States) at 20°C.

2.2.2 Electrical conductivity

20% (w/v) honey solution was prepared by dissolving 20g of honey in 100ml deionised distilled water prior to measuring the electrical conductivity using HI-38311 electrical conductivity meter (Hanna Instruments, United States).

2.2.3 Ashcontent

Ash content was indirectly determined using the measured electrical conductivity as stated in Bogdanov et al. [10]. The following equation was used:

$$X1 = (X2 - 0.143)/1.743 \tag{1}$$

Where X1 = ash value and X2 = EC (μS/cm) at 20 °C. The result is expressed as ash content in g/100g honey.

2.2.4 pH and free acidity

pH and acidity were measured using a Delta 320 pH meter (Mettler Toledo, Switzerland). 10g of honey was dissolved in 75ml of distilled water before the pH was measured. The dissolved honey solution was then titrated with 0.1M sodium hydroxide (NaOH) until pH 8.3 to obtain the free acidity value of honey.

2.2.5 Hydroxymethylfurfural (HMF)

HMF was determined as per method outlined by White [11]. 5g honey was diluted with 25ml distilled water and transferred into a 50ml volumetric flask. 0.5ml Carrez solution I was added followed by 0.5ml Carrez II solution. Distilled water was then added up to the mark. The solution was then filtered using a 0.45µm filter paper, discarding the first 10ml filtrate. 5ml honey solution was mixed with 5ml water before measuring its absorbance against reference solution of 5ml initial honey solution and 5ml 0.2% sodium bisulphite solution. The absorbance was measured at 284nm and 336nm, and the values were expressed in (mg/kg).

2.3 *Antioxidant level*

Total phenolic content (TPC) was estimated using Folin-Ciocalteu based on the method described by Khalil et al. [14] and Salleh et al. [15]. 0.2g honey was diluted in 1ml distilled water before the honey solution was mixed with 1ml Folin-Ciocalteu. 1 ml of sodium bicarbonate (10%) was added after 3 minutes. The solution was mixed and distilled water was added until 10 ml. The sample was left in the dark for 90 minutes. After incubation, the absorbance was read at 725 nm, with gallic acid used as a standard.

2.4 Storage condition

Evaluation on the effect of storage condition was conducted where moisture content, pH and TPC of honey samples were monitored monthly for six months. Samples were kept in glass jar, wrapped with aluminium foil and stored under low temperature of 4°C.

3. Results

3.1 Physicochemical analyses and antioxidant level

Physicochemical analyses were conducted to characterize the obtained samples prior to store as mentioned above. Honey samples displayed a diverse range of values, depending on the physicochemical analysis conducted, as observed in Table 1. All honey exhibited a small range of pH and ash content, while huge variations were observed in other analyses, in particular HMF content and free acidity. In several physicochemical analyses, variations were also observed for honey samples from the same type, as evident in the HMF analysis of Gelam honey and electrical conductivity of Tualang honey. Kelulut and manuka honey exhibited high antioxidant level compared to other honey samples.

3.2 Storage condition

The moisture content of honey increases while their pH level decreases the longer it is stored at low temperature (4°C), as shown in Figure 1. The moisture content of honey samples does not increase dramatically, except for Pineapple 2 honey which exhibited a 3.1% increase in moisture after six months. In general, the increase in monthly moisture content of honey at 4°C is very low (0.1 to 0.4%). pH of honey does not drop drastically at low temperature with the exception of Pineapple 2, Acacia 1 and Acacia 2 honey, which have shown a pH decrease of 0.6, 0.63 and 0.63 respectively after six months. On average, the pH level of honey decreases by 0.03 to 0.09 monthly.

TPC of honey samples increases over time when stored at low temperature (4°C), as observed in Figure 2. A remarkable increase in TPC was observed for both Kelulut 1 (40.81 mg GAE/kg) and

Gelam 2 (57.52 mg GAE/kg) honey after storage for six months. On average, the TPC of honey increases by 2.6 mg GAE/kg to 6.3 mg GAE/kg monthly.

Table 1. Physicochemical analyses and antioxidant level of honey samples from various origin before storage procedure

Honey	Moisture (%)	pH	Free acidity (meq/kg)	Ash (g/100g)	Electrical conductivity (mS/cm)	HMF (mg/kg)	TPC (mg GAE/kg)
Manuka	19.4	4.30	6.5	0.332	0.721	63.42	71.20
Kelulut 1	32.5	3.67	24.0	0.600	1.187	14.34	89.25
Kelulut 2	34.0	3.63	26.0	0.594	1.178	25.60	51.94
Acacia 1	21.0	4.03	7.5	0.438	0.906	2.09	43.75
Acacia 2	20.4	3.99	7.5	0.430	0.893	3.73	40.21
Pineapple 1	24.0	3.57	5.0	0.118	0.348	25.13	33.96
Pineapple 2	23.2	3.99	4.0	0.090	0.299	4.94	26.60
Gelam 1	27.0	3.66	6.0	0.350	0.753	5.82	30.64
Gelam 2	23.5	3.76	7.0	0.418	0.872	114.81	60.85
Tualang 1	23.3	4.21	11.0	0.248	0.576	13.46	68.09
Tualang 2	26.6	4.30	6.0	1.067	2.003	7.42	52.67

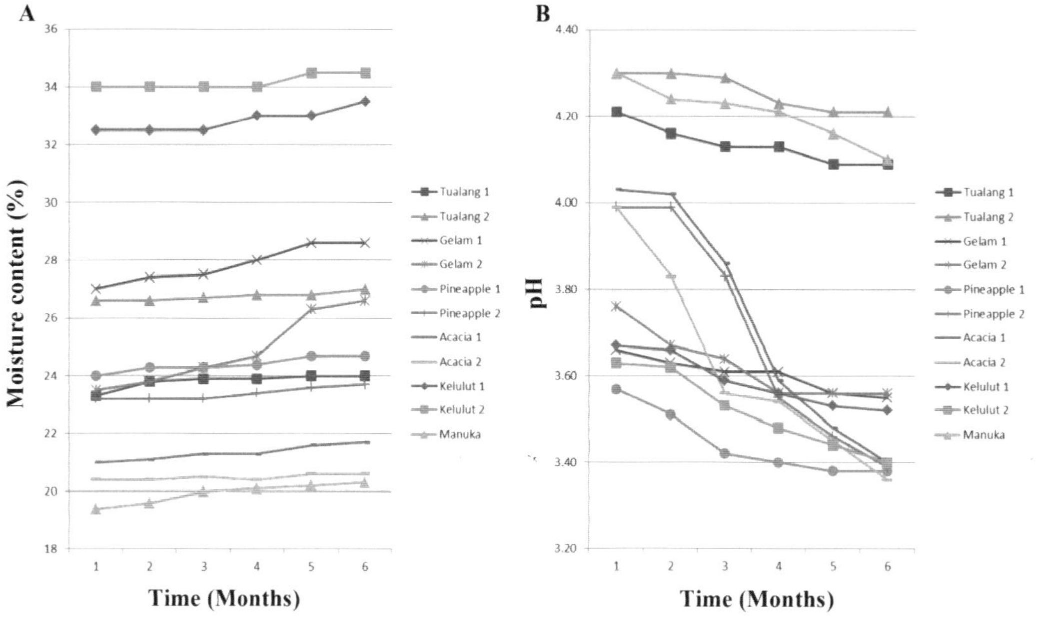

Figure 1. Selected physicochemical analyses on several types of honey stored at 4°C for six months. (A) Moisture content and (B) pH

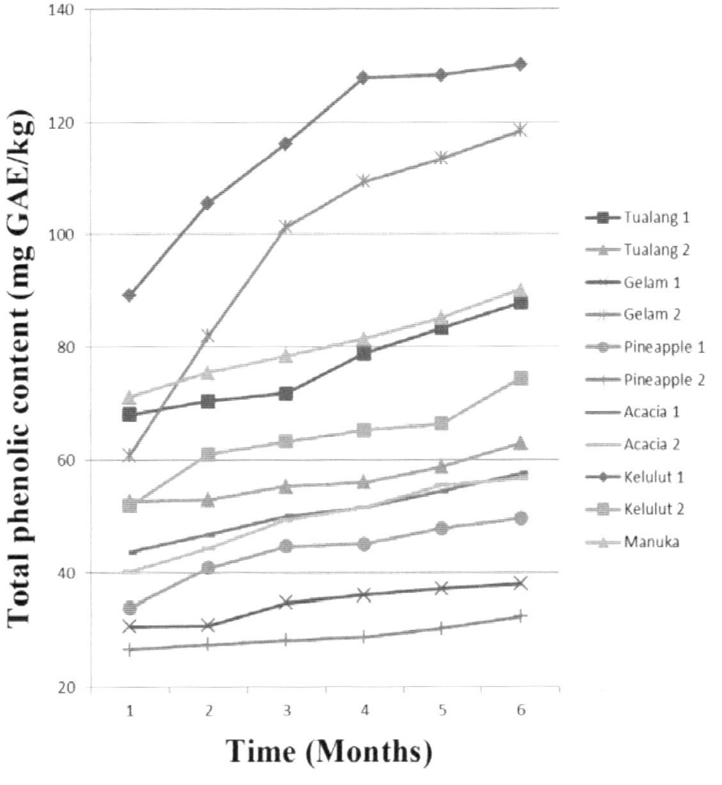

Figure 2. Antioxidant level of honey samples stored at 4°C for six months

4. Discussion

Most of the physicochemical properties (moisture, electrical conductivity, ash content, pH and free acidity) of honey samples used in the study are in accordance with the values set by the Codex Standard for Honey [12]. In terms of HMF level, all the honey samples show below than level set by the Codex (80mg/kg) except for the Gelam and Manuka honey. Both honey collected/purchased at least six months earlier than the other honey. Thus, the honey samples displayed higher level of HMF. It is difficult to obtain fresh Manuka honey since the honey is originally produced in New Zealand. Meanwhile, six month-Gelam honey was used because of lacking the honey sample during the sampling activity and can be used to compared with the Manuka honey. The finding is parallel to a current study by Shapla et al. [20], which storage can increase HMF level.

The antioxidant level of the honey, in the form of total phenolic content are lower compared to other studies on Malaysian honey [13], [14], [16]. Differences in the physicochemical properties and antioxidant levels between different honeys could be attributed to different floral sources and geographical origins. Differences in the physicochemical properties are common even for honey of the same type and at similar geographical origins [17].

The effects of storing honey at low temperature (4°C) for long periods of time were demonstrated by the changes observed in the moisture, pH and antioxidant level of honey. Contrary to honey stored at freezing temperature which experience reducing moisture content [18], honey stored at low temperature of 4°C has a slight increase in moisture. This could be due to the hygroscopic property of honey combined with the humidity present in the freezer at low temperature. Increase in moisture content above permitted level will increase the chances of honey spoilage through fermentation [18],

[19], which lowers its quality. Thus, to avoid the increase for prolong storage at low temperature, it is recommended that the storage apparatus is sealed from air, not exposed to multiple freeze-thaw cycles and presence of air in the container reduced to a minimum.

Honey pH levels decreased slowly throughout time when stored at low temperature. This might be correlated to the increasing honey moisture content. However, some honey have shown dramatic decrease in pH level (Acacia 1, Acacia 2, Pineapple 2) but only experience slight increase in moisture content. This indicated that there might be other factors that influenced the pH of honey such as the reactions between various components of honey [17] and concentrations of ions present in honey solution.

It is expected that honey's antioxidant level will decrease slowly at low temperature over time as a food is susceptible to degradation process during long storage period. Phenolics could degrade to other volatile compounds, a by-product of degradation. Floral sources and different sampling area, may affect the antioxidant level as well. However, surprisingly the antioxidant level of honey samples was found to increase throughout the six months storage period. Particularly, Gelam honey showed the highest level of antioxidant. This is in agreement with a study by Batumalaie et al. [21] that Gelam honey contained high level of antioxidant properties particularly phenolics. This may indicate that complex interactions between myriads of components in honey could affect its chemical composition which in turn affects its antioxidant level [22]. Through the reaction of enzymes or other secondary reactions, honey may degrade into new phenolic compounds or other by-products. Ultimately, it is important to note that the increase in honey antioxidant level would not last indefinitely. The antioxidant level might stop or decrease at certain time point after the six months period, which warrants further investigation in the future.

5. Conclusion

Storage of honey at low temperature above freezing point for six months will increase its moisture and antioxidant level and decrease its pH level. However, a long term study needs to be conducted to observe whether this pattern will continue indefinitely or cease to exist at a certain point in time.

6. References

[1] Moniruzzaman M, Khalil I, Sulaman SA and Gan SH 2013 *BMC Complement Altern Med.* **13** 43
[2] Khalil I, Moniruzzaman M, Boukraa L, Benhanifia M, Islam A, Islam N, Sulaiman SA and Gan SH 2012 *Molecules* **17** 11199-11215
[3] Tornuk F, Karaman S, Ozturk I, Toker OS, Tastemur B, Sagdic O, Dogan M and Kayacier A 2013 *Industrial Crops and Products* **46** 124-131
[4] Wang XH, Gheldof, N and Engeseth NJ 2004 *Journal of Food Science* **69** 96-101
[5] Moreira RFA, Maria CAB, Pietroluongo M and Trugo LC 2007 *Food Chemistry* **104** 1236-1241
[6] Costa LCV, Kaspchak E, Queiroz MB, de Almeida MM, Quast E and Quast LB 2015 *Braz J Food Technol.* **18** 155-161
[7] Conforti PA, Lupano CE, Malacalza NH, Arias V and Castells CB 2006 *Int J Food Prop.* **9** 99-107
[8] Ismail WIW 2016 *Journal of Sustainability and Science Management* **11** 70-80
[9] Samat S, Enchang FK, Hussien FN and Ismail WIW 2014 *Evidence-Based Complementary and Alternative Medicine* 1-9
[10] Bogdanov S 2009 *International Honey Commission* 1-62
[11] White JW Jr 1979 *J. Assoc. Off Anal Chem* **62** 509-514
[12] Codex Alimentarius Commission 1981 *Codex Standard* **12** 1-7

[13] Kek SP, Chin NL, Yusof YA, Tan SW and Chua LS 2014 *Agriculture and Agricultural Science Procedia* **2** 150-155

[14] Khalil MI, Sulaiman SA, Alam N, Ramli N, Mohamed M, Bai'e S and Hua GS 2012 *International Journal of Pharmacy and Pharmaceutical Sciences* **4** 214-219

[15] Salleh MAM, Eshak Z and Ismail WIW 2017 *Jurnal Teknologi* **79** 9-16

[16] Kishore RK, Sukari Halim A, Shah S and Sirajudeen KNS 2011 *Nutrition research* **31** 322-325

[17] Acquarone C, Buera P and Elizalde B 2007 *Food Chemistry* **101** 695-703

[18] Ribeiro GP, Villas-Boas JK, Spinosa WA, Prudencio SH 2018 *LWT* **90** 607-612

[19] Prica N, Zivkov-Balos M, Jaksic S, Mihaljev Z, Kartalovic B, Babic J Savic S 2014 *Arhivveterinerske medicine* **7** 99-109

[20] Shapla UM, Solayman M, Alam N, Khalil MI, Gan SH 2018 *Chemistry Central Journal* **12** 1-18

[21] Batumalaie K, Qvist R, Yusof KM, Ismail IS, Sekaran SD 2013 *Clin Exp Med* 1-11

[22] Da Silva PM, Gauche C, Gonzaga LV, Oliveira Costa AC and Fett R 2016 *Food Chemistry* **196** 309-323

Acknowledgement
This research was funded by the Research Acculturation Collaborative Effort (RACE) grant 600-RMI/RACE 16/6/2(1/2015) from the Malaysian government. Appreciation goes to the RMIC, UiTM, and to Prof. Dr. Mohd Mansor Ismail, mentor for the grant. We also would like to express our gratitude to the beekeepers and honey hunters who were involved in the honey collection process.

Physicochemical Analysis, Antioxidant and Anti Proliferation Activities of Honey, Propolis and Beebread Harvested from Stingless Bee

Wan Iryani Wan Ismail, Nurul Noratikah Hussin, Siti Nur Farhanim Mazlan, Nur Hazlin Hussin, Mohd Naim Fadhli Mohd Radzi

School of Fundamental Science, Universiti Malaysia Terengganu (UMT), 21030 Kuala Nerus, Terengganu, Malaysia

Abstract Stingless bees (Kelulut) can produce three major commercial products i.e. honey, propolis and beebread. These products are widely believed to have medicinal benefits, similar with products produced by stinging bees. However, there are very few scientific data available on the stingless bee's products to prove the claims. Thus, this study was conducted to investigate the characteristics of the products from the perspective of physicochemical analysis, activities of antioxidant and anti proliferation on cancer cells. Stingless bees honey, propolis and beebread were collected and their physicochemical and antioxidant properties were analysed prior to treatment on human breast adenocarcinoma (MCF-7) cell lines. Physicochemical analysis indicated that the samples are mostly not within the range reported by the Codex Standard for Honey. Honey, propolis and beebread exhibited antioxidant activity through the total phenolic content of 700 mg GAE/kg, 1600 mg GAE/kg and 300 mg GAE/kg respectively. Propolis has the highest antioxidant activity and inhibited MCF-7 cell growth at IC_{50} of 38.9 μg/ml. Meanwhile, stingless bee honey and beebread displayed the IC_{50} at 60 v/v and 64μg/mL respectively. The data is crucial to unveiled and prove medicinal properties and potential possessed by the stingless bee products. Subsequently, increase their commercial value in the future.

1. Introduction

Stingless bee or locally known as Kelulut is one of the native bee species available in Malaysia (Ismail, 2016). The bee can serve as a pollination agent and intriguingly it also can produce three major products, which are honey, propolis and beebread. The products, in particularly honey have high economic potential because of aesthetic taste (sweet and sour) compared to the conventional honey (sweet). In addition, many claims have been made related to the medicinal values. However, the data is still limited and is yet to be discovered at present.

Honey is classified as a functional food because of its outstanding chemical compositions including antioxidants (Bogdanov et al., 1999). Meanwhile, propolis is made up of plant resins and beeswax, which are masticated by the stingless bee using its salivary enzymes to build the hive pots. Beebread is a fermented mixture of plant pollen, honey, and saliva of bees, which is used as a major protein source to the bees. Various chemical compositions found in honey, propolis and beebread made the stingless bee's products unique with specific characterisations. The characterisation can be evaluated using physicochemical analysis such as moisture content, ash content, pH, free acidity, hydroxymethylfurfural (HMF), wax content, level of protein and carbohydrate. Method for these analyses are acknowledged globally to observe the quality of product and even used to differentiate between natural and adulterated honey (Bogdanov, 2009; Samat et al., 2018).

Many consumers believe in the medicinal value of stingless bee products. However, data on the medical properties of stingless bee products is still limited. In order to prove the claims, the antioxidant properties of all the stingless bee products in the study were tested using total phenolic content and on their anti-proliferation of cancer cell lines.

2. Materials and Methods

2.1. Sample collection

Honey and propolis were collected from species of stingless bee, *Geniotrigona thoracica*. Meanwhile, beebread was harvested from stingless bee *Heterotrigona itama* from Humaira' Honey kelulut farm in Lenggong, Perak. The samples were stored in a refrigerator at 4°C prior to use.

2.2. Sample preparation

In this study, honey sample can be used readily for the experiment. However, extraction procedures were applied on propolis and beebread prior to testing on physicochemical and antioxidant analyses, and on cancer cell lines. 50 g of ground propolis was mixed with 500 ml of 80% ethanol. Then, the mixtures were shaken at 200 rpm for 48 h at 37°C. Next, it was spn at 3000 rpm for 15 min at 10°C. Then, the filtrate was evaporated using a rotary evaporator at 175 mbar and at a temperature of 52°C at 95 rpm (Jacob et al., 2015). The propolis ethanolic extract was kept in a chiller at 4°C for further use. For the extraction process of beebread, the sample was dried before it can be crushed. Next, 20 g of sample were mixed with 80 mL of 95% ethanol and then put on a shaker at 200 rpm and 37 °C for 24 h. The top layers were decanted and centrifuged at 3000 rpm for 30 min at 20°C. After that, the supernatant was evaporated at 40°C in the rotary evaporator. The pressure and rotation were set at 156 mbar and 950 rpm respectively. The end product was weighed and stored in the dark at 8°C (Markiewicz-Zukowska et al., 2013).

2.3. Physicochemical analyses

Physicochemical analyses were conducted according to the methods defined in the IHC (Bogdanov, 2009), Samat et al., (2014) and Bogdanov et al., (1999). These include moisture, ash content, pH, free acidity and hydroxymethylfurfural (HMF) level for honey. Meanwhile for propolis, physicochemical analyses were conducted on the content of moisture, ash and wax. Physicochemical analyses of beebread were carried out on moisture and ash content; and the level of protein and carbohydrate found in the sample.

2.3.1 Moisture content

The moisture content of honey was measured using a portable digital refractometer (MISCO refractometer, United States) at 20°C. For propolis, 10g of ground propolis was heated to 100°C for 5 h in the oven. Then, the heated sample was desiccated until constant weight to measure the final weight of the sample (Cunha et al., 2004). For beebread, 1 g of sample was put in a hot-air oven at 90°C for 4 h and 48 min. Then, the samples were removed from the oven and they were allowed to reach room temperature and then weighed. The difference between dry weight and initial weight was used to calculate moisture content (Carpes et al., 2009).

2.3.2 Ash content

Ash content of honey was determined readily using electrical furnace at temperature (450-550°C) for 15 min (Bogdanov, 1999). 2g ground propolis and beebread were burnt separately at a temperature of 600°C for 1 h using the furnace (Carpes et al., 2009, Cunha et al., 2004).

2.3.3 pH and free acidity

10 g of honey was dissolved in 75 ml of distilled water before the pH was measured using a Delta 320 pH meter (Mettler Toledo, Switzerland). Then, the dissolved honey solution was titrated with 0.1 M sodium hydroxide (NaOH) until pH 8.3 to obtain the free acidity value of honey using the same device.

2.3.4 Hydroxymethylfurfural (HMF)
HMF was determined as per method outlined by White (1979). A total of 5 g honey was diluted with 25 ml distilled water and transferred prior to adding 0.5 ml of Carrez I and Carrez II solution. Distilled water was then added up to 50 ml before it was filtered using a 0.45 µm filter paper. Discard the first 10 ml filtrate. 5 ml honey solution was mixed with 5 ml water before measuring its absorbance at 284 nm and 336 nm against reference solution of 5 ml initial honey solution and 5 ml 0.2% sodium bisulphite solution. The values were then expressed in mg/kg.

2.3.5 Wax analysis
Propolis extract was left in the freezer overnight before it was filtered and its wax composition weighed based on the method described by Cunha et al., (2004).

2.3.6 Protein content
Bradford protein assay was used to determine the protein content. 100 µL of beebread sample was added to 9.9 mL distilled water. Next, 3 drops of Bradford's reagent was added. Then, the absorbance was measured at 595 nm after 10 min. Protein standard curve using bovine serum albumin (BSA) was used to determine the content of protein in the beebread sample. The protein standard curve was prepared using a protein stock solution (0.01 g/mL). The final concentrations of protein (g/mL) that were used are 0, 0.00125, 0.0025, 0.005 and 0.01 (Bradford, 1976).

2.3.7 Carbohydrate content
Somogyi-Nelson method was used to determine carbohydrate content. 100 µL of beebread sample was added to 9.9 mL of distilled water. Next, 1 mL of Nelson reagent was added. Afterwards, the solution was placed in boiling water for 20 minutes and then cooled to room temperature. Next, 1 mL arsenomolybdate reagent solution was added. The absorbance was measured at 510 nm using a spectrophotometer. Carbohydrate standard curve using glucose was used to determine the content of carbohydrate in beebread sample. The standard curve was prepared using a carbohydrate stock solution (10 mg/mL). The final concentrations of carbohydrate (mg/mL) that were used are 0, 1.25, 2.5, 5 and 10 (Carpes et al., 2009).

2.4 Antioxidant level
Total phenolic content of honey, propolis and beebread sample were estimated by using a modified spectrophotometric Folin-Ciocalteu method. Gallic acid was used as a standard by using several concentrations (1.25 mg/ml, 2.5 mg/ml, 5 mg/ml and 10 mg/ml). 2 g of honey and 200 µl of propolis and beebread respectively from stock solution was mixed with 1 ml of Folin and Ciocalteu's phenol reagent. 1 ml of 10% sodium carbonate solution was added to the mixture after 3 min. The volume of the mixture was adjusted to 10 ml by using distilled water. The reaction was left in the dark for 90 min. The absorbance of the sample was read at 725 nm (Ibrahim et al., 2015).

2.5 Anti proliferative Study
MCF-7 cell lines were purchased from ATCC (USA). The cells were treated with 0, 25, 50, 75 and 100 v/v of honey and 0, 25, 50, 75 and 100 µg/ml of propolis and beebread respectively for 72 h. Untreated cells were used as a negative control. Meanwhile, hydrogen peroxide (H_2O_2) was selected as a positive control. Anti proliferation activity was measured using MTT assay. The absorbance was read at a wavelength of 570 nm by using a microplate reader (Mohd Salleh et al., 2017).

2.6 Statistical analysis
The values were expressed as the mean ± standard error of the mean (SEM). Then, IC_{50} value from anti proliferation study was calculated using GraphPad Prism 5.0. One-way ANOVA with Tukey HSD test were used to analyse the anti proliferative results. A level of probability of $p < 0.05$ was set as statistically significant (Mohd Radzman et al., 2013).

3. Results

3.1 Physicochemical analyses and antioxidant level

Physicochemical analyses were conducted to characterize honey and propolis samples harvested from *G. thoracica* beehive, and beebread collected from *H. itama* (Table 1). The data showed that honey contained the highest level of moisture, followed by beebread and propolis. Propolis has the highest ash content while honey exhibited the lowest ash content. In terms of total phenolic content, propolis showed the highest antioxidant properties, followed by honey and beebread.

Table 1. Physicochemical analyses and antioxidant level of honey, propolis and beebread samples harvested from stingless bee

Sample	Moi-sture (%)	Ash (g/100g)	pH	Free acidity (meq/kg)	HMF (mg/kg)	Wax (%)	Protein (%)	Carbo-hydrate (%)	TPC (mg GAE/kg)
Honey	28.78 ± 3.8	0.22 ± 0.0	3.21 ± 0.0	58 ± 0.5	47 ± 1.1	nd	nd	nd	700 ± 0.0
Propolis	6.42 ± 0.0	3.85 ± 0.0	nd	nd	nd	33.7 ± 0.7	nd	nd	1600 ± 0.0
Beebread	8.10 ± 0.1	1.70 ± 0.0	nd	nd	nd	nd	47.4 ± 0.0	55.1 ± 2.4	300 ± 0.0

nd = not determined

3.2 Anti proliferation activity

Stingless bee honey demonstrated anti proliferative activity on MCF-7 cell lines at IC_{50} = 60v/v (Figure 1). Meanwhile, propolis and beebread showed the IC_{50} = 38.9μg/ml (Figure 2) and 64μg/ml (Figure 3) respectively. Based on the results, honey, propolis and bee bread showed dose dependent effects and significantly inhibited the cell cancer growth.

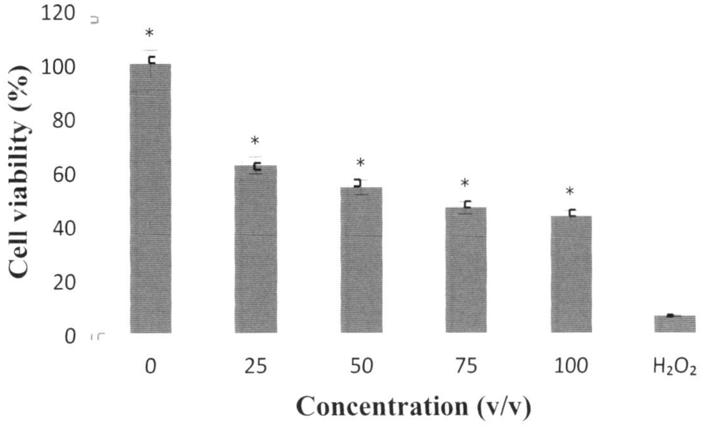

Figure 1. Anti proliferative activity of stingless bee honey from *G. thoracica* on MCF-7 cell lines after being treated with several concentrations of honey (0-100%) for 72 h. The analysis was performed using one-way ANOVA analysis (Tukey HSD test). * indicates $p < 0.05$ significant difference of the concentration towards the control. n = 3.

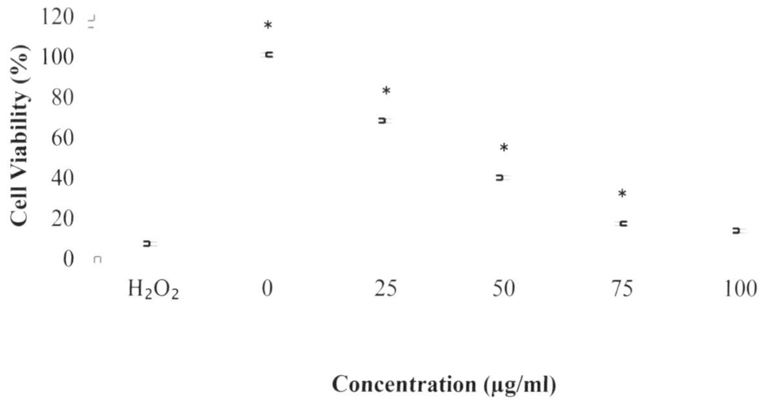

Figure 2. Anti proliferative activity of stingless bee propolis from *G. thoracica* on MCF-7 cell lines after being treated with several concentrations of propolis (0-100%) for 72 h. The analysis was performed using one-way ANOVA analysis (Tukey HSD test). * indicates $p < 0.05$ significant difference of the concentration towards the control. n = 3.

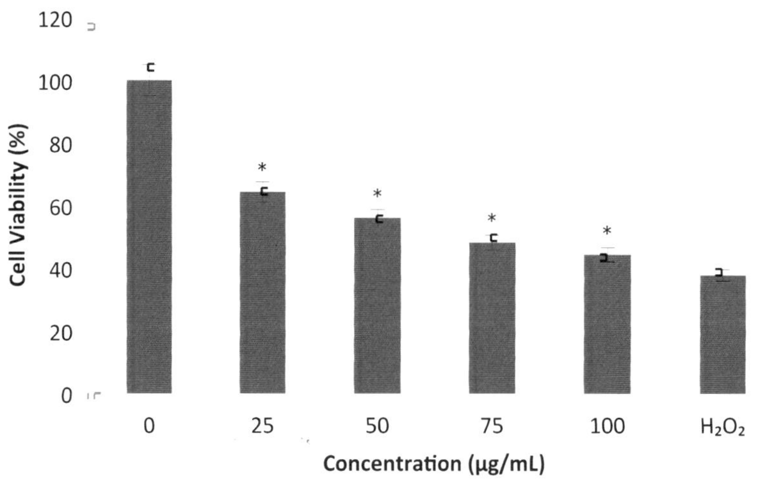

Figure 3. Anti proliferative activity of stingless bee beebread from *H. itama* on MCF-7 cell lines after being treated with several concentrations of bee bread (0-100%) for 72 h. The analysis was performed using One-way ANOVA analysis (Tukey HSD test). * indicates $p < 0.05$ significant difference of the concentration towards the control. n = 3.

4. Discussion

Most of the data obtained from physicochemical analyses on honey, propolis and beebread of stingless bee are not within the range of values set by the Codex Standard for Honey (Codex Alimentarius Commission, 1981) and other studies (Kieliszek et al., 2018 and Stan et al., 2011). These findings as expected because data set in the codex are based on honey from bee species of *Apis* (stinging bee) particularly from temperate countries. Meanwhile, our samples are collected from other bee species from tribe Meliponinae, which are a stingless bee. In addition, the stingless bee is only available in the tropical countries including Malaysia. The finding has revealed that stingless bee products have their own physicochemical characteristic and the differences need to be recognized worldwide for commercial purposes.

Each sample were subjected to different physicochemical analyses (some data were not determined-nd) as some analyses are more relevant towards certain samples, for example wax content analysis on propolis and protein and carbohydrate content on bee bread (Campos et al., 2008). The wax content of propolis used in this study is considered high (33.7%) since it is close to the 38% limit (Stan et al., 2011) and higher than the Brazilian propolis samples (Maria et al., 2016). Protein and carbohydrate content of the beebread sample are also higher compared to the 25.9% protein and 48.8% carbohydrate reported for beebread from stinging bee (Carpes et al., 2009).

Propolis has the highest antioxidant level, followed by honey and beebread. This is in agreement with another study by Bakchiche et al., 2017, which reported that propolis has higher TPC compared to honey. All samples had higher antioxidant content compared to reports from similar bee products produced by the stinging bee species *Apis mellifera* (Silva et al., 2013; Sime et al., 2015). This might be due to the differences in pollen foraging activities of singing and stingless bee products (Nurdianah et al., 2016), weather and geographical location (Markiewicz-Zukowska et al., 2013) and also differences in the solvent used for extraction (Wan et al., 2014).

Amongst the bee products, propolis with the highest antioxidant level exhibited the lowest IC_{50} on the proliferation of MCF-7 cell lines. This might be the existence of specific phenolics or flavonoids, or synergism between several compounds that are responsible for the low inhibitory concentration (IC_{50}) on MCF-7 cell lines in the propolis. However, the findings are required further investigation and validation.

5. Conclusion

Stingless bee products have unique physicochemical characters. Sample with the highest antioxidant level, propolis (1600 mg GAE/kg) has been shown to exert the lowest IC_{50} value (38.9μg/ml) on MCF-7 cell lines. Further analysis on the properties and biological activities of stingless bee products are crucial to highlight their potential benefits and at the same time promoting and increasing their commercial value in the future.

Acknowledgement

The authors would like to thank the School of Fundamental Science, UMT for providing the research and financial support. We would like to express our gratitude to the beekeepers at Humaira' Honey kelulut farm in Lenggong, Perak who were involved in the sample collection process.

References

[1] Ismail WIW 2016 *Journal of Sustainability Science and Management* **11** 70-80
[2] Bogdanov S, Lullmann C, Martin P, von der Ohe W, Russmann H, Vorwohl G, Oddo LP, Sabatini AG, Marcazzan GL, Piro R et al. 1999 *Bee World* **80** 61-69
[3] Bogdanov S 2009 *International Honey Commission* 1-62
[4] Samat S, Enchang FK, Abd Razak A, Nor Hussein F and Wan Ismail WI 2018 *Sains Malaysiana* **47** 353-365
[5] Jacob A, Parolia A, Pau A and Davamani Amalraj F 2015 *BMC Complementary and Alternative Medicine* **15** 294
[6] Markiewicz-Zukowska R, Naliwajko SK, Bartosiuk E, Moskwa J, Isidorov V, Soroczynska J and Borawska MH 2013 *Journal Apicultural Science* **57** 1339-1346
[7] Samat S, Enchang FK, Hussein FN and Ismail WIW 2014 *Evidence-Based Complementary and Alternative Medicine* **2017** 1-9
[8] Cunha IBS, Sawaya ACHF, Caetano FM, Shimizu MT, Marcucci MC, Drezza FT and Carvalho PO 2004 *Journal of the Brazilian Chemical Society* **15** 964-970
[9] Carpes ST, Mourao GB, de Alencar SM and Masson ML 2009 *Brazilian Journal of Food Technology* **12** 220-229
[10] White JW Jr 1979 *J. Assoc. Off Anal Chem* **62** 509-514
[11] Bradford MM 1976 *Anal Biochem* **72** 248-254

[12] Ibrahim N, Zakaria AJ, Ismail Z and Mohd KS 2015 *International Journal of Pharmacognosy and Phytochemical Research* **8** 156-161
[13] Mohd Salleh MA, Eshak Z and Wan Ismail WI 2017 *Jurnal Teknologi (Sciences & Engineering)* **79** 1-10
[14] Mohd-Radzman NH, Ismail WIW, Jaapar SS, Adam Z and Adam A 2013 *Evidence-Based Complementary and Alternative Medicine* **2013** 938081
[15] Codex Alimentarius Commission 1981 *Codex Standard* **12** 1-7
[16] Kieliszek M, Piwowarek K, Kot AK, Błażejak S, Chlebowska-Śmigiel A and Wolska I 2018 *Trends in Food Science & Technology* **71** 170-180
[17] Stan L, Marghitas LA and Dezmirean D 2011 *Scientific Papers Animal Science and Biotechnologies* **44** 137-140
[18] Maria SF, Nancy SB, Augusto SVF, Moura BA, Lemos SRA and Paulino N 2016 *Current Drug Discovery Technologies* **14** 21-34
[19] Bakchiche B, Habati M, Benmebarek A and Gherib A 2017 *World News of Natural Sciences* **11** 91-97
[20] Silva TMS, dos Santos FP, Evangelista-Rodrigues A, da Silva EMS, da Silva GS, de Novais JS dos Santos FAR and Camara CA 2013 *Journal of Food Composition and Analysis* **29** 10-18
[21] Sime D, Atlabachew M, Abshiro R and Zewde T 2015 *Bulletin of the Chemical Society of Ethiopia* **29** 163-172
[22] Nurdianah HF, Ahmad Firdaus AH, Eshaifol Azam O and Wan Adnan WO 2016 *International Food Research Journal* **23** 403-405
[23] Wan AWO, Nur AA, Nurdianah HF and Nik NSNMK 2014 *Asian Pacific Journal of Tropical Biomedicine* **6** 265-269

Physicochemical Analysis of Several Natural Malaysian Honeys and Adulterated Honey

Puteri Nurul Syahirah Md Dan[1], Syafizal Omar[1], Wan Iryani Wan Ismail[1,2]

[1]Faculty of Pharmacy, Universiti Teknologi MARA (UiTM) Selangor, Puncak Alam Campus, 42300 Bandar Puncak Alam, Selangor, Malaysia
[2]School of Fundamental Science, Universiti Malaysia Terengganu (UMT), 21030 Kuala Nerus, Terengganu, Malaysia

Corresponding authors emails: waniryani@umt.edu.my

Abstract: Physicochemical analysis is one of the methods used to measure the quality and authenticity of honey. Quality and authenticity of honey are crucial to ensure only a high quality of natural honey is offered to consumers and to avoid any health complications in the future. However, relatively few physicochemical data is available on the Malaysian honey. In this study, selected physicochemical analyses were conducted on several local honey. Seven honey samples i.e. Tualang honey, Rubber honey, Acacia honey, Kelulut 1 honey, Kelulut 2 honey, adulterated honey 1 and adulterated honey 2 were used and their physicochemical properties measured. Natural honey samples exhibited varying physicochemical characteristics compared to adulterated honey. The adulterated honey 1 had the highest HMF level (164.55 ± 27.14mg/kg), with a low level of pH (3.2 ± 0.02), free acidity (20.8 ± 0.17) and electrical conductivity (0.003 ± 0.00) compared to the natural honey. These preliminary data could serve as a basis for the identification and characterisation of Malaysian honey based on the physicochemical analysis. Moreover, the data can support enforcement activities by the authorities in preventing adulterated honey production and trading in the future.

1. Introduction

Honey has been used in various applications of our daily lives since ancient times (Bogdanov & Martin, 2002), with the most common one being a natural sweetener. Other profound applications can be found in the medical field such as in wound healing, as an anti-inflammatory agent and as part of a healthy functional food in general (Khalil et al., 2012; Moniruzzaman et al., 2013). However, the beneficial properties can only be exerted by consuming a natural honey of good quality. Unfortunately, at present the supply of natural honey has been marred by the overabundance of adulterated honey in the market. According to Kamaruddin et al. (2006), over 77% of honey available in the Malaysian market was adulterated. Implementation of effective enforcement by local authority to counter the excess adulterated honey has proven to be a challenge. One of the issues that impede such enforcement is a lack of physicochemical data to identify and differentiate the adulterated honey from natural honey.

Malaysia is blessed with numerous natural resources including honey. Honey produced in Malaysia can be classified based on bee species or tribe such as *Apis* (i.e. *A. dorsata, A. mellifera* and *A. cerana*) (stinging bee) or Meliponine (stingless bee; locally known as Kelulut) (Ismail,

2016). In addition, floral sources contribute to the variety of Malaysian honey such as Tualang honey, Gelam honey, Acacia honey and Rubber honey. Honey may also be influenced by the climate and its geographical origins, especially in tropical and temperate countries (Khalil et al., 2012). In order to gather the data on Malaysian honey, the study was conducted to analyse the physicochemical properties of several Malaysian honey. Concurrently, the properties of natural honey were compared with adulterated honey to observe and identify the differences between them.

2. Materials and Methods

2.1. Honey collection

Seven types of honey samples were used in this study. Five honey samples were freshly collected without any treatment from the hives of several bee species; Rubber and Acacia honey (*A. mellifera*), Tualang honey (*A. dorsata*) from Terengganu, Kelulut honey (*Heterotrigona itama*) from Kelantan (Kelulut honey 1) and Terengganu (Kelulut honey 2). Two adulterated honey samples were used in this study to compare their characteristics with the abovementioned natural honey. All samples were stored at 4°C throughout the experiment.

2.2. Physicochemical analyses

Physicochemical analyses including moisture, pH, free acidity, electrical conductivity and hydroxymethylfurfural (HMF) level were determined according to the protocols as detailed in the International Honey Commission (IHC) (2009) and Samat et al. (2017).

2.3 Moisture content

The moisture content of honey samples was determined at 20°C by using MISCO BKPR-2 digital refractometer (MISCO refractometer, United States) (IHC, 2009).

2.4 pH and free acidity

Delta 320 pH meter (Mettler Toledo, Switzerland) was used to measure the pH and free acidity of honey. Honey samples were titrated with 0.1M sodium hydroxide (NaOH) until pH 8.3 to measure their free acidity level, with the results expressed as milliequivalents acid/kilogram honey (meq/kg) (IHC, 2009).

2.5 Electrical conductivity

Electrical conductivity was carried out using HI-38311 electrical conductivity meter (Hanna Instruments, United States). A 20% (w/v) of each honey solution was prepared by dissolving 20g of honey in 100ml deionised distilled water prior to measuring the electrical conductivity using the meter (IHC, 2009).

2.6 Hydroxymethylfurfural (HMF)

5g of each honey was diluted with 25ml distilled water prior to transfer into a volumetric flask (White, 1979). 0.5ml Carrez solution I was added followed by 0.5ml Carrez II solution. Distilled water was then added up to the mark. The solution was filtered using a 0.45μm filter paper and the first 10ml filtrate was discarded. 5ml honey solution was mixed with 5ml water before its absorbance was measured against a reference solution of 5ml initial honey solution and 5ml 0.2% sodium bisulphite solution. The absorbance was measured at 284nm and 336nm, and the values were expressed in (mg/kg) according to the equation described by White (1979).

2.7 Data Analysis

Data were presented as mean ± standard error mean (SEM).

3. Results

Rubber honey displayed the highest moisture content, followed by Kelulut honey 2 and 1 (Table 1). Meanwhile, Acacia honey showed the lowest moisture content. Other honey samples including the adulterated honey demonstrated moisture content between 21.2 – 23.7%.

Table 1. Physicochemical analyses (moisture, pH, free acidity, electrical conductivity and HMF level) of several natural and adulterated honey samples. Data were measured duplicate and presented in average with SEM value.

Honey	Moisture (%)	pH	Free acidity (meq/kg)	Electrical conductivity (mS/cm)	HMF (mg/kg)
Tualang	22.5 ± 0.09	4.3 ± 0.01	16.8 ± 0.17	0.021 ± 0.00	10.5 ± 2.30
Rubber	29.3 ± 0.00	4.1 ± 0.01	18.5 ± 0.09	0.095 ± 0.00	0.60 ± 0.23
Acacia	20.2 ± 0.09	3.9 ± 0.01	18.0 ± 0.29	0.009 ± 0.00	11.62 ± 4.70
Kelulut 1	27.4 ± 0.03	3.3 ± 0.00	18.0 ± 0.29	0.068 ± 0.00	16.70 ± 2.68
Kelulut 2	28.2 ± 0.00	3.6 ± 0.01	83.0 ± 0.58	0.189 ± 0.00	0.57 ± 0.16
Adulterated 1	21.2 ± 0.03	3.2 ± 0.02	20.8 ± 0.17	0.003 ± 0.00	164.55± 27.14
Adulterated 2	23.7 ± 0.35	3.1 ± 0.06	53.2 ± 0.44	0.005 ± 0.00	126.94 ± 6.94

Adulterated honey samples had the lowest pH but are high in free acidity (Table 1). Interestingly, both kelulut honey are slightly acidic (pH of 3.3 – 3.6) compared to other natural honey but they have a wide range of free acidity (18.0 – 83.0 meq/kg). Tualang honey had the highest pH and lowest free acidity level compared to other honey. In terms of electrical conductivity, Kelulut honey 2 demonstrated the highest electrical conductivity while both adulterated honey samples showed the lowest conductivity. For HMF level, adulterated honey samples displayed very high HMF level. In contrast, the natural honey exhibited low HMF levels within the range of 0.57 – 16.70 mg/kg.

4. Discussion

Malaysian honey has a high commercial value because of its unique taste and aroma, particularly for the stingless bee (Kelulut) honey which is commonly found in the tropical countries. Based on the data obtained from the Department of Agriculture, Malaysia, the honey industry in Malaysia was estimated to be worth approximately RM17 million in 2017. Unfortunately, the quality and authenticity of the honey available in the market were unknown. A study conducted by our group revealed that long-term consumption of adulterated honey can cause various health complications on rats (Samat et al., 2018). Beneficial therapeutic effects can only be obtained from the natural honey consumption over a period of time (Samat et al., 2017). At present, enforcement by local authorities to prevent the adulterated honey from regulating in the market could not be done effectively as there is a lack of data on the local honey to identify, characterise and differentiate between natural and adulterated honey. Integration of physicochemical analysis for honey testing is seen as a good strategy as it is one of the methods which has been applied worldwide and is recommended by the International Honey Commission, a body which oversees world honey standard.

Physicochemical analysis is a set of tests to measure the quality and authenticity of honey. It includes moisture content, pH, free acidity, electrical conductivity, ash content, diastase, HMF and sugar content among others (Codex Alimentarius Commission, 2001). Multiple tests needed to be conducted on a single honey sample because honey is a complex mixture of various substances such as water, carbohydrates, proteins, enzymes, minerals and microorganisms (da Silva et al., 2016). However, selection of several parameters is adequate for the purpose of screening for a preliminary data on the differences between natural and adulterated honey. These include moisture content, pH, free acidity, electrical conductivity and HMF, which are chosen in this study.

From the results, only Acacia honey contains water (20.2 ± 0.09%) within the accepted limit set by the Codex Standard for honey (20%). Both adulterated honey and Tualang honey have slightly exceeded the limit, while other honey samples contained higher water content up to 29.3 ± 0.00%. This high water content (<30%) is considered as acceptable for honey samples harvested from tropical countries such as Malaysia, which has relatively high levels of humidity throughout the year. In addition, honey has natural hygroscopic ability to absorb water (Mandal & Mandal, 2011) especially from this humid environment. The Codex Standard is developed mostly based on the data collected from the honey of temperate countries, which experience lower humidity.

Both kelulut honey samples have low pH but are high in free acidity. This is because Kelulut honey has a more acidic taste compared to other honey produced by *Apis* bee species. The kelulut honey samples also have high electrical conductivity, which could possibly be attributed to the presence of other contents in the honey such as minerals. In contrast, both adulterated honey samples had very low electrical conductivity (Codex: ≥0.8 mS/cm). This is acceptable because adulterated honey typically only contains water and sugar. The adulterated nature of the two honey samples was supported by very high level of HMF found compared to the natural honey (<80mg/kg set by the Codex). High HMF level can be found in samples subjected to either a prolonged storage (natural honey) or heating process (adulterated honey). In this study, all natural honey samples were freshly harvested from its hive. Thus, their HMF levels are low compared to the adulterated honey.

Data obtained from the study can be used as a stepping stone for more extensive characterisation of local honey and to differentiate local natural honey from adulterated honey. Because of limitation of sample used in this study, statistical analysis cannot be performed and projected for future investigation. Determination of honey's sugar composition can also be performed as the next step for further confirmation on the quality and authenticity of honey. The initiatives recommended above are opportune for more effective enforcement in the future to impede the trading of adulterated honey. Subsequently, it will increase the user's trust in the authenticity of local honey which could contribute to the increase of local honey's production and commercial value.

5. Conclusion

Natural and adulterated honey both exhibited different characteristics, which made it possible for early screening of its quality and authenticity. Thus, the physicochemical analysis including moisture content, pH, free acidity, electrical conductivity and HMF level can be used as a preliminary step to identify and characterise the quality of natural honey and the status of adulterated honey.

6. References

[1] Bogdanov S and Martin P 2002 *Mitteilungen aus dem Gebiete der Lebensmitteluntersuchung und Hygiene* **93** 232-254
[2] Khalil I, Moniruzzaman M, Boukraa L, Benhanifia M, Islam A, Islam N, Sulaiman S A and Gan S H 2012 *Molecules* **17** 11199-11215
[3] Moniruzzaman M, Khalil I, Sulaman S A and Gan S H 2013 *BMC Complement Altern Med.* **13** 43
[4] Kamaruddin M Y, Joesima H and Mohd Nuruddin A M 2006 *1st International Conference on the Medicinal Uses of Honey* Universiti Sains Malaysia (USM) Kota Bharu Kelantan
[5] Ismail W I W 2016 *Journal of Sustainability Science and Management* **11** (2) 70-80
International Honey Commission 2009 *Harmonised methods of the international honey commission* (Liebefeld: Swiss Bee Research Centre) 63
[6] Samat S, Enchang F K, Nor Hussein F and Wan Ismail W I 2017 *Evid. Based. Complement. Alternat. Med.* **2017** 1342150
[7] White J W Jr 1979 *J. Assoc. Off Anal Chem* **62** 509-514
[8] Samat S, Enchang F K, Abd Razak A, Nor Hussin F and Wan Ismail W I 2018 *Sains Malaysiana*
47 (2) 353-365
[9] Codex Alimentarius Commission 2001 *Alinorm 41/10: Revised standard for honey,* Alinorm 1 (Rome: World Health Organisation) 19-26

IOP Conf. Series: Materials Science and Engineering **440** (2018) 012049 doi:10.1088/1757-899X/440/1/012049

[10] da Silva P M, Gauche C, Gonzaga L V, Costa A C O and Fett R 2016 *Food Chemistry* **196** 309-323
[11] Mandal M D and Mandal S 2011 *Asian Pac. J. Trop. Biomed.* **1** (2) 154-160

Acknowledgement
The authors would like to thank the Faculty of Pharmacy and RMIC, UiTM Selangor for providing the research and financial support. We would like to express our gratitude to the beekeepers and honey hunters who were involved in the honey collection process.

IOP Conf. Series: Materials Science and Engineering **440** (2018) 012050 doi:10.1088/1757-899X/440/1/012050

Early Investigation on the Effects of Stevioside on Differentiation Process of Fibroblast into Adipocytes

Muhammad Farhi Md Yusoff[1] and Wan Iryani Wan Ismail[1,2]

[1]Faculty of Pharmacy, Universiti Teknologi MARA (UiTM) Selangor, Puncak Alam Campus, 42300 Bandar Puncak Alam, Selangor, Malaysia
[2]School of Fundamental Science, Universiti Malaysia Terengganu (UMT), 21030 Kuala Nerus, Terengganu, Malaysia

Corresponding author email: waniryani@umt.edu.my

Abstract. Obesity is one of major health problems occurred worldwide including Malaysia. Unfortunately, no suitable treatment or mechanism can be used to solve the problem effectively. In order to find the most appropriate treatment, stevioside was chosen to investigate its effects in preventing obesity. Stevioside was used to observe its effect in four stages of differentiation process (lipid formation) 3T3-L1 murine pre-adipocytes (fibroblasts) into adipocytes. The process was observed using Oil red-O staining and measured using a spectrophotometer at the wavelength of 520 nm. From the results, stevioside was significantly inhibited the differentiation process at stage 2. Stevioside may interfere with the other supplements added in the cell culture media, which triggered the inhibition. However, the mechanism was unclear and required further studies.

1. Introduction

Obesity remains the major health problem in Malaysia (Lim, 2016). Until now, none effective treatments are available to solve the problem. The best practices are by taking an appropriate and healthy diet and do a regular and consistent exercise. Beside the development in preventing, managing and controlling obesity, various researches are being conducted to find the most possible methods and treatment to resolve the challenge. One of the efforts is by elucidating a potential of stevioside, a natural sweetener, in preventing the excess gain.

Stevioside is a bioactive compound extracted from the leaf of *Stevia rebaudiana*. The herb is widely available in the northeastern region of Paraguay, Brazil, Argentina, Europe and Asia including in Malaysia (Tavarini & Angelini, 2013). Stevioside has been used as a natural sweetener to replace table sugar, which is known to cause various health complications. Meanwhile, the natural sweetener is zero calorie and has been used traditionally to treat diabetes in Paraguay (Mohd-Radzman et al., 2013a). However, its effect in preventing obesity is still unknown and will be evaluated in this study.

Physiologically, obesity occurs when size and volume of adipocytes in adipose tissue are increasing unlimited because of high glucose uptake and lipogenesis (González-Muniesa et al., 2017). In order to control the obesity from the beginning, an effective treatment may be evaluated its effects on adipocyte formation. Adipocyte formation involves differentiation procedure from fibroblast cells to adipocytes. In order to access potential stevioside in controlling obesity, differentiation process of 3T3-L1 murine adipocytes was measured with and without the stevioside treatment.

2. Materials and Methods

2.1. Sample preparation

Stevioside was purchased from Sigma-Aldrich Co. (Germany) and prepared in the concentration of 30μM (Mohd-Radzman et al., 2013) prior to adding in the four different stages of adipocyte differentiation i.e. stage 1, 2, 3, 4 (Ismail et al., 2013 and Mohd-Radzman et al., 2013b).

2.2. Cell culture and differentiation

3T3-L1 pre-adipocytes or also known as fibroblasts were cultured in Dulbecco's Modified Eagle's Media (DMEM) according to Ismail *et al.,* 2013 and Mohd-Radzman et al., 2013. For the differentiation procedures, the cells were cultured in the DMEM supplemented with insulin, dexamethasone (DMX) and 3-isobutyl-1-methyl-xanthine (IBMX).

2.3. Oil red-O staining

The cells were washed with phosphate-buffered saline (PBS) before being fixed with a solution of 10% formaldehyde in PBS for overnight. The solution was discarded prior to introduce the cells with the Oil Red-O dye for 10 minutes in room temperature. Then, the dye was removed with 100% isopropanol before the cells were measured using a spectrophotometer at 520 nm (Mohd-Radzman et al., 2013).

2.4 Statistical analysis

Data were presented as mean ± standard error mean (SEM) before proceeded with one-way analysis of variance (ANOVA) followed by Tukey's post-hoc tests. The analysis was conducted using the Sigma Plot version 16 software. Statistically different means were acknowledged at $p < 0.05$.

3. Results

3T3-L1 pre-adipocytes (control) were successfully and significantly differentiated into adipocytes (untreated) (Figure 1). After the cells treated with stevioside at every stage of adipocyte differentiation, the lipid content in the adipocytes was changed compared to the untreated cells. Stevioside was found to inhibit adipogenesis process significantly at stage 2 compared to the other stages. Meanwhile, the compound found to stimulate the fat formation in stage 1, 3 and 4.

4. Discussion

3T3-L1 adipocytes were used in the study because the cells are well-known as a cell culture model to study obesity and diabetes (Knutson and Balba, 1997). In addition, the cells have the ability to imitate fat cell formation in animal and human. Basically, adipocytes are the major component in adipose tissue; our lipid storage. Investigation and observation on the cells before and after some treatments may give a clue in controlling excess weight gain.

Fat cell formation or also known as differentiation procedure can be conducted *in-vitro* through four different stages (Ismail et al., 2013). Each stage requires different supplements to support the growth. In stage 1, the cells were cultured in medium 1, which contained DMEM with newborn calf serum (NCS) (10%), L-glutamine (2 mM), antibiotic/antimycotic (1% of 100x) and gentamicin (0.1%) to maintain 3T3-L1 fibroblast growth. DMEM was used as a base medium. Addition of NCS in the medium because it has a low level of insulin-like growth factor (IGF), which allows cells to grow as fibroblasts and prevents them from differentiating (Knutson and Balba, 1997). However, with additional of stevioside, the cells seem able to differentiate into adipocytes and hinder the effect of NCS.

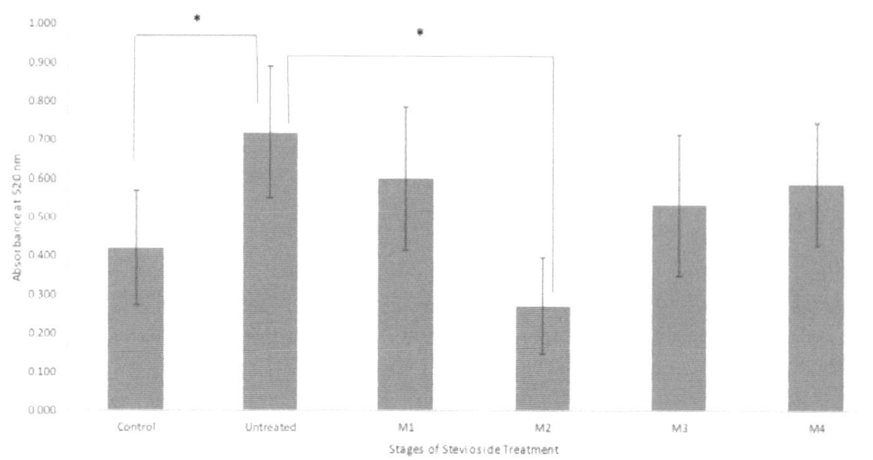

Figure 1. Optical density of lipid formation with and without treatment with stevioside (30μM) after stained with Oil Red-O dye and measured using spectrophotometer at 520 nm. Control: 3T3-L1 pre- adipocytes (fibroblast), Untreated: 3T3-L1 adipocytes after differentiation process without stevioside's treatment, M1: 3T3-L1 pre-adipocytes in stage 1 of lipid formation, M2: 3T3-L1 pre-adipocytes in stage 2 of lipid formation, M3: 3T3-L1 pre-adipocytes in stage 3 of lipid formation and M4: 3T3-L1 pre-adipocytes in stage 4 of lipid formation. Mean ± SEM (n = 6). Statistically significant differences with $p < 0.05$ (ANOVA and Tukey's posthoc test).

In stage 2, the cells were cultured in medium 2, which is similar to medium 1 except NCS was replaced with FCS (fetal calf serum) (10 %) and addition of DMX (0.25 μM), IBMX (0.5 mM) and insulin (166 nM). FCS has a high level of IGF, which stimulates the differentiation of pre-adipocytes into adipocytes. Meanwhile, DMX and IBMX were used to induce differentiation of 3T3-L1 pre- adipocytes through the activation of transcription factors, the CCAAT/enhancer-binding proteins (C/EBPs) and the PPARγ. Insulin was also added to the culture medium to accelerate the differentiation procedure in the cells (Knutson and Balba, 1997). The most effective differentiation process should occur at the stage. Interestingly, stevioside is able to inhibit the lipid formation hereby decreasing the differentiation process (Figure 1). The possibility of the mechanism is the stevioside may interfere with effects of FCS added in the medium. FCS contains a high level of IGF, which plays important role in adipocyte formation and has proteins sequence similar to insulin (Ismail et al., 2013). Meanwhile, stevioside has similar function to insulin (umpublished data), which can mimic the insulin hormone. The existence of both stevioside and IGF may disrupt the role of IGF, thus inhibited the fat formation. However, the notion is required further investigation.

In stage 3 and 4, the cells were grown respectively in medium 3 and 4. Chemical compositions in the medium 3 was similar to components in medium 2 except without DMX and IBMX addition. Meanwhile, medium 4 was prepared as according to medium 3, but without insulin to maintain the adipocyte differentiation (Knutson and Balba, 1997). Stevioside is found to stimulate the differentiation process at both stages. According to the results, the potential of stevioside in controlling obesity is still unclear. The finding shows the ability of the compound in inhibiting lipid accumulation at stage 2 compared to the other stages. However, the reaction between chemical and supplements used in the media may interfere with the results. Thus, more studies need to be conducted such as by using an obesity animal model with and without the stevioside consumption to observe its effects as an anti-obesity agent.

IOP Conf. Series: Materials Science and Engineering **440** (2018) 012050 doi:10.1088/1757-899X/440/1/012050

5. Conclusion
Potential stevioside in controlling obesity is still uncertain even it shows positive results at stage 2 of the differentiation procedures. Other studies should be done to confirm the investigation.

6. References
[1] Lim K G 2016 *Med J Malaysia* **71** (1) 1-19
[2] Tavarini S and Angelini L G 2013 *Journal of the Science of Food and Agriculture* **93** (9) 2121– 2212
[3] Mohd-Radzman N H, Ismail W I W, Adam Z, Jaapar S S and Adam A 2013a *Evid Based Complement Alternat Med* **2013** 1-10
[4] González-Muniesa P, Mártinez-González M A, Hu F B, Després J P, Matsuzawa Y, Loos R J F, Moreno L A, Bray G A and Martinez J A 2017 *Nature Reviews Disease Primers* **3** 17034
[5] Ismail W I W, King J A, Anwar K and Pillay TS 2013 *J Cell Biochem* **114** 1729-1737
[6] Mohd-Radzman N H, Ismail W I W, Jaapar S S, Adam Z and Adam A 2013b *Evid Based Complement Alternat Med* **2013** 1-8
[7] Knutson V P and Balba Y 1997 *In Vitro Cellular & Developmental Biology - Animal* **33**77-81

Acknowledgement
The authors would like to thank the Faculty of Pharmacy and RMIC, UiTM Selangor for providing the research and financial support.

AUTHOR INDEX

AUTHOR INDEX